AF356544

# Advances in UAV Detection, Classification and Tracking

# Advances in UAV Detection, Classification and Tracking

Editors

**Daobo Wang**
**Zain Anwar Ali**


MDPI • Basel • Beijing • Wuhan • Barcelona • Belgrade • Manchester • Tokyo • Cluj • Tianjin

*Editors*
Daobo Wang
College of Automation
Engineering
Nanjing University of
Aeronautics & Astronautics
Nanjing
China

Zain Anwar Ali
Electronic Engineering
Sir Syed University of
Engineering & Technology
Karachi
Pakistan

*Editorial Office*
MDPI
St. Alban-Anlage 66
4052 Basel, Switzerland

This is a reprint of articles from the Special Issue published online in the open access journal *Drones* (ISSN 2504-446X) (available at: www.mdpi.com/journal/drones/special_issues/UAV_Detection_Classification_Tracking).

For citation purposes, cite each article independently as indicated on the article page online and as indicated below:

LastName, A.A.; LastName, B.B.; LastName, C.C. Article Title. *Journal Name* **Year**, *Volume Number*, Page Range.

**ISBN 978-3-0365-7561-2 (Hbk)**
**ISBN 978-3-0365-7560-5 (PDF)**

# Contents

# About the Editors

**Daobo Wang**

Prof. Dr. Wang Dao Bo is a distinguished academic and researcher in the field of automation engineering. He is currently affiliated with the College of Automation Engineering at Nanjing University of Aeronautics and Astronautics (NUAA), located in Nanjing, China.

Prof. Wang has made significant contributions to the development of control theory, system identification, and fault diagnosis of complex systems. He has authored or co-authored over 200 scientific papers and several books on these topics.

In addition to his research work, Prof. Wang has also served in various leadership roles at NUAA, including as the dean of the College of Automation Engineering. He is also an elected member of the Chinese Academy of Engineering, which is a prestigious honor in the academic community.

Overall, Prof. Wang Dao Bo is a respected and accomplished scholar in the field of automation engineering, with a long-standing record of academic excellence and contributions to the field.

**Zain Anwar Ali**

Engr. Dr. Zain Anwar Ali received a B.S. degree in Electronic Engineering from Sir Syed University of Engineering and Technology, Karachi, Pakistan, in 2009. In the same year, he joined Sir Syed UET, as a Research Assistant in the Electronic Engineering department, and the next year in 2011, he was promoted to the post of junior lecturer due to his hard work and research contributions. He did his master's in Industrial Control and Automation from the Hamdard University of Engineering in 2012 and secured second position in his batch. After that, he was promoted to the post of Lecturer in 2012. Later, in September 2013, he joined Nanjing University of Aeronautics and Astronautics (NUAA) as a Ph.D. research scholar and joined Nanjing Strong flight electronics and machinery LTD to complete his Ph.D. experiment work there. In the year 2017, he completed his Ph.D. in the field of Control Theory and Control Engineering NUAA.

 *drones*

*Editorial*

# Editorial of Special Issue "Advances in UAV Detection, Classification and Tracking"

Daobo Wang [1] and Zain Anwar Ali [2,3,*]

[1] College of Automation Engineering, Nanjing University of Aeronautics and Astronautics, Nanjing 210016, China
[2] School of Physics and Electronic Engineering, JiaYing University, Meizhou 514015, China
[3] Department of Electronic Engineering, Sir Syed University of Engineering and Technology, Karachi 75300, Pakistan
* Correspondence: zainanwar86@hotmail.com

**Citation:** Wang, D.; Ali, Z.A. Editorial of Special Issue "Advances in UAV Detection, Classification and Tracking". *Drones* **2023**, *7*, 195. https://doi.org/10.3390/drones7030195

Received: 1 March 2023
Accepted: 10 March 2023
Published: 14 March 2023

This is an editorial for a Special Issue of *Drones* titled "Advances in UAV Detection, Classification and Tracking". The main aim of this Special Issue is to promote global collaboration and knowledge transfer between researchers. This Special Issue also includes the top 17 out of 42 selected papers received from academicians, researchers, and students in the field. These papers mainly emphasize recent trends in drone research.

The first paper is titled "Research on Modeling and Fault-Tolerant Control of Distributed Electric Pro-pulsion Aircraft." This study proposes a very promising distributed electric propulsion (DEP) system with high propulsion efficiency and low fuel consumption. The redundant thrusters of DEP aircraft increase the risk of faults in the propulsion system, so it is necessary to study fault-tolerant controls to ensure flight safety. Little research has been performed on coordinated thrust control, and the research on fault-tolerant controls for DEP systems is also in the preliminary stage. In this study, a mathematical model of a DEP aircraft was built. Aimed at the lateral and longitudinal control of DEP aircraft, a coordinated thrust-control method based on the control of total energy and total heading was designed. Furthermore, a fault-tolerant control strategy and control method were developed for faults in the propulsion system. Simulation results showed that the controller could control the thrust at the pre-fault level. The correctness and effectiveness of the coordinated thrust-control method designed and the fault-tolerant control method for DEP aircraft were theoretically verified. This study provides a theoretical basis for the future engineering application and development of control systems for DEP aircraft [1].

The second paper in this Special Issue is titled, "Design and Implementation of Sensor Platform for UAV-Based Target Tracking and Obstacle Avoidance." Small-scale unmanned aerial vehicles are currently being deployed in urban areas for missions such as ground target tracking, crime scene monitoring, and traffic management. Aerial vehicles deployed in such cluttered environments are required to demonstrate robust, autonomous navigation and have both target-tracking and obstacle-avoidance capabilities. To this end, this work presents a simply designed but effective steerable sensor platform and implementation techniques for both obstacle avoidance and target tracking. The proposed platform is a two-axis gimbal system capable of roll and pitch/yaw. A mathematical model was developed to govern the dynamics of this platform. The performance of the platform was validated using a software-in-the-loop simulation. The simulation's results showed that the platform can be effectively steered to all regions of interest except in a backwards direction. Due to its design layout and mount location, the platform can engage sensors for obstacle avoidance and target tracking as per UAV requirements. Moreover, steering the platform in any direction does not induce aerodynamic instability with respect to the unmanned aerial vehicle [2].

The third paper Is titled "Mathematical Modeling and Stability Analysis of Tiltrotor Aircraft." Mathematical modeling is the key problem in developing a tiltrotor. Therefore, this paper proposes a dividing modeling method that divides a tiltrotor into five parts (rotor, wing, fuselage, horizontal tail, and vertical fin) and develops aerodynamic models for each of them. In this way, the force and moment generated by each part can be obtained. First, a dynamic model of the rotor and its flapping angle expression was developed using the blade element theory. Using the mature lifting line theory, dynamic models of the wings, fuselage, horizontal tail, and vertical fin were then built. The rotors' dynamic interference on the wings and the nacelle tilt variations against the center of gravity and moment of inertia were taken into account. A non-linear tiltrotor simulation model was built in a MATLAB/Simulink simulation environment, and the Trim command was then applied to trim the tiltrotor. Finally, the XV-15 tiltrotor was used as an example to validate the rationality of the developed model. In the end, the non-linear simulation model was linearized to obtain a state–space matrix, thus preforming a stability analysis of the tiltrotor [3].

The fourth paper, which compiled research on motion-planning algorithms for UAVs, is titled "Optimization Methods Applied to Motion Planning of Unmanned Aerial Vehicles: A Review." A flying robot is a system that can fly off and touch down to execute specific tasks. These flying robots are currently capable of flying without human control and can make situation-appropriate decisions with the help of onboard sensors and controllers. Among flying robots, unmanned aerial vehicles (UAVs) are highly attractive and applicable for military and civilian purposes. These UAV applications require motion-planning and collision-avoidance protocols to achieve improved robustness and a faster convergence rate for meeting their targets. Furthermore, optimization algorithms improve the performance of the system and minimize convergence errors. In this survey, diverse scholarly articles were gathered to highlight motion planning for UAVs using bio-inspired algorithms. This study will assist researchers in understanding the latest research performed on UAV motion planning with various optimization techniques. Moreover, this review presents the contributions and limitations of every article to demonstrate the effectiveness of the proposed work [4].

In the fifth paper, the authors present a study titled "The Development of a Visual Tracking System for a Drone to Follow an Omnidirectional Mobile Robot." This research developed a UAV visual tracking system that guides a drone in tracking a mobile robot and accurately landing on it when it stops moving. Two different-color LEDs were installed on the bottom of the drone. The visual tracking system on the mobile robot can detect the heading angle and the distance between the drone and the mobile robot. The heading angle and flight velocity in the pitch and roll direction of the drone were modified by PID controls so that the flying speed and angle are more accurate and the drone can land quickly. The PID tuning parameters were also adjusted according to the height of the drone. The system embedded in the mobile robot, which is equipped with Linux Ubuntu and processes images with OpenCV, can send a control command (SDK 2.0) to the Tello EDU drone via WIFI by using the UDP Protocol. The drone can auto-track the mobile robot. After the mobile robot stops moving, the drone can land on top of the mobile robot. Experimental results indicate that the drone can take off from the top of the mobile robot, visually track the mobile robot, and finally land on the top of the mobile robot accurately [5].

The sixth paper, "A Multi-Colony Social Learning Approach for the Self-Organization of a Swarm of UAVs", offers an improved method for the self-organization of a swarm of UAVs based on a social learning approach. To begin, the authors used three different colonies and three best members, i.e., unmanned aerial vehicles (UAVs), that are randomly placed in the colonies. This study used max–min ant colony optimization (MMACO) in conjunction with a social learning mechanism to plan an optimized path for an individual colony. A multi-agent system (MAS) chooses the most optimal UAV as the leader of each colony and selects the remaining UAVs as agents, which helps organize the randomly positioned UAVs into three different formations. The algorithm then synchronizes and connects the three colonies into a swarm and controls it using dynamic leader selection. The major contribution of this study was to hybridize two different approaches to produce a more optimized, efficient, and effective strategy. The results verified that the proposed algorithm completed the given objectives. This study also compared the designed method with the non-dominated sorting genetic algorithm II (NSGA-II) to prove that the new method offers better convergence and reaches the target via a shorter route than NSGA-II [6].

The seventh paper, which proposes a triangular topological sequence for UAVs, is titled "Multi-Target Association for UAVs Based on Triangular Topological Sequence." Thanks to their wide coverage and multi-dimensional perception, multi-UAV cooperative systems are highly regarded in the field of cooperative multi-target localization and tracking. However, due to the similarity of target visual characteristics and the limitations of UAV sensor resolution, it is difficult for UAVs to correctly distinguish visually similar targets. Incorrect correlation matching between targets results in the incorrect localization and tracking of multiple targets by multiple UAVs. In order to solve the association problem of targets with similar visual characteristics and to reduce the localization and tracking errors caused by target association errors based on the relative positions of the targets, the paper proposes a globally consistent target association algorithm for multiple UAV vision sensors based on triangular topological sequences. In contrast to Siamese neural networks and trajectory correlations, this algorithm uses the relative position relationship between targets to distinguish and correlate targets with similar visual features and trajectories. The sequence of neighboring target triangles is constructed using the relative position relationship to produce a specific triangular network. Moreover, this paper proposes a method for calculating the similarity of topological sequences with similar transformation invariances and a two-step optimal association method that considers global objective association consistency. Experimental flight results indicated that the algorithm achieves an association accuracy of 84.63%, and the two-step association is 12.83% more accurate than a single-step association. By conducting this research, the multi-target association problem of similar or even identical visual characteristics can be solved via cooperative surveillance and using multiple UAVs to track suspicious vehicles on the ground [7].

The eighth paper is titled "Drones Classification by the Use of a Multifunctional Radar and Micro-Doppler Analysis." The use of radars to classify targets has received great interest in recent years, particularly for defense and military applications in which the development of sensor systems for identifying and classifying threatening targets is a mandatory requirement. In the specific case of drones, several classification techniques have already been proposed. Until recently, a micro-Doppler analysis in conjunction with machine learning tools was considered the most effective technique. The micro-Doppler signatures of targets are usually represented in the form of a spectrogram, which is a time–frequency diagram obtained by performing a short-time Fourier transform (STFT) on a radar return signal. Moreover, it is often possible to extract useful information from a target's spectrogram that can also be used in a classification task. The main aim of this paper is to compare different methods of exploiting a drone's micro-Doppler analysis on different stages of a multifunctional radar. Three different classification approaches were compared: a classic spectrogram-based classification; spectrum-based classification in which the received signal from the target is picked up after the moving target detector (MTD); and feature-based classification in which the received signal from the target undergoes a detection step after the MTD to extract and use discriminating features as input for the classifier. A theoretical model for the radar return signals of different types of drones and aerial targets was developed to compare the three approaches. This model was validated via a comparison with real recorded data, and it was used to simulate the targets. The results showed that the third approach (feature-based) demonstrated improved performance. Moreover, it also required less modification and less processing power when using a modern, multifunctional radar because it is capable of reusing most of the processing facilities that are already present [8].

The ninth paper is "Anti-Occlusion UAV Tracking Algorithm with a Low-Altitude Complex Background by Integrating Attention Mechanism." In recent years, the increasing number of unmanned aerial vehicles (UAVs) in low-altitude airspace has introduced not only convenience to work and life but also great threats and challenges. There are common problems in the process of UAV detection and tracking, such as target deformation, target occlusion, and the submersion of targets by complex background clutter. This paper proposes an anti-occlusion UAV tracking algorithm for low-altitude, complex backgrounds that integrates an attention mechanism to solve the problems of complex backgrounds and occlusions. The algorithm process is as follows: first, extracted features are enhanced using the SeNet attention mechanism. Second, an occlusion-sensing module is used to determine whether the target is occluded. If the target is not occluded, tracking continues. Otherwise, the LSTM trajectory-prediction network is used to predict the UAV position in subsequent frames from the UAV flight trajectory prior to occlusion. This study was verified using the OTB-100, GOT-10k, and integrated UAV datasets. The accuracy and success rates of the integrated UAV datasets were 79% and 50.5%, respectively, 10.6% and 4.9% higher than the rates of the SiamCAM algorithm. Experimental results showed that the algorithm could robustly track a small UAV in a low-altitude, complex background [9].

The tenth paper is titled, "A Modified YOLOv4 Deep Learning Network for Vision-Based UAV Recognition." The use of drones in various applications has increased, as has their popularity among the general public. As a result, the possibility of drone misuse and their unauthorized intrusion into important places, such as airports and power plants, are increasing. This threatens public safety. Therefore, the accurate and rapid recognition of drone types is important for preventing their misuse and security problems caused by unauthorized drone access. Performing this operation from visible images is always associated with challenges, such as the drone's small size, confusion with birds, the presence of hidden areas, and crowded backgrounds. In this paper, a novel and accurate technique with a change in the YOLOv4 network is presented to recognize four types of drones (multirotors, fixed-wing, helicopters, and VTOLs) and distinguish them from birds using a set of 26,000 visible images. In this network, more precise and detailed semantic features could be extracted by changing the number of convolutional layers. The performance of the basic YOLOv4 network was also evaluated on the same dataset, and the proposed model performed better than the basic network in solving the challenges. The proposed model also achieved automated, vision-based recognition with a loss of 0.58 in the training phase and an 83% F1-score, 83% accuracy, 83% mean average precision (mAP), and 84% intersection over union (IoU) in the testing phase. These results represent a slight improvement of 4% in these evaluation criteria over the basic YOLOv4 model [10].

The 11th paper is titled "Using Classify-While-Scan (CWS) Technology to Enhance Unmanned Air Traffic Management (UTM)." Drone detection radar systems have been verified to support unmanned air traffic management (UTM). In this paper, the authors propose the use of classify-while-scan (CWS) technology to improve the detection performance of drone-detection radar systems and to enhance UTM applications. The CWS recognizes radar data from each radar cell in the radar beam using an advanced automatic target recognition (ATR) algorithm. It then integrates the recognized results into the tracking unit to obtain real-time situational-awareness results for the entire surveillance area. Real X-band radar data, collected in a coastal environment, demonstrated a significant advancement in a powerful situational-awareness scenario in which birds were chasing a ship to feed on fish. The CWS technology turns drone-detection radars into a sense-and-alert platform that revolutionizes UTM systems by reducing the detection unit's detection response time (DRT) [11].

The 12th paper, "ARSD: An Adaptive Region Selection Object Detection Framework for UAV Images", proposes an object detection framework for UAVs. The performance of object detection has greatly improved due to the rapid development of deep learning. However, object detection in high-resolution images from unmanned aerial vehicles images remains a challenging problem for three main reasons: (1) the objects in aerial images have different scales and are usually small; (2) the images are high-resolution, but state-of-the-art object-detection networks are of a fixed size; (3) the objects are not evenly distributed in aerial images. To solve these problems, the authors proposed a two-stage adaptive region selection detection framework. An overall region detection network was first applied to coarsely localize the object. A fixed-point, density-based target-clustering algorithm and an adaptive selection algorithm were then designed to select object-dense sub-regions. The object-dense sub-regions were sent to a key region detection network where the results were fused with the results from the first stage. Extensive experiments and comprehensive evaluations on the VisDrone2021-DET benchmark datasets demonstrated the effectiveness and adaptiveness of the proposed framework. Experimental results showed that the proposed framework outperformed the existing baseline methods by 2.1% without additional time consumption in terms of the mean average precision (mAP) [12].

The 13th paper is titled "Comprehensive Review of UAV Detection, Security, and Communication Advancements to Prevent Threats." Over time, unmanned aerial vehicles (UAVs), also known as drones, have been used in very different ways. Advancements in key UAV areas include detection (including radio frequency and radar), classification (including micro-, mini-, close-, short-, and medium-range; medium-range endurance; low-altitude, deep-penetration; low-altitude, long-endurance; and medium-altitude, long-endurance), tracking (including lateral tracking, vertical tracking, a moving aerial pan with a moving target, and a moving aerial tilt with a moving target), and so forth. Even with these improvements and advantages, security and privacy can still be ensured by researching a number of key aspects of unmanned aerial vehicles, such as jamming a UAV's control signals and redirecting them for a high-assault activity. This review article examined the privacy issues related to drone standards and regulations. The manuscript provides a comprehensive answer to these limitations. In addition to updated information on current legislation and the many classes that can be used to establish communication between a ground control room and an unmanned aerial vehicle, this article provides a basic overview of unmanned aerial vehicles. This review provides readers with an understanding of UAV shortcomings, recent advancements, and strategies for addressing security issues, assaults, and limitations. The open research areas described in this manuscript can be utilized to create novel methods for strengthening the security and privacy of unmanned aerial vehicles [13].

The 14th paper is "Detection of Micro-Doppler Signals of Drones Using Radar Systems with Different Radar Dwell Times." Not all drone radar dwell times are suitable for detecting the micro-Doppler signals (or jet engine modulation, JEM) produced by rotating blades in the radar signals of drones. Theoretically, any X-band drone radar system can detect the micro-Doppler effects of blades due to the micro-Doppler effect and the partial resonance effect. However, the authors of this paper analyzed radar data from three radar systems with different radar dwell times and similar resolutions for frequency and velocity. These systems, Radar$-\alpha$, Radar$-\beta$, and Radar$-\gamma$, had radar dwell times of 2.7 ms, 20 ms, and 89 ms, respectively. The results indicated that Radar$-\beta$ is the best radar for detecting the micro-Doppler signals (i.e., JEM signals) produced by the rotating blades of DJI Phantom 4, a quadrotor drone. Radar$-\beta$ achieved the best results because the detection probability for JEM signals was almost 100% with approximately two peaks that had similar magnitudes to the body Doppler. In contrast, Radar$-\alpha$ could barely detect any micro-Doppler effects, and Radar$-\gamma$ detected only weak micro-Doppler signals at a magnitude of only 10% of the body Doppler's magnitude. A proper radar dwell time is the key to micro-Doppler detection. This research provides an idea for designing a cognitive micro-Doppler radar by changing the radar dwell time to detect and track the micro-Doppler signals of drones [14].

The 15th paper is titled, "Elliptical Multi-Orbit Circumnavigation Control of UAVS in Three-Dimensional Space Depending on Angle Information Only." In order to analyze the circumnavigation tracking problem in a complex, three-dimensional space, this paper proposes a UAV group circumnavigation control strategy in which the UAV circumnavigation orbit is an ellipse for which its size can be adjusted arbitrarily. At the same time, the UAV group can be assigned to multiple orbits for tracking. The UAVs only have the target's angle information, and the target's position information can be obtained by using the angle information and the proposed three-dimensional estimator, thereby establishing an ideal relative velocity equation. By constructing the error dynamic equation between the actual relative velocity and the ideal relative velocity, the three-dimensional circumnavigation problem is transformed into a velocity-tracking problem. Since UAVs are easily disturbed by external factors during flight, a sliding mode control was used to improve the system's robustness. Finally, the effectiveness of the control law and its robustness to unexpected situations were verified via simulations [15].

The 16th paper is titled, "Small-Object Detection for UAV-Based Images Using a Distance Metric Method." Object detection is important in unmanned aerial vehicle (UAV) reconnaissance missions. However, since UAVs fly at a high altitude to obtain a large reconnaissance view, the objects captured often have a small pixel size and their categories have high uncertainties. Given the limited computing capability of UAVs, large detectors based on convolutional neural networks (CNNs) have difficulty achieving real-time detection performance. To address these problems, the authors designed a small-object detector for UAV-based images in this paper. They modified the backbone of YOLOv4 according to the characteristics of small-object detection and improved the performance of small-object positioning by modifying the positioning loss function. Using the distance metric method, the proposed detector can classify trained and untrained objects by using the objects' features. Furthermore, the authors designed two data-augmentation strategies to enhance the diversity of the training set. The method was evaluated on a small-object dataset and obtained 61.00% on trained objects and 41.00% on untrained objects with 77 frames per second (FPS). Flight experiments confirmed the utility of the approach on small UAVs, which demonstrated a satisfying detection performance and real-time inference speed [16].

The 17th paper is "Simultaneous Astronaut Accompanying and Visual Navigation in Semi-Structured and Dynamic Intravehicular Environment." The application of intravehicular robotic assistants (IRA) can save valuable working hours for astronauts in space stations. There are various types of IRAs, such as those that accompany drones working in microgravity and a dexterous humanoid robot for collaborative operations. In either case, the ability to navigate and work alongside human astronauts lays the foundation for IRA deployment. To address this problem, this paper proposes a framework of simultaneous astronaut accompaniment and visual navigation. The framework contains a customized astronaut detector, an intravehicular navigation system, and a probabilistic model for astronaut visual tracking and motion prediction. The customized detector was designed to be lightweight. It achieved superior performance (AP@0.5 of 99.36%) for astronaut detection in diverse postures and orientations during intravehicular activities. A map-based visual navigation method was proposed for accurate localization with 6 DoF (1~2 cm, 0.5°) in semi-structured environments. To ensure navigation robustness in dynamic scenes, the feature points within the detected bounding boxes were filtered out. A probabilistic model was formulated based on the map-based navigation system and the customized astronaut detector. Both trajectory correlation and geometric similarity clues were incorporated into the model for stable visual tracking and the trajectory estimation of the astronaut. The overall framework enables the robotic assistant to track and identify the astronaut efficiently during intravehicular activities and can provide foresighted services while moving. The overall performance and superiority of the proposed framework were verified by conducting extensive ground experiments in a space station mockup [17].

**Funding:** This research received no external funding.

**Conflicts of Interest:** The authors declare no conflict of interest.

## References

1.  Li, J.; Yang, J.; Zhang, H. Research on Modeling and Fault-Tolerant Control of Distributed Electric Propulsion Aircraft. *Drones* **2022**, *6*, 78. [CrossRef]
2.  Tullu, A.; Hassanalian, M.; Hwang, H.-Y. Design and Implementation of Sensor Platform for UAV-Based Target Tracking and Obstacle Avoidance. *Drones* **2022**, *6*, 89. [CrossRef]

3. Sheng, H.; Zhang, C.; Xiang, Y. Mathematical Modeling and Stability Analysis of Tiltrotor Aircraft. *Drones* **2022**, *6*, 92. [CrossRef]
4. Israr, A.; Ali, Z.A.; Alkhammash, E.H.; Jussila, J.J. Optimization Methods Applied to Motion Planning of Unmanned Aerial Vehicles: A Review. *Drones* **2022**, *6*, 126. [CrossRef]
5. Zou, J.-T.; Dai, X.-Y. The Development of a Visual Tracking System for a Drone to Follow an Omnidirectional Mobile Robot. *Drones* **2022**, *6*, 113. [CrossRef]
6. Shafiq, M.; Ali, Z.A.; Israr, A.; Alkhammash, E.H.; Hadjouni, M. A Multi-Colony Social Learning Approach for the Self-Organization of a Swarm of UAVs. *Drones* **2022**, *6*, 104. [CrossRef]
7. Li, X.; Wu, L.; Niu, Y.; Ma, A. Multi-Target Association for UAVs Based on Triangular Topological Sequence. *Drones* **2022**, *6*, 119. [CrossRef]
8. Leonardi, M.; Ligresti, G.; Piracci, E. Drones Classification by the Use of a Multifunctional Radar and Micro-Doppler Analysis. *Drones* **2022**, *6*, 124. [CrossRef]
9. Wang, C.; Shi, Z.; Meng, L.; Wang, J.; Wang, T.; Gao, Q.; Wang, E. Anti-Occlusion UAV Tracking Algorithm with a Low-Altitude Complex Background by Integrating Attention Mechanism. *Drones* **2022**, *6*, 149. [CrossRef]
10. Dadrass Javan, F.; Samadzadegan, F.; Gholamshahi, M.; Ashatari Mahini, F. A Modified YOLOv4 Deep Learning Network for Vision-Based UAV Recognition. *Drones* **2022**, *6*, 160. [CrossRef]
11. Gong, J.; Li, D.; Yan, J.; Hu, H.; Kong, D. Using Classify-While-Scan (CWS) Technology to Enhance Unmanned Air Traffic Management (UTM). *Drones* **2022**, *6*, 224. [CrossRef]
12. Wan, Y.; Zhong, Y.; Huang, Y.; Han, Y.; Cui, Y.; Yang, Q.; Li, Z.; Yuan, Z.; Li, Q. ARSD: An Adaptive Region Selection Object Detection Framework for UAV Images. *Drones* **2022**, *6*, 228. [CrossRef]
13. Abro, G.E.M.; Zulkifli, S.A.B.M.; Masood, R.J.; Asirvadam, V.S.; Laouti, A. Comprehensive Review of UAV Detection, Security, and Communication Advancements to Prevent Threats. *Drones* **2022**, *6*, 284. [CrossRef]
14. Gong, J.; Yan, J.; Li, D.; Kong, D. Detection of Micro-Doppler Signals of Drones Using Radar Systems with Different Radar Dwell Times. *Drones* **2022**, *6*, 262. [CrossRef]
15. Wang, Z.; Luo, Y. Elliptical Multi-Orbit Circumnavigation Control of UAVS in Three-Dimensional Space Depending on Angle Information Only. *Drones* **2022**, *6*, 296. [CrossRef]
16. Zhou, H.; Ma, A.; Niu, Y.; Ma, Z. Small-Object Detection for UAV-Based Images Using a Distance Metric Method. *Drones* **2022**, *6*, 308. [CrossRef]
17. Zhang, Q.; Fan, L.; Zhang, Y. Simultaneous Astronaut Accompanying and Visual Navigation in Semi-Structured and Dynamic Intravehicular Environment. *Drones* **2022**, *6*, 397. [CrossRef]

*Article*

# Simultaneous Astronaut Accompanying and Visual Navigation in Semi-Structured and Dynamic Intravehicular Environment

Qi Zhang [1,*], Li Fan [2,3] and Yulin Zhang [2,3]

1. College of Aerospace Science and Engineering, National University of Defense Technology, Changsha 410073, China
2. College of Control Science and Engineering, Zhejiang University, Hangzhou 310000, China; fanli77@zju.edu.cn (L.F.); zhangyulin@zju.edu.cn (Y.Z.)
3. Huzhou Institute, Zhejiang University, Huzhou 313000, China
* Correspondence: zhangqi9241@alumni.nudt.edu.cn

**Abstract:** The application of intravehicular robotic assistants (IRA) can save valuable working hours for astronauts in space stations. There are various types of IRA, such as an accompanying drone working in microgravity and a dexterous humanoid robot for collaborative operations. In either case, the ability to navigate and work along with human astronauts lays the foundation for their deployment. To address this problem, this paper proposes the framework of simultaneous astronaut accompanying and visual navigation. The framework contains a customized astronaut detector, an intravehicular navigation system, and a probabilistic model for astronaut visual tracking and motion prediction. The customized detector is designed to be lightweight and has achieved superior performance (AP@0.5 of 99.36%) for astronaut detection in diverse postures and orientations during intravehicular activities. A map-based visual navigation method is proposed for accurate and 6DoF localization (1~2 cm, 0.5°) in semi-structured environments. To ensure the robustness of navigation in dynamic scenes, feature points within the detected bounding boxes are filtered out. The probabilistic model is formulated based on the map-based navigation system and the customized astronaut detector. Both trajectory correlation and geometric similarity clues are incorporated into the model for stable visual tracking and trajectory estimation of the astronaut. The overall framework enables the robotic assistant to track and distinguish the served astronaut efficiently during intravehicular activities and to provide foresighted service while in locomotion. The overall performance and superiority of the proposed framework are verified through extensive ground experiments in a space-station mockup.

**Keywords:** astronaut detection; astronaut accompanying; intravehicular visual navigation; semi-structured environment; dynamic scenes

**Citation:** Zhang, Q.; Fan, L.; Zhang, Y. Simultaneous Astronaut Accompanying and Visual Navigation in Semi-Structured and Dynamic Intravehicular Environment. *Drones* **2022**, *6*, 397. https://doi.org/10.3390/drones6120397

Academic Editors: Daobo Wang and Zain Anwar Ali

Received: 17 October 2022
Accepted: 3 December 2022
Published: 6 December 2022

**Publisher's Note:** MDPI stays neutral with regard to jurisdictional claims in published maps and institutional affiliations.

## 1. Introduction

Human resources in space are scarce and expensive due to launch costs and risks. There is evidence that astronauts will become increasingly physically and cognitively challenged as missions become longer and more varied [1]. The application of artificial intelligence and the use of robotic assistants allow astronauts to focus on more valuable and challenging tasks during both intravehicular and extravehicular activities [2–4]. Up to now, several robotic assistants of various types and functionalities have been developed to improve astronauts' onboard efficiency and help perform regular maintenance tasks such as thermal inspection [5] and on-orbit assembly [6]. These robots include free-flying drones designed to operate in microgravity such as Astrobee [7], Int-Ball [8], CIMON [9], IFPS [10], BIT [11], and more powerful humanoid assistants such as Robonaut2 [12] from NASA and Skybot F-850 [13] proposed by Roscosmos. Although different designs and principles are adopted, robust intravehicular navigation and the ability to work along with human astronauts constitutes the basis for their onboard deployment.

Firstly, to provide immediate service, the robotic assistant should be able to detect and track the served astronaut with high accuracy and efficiency. The recent advances in deep learning have made it possible to solve this problem. Extensive research has been carried out in terms of pedestrian detection [14] and object detection [15,16] by predecessors based on computer-vision techniques. However, the problem of astronaut detection and tracking has some distinctive characteristics due to the particular onboard working environment. On one hand, astronauts can wear similar uniforms and present diverse postures and orientations during intravehicular activities. This can cause problems for general-purpose detectors that are designed and trained for daily life scenes. On the other hand, the relatively fixed and stable background, and the limited range of motion in the space station are beneficial to customizing the astronaut detector. In terms of astronaut motion tracking and prediction, the problem cannot be simply resolved by calibrating intrinsic parameters as with a fixed surveillance camera. The robotic assistant can move and rotate at all times in the space station. Both the movement of the robot and the motion of the served astronauts will change the projected trajectories on the image plane. The trajectories must be decoupled so that the robot can distinguish the actual movement of the served astronauts. Our previous works [17,18] have mainly focused on the astronaut detection and tracking problem from a simplified fixed point of view.

An effective way to decouple motion is to incorporate the robot's 6DoF localization result so that the measured 3D positions of the astronauts can be transformed into the inertial world frame of the space station. Many approaches can be applied to achieve 6DoF localization in the space station. SPHERES [19] is a free-flying research platform propelled by cold-gas thrusters in the International Space Station (ISS). A set of ultrasonic beacons are mounted around the experimental area to provide localization with high efficiency, which resembles a regional satellite navigation system. However, the system can only provide the positioning service within a cubic area of 2 m and may suffer from the problem of signal occlusion and multi-path artifact [20]. The beacon-based approach is more suitable for experiments than service-robot applications. Int-Ball [8] is a spherical camera drone that can record HD videos under remote control currently deployed in the Japanese Experiment Module (JEM). It aims to realize zero photographing time by onboard crew members. Two stereoscopic markers are mounted on the airlock port and the entrance side for in-cabin localization. The accuracy of the marker-based method depends heavily on observation distance. When the robot is far away from the marker, the localization accuracy drops sharply. Moreover, if the robot conducts large-attitude maneuvering, the makers will move out of the robot's vision, and an auxiliary localization system has to take over.

From our perspective, the robotic assistant should not rely on any marker or auxiliary device other than its proprietary sensors for intravehicular navigation. This ideology aims to make the robot's navigation system an independent module and to enhance its adaptability to environmental changes. The space station is an artificial facility with abundant visual clues, which can provide ample references for localization. Astrobee [7] is a new generation of robotic assistants propelled by electric fans in the ISS. It adopts a map-based visual navigation system which does not rely on any external device. An intravehicular map of the ISS is constructed to assist the 6DoF localization of the robot [21]. The team has also studied the impact of light-intensity variations on the map-based navigation system [22]. However, they did not consider the coexistence of human astronauts and the problem of dynamic scenes introduced by various intravehicular activities. These problems are crucial for IRA to work in the manned space station and to provide satisfactory assistance.

To resolve the problem, this paper proposes the framework of simultaneous astronaut-accompanying and in-cabin visual navigation. The semi-structured environment of the space station is utilized to build various registered maps to assist intravehicular localization. Astronauts are detected and tracked in real time with a customized astronaut detector. To enhance the robustness of navigation in dynamic scenes, map matches within the bounding boxes of astronauts are filtered out. The computational workload is evenly distributed within a multi-thread computing architecture so that real-time performance can be achieved. Based on the robust localization and the customized astronaut detector, a probabilistic model is proposed for astronaut visual tracking and short-term motion prediction, which is crucial for the robot to accompany the served astronaut in the space station and to provide immediate assistance. Table 1 compares our proposed approach and existing methods in the literature. The incorporation of the intravehicular navigation system enables astronaut visual tracking and trajectory prediction from a moving point of view, which is one of the unique contributions of this paper.

**Table 1.** Comparison between our proposed approach and existing methods in the literature.

| Robotic Assistant | Navigation Method | Accuracy | Additional Devices | Dynamic Scene | Drawbacks |
| --- | --- | --- | --- | --- | --- |
| SPHERES [19] | radio-based | 0.5 cm/2.5° | ultrasonic beacons | yes | limited workspace |
| Astrobee [7] | map-based | 5~12 cm/1°~6° | not required | no | for static scene |
| Int-Ball [8] | marker-based | 2 cm/3° | marker | yes | limited field of view |
| CIMON [9] | vision-based | / | / | / | / |
| IFPS [10] | map-based | 1~2 cm/0.5° | not required | no | for static scene |
| Robonaut2 [12] | / | / | / | / | / |
| Skybot F-850 [13] | / | / | / | / | / |
| Proposed | map-based | 1~2 cm/0.5° | not required | yes | / |

The rest of this paper is organized as follows. In Section 2, the problem of astronaut detection in diverse postures and orientations is discussed. In Section 3, we focus on the problem of map-based intravehicular navigation in both static and dynamic environments. In Section 4, the astronaut visual tracking and short-term motion prediction model is presented. Experiments to evaluate the overall design and comparative analyses are discussed in Section 5. Finally, we summarize in Section 6.

## 2. Astronaut Detection in Diverse Postures and Orientations

In this section, we address the problem of astronaut detection during intravehicular activities, which is an important component of the overall framework. A lightweight and customized network is designed for astronaut detection in diverse postures and orientations, which achieved superior performance after fine-tuning with a homemade dataset.

### 2.1. Design of the Customized Astronaut-Detection Network

The special intravehicular working environment has introduced some new features to the astronaut-detection problem, which can be summarized as

(1)     Astronauts can present diverse postures and orientations during intravehicular activities, such as standing upside down and climbing with handrails.
(2)     Astronauts may wear similar uniforms, which are hard to distinguish.
(3)     Images can be taken from any position or orientation by IRA in microgravity.
(4)     It is possible to simplify the astronaut detector while maintaining satisfactory performance by utilizing the relatively fixed and stable background and the limited range of motion in the space station.
(5)     There is a limited number of crew members onboard the space station at the same time.

To achieve satisfactory performance, the astronaut-detection network should be equipped to cope with the above features and be lightweight enough to provide real-time detections. Anchor-based and one-shot object-detection methods [15,23], such as the Yolo network, are widely used in pedestrian detection for their balance between accuracy and efficiency. However, these networks do not perform well in the astronaut-detection task. Many false and missed detections can be found in their results. This poor performance is due to the fact that the structures of those networks are designed for general-purpose applications and the parameters are trained with daily life examples. There lies a gap between the networks' expertise and the actual application scenarios.

To fill the gap, we proposed a lightweight and customized astronaut-detection network based on the anchor-based technique. The main structure of the network is illustrated in Figure 1, where some repetitive layers are collapsed for better understanding. Input to the network is the color image taken by the robot with a resolution of 640 × 480. Layers in blue are feature-extraction modules characterized by abundant residual blocks [24], which can mitigate the notorious issue of vanishing and exploding gradients. The raw pixels are gradually compressed to the feature maps of 80 × 80, 40 × 40, and 20 × 20, respectively. Layers in the dashed box apply structures of feature pyramid network (FPN) and path aggregation network (PAN) [25] to accelerate feature fusion in different scales. The residual blocks, FPN, and PAN structures have introduced abundant cross-layer connections, which improves the network's overall fitting capacity. Green layers on the right-hand side are the anchor-based detection heads that output the final detection results after non-maximum suppression.

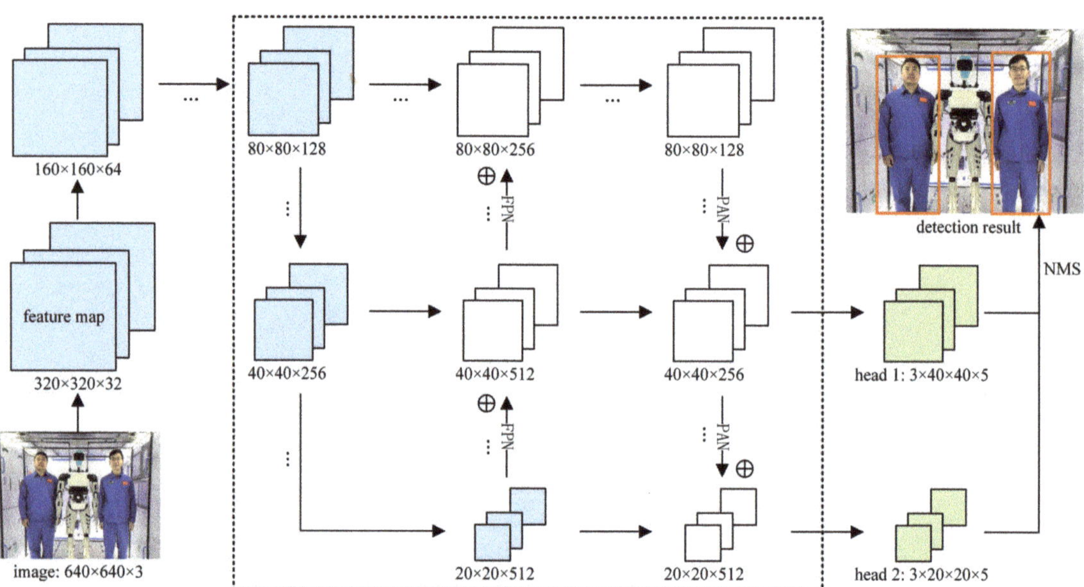


**Figure 1.** Architecture of the lightweight and customized astronaut-detection network. Layers in blue are the feature-extraction modules. Layers in the dashed box are characterized by abundant cross-layer connections for feature fusion. Layers in green are the two anchor-based detection heads.

Considering the limited number of served astronauts and their possible scales on the images, only two detection heads are designed, which also reduces the parameters and improves the network's real-time performance. The detection head with a 20 × 20 grid system is mainly responsible for astronaut detection in proximity, while the other head with a 40 × 40 grid system mainly provides smaller scale detections when astronauts are far away. As shown in Table 2, three reference bounding boxes of different sizes and shapes are designed for each anchor to adapt to the diverse postures and orientations of astronauts during intravehicular activities. A set of correction parameters ($\Delta x$, $\Delta y$, $\sigma_w$, and $\sigma_h$) are estimated with respect to the most similar reference boxes to characterize the final detection, as shown in Figure 2. Each reference box also outputs the confidence $p$ of the detection. The two detection heads provide a total of 6000 reference boxes, which is sufficient to cover all possible scenarios in the space station. To summarize, the astronaut-detection problem is modeled as a regression problem fitted by a lightweight and customized convolutional network with 7.02 million trainable parameters.

**Table 2.** Detection-head specifications of the lightweight and customized astronaut detector.

| Detection Head | Grid System | Prior Bounding Boxes | Ratio | Predictions for Each Anchor |
|:---:|:---:|:---:|:---:|:---:|
| 1 | 40 × 40 | [100, 200]<br>[200, 100]<br>[150, 150] | 1/2<br>2/1<br>1/1 | $[\Delta x, \Delta y, \sigma_w, \sigma_h, p_i] \times 3$ |
| 2 | 20 × 20 | [200, 400]<br>[400, 200]<br>[300, 300] | 1/2<br>2/1<br>1/1 | $[\Delta x, \Delta y, \sigma_w, \sigma_h, p_i] \times 3$ |

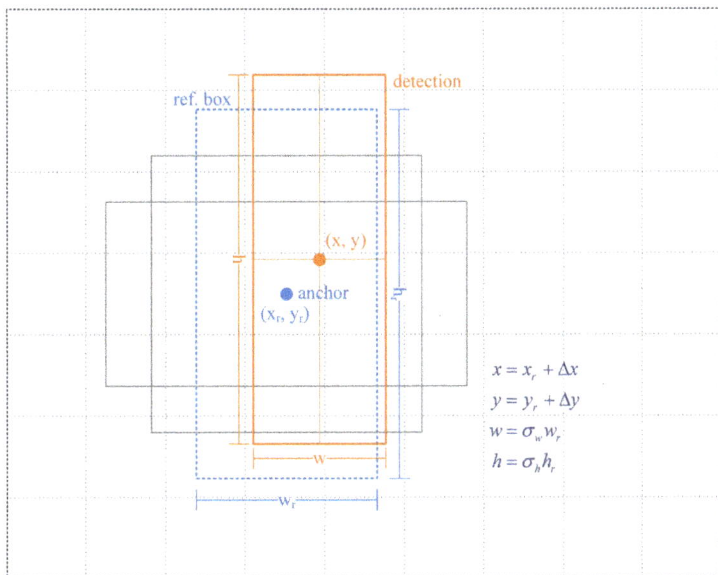


**Figure 2.** A set of correction parameters is estimated with respect to the most similar reference boxes to characterize the final detection.

### 2.2. Astronaut-Detection Dataset for Network Fine Tuning

The proposed network cannot maximize its performance before training with an appropriate dataset. General-purpose datasets such as the COCO [26] and the CrowdHuman [27] dataset mismatch the requirements. The incompatibility may be found in the crowdedness of people, the diversity of people's postures and orientations, and the scale of the projections, etc. Even though various data-augmentation techniques can be employed in the training process, it is difficult to mitigate the mismatches between the daily life scenes and the actual working scenarios in the space station.

To address the problem, we built a space-station mockup of high fidelity on the ground, and created a customized dataset for astronaut detection and visual tracking. Volunteers are invited to imitate the intravehicular activities of astronauts in the space-station mockup. During data collection, we constantly moved and rotated the camera so that bodies in the captured images show diverse perspectives. As shown in Figure 3, the proposed dataset incorporated a variety of scenes such as diverse postures and orientations of astronauts, partially observable human bodies, illumination variations, and motion blur. In total, 17,824 labeled images were collected, where 12,000 were used as the training dataset while the remaining 5824 were used as the testing dataset.

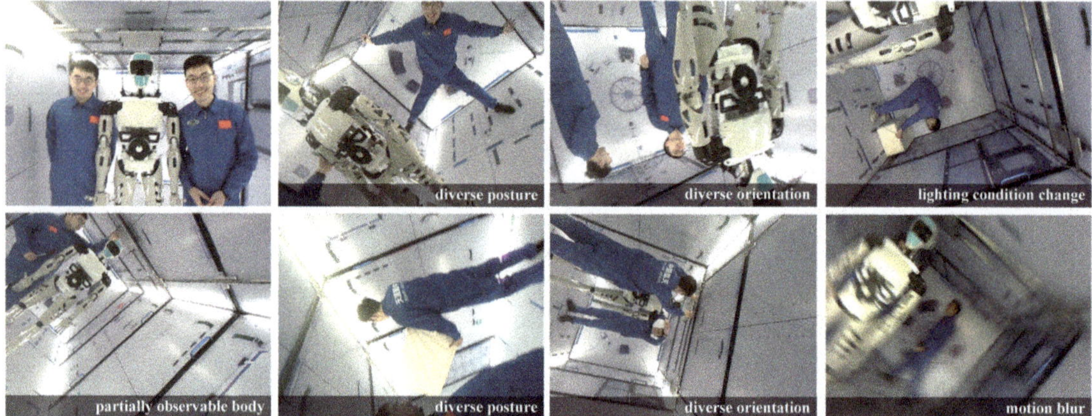


**Figure 3.** Examples in the customized astronaut-detection dataset.

### 2.3. Network Pre-Training and Fine Tuning

The astronaut detector is trained in two steps. In the pre-training phase, the network is fed with a cropped COCO dataset for 300 epochs. The cropped dataset is made by discarding crowd labels and labels that are too small or not human from the COCO 2017 dataset. The pre-training process will improve the detector's generalization ability and reduce the risk of over fitting by incorporating large numbers of samples. In the second step, the pretrained model is fine-tuned with the customized astronaut-detection dataset for 100 epochs to obtain the final detector with superior accuracy. The objective function is kept the same in both steps and is formulated as a weighted sum of the confidence loss and the bounding-box regression loss.

$$Loss = \frac{1}{A} \sum_{i=1}^{A} \sum_{j=1}^{G} \left( L_{conf}\left(p_i, \hat{p}_{ij}\right) + \lambda \hat{c}_{ij} L_{loc}\left(l_i, \hat{l}_{ij}\right) \right) \tag{1}$$

where $L_{conf}(\cdot)$ is the cross-entropy confidence loss, $L_{loc}(\cdot)$ is the bounding-box regression loss related to the prediction $l_i$ and the matched target $\hat{l}_{ij}$ where the CIOU [28] criterion is adopted, $\hat{c}_{ij}$ is 1 if the match exists, $\lambda$ is the weight parameter set to 1, $G$ is the number of ground truth label, and $A$ is the total number (6000) of reference bounding boxes.

After the two-step training, the proposed detector achieved superior detection accuracy (better than 99%) and recall rate (better than 99%) in the testing dataset, which outperforms the general-purpose detector and the pre-trained detector. Detailed analyses will be discussed in Section 5.

The proposed astronaut detector will play an important role in the robust intravehicular visual navigation in the manned space station to be discussed in Section 3, and support the astronaut visual tracking and motion prediction to be discussed in Section 4.

## 3. Visual Navigation in Semi-Structured and Dynamic Environments

In this section, we focus on the problem of robust visual navigation in the semi-structured and dynamic intravehicular environment, which is the other component of the overall framework. The problem is addressed using a map-based visual navigation technique that does not rely on any maker or additional device. The semi-structured environment makes it unnecessary to use a SLAM-like approach to explore unknown areas, and a map-based method is more practical and reliable. Moreover, compared with possible long-term environmental changes, the ability to cope with instant dynamic factors introduced by various intravehicular activities is more important.

*3.1. Map-Based Navigation in Semi-Structured Environments*

A proprietary RGB-D camera is used as the only sensor for mapping and intravehicular navigation. The RGB-D camera can not only provide color images with rich semantic information, but also the depth value of each pixel, which can improve the perception of distance and eliminate scale uncertainties.

(A)  Construction of the visual navigation map

In the mapping phase, the RGB-D camera is used to collect a video stream inside the space-station mockup from various positions and orientations. The collected data covered the entire space so that few blind areas are introduced. Based on the video stream, three main steps are utilized to build the final maps for intravehicular navigation.

(1)    Build initial map using standard visual SLAM technique.
(2)    Maps are optimized to minimize the distortion and the overall reprojection error.
(3)    The optimized maps are registered to the space station with a set of known points.

The initial map can be constructed using the Structure from Motion (SFM) technique [29] or standard visual SLAM technique. In our case, a widely used keyframe-based SLAM method [30] is adapted to build the initial point cloud map of the space station. The very first image frame is set as the map's origin temporarily. The point-cloud map contains plenty of distinguishable map points for localization and keyframes to reduce redundancy and assist feature matching. By searching enough associated map points in the current image, the robot can obtain its 6DoF pose with respect to the map.

In the second step, the map is optimized several times to minimize the overall measurement error, so that the map's distortion can be reduced as much as possible, and higher navigation accuracy can be achieved. The optimization problem is summarized as the minimization of reprojection error of associated map points in all keyframes.

$$\left\{ \mathbf{X}^j, \mathbf{R}_k, \mathbf{t}_k \right\} = \arg\min \sum_{k=1}^{K} \sum_{j=1}^{M} \rho \left( c_k^j \left\| \mathbf{x}_k^j - \pi \left( \mathbf{R}_k \mathbf{X}^j + \mathbf{t}_k \right) \right\|^2 \right) \tag{2}$$

where $\mathbf{X}^j$ is the coordinate of the $j$th map point, $\mathbf{R}_k$ and $\mathbf{t}_k$ are the rotational matrix and translational vector of the $k$th keyframe, $\pi(\cdot)$ is the camera projection function with known intrinsic parameters, $\mathbf{x}_k^j$ is the pixel coordinate of the matched feature point in the $k$th keyframe with respect to the $j$th map point, and $c_k^j$ is 1 if the match exists. $\rho(\cdot)$ is the robust Huber cost function to reduce the impact of error matches.

The constrained space in the space station allows for minimal distortion of the maps after global optimization as compared with applications in large-scale scenes. In the third step, a set of points with known coordinates are utilized to transform the optimized map to the world frame of the space station. Various types of maps can be constructed accordingly for different purposes such as localization, obstacle avoidance and communication [31]. Figure 4 presents three typical maps registered to the space-station mockup. All maps have an internal dimension of $2 \times 4 \times 2$ m. The point-cloud map shown in Figure 4a contains, in total, 12,064 map points with distinctive features, and 209 keyframes which are used to accelerate feature matching for pose initialization and re-localization. Figure 4b,c illustrates the dense point-cloud map and the octomap [32] constructed concurrently with the sparse point-cloud map. The clear definition and the straight contours of the mockup in the dense point cloud and the distinguishable handrails in the octomap proved the high accuracy of the maps after global optimization (2), which guarantees the accuracy of the map-based navigation system.

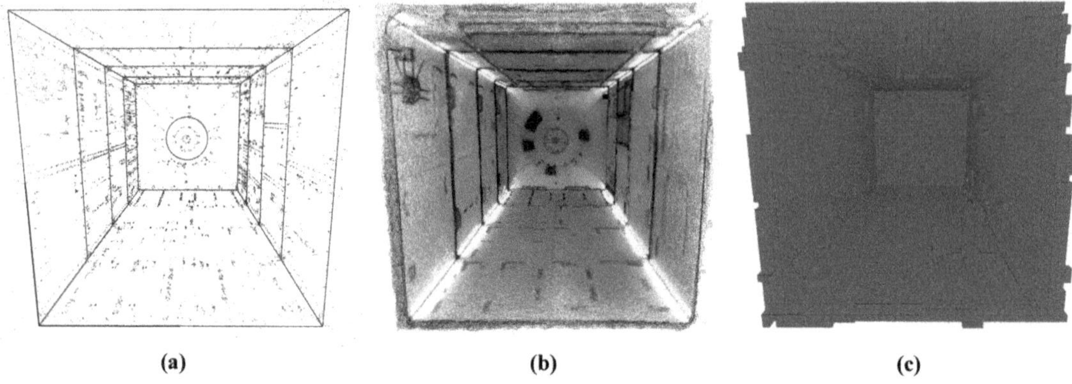

**Figure 4.** Various maps constructed and registered to the space-station mockup. (**a**) The (sparse) point-cloud map. (**b**) The dense point-cloud map. (**c**) Octomap for obstacle avoidance and motion planning.

(B)  Map-based localization and orientation

With prebuilt maps, two steps are carried out for intravehicular localization and orientation. In the first step, the robot tries to obtain an initial estimate of its 6DoF pose from scratch. This is achieved by comparing the current image with each similar keyframe in the sparse point-cloud map. Initial pose will be recovered using a PnP solver when enough 2D–3D matches are associated. With an initial pose estimation, local map points are then projected to the current image to search more 2D–3D matches for pose-only optimization, which will provide a more accurate localization result. The pose-only optimization problem can be summarized as the minimization of reprojection error with a static map.

$$\{\mathbf{R}, \mathbf{t}\} = \arg\min \sum_{j=1}^{M} \rho\left( c^j \left\| \mathbf{x}^j - \pi\left( \mathbf{R}\mathbf{X}^j + \mathbf{t} \right) \right\|^2 \right) \tag{3}$$

where $\mathbf{R}$ and $\mathbf{t}$ are the rotational matrix and translational vector of the robot with respect to the world frame, $\mathbf{x}^j$ is the pixel coordinate of the matched feature point with respect to the $j$th map point, and $c^j$ is 1 if the match exists.

When the robot succeeds to localize itself for several consecutive frames after initialization or re-localization, frame-to-frame velocity is utilized to provide the initial guess to search map points, which saves time and helps improve computational efficiency.

*3.2. Robust Navigation during Human–Robot Collaboration*

The robotic assistant usually works side by side with human astronauts to provide immediate service. In the constrained intravehicular environment, astronauts can take a field of vision in front of the robot. Astronauts' various intravehicular activities can also occlude the map points and introduce dynamic disturbance to the navigation system. In such a condition, the robot may be confused to search enough map points for stable in-cabin localization. The poor localization will, in turn, create uncertainties for the robot to accomplish various onboard tasks.

To address the problem, we proposed the integrated framework of simultaneous astronaut accompanying and visual navigation in the dynamic and semi-structured intravehicular environment. The framework can not only solve the problem of robust navigation in dynamic scenes during human–robot collaboration, but also assist in tracking and predicting the short-motion of the served astronaut to provide more satisfactory and foresighted assistance.

As shown in Figure 5, the framework adopts a multi-thread computing architecture to ensure real-time performance. The main thread in the red dashed box is mainly responsible for image pre-processing and in-cabin visual navigation, whereas the sub thread in the blue dashed box is mainly responsible for astronaut detection, visual tracking, and trajectory estimation. Specifically, while the main thread is working on frame registration and feature extraction, the sub thread tries to detect astronauts in the meantime. The first-round information exchange between the two threads is carried out at this point. Then, feature points within the detected bounding boxes are filtered out to avoid large areas of disturbances to the visual navigation system.

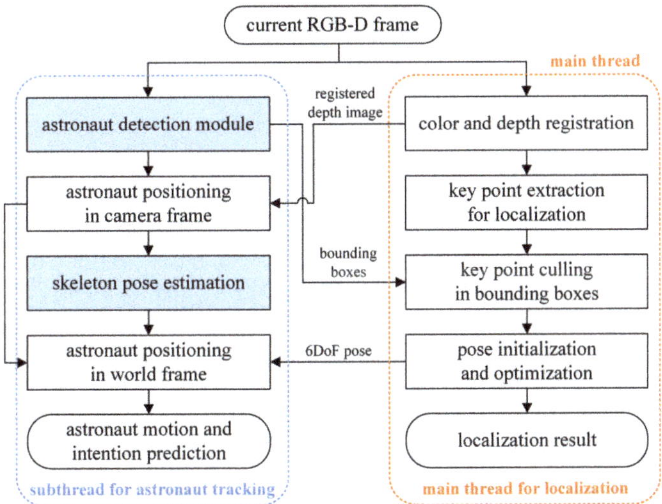


**Figure 5.** Computational work flow of the integrated framework of simultaneously astronaut accompanying and visual navigation in the semi-structured and dynamic intravehicular environment.

While the main thread is working on 6DoF pose initialization and optimization, the sub thread is idle and can perform some computations such as astronaut skeleton extraction. The second-round information exchange i8s carried out once the main thread has obtained the localization result. The optimized 6DoF pose together with the detected bounding boxes are utilized for astronaut visual tracking and motion prediction in the sub thread, which will be discussed in Section 4.

The computational burden of the proposed framework is evenly distributed where the main thread uses mainly CPU resources and the sub thread consumes mainly GPU resources. The overall algorithm is tested to run at over 30 Hz with a GS66 laptop (low-power i9@2.4GHz processor and RTX 2080 GPU for notebook).

## 4. Astronaut Visual Tracking and Motion Prediction

Astronaut visual tracking and motion prediction help the robot track and identify the served astronaut and provide immediate assistance when required. The solution to the problem is based on the research into astronaut detection in Section 2 and the research into robust intravehicular navigation in Section 3.

Specifically, the astronaut visual-tracking problem is to detect and track the movement of a certain target astronaut in a sequence of images, which is formulated as a maximum a-posteriori (MAP) estimation problem as

$$i = \arg\max P^k\left(p \mid \beta_i^k, \beta_t^{k-1}\right), i = 1, 2, \ldots, M \tag{4}$$

where $M$ is the number of detected astronauts in the current (or $k$th) frame; $\beta_i^k$ and $\beta_i^{k-1}$ are the $i$th bounding box in the current frame and the target to be matched in the previous frame, respectively; and $P^k(\cdot \mid \beta_i^k, \beta_t^{k-1})$ defines the probability of the match. We seek to find the bounding box with the largest posterior probability. If no bounding box is matched for a long time or the wrong bounding box is selected, the tracking task fails.

The posterior probability in Equation (4) is determined by a variety of factors. For example, when the 3D position of $\beta_i^k$ is close to the predicted trajectory of the served astronaut, or the bounding boxes overlap, there is a high match probability. On the contrary, when the 3D position of $\beta_i^k$ deviates from the predicted trajectory or the geometry mismatches, the probability will be small. According to the above factors, the overall posterior probability is decomposed into the trajectory correlation probability $P^k_{\text{predicton}}$, and the geometric correlation probability $P^k_{\text{geometry}}$ and other clues $P^k_{\text{others}}$ include identity identification probability.

$$
\begin{aligned}
i &= \arg\max P^k\left(p \mid \beta_i^k, \beta_t^{k-1}\right) \\
&= \arg\max P^k_{\text{predicton}}\left(p \mid \beta_i^k\right) P^k_{\text{geometry}}\left(p \mid \beta_i^k, \beta_t^{k-1}\right) P^k_{\text{others}}\left(p \mid \beta_i^k, \beta_t^{k-1}\right)
\end{aligned}
\tag{5}
$$

(A)  Matching with predicted trajectory

The served astronaut's trajectory can be estimated and predicted using the astronaut detection result in the image flow and the robot's 6DoF localization information.

Firstly, the 3D position of the astronaut in the robot body frame $[x_b, y_b, z_b]^T$ is obtained using the camera's intrinsic parameters. The coordinates are averaged over a set of points within a small central area in the bounding box to reduce the measurement error. Then, by incorporating the 6DoF pose $\{\mathbf{R}, \mathbf{t}\}$ of the robot (3), the astronaut's 3D position can be transformed to be represented in the world frame of the space station $[x_w, y_w, z_w]^T$.

$$
\begin{bmatrix} x_w \\ y_w \\ z_w \end{bmatrix} = \mathbf{R}^T \left( \begin{bmatrix} x_b \\ y_b \\ z_b \end{bmatrix} - \mathbf{t} \right)
\tag{6}
$$

The space station usually keeps three axes stabilized to the earth and orbits every 1.5 h. We assume the space station to be an inertial system when modeling the instant motion of astronauts. The motion is formulated as a constant acceleration model for simplicity. For example, when the astronaut moves freely in microgravity, a constant speed can be estimated and the acceleration is zero. When the astronaut contacts with the surroundings, a time-varying acceleration can be estimated by introducing a relatively large acceleration noise in the model. The above motion model and corresponding measurement model are defined as

$$
\begin{aligned}
x_w^k &= A x_w^{k-1} + w \\
z^k &= H x_w^k + v
\end{aligned}
\tag{7}
$$

where $A \in \mathbb{R}^{9\times9}$ is the state transition matrix that determines the relationship between the current state $x_w^k \in \mathbb{R}^9$ and the previous state $x_w^{k-1}$, vector $z^k$ is the measured 3D position of the astronaut represented in the world frame, $H \in \mathbb{R}^{3\times9}$ is the measurement matrix, $w$ is the time-invariant process noise to characterize the error of the simplified motion model, and $v$ is the time-invariant measurement noise determined by the positioning accuracy of the served astronaut. The process and measurement noise are assumed to be white Gaussian with zero means and covariance matrices of $Q$ and $R$, respectively.

The nine-dimentional state vector contains the estimated position, velocity, and acceleration of the served astronaut represented in the world frame as

$$
x_w^k = \begin{bmatrix} x_w^k & y_w^k & z_w^k & v_{x,w}^k & v_{y,w}^k & v_{z,w}^k & a_{x,w}^k & a_{y,w}^k & a_{z,w}^k \end{bmatrix} \in \mathbb{R}^9
\tag{8}
$$

The trajectory of the served astronaut can be predicted with the above constant acceleration model. The update interval is kept the same as the frequency of the overall astronaut-detection and visual-navigation framework at 30 Hz.

$$
\begin{aligned}
x_w^{k-} &= A x_w^{k-1} \\
P^{k-} &= A P^{k-1} A^T + Q
\end{aligned}
\tag{9}
$$

where $P$ is the state covariance matrix. We propagate Equation (9) for a few seconds to predict the short-term motion of the served astronaut.

With estimated trajectories, a comparison is made between the prediction and each bounding box in the current image frame. There would be a high correlation probability if the 3D position of a certain bounding box is close to the predicted trajectory of the target astronaut. The trajectory correlation probability is defined as

$$
P_{\text{predicton}}^k \left( p \mid \beta_i^k \right) = e^{-\alpha_0 \left\| z_i^k - x_w^{k-}(1:3) \right\|}
\tag{10}
$$

where $z_i^k$ is the measured position of the astronaut in the $i$th bounding box and $\alpha_0$ is a non-negative parameter.

Once the match is verified together with other criteria, the measurement will be used to correct the motion model of the target astronaut.

$$
\begin{aligned}
K^k &= P^{k-} H^T (H P^{k-} H^T + R)^{-1} \\
x_w^k &= x_w^{k-} + K^k (z^k - H x_w^{k-}) \\
P^k &= (I - K^k H) P^{k-}
\end{aligned}
\tag{11}
$$

(B)    Matching with geometric similarity

Besides trajectory correlation, the geometric similarity of the bounding boxes can also provide valuable information for visual tracking. The overall algorithm runs at 30 Hz, and, thus, we assume few changes between consecutive frames to be introduced. Many criteria can be used to characterize the similarity between bounding boxes. We selected the most straightforward IOU (intersection over union) criterion. When a certain bounding box in the current image frame overlaps heavily with the target in the previous frame, there would be a high matching probability. The geometric correlation probability is defined as

$$
P_{\text{geometry}}^k \left( p \mid \beta_i^k, \beta_t^{k-1} \right) = e^{-\alpha_1 (1 - \mathbf{IOU})}
\tag{12}
$$

where $\alpha_1$ is a non-negative parameter, and a larger $\alpha_1$ will give more weight to the IOU criterion.

(C)    Matching with other clues

Some other clues can also be incorporated to assist astronaut visual tracking. For example, face recognition is helpful for initial identity confirmation and tracking recovery after long-time loss. The corresponding posterior probability is formulated as

$$
P_{\text{others}}^k \left( p \mid \beta_i^k, \beta_t^{k-1} \right) = \begin{cases} 1.0, & \text{matched} \\ 0.5, & \text{not sure} \\ 0.0, & \text{not matched} \end{cases}
\tag{13}
$$

During the experiments, we only applied the trajectory and geometric correlation probabilities into the framework. The face-recognition part is out of the scope of this paper, and can be referenced from our previous work [18].

## 5. Experimental Results and Discussion

Experiments were carried out to evaluate each component of the proposed framework in Section 5.1 (astronaut detection) and Section 5.2 (visual navigation) respectively. The overall performance is verified and discussed in Section 5.3.

### 5.1. Evaluation of the Customized Astronaut Detector

The performance of the proposed astronaut-detection network is evaluated in the testing dataset (5824 images) collected in the space-station mockup. As shown in Figure 6, the fine-tuned detector shows an AP@0.5 of 99.36%, which outperforms the general-purpose detection network (85.06%) and the pretrained detector (90.78%). The superior performance of the astronaut detector benefited from the customized network structure designed for intravehicular applications and the proposed astronaut-detection dataset to mitigate possible domain inconsistency.

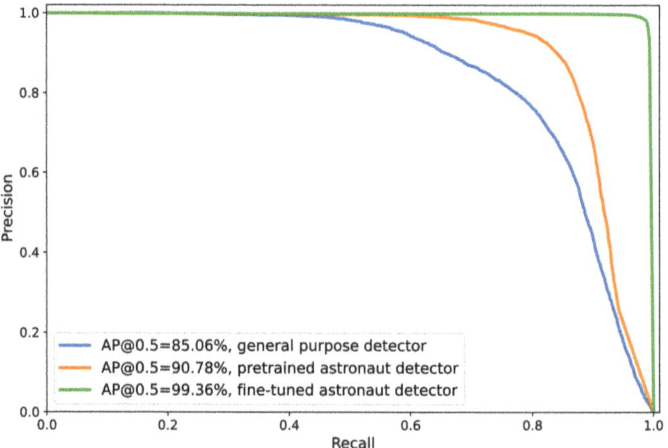


**Figure 6.** Comparison of the precision-recall curves of three detectors in the task of astronaut detection in the space-station mockup.

Figure 7 presents some typical results for comparative studies. All three networks achieved satisfactory detection when volunteers are close to the upright posture. The general-purpose network may give some false bounding boxes that do not exist in the mockup, such as a clock. When astronauts' body postures or orientations are significantly different from daily life scenes on the ground, both the general-purpose detector and the pretrained detector degrade. A large number of missed detections and poor detections can be found. On the other hand, the fine-tuned astronaut detector still guarantees its performance when dealing with the challenging task. It is worth mentioning that all networks showed satisfactory performance to cope with illumination variation and motion blur without implementing image-enhancement algorithms [33].

The proposed astronaut detector showed superior performance to cope with the rich body postures and orientations. The estimated pixel coordinates of the bounding boxes are also more accurate than its competitors. The proposed detector runs at over 80 Hz on a GS66 laptop, proving the sufficiency for real-time performance.

### 5.2. Evaluation of Map-Based Navigation in Semi-Structured and Dynamic Environments

Experiments were conducted in the mockup to test the accuracy and robustness of the proposed map-based navigation system in both static and dynamic scenarios. As shown in Figure 8a, the mockup has an internal dimension of 2 × 4 × 2 m and has high fidelity to a real space station. The handrails, buttons, experiment cabinets, and airlock provided stable visual references for visual navigation.

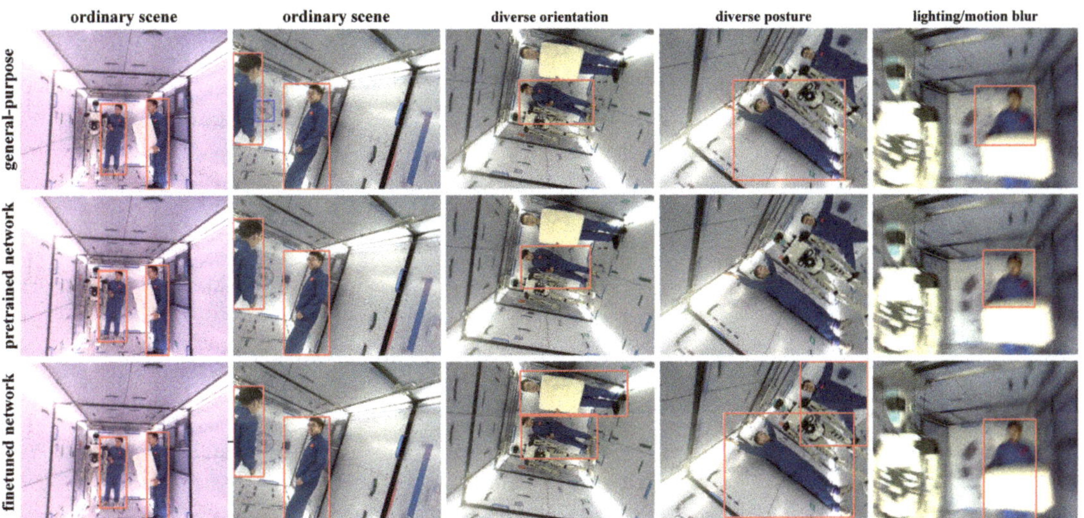

**Figure 7.** Astronaut-detection performance of the general-purpose network, pretrained network and the fine-tuned astronaut detector on the testing dataset.

**Figure 8.** The ground experimental environment. (**a**) The space-station mockup of high fidelity. (**b**) The humanoid robotic assistant Taikobot used in the experiment.

(A)  Performance in static environment

During the experiment, the RGB-D camera is moved and rotated constantly to collect video streams in the mockup. Four large (60 × 60 cm) Aruco markers [34] are fixed to the back of the camera to provide reference trajectories for comparison. Figure 9 presents the results in a static environment. As shown in Figure 9a,b, we performed a large range of motion in all six translational and rotational directions consecutively. The estimated 6DoF pose almost coincides with the reference trajectories. Figure 9c presents the corresponding error curves. The average positional error is less than 1cm (the maximum error does not exceed 2 cm) and the average three-axis angular error is less than 0.5°. The camera's overall trajectory during the experiment is shown in Figure 9d. Two other random trajectories were also collected and analyzed as shown in Figure 10 where identical performance was achieved, proving the feasibility of the proposed navigation method.

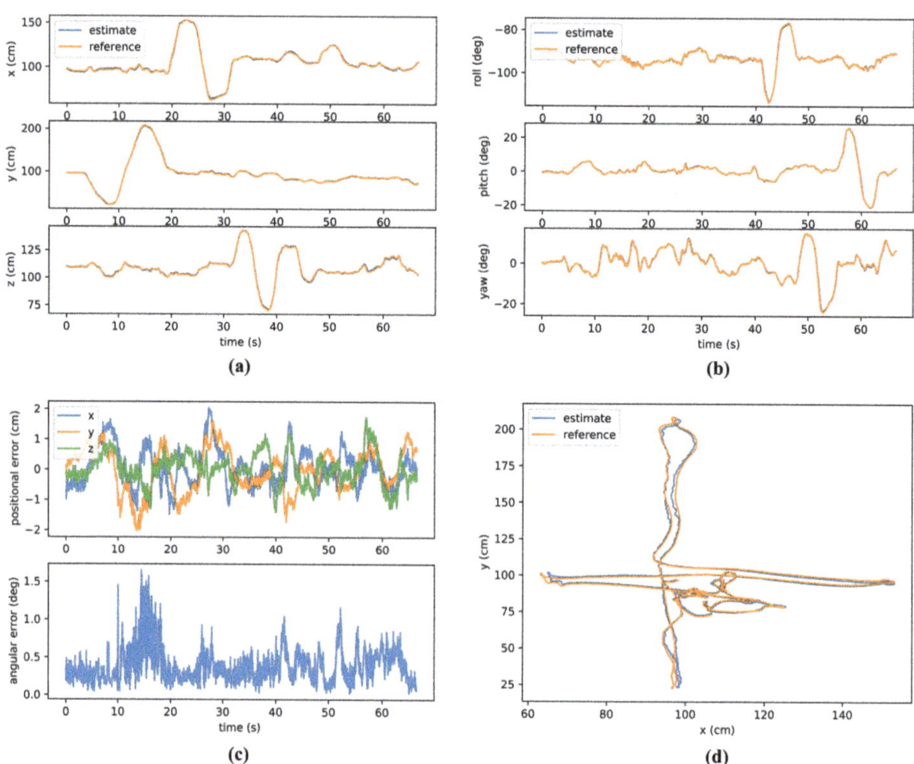

**Figure 9.** Localization and orientation performance of the proposed map-based navigation system in static environment. (**a**) Position curves in world frame. (**b**) Euler angle curves with respect to the world frame. (**c**) Positional error and three-axis angular error. (**d**) The estimated trajectories in the XY plane of the space-station mockup.

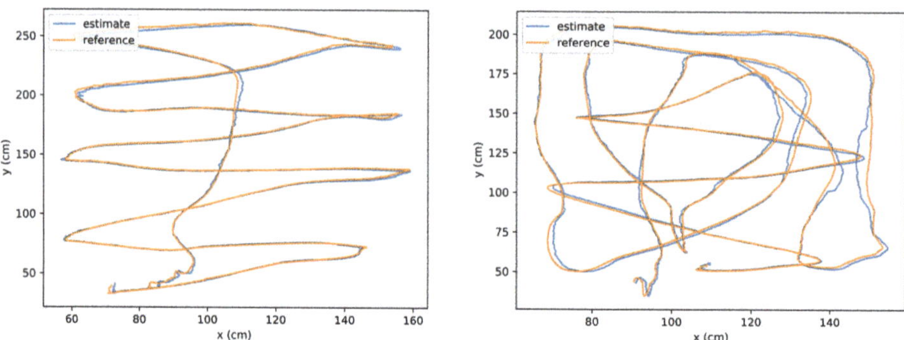

**Figure 10.** Two random trajectories tested in the space-station mockup (static environment).

(B)    Performance in dynamic environment

Next, we evaluate the map-based navigation in dynamic scenes when the robot works along with human astronauts. As shown in Figure 8b, the humanoid robotic assistant

Taikobot [35] we developed previously is used this time. The RGB-D camera mounted in the head of Taikobot is used both for astronaut detection and intravehicular navigation. During the experiment, the robot moves along with a volunteer astronaut in the mockup to provide immediate assistance. The astronaut can occasionally require a large field of vision in front of the robot during intravehicular activities. As shown in Figure 11a,b, the robot navigates robustly and smoothly in the dynamic environment with the proposed framework. Based on the stable localization result of the robot, the trajectories of the served astronaut are also estimated and predicted in the meantime, which will be discussed in Section 5.3.

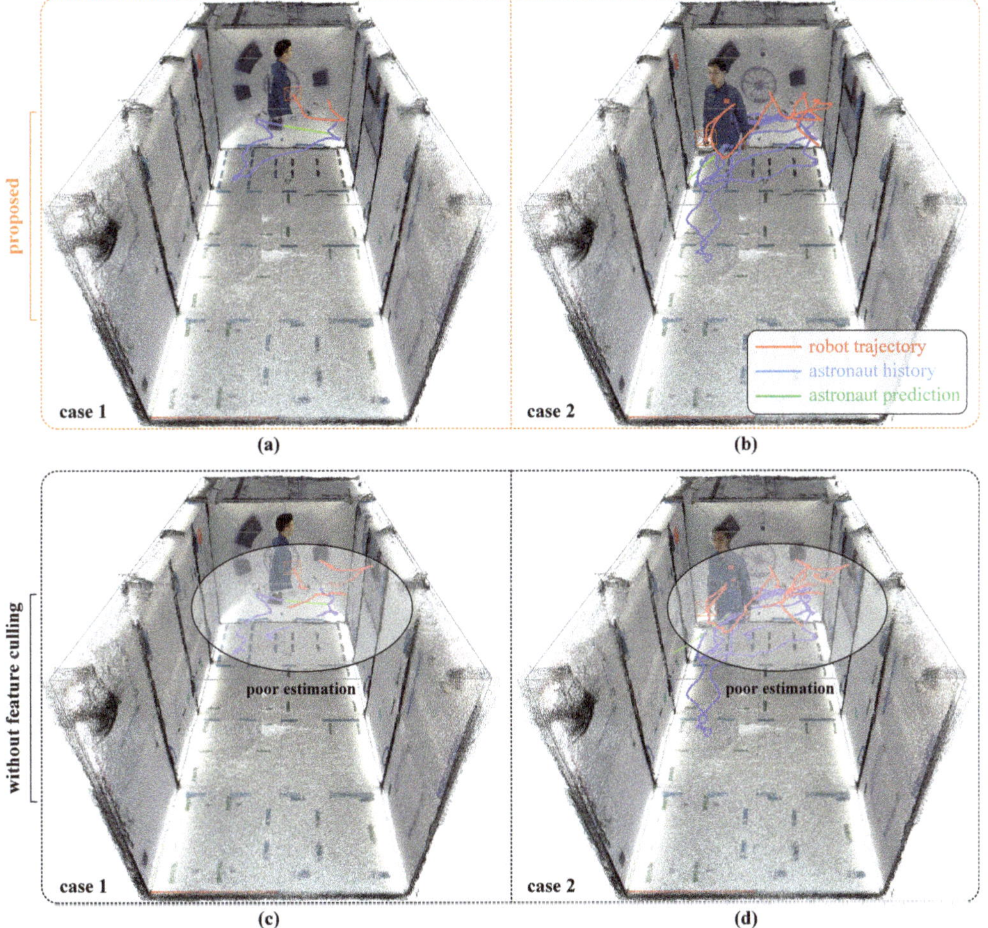


**Figure 11.** Localization performance of the map-based navigation system in dynamic environment. Red lines are the estimated trajectories of the robotic assistant. Blue and green lines are the estimated and predicted trajectories of the served astronaut, respectively. (**a**,**b**) performance of the proposed framework. (**c**,**d**) performance without feature culling.

By comparison, when we remove the feature-culling module in the framework, the robot becomes lost several times with the same data input as shown in Figure 11c,d. The degradation in the navigation system is caused by the dynamic feature points detected on the served astronauts, which makes it difficult for the robot to locate sufficient references for stable in-cabin navigation. As we can see, the poor localization result also led to poor trajectory estimation of the astronaut.

*5.3. Verification of Simultaneous Astronaut Accompanying and Visual Navigation*

Based on the robust intravehicular navigation system and the customized astronaut detector, the trajectory of the served astronaut can be identified, estimated and predicted efficiently.

Firstly, we present the results when only one astronaut is served. Figure 12 gives two typical scenarios where the robot moves along with one astronaut in the mockup. The red and green curves are the measured and predicted trajectories of the served astronaut, respectively. The blue curves are the estimated trajectories of the robot by (3). During the experiments, the astronaut was kept within the robot's perspective. In both scenarios, the robot can navigate smoothly in the dynamic scenes, and the astronaut is tracked stably in the image flow at all times. The predicted trajectories of the astronaut are identical to the measurements. By applying the proposed motion model, the predictions are also smoothed compared with the raw measurements.

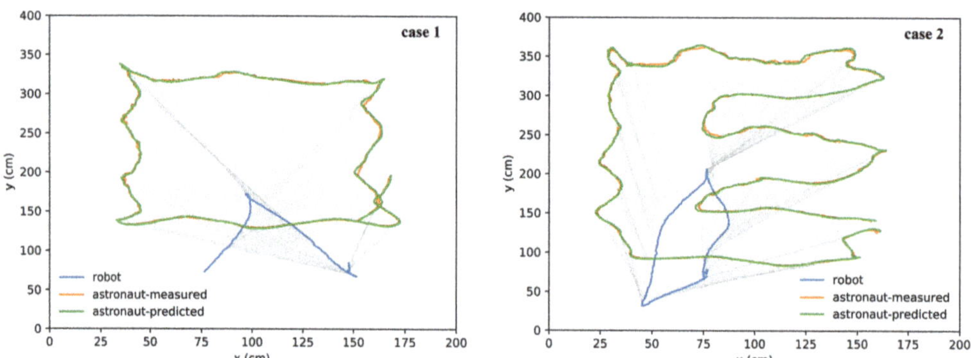


**Figure 12.** Experimental results of simultaneous astronaut tracking and visual navigation when the robotic assistant accompanies one astronaut.

When multiple astronauts coexist, the robot is able to track a certain astronaut to provide a customized service. The task is more challenging compared with the previous examples. Figure 13 presents the results of two typical scenarios where the robot works along with two astronauts at the same time. We take case 4 to discuss in detail. As shown in the picture series in Figure 13, when the robot has confirmed (red bounding box) the target astronaut (astronaut A), the other astronaut (astronaut B) enters in the field of view of the robot. The robot can distinguish the target astronaut from astronaut B by utilizing the trajectory correlation and geometric similarity criteria (5). The robot tracks the target astronaut robustly even though the two astronauts move closely and overlap. The most challenging part occurs when astronaut B moves in between the robot and the target astronaut. When astronaut A is completely obscured from the robot, tracking loss is inevitable. However, when astronaut A reappears in the image, the robot recovers the tracking immediately. It is worth mentioning that only the trajectory and geometry criteria are used in the tracking process, which have minimal computing burden. Other criteria such as face recognition can also be incorporated into the framework for tracking recovery after long-time loss.

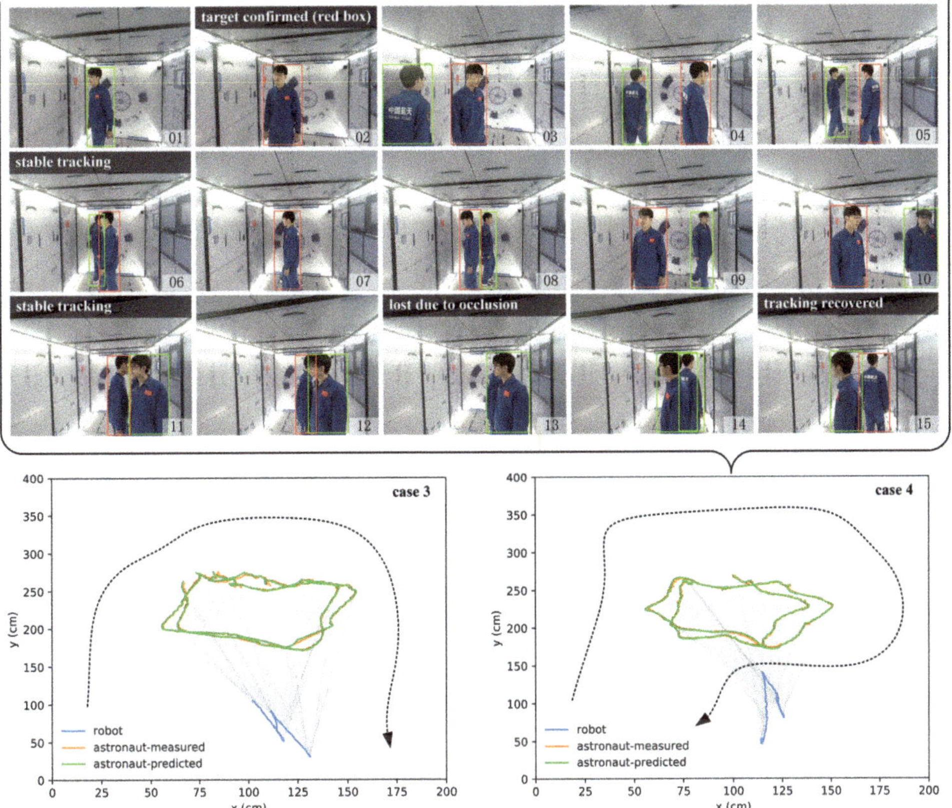


**Figure 13.** Experimental results of simultaneous astronaut tracking and visual navigation when multiple astronauts coexist in the space-station mockup. The red bounding boxes in the sequentially numbered pictures denote the target astronaut. The dotted curves in the two sketches denote the routes of astronaut B.

## 6. Conclusions

This paper proposed the framework of simultaneous astronaut accompanying and visual navigation in the semi-structured and dynamic intravehicular environment. In terms of the intravehicular navigation problem of IRA, the proposed map-based visual-navigation framework is able to provide real-time and accurate 6DoF localization results even in dynamic scenes during human–robot interaction. Moreover, compared with the other map-based localization methods of IRA in the literature, we achieved superior accuracy (1~2 cm, 0.5°). In terms of the astronaut visual tracking and short-term motion-prediction problem, the proposed MAP model with geometric similarity and trajectory correlation hints enables IRA to distinguish and accompany the served astronaut with minimal calculation from a moving point of view. The overall framework provided a feasible solution to address the problem of intravehicular robotic navigation and astronaut–robot coordination in the manned and constrained space station.

**Author Contributions:** Conceptualization, Q.Z. and Y.Z.; methodology, Q.Z. and Y.Z.; software, Q.Z.; validation, Q.Z.; formal analysis, Q.Z.; investigation, Q.Z.; resources, L.F. and Y.Z.; data curation, Q.Z. and L.F.; writing—original draft preparation, Q.Z.; writing—review and editing, L.F. and Y.Z.; visualization, Q.Z. and L.F.; supervision, L.F. and Y.Z.; project administration, L.F. and Y.Z.; funding acquisition, Y.Z. All authors have read and agreed to the published version of the manuscript.

**Funding:** This work is supported by Huzhou Institute of Zhejiang University under the Huzhou Distinguished Scholar Program (ZJIHI—KY0016).

**Institutional Review Board Statement:** Not applicable.

**Informed Consent Statement:** Not applicable.

**Data Availability Statement:** The data presented in this study are available on request from the corresponding author. The data are not publicly available due to intellectual-property protection.

**Conflicts of Interest:** The authors declare no conflict of interest.

## Abbreviations

| | |
|---|---|
| IRA | Intravehicular robotic assistants |
| CIMON | Crew interactive mobile companion |
| IFPS | Intelligent formation personal satellite |
| SPHERES | Synchronized position hold engage and reorient experimental satellite |
| ISS | Internationall Space Station |
| JEM | Japanese experiment module |
| FPN | Feature pyramid network |
| PAN | Path aggregation network |
| IOU | Intersection over union |
| CIOU | Complete intersection over union |
| COCO | Common object in context |
| RGB-D | Red green blue-depth |
| SFM | Structure from motion |
| SLAM | Simultaneous localization and mapping |
| PnP | Perspective-n-point |
| AP | Average precision |
| MAP | Maximum a posteriori |


## References

1. Sgobba, T.; Kanki, B.; Clervoy, J.F. *Space Safety and Human Performance*, 1st ed.; Butterworth-Heinemann: Oxford, UK, 2018; pp. 357–376. Available online: https://www.elsevier.com/books/space-safety-and-human-performance/sgobba/978-0-08-1 01869-9 (accessed on 10 October 2022).
2. Russo, A.; Lax, G. Using artificial intelligence for space challenges: A survey. *Appl. Sci.* **2022**, *12*, 5106. [CrossRef]
3. Miller, M.J.; McGuire, K.M.; Feigh, K.M. Information flow model of human extravehicular activity operations. In Proceedings of the 2015 IEEE Aerospace Conference, Big Sky, MT, USA, 7–14 March 2015.
4. Miller, M.J. Decision support system development for human extravehicular activity. Ph.D. Thesis, Georgia Institute of Technology, Atlanta, GA, USA, 2017.
5. Akbulut, M.; Ertas, A.H. Establishing reduced thermal mathematical model (RTMM) for a space equipment: An integrative review. *Aircr. Eng. Aerosp. Technol.* **2022**, *94*, 1009–1018. [CrossRef]
6. Li, D.; Zhong, L.; Zhu, W.; Xu, Z.; Tang, Q.; Zhan, W. A survey of space robotic technologies for on-Orbit assembly. *Space Sci. Technol.* **2022**, *2022*, 9849170. [CrossRef]
7. Smith, T.; Barlow, J.; Bualat, M. Astrobee: A new platform for free-flying robotics on the international space station. In Proceedings of the 13th International Symposium on Artificial Intelligence, Robotics, and Automation in Space, Beijing, China, 20–22 June 2016.
8. Mitani, S.; Goto, M.; Konomura, R. Int-ball: Crew-supportive autonomous mobile camera robot on ISS/JEM. In Proceedings of the 2019 IEEE Aerospace Conference, Yellowstone Conference Center, Big Sky, MT, USA, 2–9 March 2019.
9. Experiment CIMON—Astronaut Assistance System. Available online: https://www.dlr.de/content/en/articles/missions-projects/horizons/experimente-horizons-cimon.html (accessed on 10 October 2022).
10. Zhang, R.; Wang, Z.K.; Zhang, Y.L. A person-following nanosatellite for in-cabin astronaut assistance: System design and deep-learning-based astronaut visual tracking implementation. *Acta Astronaut.* **2019**, *162*, 121–134. [CrossRef]
11. Liu, Y.Q.; Li, L.; Ceccarelli, M.; Li, H.; Huang, Q.; Wang, X. Design and testing of BIT flying robot. In Proceedings of the 23rd CISM IFToMM Symposium, Online, 20–24 September 2020. Available online: http://doi.org/10.1007/978-3-030-58380-4_9 (accessed on 10 October 2022).
12. NASA Facts Robonaut 2, Technical Report. Available online: https://www.nasa.gov/sites/default/files/files/Robonaut2_508 .pdf (accessed on 10 October 2022).
13. Meet Skybot F-850, the Humanoid Robot Russia Is Launching into Space. Available online: https://www.space.com/russia-launching-humanoid-robot-into-space.html (accessed on 10 October 2022).

14. Chen, L.; Lin, S.; Lu, X.; Cao, D.; Wu, H.; Guo, C.; Liu, C.; Wang, F. Deep neural network based vehicle and pedestrian detection for autonomous driving: A survey. *IEEE Trans. Intell. Transp. Syst.* **2021**, *22*, 3234–3246. [CrossRef]
15. Bochkovskiy, A.; Wang, C.Y.; Liao, H. YOLOv4: Optimal speed and accuracy of object detection. *arXiv* **2020**, arXiv:2004.10934.
16. Avdelidis, N.P.; Tsourdos, A.; Lafiosca, P.; Plaster, R.; Plaster, A.; Droznika, M. Defects recognition algorithm development from visual UAV inspections. *Sensors* **2022**, *22*, 4682. [CrossRef] [PubMed]
17. Zhang, R.; Wang, Z.K.; Zhang, Y.L. Astronaut visual tracking of flying assistant robot in space station based on deep learning and probabilistic model. *Int. J. Aerosp. Eng.* **2018**, *2018*, 6357185. [CrossRef]
18. Zhang, R.; Zhang, Y.L.; Zhang, X.Y. Tracking in-cabin astronauts Using deep learning and head motion clues. *Int. J. Aerosp. Eng.* **2018**, *9*, 2680–2693. [CrossRef]
19. Saenz-Otero, A.; Miller, D.W. Initial SPHERES operations aboard the International Space Station. In Proceedings of the 6th IAA Symposium on Small Satellites for Earth Observation, Berlin, Germany, 23–26 April 2008.
20. Prochniewicz, D.; Grzymala, M. Analysis of the impact of multipath on Galileo system measurements. *Remote Sens.* **2021**, *13*, 2295. [CrossRef]
21. Coltin, B.; Fusco, J.; Moratto, Z.; Alexandrov, O.; Nakamura, R. Localization from visual landmarks on a free-flying robot. In Proceedings of the 2016 IEEE/RSJ International Conference on Intelligent Robots and Systems, Seoul, Republic of Korea, 8–9 October 2016.
22. Kim, P.; Coltin, B.; Alexandrov, O. Robust visual localization in changing lighting conditions. In Proceedings of the 2017 IEEE International Conference on Robotics and Automation, Marina Bay Sands, Singapore, 29 May–3 June 2017.
23. Xiao, Z.; Wang, K.; Wan, Q.; Tan, X.; Xu, C.; Xia, F. A2S-Det: Efficiency anchor matching in aerial image oriented object detection. *Remote Sens.* **2021**, *13*, 73. [CrossRef]
24. He, K.; Zhang, X.; Ren, S.; Sun, J. Deep residual learning for image recognition, In Proceedings of the 2016 IEEE Conference on Computer Vision and Pattern Recognition, Las Vegas, NV, USA, 26 June–1 July 2016.
25. Liu, S.; Qi, L.; Qin, H. Shi, J.; Jia, J. Path aggregation network for instance segmentation. In Proceedings of the 2018 IEEE Conference on Computer Vision and Pattern Recognition, Salt Lake City, UT, USA, 19–21 June 2018.
26. COCO Common Objects in Context. Available online: https://cocodataset.org/ (accessed on 10 October 2022).
27. Shao, S.; Zhao, Z.; Li, B.; Xiao, T.; Yu, G.; Zhang, X.; Sun, J. CrowdHuman: A benchmark for detecting human in a crowd. *arXiv* **2018**, arXiv:1805.00123.
28. Zheng, Z.; Wang, P.; Ren, D.; Liu, W.; Ye, R.; Hu, Q.; Zuo, W. Enhancing geometric factors in model learning and inference for object detection and instance segmentation. *IEEE Trans. Cybern.* **2020**, *52*, 8574–8586. [CrossRef] [PubMed]
29. Jiang, S.; Jiang, C.; Jiang, W. Efficient structure from motion for large-scale UAV images: A review and a comparison of SfM tools. *ISPRS J. Photogramm. Remote. Sens.* **2020**, *167*, 230–251. [CrossRef]
30. Mur-Artal, R.; Tardos, J.D. ORB-SLAM2: An open-source SLAM system for monocular, stereo and RGB-D cameras. *IEEE Trans. Robot.* **2017**, *33*, 1255–1262. [CrossRef]
31. Koletsis, E.; Cartwright, W.; Chrisman, N. Identifying approaches to usability evaluation. In Proceedings of the 2014 Geospatial Science Research Symposium, Melbourne, Australia, 2–3 December 2014.
32. Hornung, A.; Wurm, K.M.; Bennewitz, M.; Stachniss, C.; Burgard, W. OctoMap: An efficient probabilistic 3D mapping framework based on octrees. *Auton. Robot.* **2013**, *34*, 189–206. [CrossRef]
33. Irmak, E.; Ertas, A.H. A review of robust image enhancement algorithms and their applications. In Proceedings of the 2016 IEEE Smart Energy Grid Engineering Conference, Oshawa, ON, Canada, 21–24 August 2016.
34. Romero-Ramirez, F.J.; Muñoz-Salinas, R.; Medina-Carnicer, R. Speeded up detection of squared fiducial markers. *Image Vis. Comput.* **2018**, *76*, 38–47. [CrossRef]
35. Zhang, Q.; Zhao, C.; Fan F.; Zhang Y. Taikobot: A full-size and free-flying humanoid robot for intravehicular astronaut assistance and spacecraft housekeeping. *Machines* **2022**, *10*, 933. [CrossRef]

*Article*

# Small-Object Detection for UAV-Based Images Using a Distance Metric Method

Helu Zhou [1,2,†], Aitong Ma [1,†], Yifeng Niu [1] and Zhaowei Ma [1,*]

1    College of Intelligence Science and Technology, National University of Defense Technology, Changsha 410078, China
2    Aerocraft Technology Branch of Hunan Aerospace Co., Ltd., Changsha 410200, China
*    Correspondence: mazhaowei1989@126.com
†    These authors contributed equally to this work.

**Abstract:** Object detection is important in unmanned aerial vehicle (UAV) reconnaissance missions. However, since a UAV flies at a high altitude to gain a large reconnaissance view, the captured objects often have small pixel sizes and their categories have high uncertainty. Given the limited computing capability on UAVs, large detectors based on convolutional neural networks (CNNs) have difficulty obtaining real-time detection performance. To address these problems, we designed a small-object detector for UAV-based images in this paper. We modified the backbone of YOLOv4 according to the characteristics of small-object detection. We improved the performance of small-object positioning by modifying the positioning loss function. Using the distance metric method, the proposed detector can classify trained and untrained objects through object features. Furthermore, we designed two data augmentation strategies to enhance the diversity of the training set. We evaluated our method on a collected small-object dataset; the proposed method obtained 61.00% $mAP_{50}$ on trained objects and 41.00% $mAP_{50}$ on untrained objects with 77 frames per second (FPS). Flight experiments confirmed the utility of our approach on small UAVs, with satisfying detection performance and real-time inference speed.

**Keywords:** small-object detection; backbone design; object positioning; object classification; UAV flight experiment

**Citation:** Zhou, H.; Ma, A.; Niu, Y.; Ma, Z. Small-Object Detection for UAV-Based Images Using a Distance Metric Method. *Drones* **2022**, *6*, 308. https://doi.org/10.3390/drones6100308

Academic Editor: Diego González-Aguilera

Received: 11 September 2022
Accepted: 17 October 2022
Published: 20 October 2022

**Publisher's Note:** MDPI stays neutral with regard to jurisdictional claims in published maps and institutional affiliations.

## 1. Introduction

Nowadays, Unmanned aerial vehicles (UAVs) play an important role in civil and military fields, such as system mapping [1], low-attitude remote sensing [2], collaborative reconnaissance [3], and others. In many applications, reconnaissance tasks are mostly based on UAV airborne vision. In this case, the detection and recognition of ground targets is an important demand. However, when the UAV flies at high altitudes, the captured object occupies a relatively small pixel scale in UAV airborne images. It is a challenge to detect such small objects in complex large scenes. Additionally, due to the limited computing resources in UAVs, many large-scale detection models based on server and cloud computing are not suitable for online real-time detection of small unmanned aerial vehicles. In this case, achieving fast and accurate small-object detection using the onboard computer becomes challenging. This paper mainly focuses on the detection of small objects in UAV reconnaissance images.

Combining with the flight characteristics of small UAVs and the computing capability of onboard processors, this paper selects a neural-network-based model as the basic detection model. To the best of our knowledge, most of the current detection algorithms for UAVs use one-stage detectors [4]. One of the state-of-the-art detectors among the one-stage detectors is YOLOv4 [5]. The YOLOv4 object detector integrates various classic ideas [6–9] in the field of object detection and works at a faster speed and higher accuracy than other alternative detectors. We choose YOLOv4 as the benchmark detector. The YOLOv4 detector was proposed and trained using a common dataset [10] covering various objects. However, the objects in the field of UAV reconnaissance that we are concerned with are limited in category, such as cars, aircraft, and ships. There are many instances of subdividing these limited categories of targets, but current object detection training sets rarely care about all types of objects. Therefore, objects that have not appeared in the training set are difficult to recognize in the UAV reconnaissance image during the inference stage, which is also a major challenge for UAV object detection. Generally, these objects are small in the vision of UAVs. When the flight altitude of the UAV is different, the image pixels of the same object are also different. Since YOLOv4 is a multi-scale detector, we improve YOLOv4 to make it more suitable for small-object detection in UAV reconnaissance images. Furthermore, the images of the same scene obtained by the UAV are different under different flight weather conditions.

To solve the above challenges, we proposed a small-object detection method that is applied to UAV reconnaissance. Our contributions are described as follows:

1. We propose two data augmentation methods to improve the generalization of the algorithm on the scene;

2. We design a backbone network that is suitable for small-object detection and modify the positioning loss function of the one-stage detector to improve detection accuracy;

3. We design a metric-based object classification method to classify objects into subclasses and detect objects that do not appear in the training phase, in other words, untrained objects.

The remainder of this manuscript is structured as follows. Section 2 introduces some related works of object detection algorithms. Section 3 formulates the detector structure for UAV untrained small-object detection and introduces the improved algorithm. Experimental results are presented in Section 4 to validate the effectiveness of our proposed method. Section 5 concludes this paper and envisages some future work.

## 2. Related Works

### 2.1. Small-Object Detection Algorithm

Most of the state-of-the-art detectors are based on deep-learning methods [11]. These methods mainly include two-stage detectors, one-stage detectors, and anchor-free detectors. Two-stage detectors first extract possible object locations and then perform classification and relocation. The classic two-stage detectors include spatial pyramid pooling networks (SPPNet) [9], faster region-CNN (RCNN) [12], etc. The one-stage detectors perform classification and positioning at the same time. Some effective one-stage detectors mainly include the single shot multi-box detector (SSD) [13], You Only Look Once (YOLO) series [5,8,14,15], etc. Anchor-free detectors include CenterNet [16], ExtremeNet [17], etc. These methods do not rely on predefined anchors to detect objects. In addition, some scholars have introduced transformers into the object detection field, such as detection with transformers (DETR) [18] and vision transformer faster RCNN (ViT FRCNN) [19], which have also achieved good results. However, the detection objects of the general object detectors are multi-scale. They are not designed for small-object detection specifically.

Small-object detection algorithms can be mainly divided into two kinds. One is to improve the detection performance of small objects with multiple scales in a video or image sequence. The other is to improve the detection performance of small objects with only a scale in an image. The improved detection methods of small objects with multiple scales mainly include feature pyramids, data augmentation, and changing training strategies. In 2017, Lin et al. proposed feature pyramid networks (FPN) [20], which improves the detection performance effect of small objects by fusing high-level and low-level features to generate multi-scale feature layers. In 2019, M. Kisantal et al. proposed two data augmentation methods [21] for small objects to increase the frequency of small objects in training images. B. Singh et al. designed scale normalization for image pyramids (SNIP) [22]. SNIP selectively backpropagates the gradients of objects of different sizes, and trains and tests images of different resolutions, respectively. The research object of these methods is multi-scale objects, which cannot make full use of the characteristics of small objects. The approaches that only detect small objects with a scale are mainly of three kinds: designing networks, using context information, and generating super-resolution images. L. Sommer et al. proposed a very shallow network for detecting objects in aerial images [23]. Small-object detection based on network design is few and immature. J. Li et al. proposed a new perceptual GAN network [24] to improve the resolution of small objects to improve the detection performance. In this paper, we focus on algorithms that are suitable for small-object detection in UAV reconnaissance images.

### 2.2. Object Detection Algorithm for UAV

The object detection algorithms used in UAVs are mainly designed based on the requirements of the task scenarios. M. Liu et al. proposed an improved detection algorithm based on YOLOv3 [8]. The algorithm first optimizes the Res-Block and then improves the darknet structure by increasing the convolution operation. Y. Li et al. proposed a multi-block SSD (MBSSD) mechanism [25] for railway scene monitored by UAVs. MBSSD uses transfer learning to solve the problem of insufficient training samples and improve accuracy. Y. Liu et al. proposed multi-branch parallel feature pyramid networks (MPFPN) [26] and used a supervised spatial attention module (SSAM) to focus the model on object information. MPFPN conducted experiments on a UAV public dataset named VisDrone-DET [27] to prove its effectiveness. These algorithms are combined with UAV application scenarios, but are all based on classic methods. They cannot work well when inferencing against untrained small objects.

### 3. Proposed Method

In this paper, we focus on small objects in images from the aerial view of UAVs. The proportion of object pixels in the image is less than 0.1% and the objects to be detected include objects that have not appeared in the training set. Current deep-learning-based object detection algorithms depend on a large amount of data. Therefore, we design an approach to expand the dataset in the proposed detection framework. In addition to the classic methods such as rotation, cropping, and color conversion, we propose two data augmentation methods—background replacement and noise adding—to improve the generalization of the detection algorithm. Background replacement gives the training images more negative samples to make the training set have a richer background. Noise adding can prevent the training model from overfitting the object.

After preprocessing the training images, the image features need to be obtained through the backbone network. We choose YOLOv4 [5], which has a good trade-off between speed and accuracy, as the benchmark algorithm for research. Common backbones in object detection algorithms are used to detect multi-scale objects. The receptive field of the backbone module is extremely large. In YOLOv4, the input of the backbone called CSPDarknet53 [7] is $725 \times 725$. However, the number of pixels of the small object in the image generally does not exceed $32 \times 32$. For small-object detection, the backbone does not need such a large receptive field. Therefore, the common backbone needs to be modified to apply to small object detection.

The YOLO series algorithms have three outputs, whether it is an object, which category it belongs to, and the bounding box coordinates of the object. These outputs are calculated by the same feature map. However, the convergence direction of the object positioning and object classification is not consistent. For example, the same object may have different coordinate positions, but the same coordinate position may have different objects. From this perspective, object positioning and object classification cannot use the same features. One of the research ideas of the proposed algorithm is to use different methods to deal with object positioning and object classification separately. This can avoid influence between positioning and classification.

As the input and output sizes of convolution neural networks (CNNs) are determined, the YOLO series algorithms can only detect a fixed number of objects. In order to recognize untrained categories of objects, we extract the object features from feature maps and design a metric-based method to classify objects. When the algorithm is trained, the classification loss and the positioning loss are backpropagated at the same time.

The overall structure of the proposed detector is shown in Figure 1. The data augmentation module uses methods such as background replacement and noise adding to change the training data to obtain better training performance. The backbone network extracts the multi-layer features in the image for object positioning and object classification. The object positioning module uses high-level features to obtain the center point coordinates, width and height of the object. The object classification module uses the positioning results to extract object features and then judges the category of the object by distance measurement. Finally, the detection result are obtained by combining the positioning and classification calculation.

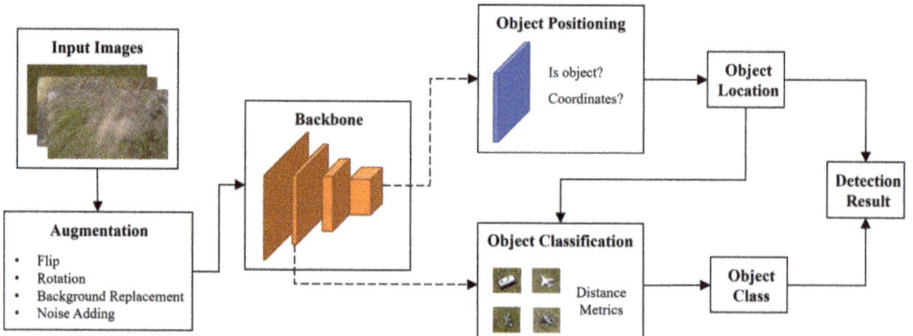


**Figure 1.** Structure of the proposed small-object detection method, including data augmentation, backbone network and object positioning and object classification modules.

### 3.1. Image Augmentation

We analyze two main reasons for insufficient datasets. One is that the background in the image is relatively monotonous and the other is that the object state is relatively monotonous. Aimed at these two reasons, we propose two methods to increase the image number of the database, as shown in Figure 2.

The purpose of background replacement is to reduce the impact of background singularity in UAV-based images. We randomly crop some areas in images that are not in the training set to cover areas in training images that do not contain the object. This can increase the diversity of negative samples, making the model eliminate the interference of irrelevant factors.

The output result of the object detection is a rectangular box surrounding the object. However, the object is generally not a standard rectangle, which means that the detected rectangular box will contain some information that does not belong to the object. If the object location does not change much, it is very likely to overfit the background information near the object, which is not conducive to the generalization of the detector. Generally speaking, invalid information will appear at the edge of the rectangular box. Therefore, we design a noise-adding augmentation strategy. We randomly select pixels in the image to cover the pixels near the edge of the rectangular box containing the object. Since we cannot accurately determine whether a pixel belongs to the object or background, we fill the pixels along the bounding box and the pixel block used as noise contains no more than 10 pixels, considering that the object is in the center of the detection box. The pixels near the object are changed by randomly adding noise to improve the generalization of the detector. The formula of background replacement is expressed as follows:

$$\tilde{x} = M \odot x_A + (1 - M) \odot z_B$$
$$\tilde{y} = y_A$$

$$(1)$$

where $M \in \{0,1\}^{W \times H}$ represents the part of the image that needs to be filled and $\odot$ means pixel by pixel multiplication. $x_A$ and $x_B$ denote two samples in the training set. $y_A$ and $y_B$ represent the label corresponding to the training samples, and $z_B$ is an image in the background image set.

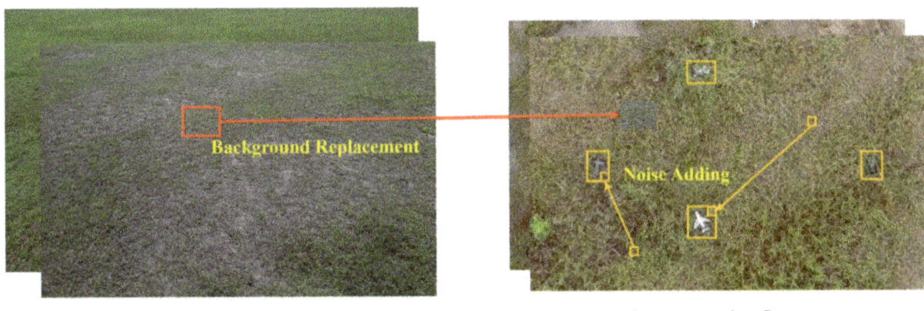


**Figure 2.** Two data augmentation methods. The red line represents the background replacement and the yellow lines represent noise adding.

*3.2. Backbone Design*

YOLOv4 proves that CSPDarknet53 is the relatively optimal model. We modify CSPDarknet53 to make it suitable for small-object detection. The comparison between the modified backbone in this paper and the original backbone is shown in the Figure 3.

Compared with the original network, the modified network reduces the receptive field and improves the input image resolution without increasing the computational complexity. Since the research object focuses on small-scale objects, there is no need to consider large-scale objects. The modified backbone deletes the network layers used to predict large-scale objects, which reduces the network depth by half. We call it DCSPDarknet53.

In order to reduce the computational complexity of deep learning, the resolution of the input image is usually downsampled. However, low image resolution will make it difficult to correctly classify and locate small objects. Therefore, a convolutional layer is added in the front of the network to calculate higher-resolution images. We call it as ADCSPDarknet53. At the cost of a small amount of calculation speed, the detection accuracy is improved on a large scale. The specific network structure of our proposed backbone network ADCSPDarknet53 is shown in Figure 4.

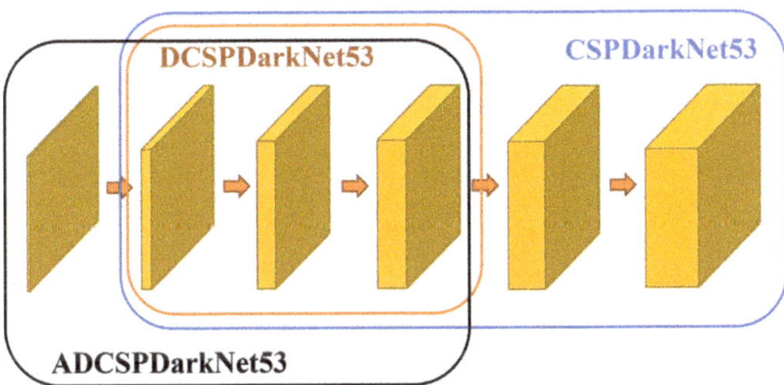


**Figure 3.** Comparison of backbones. CSPDarknet53 is the backbone of YOLOv4. DCSPDarknet53 is a backbone for small objects. ADCSPDarknet53 is a backbone that increases the downsampling network layer.

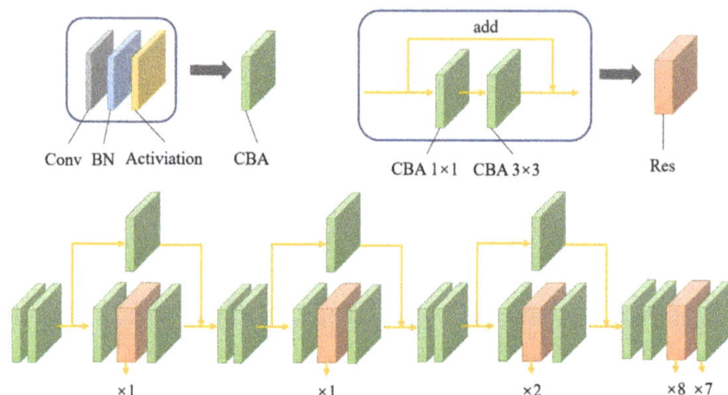



**Figure 4.** The structure of backbone network ADCSPDarknet53. Conv means convolution layer. BN denotes batch normalization. Activation layer uses ReLU function.

### 3.3. Object Positioning

The object positioning algorithm is improved based on the YOLO series. YOLOv5 [28] uses a positive sample expansion method. In addition to the original positive sample, two anchor points close to the object center are also selected as positive samples. The calculation formula is expressed as follow:

$$
P = \left\{ \begin{array}{c} p, \\ \text{if } (p \cdot x - \lfloor p \cdot x \rfloor \leq 0.5) : p + (-1, 0) \\ \text{if } (p \cdot x - \lfloor p \cdot x \rfloor > 0.5) : p + (1, 0) \\ \text{if } (p \cdot y - \lfloor p \cdot y \rfloor \leq 0.5) : p + (0, -1) \\ \text{if } (p \cdot y - \lfloor p \cdot y \rfloor > 0.5) : p + (0, 1) \end{array} \right\} \tag{2}
$$

where $P$ represents the expanded positive sample coordinate set and $p$ means the original positive sample coordinate. For example, in Figure 5, the gray plane is predicted by the grid where the gray plane's center point is located. After the expansion, the gray plane is also predicted by the grid where the red dots are located.

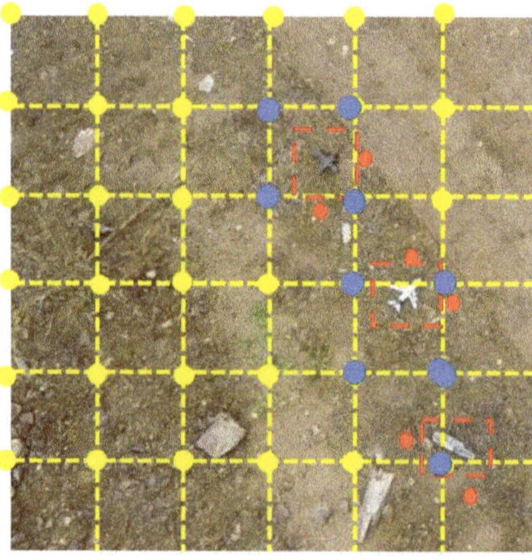

**Figure 5.** Selection of positive sample. The yellow dots are anchor points. The red dots are positive samples expanded by YOLOv5. The blue dots are positive samples proposed in this article.

The distance between each anchor point is not far in the last feature layer, which means that the object may contain multiple anchor points. It is not appropriate to select the closest anchor point as a positive sample and define other anchor points as negative samples. Therefore, we revise the selection method of positive samples. We calculate the four anchor points around the object center point as positive samples, as shown by the blue dots in Figure 5.

The YOLO series algorithms define the probability that the anchor contains the object as 1. Since each anchor point has a different distance from the object, the strategy of defining all as 1 cannot reflect the difference between different anchor points. Therefore, we use the Euclidean distance between the anchor point and the object center point as the metric for the probability that the anchor contains the object. As the positive sample must contain the object, the probability of containing objects of the positive sample anchor point cannot be 0. According to the design principle, the function of calculating the probability is shown as follows:

$$p_{obj} = 1 - \left( (x_a - x_t)^2 + (y_a - y_t)^2 \right)/4 \tag{3}$$

### 3.4. Untrained Sub-Class Object Detection

Generally speaking, top-level features are beneficial to object positioning and bottom-level features are beneficial to object classification. In order to separate the process of object localization and object classification to reduce the distractions between these two and make better use of the features extracted from the backbone network, we select features from the middle layers of the backbone for object classification. Through object positioning, we can obtain the coordinates of the object. Using these coordinates, the feature vector of the object can be extracted from feature maps and then can be used to classify the object.

In UAV airborne images, it is common to think of objects in terms of large classes, such as aircraft, cars, buildings, pedestrians, etc. However, specific objects such as black cars and white cars are difficult to determine. To address this sub-class classification problem, we divide the object classification process into the rough classification process and the fine classification process. Rough classification mainly distinguishes objects with large differences in appearance, such as aircraft and cars. Fine classification mainly distinguishes objects that have similar appearance characteristics, but belong to different classes, such as black cars, white cars, etc.

In this paper, a measurement method based on Euclidean distance is used to classify the object. The advantage is that it does not need to fix the object class. By using this metric learning, the algorithm can identify the potential objects in the scene that do not appear in training process, in other words, untrained object. The training goal of object classification is to make the object features of the same class as close as possible and to make the object features of different classes as far as possible. After extracting the object features, three objects are randomly selected from all objects, in which two classes are the same. We use the triple loss [29] as the loss function of object classification. The loss function is defined as:

$$\text{loss}_{cls} = \max(d(a_1, a_2) - d(a_1, b) + \text{thre}, 0) \tag{4}$$

where $a_1$ and $a_2$ represent objects that belong to the same class. $b$ means the object that is different from $a_1$ and $a_2$ and *thre* is the expected distance between objects of different classes. The rough classification calculates the loss value between the object classes, while the fine classification calculates the loss value between the object sub-classes. The *thre* of the fine classification process is lower than the *thre* of the rough classification.

In the testing process, we input the labeled images with all objects into the trained model to obtain image features and then extract the feature vector of the object according to the object position to construct a classification database. The classification database is used to classify the object in the test image. The flowchart of object classification is shown in Figure 6. First, the rough classification database is used to determine the object class and then the fine classification database is used to determine the object sub-class.

The principle of object classification is to classify the object into the class closest to the object. If the distance between the object and any category in the database is greater than the threshold, the object is considered to belong to an unknown class that has not appeared in the database. If the distance between the object and the class closest to it is less than the threshold, the object is considered to belong to that class.

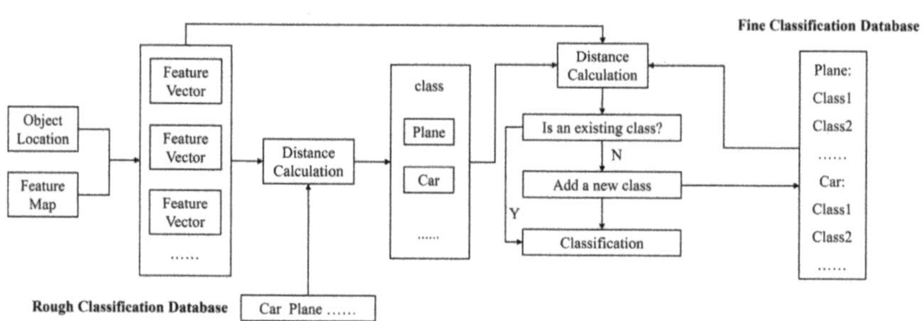

**Figure 6.** The flowchart of object classification for fine and rough classification.

In summary, we designed two data augmentations—background replacement and noise adding—to increase the background diversity of the dataset. Based on the information flow of small objects through convolution layers, we modified the detector backbone CSPDarkNet53 to ADCSPDarknet53 to obtain a larger feature map for small-object detection as well as to reduce the computation cost. For object positioning, we selected the four anchor points around the object center point as positive samples and modified the function for calculating the objectness probability, which can increase the positioning accuracy of small objects. For object classification, we combined information from shallow feature maps and positioning results to perform rough and fine classification processes to obtain more accurate classification results and identity untrained sub-class small objects.

## 4. Experiments

To evaluate the small-object detection and classification algorithm proposed in this paper, we first constructed a dataset consists of small objects. Then, we performed experiments on trained and untrained small objects to compare localization and classification performance. Finally, we conducted flight experiments to test the detection performance and real-time inference of the proposed algorithm on small UAVs.

### 4.1. Dataset of Small Objects and Evaluation Metrics

We choose to detect small objects in the visual field of UAVs to evaluate our algorithm. In order to obtain as much target data as possible, we built a scaled-down experimental environment to collect our dataset. The UAV we used to collect the dataset was a DJI Mini2. To obtain various data, we used some small target models as objects to be detected. The target models were between 15 cm–25 cm in length and 5 cm–20 cm in width. When taking the image data, the flight altitude of the UAV was controlled between 8–10 m to simulate the dataset captured at high altitudes. As the resolution of captured images is $1920 \times 1080$ pixels, the pixel ratio of the object to be detected in the image is less than 0.1% as shown in Figure 7a. Eight types of objects were selected to construct the dataset. There are two object classes in the dataset, car and plane. Each class has four sub-classes, which are listed in Figure 7b.

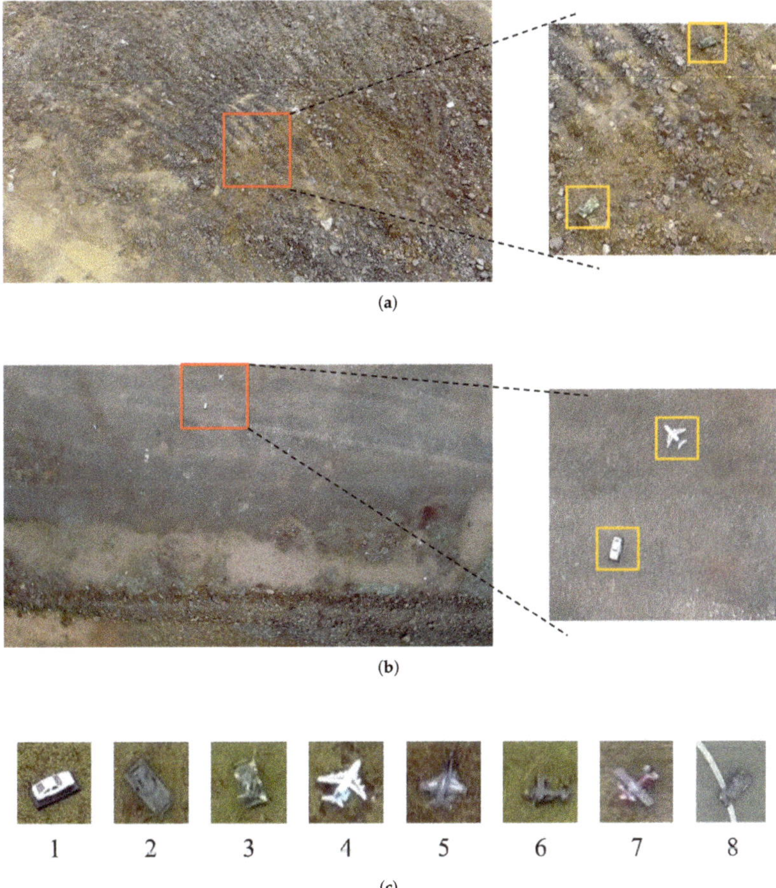

**Figure 7.** (**a**,**b**) are example images in the collected small-object dataset. The images on the left show the pixel proportion of the object to be detected in the whole image, and the patches on the right are a zoomed-in version of the red box on the left images. (**c**) shows eight objects in the dataset. 1, 2 and 3 are three trained cars belonging to different sub-classes. 4, 5 and 6 are three trained planes belonging to different sub-classes. 7 is an untrained plane and 8 is an untrained car. 7 and 8 are objects that did not appear in the training set and validation set.

The collected dataset are split into training set, validation set and testing set, with 977 images, 100 images and 195 images, respectively. For the object-detection task, the label file contains four object positioning coordinates and an object class label. For the sub-class object classification task, the label file contains two object class labels, in which the first label denotes the object class and the second one is the sub-class. In order to evaluate the detection performance of the proposed algorithm on untrained small objects, the designs of the testing set are slightly different from the training set and validation set. The training set and the validation set contain six types of objects, including three types of cars and three types of planes. In addition to these six types of objects, one type of car and one type of plane are added to the testing set.

For evaluation, we use the general indicators in the field of object detection [30] to evaluate the performance of the proposed algorithm, including mean average precision (mAP), mean average precision at $IoU = 0.50$ ($mAP_{50}$) and frames per second (FPS). Intersection over union (IoU) evaluates the overlap between the ground truth bounding boxes and the predicted bounding box. Based on IoU and the predicted object category confidence scores, mAP is applied to measure the detection performance with classification and localization. $mAP_{50}$ is calculated with IoU $\geq$ 0.50. These two metrics are performed under the COCO criteria [10]. FPS is used to evaluate the running speed of the algorithm on certain computation platforms. The FPS calculation method is the same as in YOLOv4.

## 4.2. Implementation Details

We chose YOLOv4 as our baseline model since the proposed network is improved based on YOLOv4. For comparison of object detection, we implemented several object-detection models on the collected dataset, including Faster RCNN (with VGG16 [31] as backbone), SSD [13], FCOS [31], PPYOLO [32], PPYOLOv2 [33], PPYOLOE [34] and PicoDet [35]. Among them, Faster RCNN is a two-stage object detector and the rest are one-stage detectors. FCOS, PPYOLOE and PicoDet are anchor-free detectors, and PicoDet is designed for mobile devices. All the detectors have the same size of input image (608 $\times$ 608) and were trained from the pretrained model on the COCO dataset [10] to obtain faster and better convergence. The training epochs were set to 300 with initial learning rate 0.001 and Adam optimizer. Then, learning rate decayed by 0.1 times at the 150th epoch and 250th epoch. The batch size was set to be 16. For the special sub-class classification part of our method, the *thre* of the rough classification was 10 and the *thre* of the fine classification was 2. All the training and testing experiments were conducted on one NVIDIA Quadro GV100 GPU.

## 4.3. Experiment Results

### 4.3.1. Small Object Detection

We compare the detection performance on small objects with several existing object detectors. The results are listed in Table 1, from which we can see that our proposed method gives the highest average precision (33.80%) compared to the others, as well as running at the highest speed (77 FPS). It also achieves the third highest $mAP_{50}$, with 61.00% among ten models. These improvements can be attributed to the following aspects: (1) the modified backbone network focuses more on the small objects and discards the deep layers, which have little effect on detecting small objects. Meanwhile, the interference of extra parameters on network learning is reduced. (2) Metric-based learning improves the ability of the network to classify objects. (3) The proposed object positioning method increase the number of positive samples and thus improves small-object localization abilities. (4) The modified backbone network reduces computation significantly, which makes the network run at a faster speed and achieves the performance of real-time detection. Figure 8 shows some detection results of the collected dataset using YOLOv4 and our algorithm. Our algorithm has stronger ability to detect small objects.

**Table 1.** Experiment results of small object detection.

| Method | FPS | $mAP_{50}$ | mAP |
|---|---|---|---|
| PPYOLOE_s | 56 | 33.40% | 12.70% |
| PPYOLOE_l | 34 | 52.50% | 22.40% |
| PicoDet_s | 68 | 21.70% | 7.50% |
| FasterRCNN | 18 | 29.40% | 10.80% |
| SSD | 51 | 34.80% | 10.80% |
| PPYOLO | 54 | 63.70% | 27.50% |
| PPYOLOv2 | 36 | 62.00% | 25.60% |
| FCOS | 26 | 21.20% | 8.80% |
| YOLOv4 | 71 | 39.60% | 21.70% |
| Ours(ADCSPDarknet53) | **77** | 61.00% | **33.80%** |

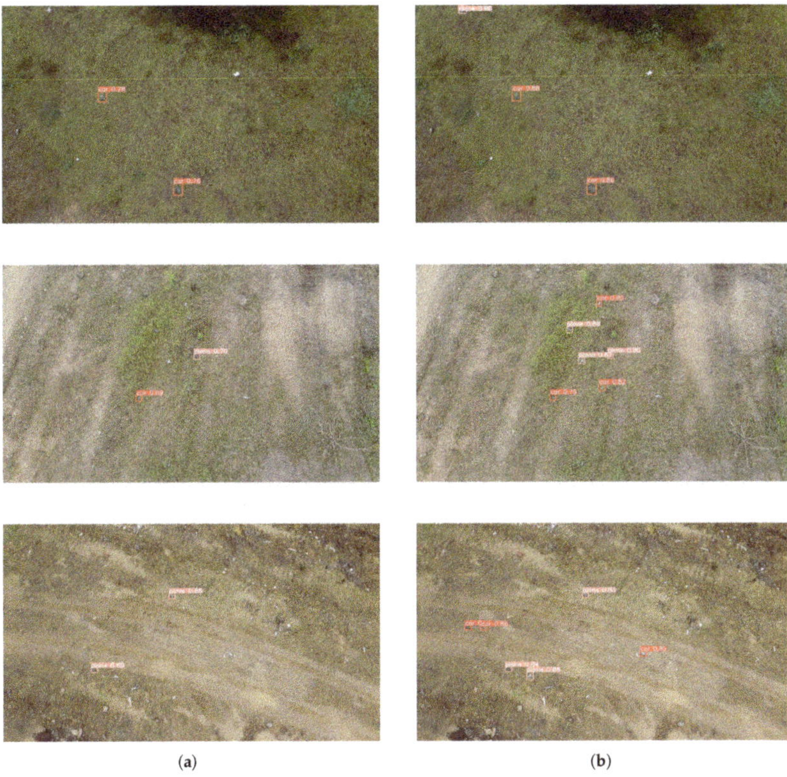

(a)             (b)

**Figure 8.** Small-object detection experiment with other algorithms. (**a**) YOLOv4 algorithm detection results; (**b**) our algorithm detection results.

### 4.3.2. Untrained Sub-Class Object Classification

In our proposed method, object classification includes the process of rough object classification and fine object classification. The designed classification method has two advantages. One is that it can classify untrained objects, and the other is that it avoids the mutual influence of classification and positioning. To illustrate both points, we conduct multiple comparative experiments. The experimental results of object classification are shown in Table 2. Experiments 1, 2 and 3 represent detection results of YOLOv4 under different conditions. Experiments 4 and 5 are the detection results of the same rough classification model under different test categories. Experiments 6, 7 and 8 are the detection results of the same fine classification model under different test categories.

Comparing Experiments 1 and 3, the accuracy of object detection decreases from 88.30% to 65.60% when objects are classified during training, which shows the interference between object classification and localization. Through Experiments 3 and 4, it can be found that metric-based learning is beneficial to improve the result of object detection, as the $mAP_{50}$ increases from 65.60% to 88.30%. It can be demonstrated by Experiments 1, 3 and 6 that the detection performance of the object is worse for more categories. In Experiment 8, we can find that the untrained car and the untrained plane can be detected with 49.4% $mAP_{50}$ and 34.1% $mAP_{50}$, respectively. Although the metric-based method is not very accurate in detecting untrained objects, it can still locate untrained objects and distinguish them from trained objects.

**Table 2.** Experiment results of untrained object classification.

| Experiment | Model | Number of Training Classes | Number of Testing Classes | $mAP_{50}$ (Trained) | $mAP_{50}$ (Untrained) |
|---|---|---|---|---|---|
| 1 | YOLOv4 | 1 | 1 | 88.30% | - |
| 2 | YOLOv4 | 2 | 2 | 51.10% | - |
| 3 | YOLOv4 | 2 | 1 | 65.60% | - |
| 4 | rough classification | 2 | 1 | 83.80% | - |
| 5 | rough classification | 2 | 2 | 64.70% | - |
| 6 | fine classification | 9 | 1 | 78.30% | - |
| 7 | fine classification | 9 | 2 | 57.70% | - |
| 8 | fine classification | 9 | 9 | 41.900% | aircraft: 34.10% car: 49.40% |

Figure 9 shows the visualization results of untrained object classification using YOLOv4 and our algorithm. For the untrained sub-class objects, YOLOv4 will give incorrect classification results, but the proposed algorithm will add these untrained objects to new sub-classes using metric-based learning.

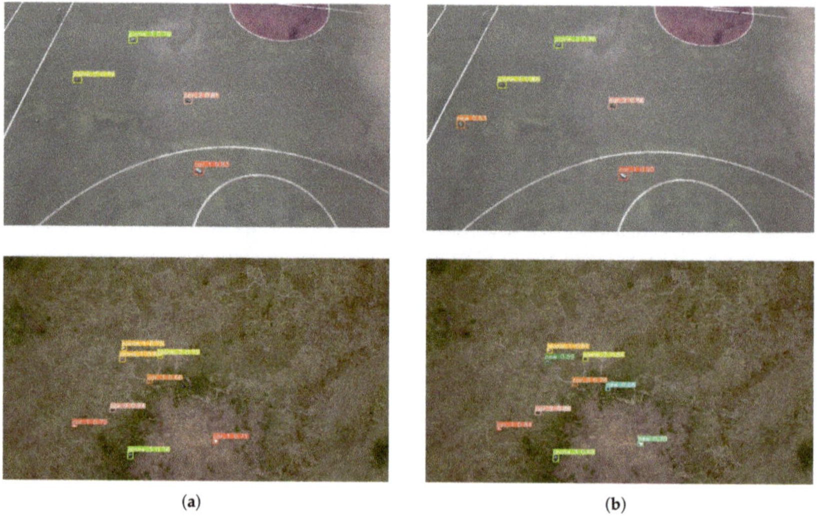

(a)       (b)

**Figure 9.** Untrained object classification experiment with other algorithms. (**a**) YOLOv4 algorithm detection results; (**b**) our algorithm detection results. Different colors of bounding boxes mean different sub-classes, and the recognized untrained sub-class objects are labeled with 'new'.

### 4.3.3. Ablation Study

To analyze the effectiveness of our proposed method, we conducted an ablation study on data augmentation, backbone design and object positioning.

**Data Augmentation.** Based on YOLOv4, we analyze the results of two data augmentation methods. It can be seen from Table 3 that the two types of data augmentation can improve detection performance. The results show that replacing part of the image background can lead the network to learn more combinations of patterns and effectively increase the diversity of the dataset. Adding noise around the object reduces the overfitting to special backgrounds. However, the effect of running the two methods at the same time is not as good as the effect of running the two methods separately. This is mainly because the data augmentation method we propose introduces significant noise while increasing the diversity of the dataset. Applying both methods at the same time may cause too much noise and cannot obtain better performance.

**Table 3.** Experiment results of data augmentation.

| Image Input Size | Replace Background | Add Noise | $mAP_{50}$ | mAP |
|---|---|---|---|---|
| $608 \times 608$ | N | N | 36.50% | 19.10% |
| $608 \times 608$ | Y | N | 44.60% | 24.90% |
| $608 \times 608$ | N | Y | 41.70% | 20.70% |
| $608 \times 608$ | Y | Y | 39.60% | 22.40% |

**Backbone Design.** The backbone design consists of two steps. First, the DCSPDarknet53 deletes the deep network layers used to detect large-scale objects in the original detection backbone CSPDarknet53 to reduce the influence of the deep network on the detection of small objects. Then, ADCSPDarknet53 adds a network layer for downsampling at the front end of the network, so as to obtain better detection results while increasing the computational complexity as little as possible. The experiment results are shown in Table 4. Compared to the CSPDarknet53-based detector, there is little increase in $mAP_{50}$ and $mAP$ of the DCSPDarknet53-based detector in small-object detection, but the calculation speed is more than double, which proves that small-object detection can use a high-resolution network with fewer layers. As for the ADCSPDarknet53-based detector, the $mAP_{50}$ is increased by 18% and the $mAP$ is increased by 11% compared to the original detector. Although the FPS drops, it still runs faster than the CSPDarknet53-based detector. In the actual scene, the image size and backbone can be adjusted as needed.

**Table 4.** Experiment results of backbone design.

| Backbone | Image Input Size | FPS | $mAP_{50}$ | mAP |
|---|---|---|---|---|
| CSPDarknet53 | $608 \times 608$ | 71 | 39.60% | 21.70% |
| DCSPDarknet53 | $608 \times 608$ | 166 | 42.90% | 23.90% |
| ADCSPDarknet53 | $1216 \times 1216$ | 77 | 57.60% | 32.70% |

**Object Positioning.** We modified the loss function of object positioning; Table 5 shows the detection evaluation of detectors with and without loss function modification. After modifying the loss function of object positioning, the $mAP_{50}$ of the CSPDarknet53-based detector is increased by 10.7% and the $mAP$ is increased by 2.9%. For the ADCSPDarknet53-based detector, the $mAP_{50}$ increased by 3.4% and the $mAP$ increased by 1.1%. This proves the effectiveness of the modified loss function, which can select positive samples that can represent the ground truth more accurately from many candidate samples. The training process of object positioning is shown in Figure 10. With the modified object positioning loss function, the training process is more stable.

**Table 5.** Experiment results of object positioning.

| Backbone | Loss Function Modification | $mAP_{50}$ | mAP |
|---|---|---|---|
| CSPDarknet53 | N | 39.60% | 21.70% |
| CSPDarknet53 | Y | 50.30% | 24.60% |
| ADCSPDarknet53 | N | 57.60% | 32.70% |
| ADCSPDarknet53 | Y | 61.00% | 33.80% |

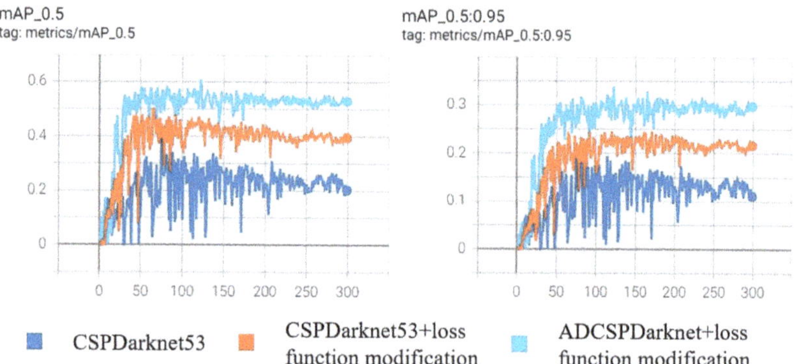



**Figure 10.** Training process of object positioning.

*4.4. Flight Experiments*

In order to verify the effectiveness of the proposed algorithm in the actual UAV application scenario, we built a UAV flight experiment system and deployed the proposed algorithm on a small drone.

### 4.4.1. Experiment Settings

The drone we used as the experiment platform was a Matrice M210v2 drone manufactured by DJI Innovations. Its overall dimensions are $883 \times 886 \times 398$ mm with a maximum takeoff weight 6.14 kg. The onboard optical camera was a DJI innovation company Chansi X5s camera equipped with a DJI MFT 15 mm/1.7 ASPH lens. It was fixed to the drone body through a 3-DOF gimbal and its posture can be controlled with a remote controller. The resolution of the images taken by the camera was set to be $1920 \times 1080$ pixels. To implement our proposed algorithm on the drone, we deployed a Nvidia Jetson Xavier NX processor on the drone for real-time processing. In order to ensure the safe outdoors flight of the drone, we also set the real-time kinematic (RTK) global navigation satellite system on the drone for the positioning. The flight experiment system is shown in Figure 11.

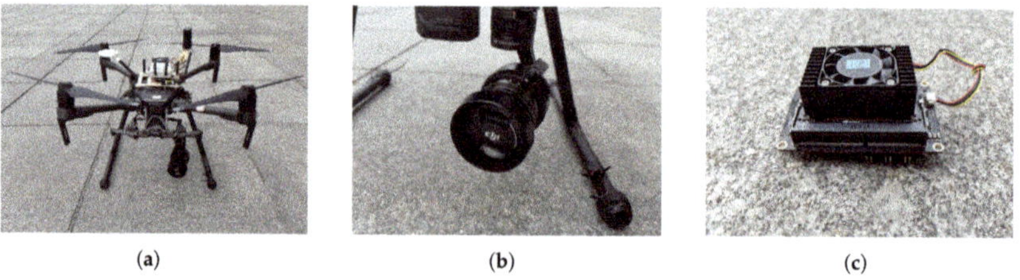



**Figure 11.** Hardware system for flight experiments. (**a**) DJI Matrice M210v2 drone; (**b**) DJI camera; (**c**) Nvidia Jetson Xavier NX.

The proposed algorithm is implemented by PyTorch 1.10 on Nvidia Jetson Xavier NX's GPU with a computational capacity of 7.2. All programs run on Robot Operating System (ROS) systems. While the drone is flying, we use the Rosbag tool to record the on-board processing data, such as the real-time detection image results. Once the drone is back on the ground, we can use the Rosbag's playback function to check how well the algorithm works.

We set up two flight scenarios to validate our algorithm with trained and untrained objects. In one detection scenario, the objects to be detected were the small models used for creating the above dataset, but with new backgrounds. In this case, the flight altitude of the drone was set to 10 m to stay consistent with the dataset. The purpose of this detection scenario is to verify the generalization performance of the learned model in practical application scenarios. In another detection scenario, the model detected real vehicles with flight altitude of 95 m. We used seven different types of vehicles to test the classification and localization ability of the model for object classes with high inter-class similarity. In this case, we collected new data but only six types were labeled and appear in the training set. Then the model is retrained and tested for detection performance and speed. The two scenarios settings are shown in Figure 12.

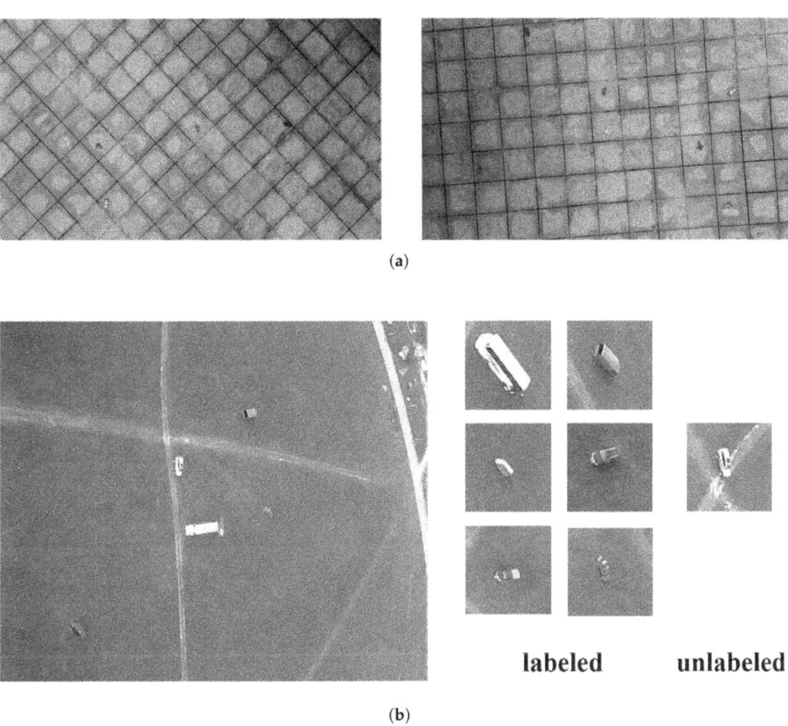


(a)

(b)

**Figure 12.** (**a**) The first detection scenario is set to have the same objects as the collected dataset, but with different background. (**b**) The second detection scenario uses real vehicles as objects. The figure on the left shows part of the drone's field of view, and the right images show different types of vehicles, with six labeled types and one unlabeled.

### 4.4.2. Results

Some qualitative detection results are shown in Figure 13. In Figure 13a, our proposed algorithm can detect small objects in the visual field without being influenced by the changing background. This is because the data enhancement method we used effectively prevents the model from overfitting to the background during the training process. In Figure 13b, the learned objects are detected by the proposed detector. In addition, the potential target (the unlabeled one) is also identified by the network and classified into a new class according to metric learning.

It is worth noting that small-object detection during flight faces additional challenges, such as camera vibration and motion caused by flight. Camera motion in the imaging process leads to the blurring of objects and damages the features. In this case, our algorithm can still detect small objects in the airborne visual field accurately. Our proposed algorithm not only extracts the features of small objects with CNNs, but also distinguishes the inter-class and intra-class differences of objects by measuring the distance metric. This more powerful feature extraction method helps reduce the effect of motion blur.

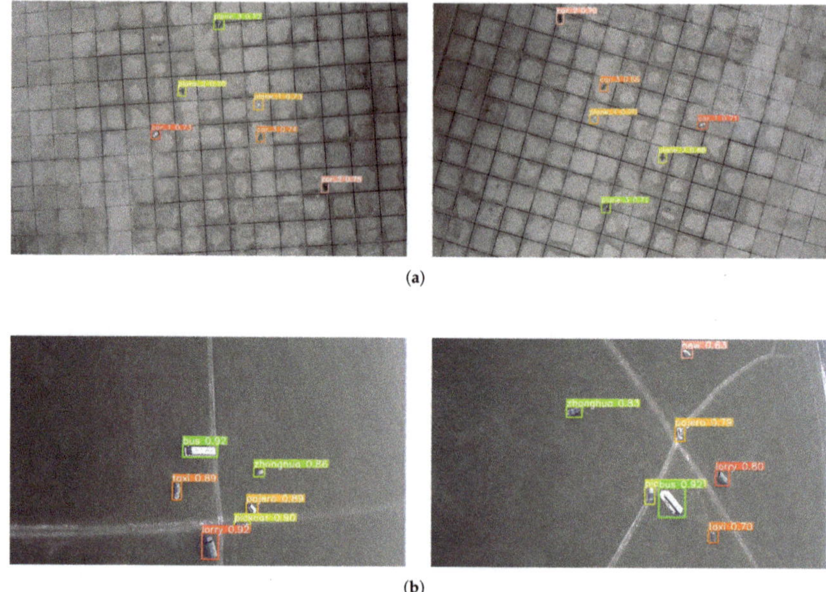

**Figure 13.** (**a**) Detection results on small model objects with different backgrounds. (**b**) Detection results on seven types of vehicles.

We also checked the real-time performance of our proposed method. We computed the runtime of the algorithm on the edge GPU (Nvidia Xavier NX) using different input image resolutions, including the pre-processing phase, model inference phase, and post-processing phase. The results are listed in Table 6. For images with input sizes of $416 \times 416$, $608 \times 608$ and $640 \times 640$, our algorithm reaches an average speed of 22 FPS. The small runtime difference mainly comes from the pre-processing phase and post-processing phase, since these two phases are run on the CPU. The parallel computing capability of the GPU makes the inference time of the model almost the same. However, for images with input size of $1216 \times 1216$, it takes more than twice as long to process a single frame. As there are four times as many pixels in the image, more time is needed to perform normalization for each pixel and permute the image channels in the pre-processing stage. For model inference, the larger input image size makes the feature map in the network larger, with more activation needs to compute, which accounts for the increase of time usage [36]. During the post-processing phase, more candidate detection boxes need to be computationally suppressed.

In our flight test, we used an image input size of $608 \times 608$. Without any acceleration library, our algorithm can achieve real-time performance, which is sufficient for reconnaissance missions on UAVs. Some runtime optimizations can be made, for example, using the TensorRT [37] library to accelerate the model inference, or by improving the code efficiency in pre-processing and post-processing stages.

**Table 6.** Experiment results of real-time performance.

| Image Input Size | Average Pre-Processing Time (ms) | Average Inference Time (ms) | Average Post-Processing Time (ms) | Average Total Process Time (ms) | FPS |
| --- | --- | --- | --- | --- | --- |
| 416 × 416 | 1.6 | 40.8 | 1.3 | 43.7 | 22.88 |
| 608 × 608 | 2.1 | 41.0 | 1.3 | 44.4 | 22.52 |
| 640 × 640 | 2.2 | 41.3 | 1.4 | 44.9 | 22.27 |
| 1216 × 1216 | 3.1 | 51.9 | 49.3 | 104.3 | 9.59 |

## 5. Conclusions

Aimed at challenges such as small-scale objects, untrained objects during inference, and real-time performance requirements, we designed a detector to detect small objects in UAV reconnaissance images. To conduct our research, we collected a small object dataset from the perspective of a high-flying UAV. We proposed two data augmentation methods, background replacement and noise adding, which improve the background diversity of the collected dataset. For the backbone design, we designed ADCSPDarkent53 based on the characteristics of small objects and evaluated the improved backbone on accuracy and speed. For object positioning, we modified the positioning loss function, which greatly improved detection accuracy. For object classification, a metric-based classification method was proposed to solve the problem of untrained sub-class object classification. Experiments on UAV-captured images and flight tests show the effectiveness and applicable scope of the proposed small-object detector. In the next step, improvements can be made in terms of dataset construction, feature selection and metric function design.

**Author Contributions:** Conceptualization, H.Z.; methodology, H.Z.; software, H.Z. and A.M.; validation, A.M.; formal analysis, H.Z. and A.M.; investigation, H.Z.; resources, Y.N.; data curation, H.Z. and A.M.; writing—original draft preparation, H.Z.; writing—review and editing, A.M.; visualization, H.Z. and A.M.; supervision, Y.N.; project administration, Z.M.; funding acquisition, Y.N. and Z.M. All authors have read and agreed to the published version of the manuscript.

**Funding:** This paper was supported by National Natural Science Foundation of China (No. 61876187) and Natural Science Foundation of Hunan Province (No. S2022JJQNJJ2084 and No. 2021JJ20054).

**Acknowledgments:** Thanks to the Unmanned Aerial Vehicles Teaching and Research Department, Institute of Unmanned Systems, College of Intelligence Science and Technology, National University of Defense Technology for providing the experimental platform.

**Conflicts of Interest:** The authors declare no conflict of interest.

## References

1. Belmonte, L.M.; Morales, R.; Fernández-Caballero, A. Computer Vision in Autonomous Unmanned Aerial Vehicles—A Systematic Mapping Study. *Appl. Sci.* **2019**, *9*, 3196. [CrossRef]
2. Zhang, H.; Wang, L.; Tian, T.; Yin, J. A Review of Unmanned Aerial Vehicle Low-Altitude Remote Sensing (UAV-LARS) Use in Agricultural Monitoring in China. *Remote Sens.* **2021**, *13*, 1221. [CrossRef]
3. Zheng, Y.J.; Du, Y.C.; Ling, H.F.; Sheng, W.G.; Chen, S.Y. Evolutionary Collaborative Human-UAV Search for Escaped Criminals. *IEEE Trans. Evol. Comput.* **2020**, *24*, 217–231. [CrossRef]
4. Wu, X.; Li, W.; Hong, D.; Tao, R.; Du, Q. Deep Learning for Unmanned Aerial Vehicle-Based Object Detection and Tracking: A survey. *IEEE Geosci. Remote Sens. Mag.* **2022**, *10*, 91–124. [CrossRef]
5. Alexey, B.; Wang, C.; Mark Liao, H. Optimal speed and accuracy of object detection. *arXiv* **2020**, arXiv:2004.10934.
6. Liu, S.; Qi, L.; Qin, H.; Shi, J.; Jia, J. Path Aggregation Network for Instance Segmentation. In Proceedings of the IEEE Conference on Computer Vision and Pattern Recognition (CVPR), Salt Lake City, UT, USA, 18–23 June 2018.
7. Wang, C.Y.; Liao, H.Y.M.; Wu, Y.H.; Chen, P.Y.; Hsieh, J.W.; Yeh, I.H. CSPNet: A New Backbone That Can Enhance Learning Capability of CNN. In Proceedings of the IEEE/CVF Conference on Computer Vision and Pattern Recognition (CVPR) Workshops, Seattle, WA, USA, 14–19 June 2020.
8. Redmon, J.; Farhadi, A. Yolov3: An incremental improvement. *arXiv* **2018**, arXiv:1804.02767.

9. He, K.; Zhang, X.; Ren, S.; Sun, J. Spatial Pyramid Pooling in Deep Convolutional Networks for Visual Recognition. *IEEE Trans. Pattern Anal. Mach. Intell.* **2015**, *37*, 1904–1916. [CrossRef] [PubMed]

10. Lin, T.Y.; Maire, M.; Belongie, S.; Hays, J.; Perona, P.; Ramanan, D.; Dollár, P.; Zitnick, C.L. Microsoft COCO: Common Objects in Context. In *Proceedings of the Computer Vision—ECCV 2014*; Fleet, D., Pajdla, T., Schiele, B., Tuytelaars, T., Eds.; Springer International Publishing: Cham, Switzerland, 2014; pp. 740–755.

11. Tong, K.; Wu, Y.; Zhou, F. Recent advances in small object detection based on deep learning: A review. *Image Vis. Comput.* **2020**, *97*, 103910. [CrossRef]

12. Ren, S.; He, K.; Girshick, R.; Sun, J. Faster r-cnn: Towards real-time object detection with region proposal networks. *Adv. Neural Inf. Process. Syst.* **2015**, *28*, 91–99. [CrossRef] [PubMed]

13. Liu, W.; Anguelov, D.; Erhan, D.; Szegedy, C.; Reed, S.; Fu, C.Y.; Berg, A.C. SSD: Single shot multibox detector. In Proceedings of the European Conference on Computer Vision, Amsterdam, The Netherlands, 11–14 October 2016; Springer: Cham, Switzerland, 2016; pp. 21–37.

14. Redmon, J.; Divvala, S.; Girshick, R.; Farhadi, A. You only look once: Unified, real-time object detection. In Proceedings of the IEEE Conference on Computer Vision and Pattern Recognition, Las Vegas, NV, USA, 26 June–1 July 2016; pp. 779–788.

15. Redmon, J.; Farhadi, A. YOLO9000: Better, faster, stronger. In Proceedings of the IEEE Conference on Computer Vision and Pattern Recognition, Honolulu, HI, USA, 21–26 July 2017; pp. 7263–7271.

16. Zhou, X.; Wang, D.; Krähenbühl, P. Objects as points. *arXiv* **2019**, arXiv:1904.07850.

17. Zhou, X.; Zhuo, J.; Krahenbuhl, P. Bottom-up object detection by grouping extreme and center points. In Proceedings of the IEEE/CVF conference on computer vision and pattern recognition, Long Beach, CA, USA, 15–20 June 2019; pp. 850–859.

18. Carion, N.; Massa, F.; Synnaeve, G.; Usunier, N.; Kirillov, A.; Zagoruyko, S. End-to-end object detection with transformers. In Proceedings of the European Conference on Computer Vision, Glasgow, UK, 23–28 August 2020; Springer: Cham, Switzerland, 2020; pp. 213–229.

19. Beal, J.; Kim, E.; Tzeng, E.; Park, D.H.; Zhai, A.; Kislyuk, D. Toward transformer-based object detection. *arXiv* **2020**, arXiv:2012.09958.

20. Lin, T.Y.; Dollár, P.; Girshick, R.; He, K.; Hariharan, B.; Belongie, S. Feature pyramid networks for object detection. In Proceedings of the IEEE Conference on Computer Vision and Pattern Recognition, Honolulu, HI, USA, 21–26 July 2017; pp. 2117–2125.

21. Kisantal, M.; Wojna, Z.; Murawski, J.; Naruniec, J.; Cho, K. Augmentation for small object detection. *arXiv* **2019**, arXiv:1902.07296.

22. Singh, B.; Davis, L.S. An analysis of scale invariance in object detection snip. In Proceedings of the IEEE Conference on Computer Vision and Pattern Recognition, Salt Lake City, UT, USA, 18–23 June 2018; pp. 3578–3587.

23. Schumann, A.; Sommer, L.; Klatte, J.; Schuchert, T.; Beyerer, J. Deep cross-domain flying object classification for robust UAV detection. In Proceedings of the 2017 14th IEEE International Conference on Advanced Video and Signal Based Surveillance (AVSS), Lecce, Italy, 29 August–1 September 2017; pp. 1–6.

24. Li, J.; Liang, X.; Wei, Y.; Xu, T.; Feng, J.; Yan, S. Perceptual generative adversarial networks for small object detection. In Proceedings of the IEEE Conference on Computer Vision and Pattern Recognition, Honolulu, HI, USA, 21–26 July 2017; pp. 1222–1230.

25. Yundong, L.; Han, D.; Hongguang, L.; Zhang, X.; Zhang, B.; Zhifeng, X. Multi-block SSD based on small object detection for UAV railway scene surveillance. *Chin. J. Aeronaut.* **2020**, *33*, 1747–1755.

26. Liu, Y.; Yang, F.; Hu, P. Small-object detection in UAV-captured images via multi-branch parallel feature pyramid networks. *IEEE Access* **2020**, *8*, 145740–145750. [CrossRef]

27. Du, D.; Zhu, P.; Wen, L.; Bian, X.; Lin, H.; Hu, Q.; Peng, T.; Zheng, J.; Wang, X.; Zhang, Y.; et al. VisDrone-DET2019: The vision meets drone object detection in image challenge results. In Proceedings of the IEEE/CVF International Conference on Computer Vision Workshops, Seoul, Korea, 27–28 October 2019.

28. Glenn, J.; Ayush, C.; Jirka, B.; Alex, S.; Yonghye, K.; Jiacong, F.; Tao, X.; Kalen, M.; Yifu, Z.; Colin, W.; et al. Ultralytics/Yolov5. Available online: https://github.com/ultralytics/yolov5 (accessed on 1 April 2021).

29. Schroff, F.; Kalenichenko, D.; Philbin, J. Facenet: A unified embedding for face recognition and clustering. In Proceedings of the IEEE Conference on Computer Vision and Pattern Recognition, Boston, MA, USA, 7–12 June 2015; pp. 815–823.

30. Padilla, R.; Netto, S.L.; Da Silva, E.A. A survey on performance metrics for object-detection algorithms. In Proceedings of the 2020 international conference on systems, signals and image processing (IWSSIP), Niteroi, Brazil, 1–3 July 2020; pp. 237–242.

31. Tian, Z.; Shen, C.; Chen, H.; He, T. Fcos: Fully convolutional one-stage object detection. In Proceedings of the IEEE/CVF International Conference on Computer Vision, Seoul, Korea, 27–28 October 2019; pp. 9627–9636.

32. Long, X.; Deng, K.; Wang, G.; Zhang, Y.; Dang, Q.; Gao, Y.; Shen, H.; Ren, J.; Han, S.; Ding, E.; et al. PP-YOLO: An effective and efficient implementation of object detector. *arXiv* **2020**, arXiv:2007.12099.

33. Huang, X.; Wang, X.; Lv, W.; Bai, X.; Long, X.; Deng, K.; Dang, Q.; Han, S.; Liu, Q.; Hu, X.; et al. PP-YOLOv2: A practical object detector. *arXiv* **2021**, arXiv:2104.10419.

34. Xu, S.; Wang, X.; Lv, W.; Chang, Q.; Cui, C.; Deng, K.; Wang, G.; Dang, Q.; Wei, S.; Du, Y.; et al. PP-YOLOE: An evolved version of YOLO. *arXiv* **2022**, arXiv:2203.16250.

35. Yu, G.; Chang, Q.; Lv, W.; Xu, C.; Cui, C.; Ji, W.; Dang, Q.; Deng, K.; Wang, G.; Du, Y.; et al. PP-PicoDet: A Better Real-Time Object Detector on Mobile Devices. *arXiv* **2021**, arXiv:2111.00902.

36. Dollar, P.; Singh, M.; Girshick, R. Fast and Accurate Model Scaling. In Proceedings of the IEEE/CVF Conference on Computer Vision and Pattern Recognition (CVPR), Nashville, TN, USA, 20–25 June 2021; pp. 924–932.
37. Rajeev, R.; Kevin, C.; Xiaodong, H.; Samurdhi, K.; Shuyue, L.; Ryan, M.; Gwena, C.; Vinh, N.; Boris, F.; Paul, B.; et al. TensorRT Open Source Software. Available online: https://github.com/NVIDIA/TensorRT (accessed on 2 October 2022).

 *drones*

*Article*

# Elliptical Multi-Orbit Circumnavigation Control of UAVS in Three-Dimensional Space Depending on Angle Information Only

Zhen Wang [1,2] and Yanhong Luo [1,2,*]

[1]  The State Key Laboratory of Synthetical Automation for Process Industries, Northeastern University, Shenyang 110004, China
[2]  School of Information Science and Engineering, Northeastern University, Shenyang 110004, China
*  Correspondence: luoyanhong@ise.neu.edu.cn; Tel.: +86-024-8368-6470

**Abstract:** In order to analyze the circumnavigation tracking problem in complex three-dimensional space, in this paper, we propose a UAV group circumnavigation control strategy, in which the UAV circumnavigation orbit is an ellipse whose size can be adjusted arbitrarily; at the same time, the UAV group can be assigned to multiple orbits for tracking. The UAVs only have the angle information of the target, and the position information of the target can be obtained by using the angle information and the proposed three-dimensional estimator, thereby establishing an ideal relative velocity equation. By constructing the error dynamic equation between the actual relative velocity and the ideal relative velocity, the circumnavigation problem in three-dimensional space is transformed into a velocity tracking problem. Since the UAVs are easily disturbed by external factors during flight, the sliding mode control is used to improve the robustness of the system. Finally, the effectiveness of the control law and its robustness to unexpected situations are verified by simulation.

**Keywords:** three-dimensional circumnavigation control; elliptical multi-orbit; UAV group

**Citation:** Wang, Z.; Luo, Y. Elliptical Multi-Orbit Circumnavigation Control of UAVS in Three-Dimensional Space Depending on Angle Information Only. *Drones* **2022**, *6*, 296. https://doi.org/10.3390/drones6100296

Academic Editors: Daobo Wang and Zain Anwar Ali

Received: 30 August 2022
Accepted: 6 October 2022
Published: 10 October 2022



## 1. Introduction

In the past decade, the problem of UAV control has attracted more and more attention [1–3]. Since the multi-agent control problem has received continuous attention [4,5], multi-UAV control is also a hot area of research [6–8]. The multi-UAV circumnavigation control problem is a field of application for multi-UAV control, which refers to a group of UAVs surrounding the target in a prescribed formation to perform tasks such as monitoring [9], tracking and rounding up [10].

The circumnavigation problem has gradually developed from a single-agent circumnavigation static target [11–13] to a single-agent circumnavigation dynamic target [14–16], and up to the current multi-agent circumnavigation dynamic target [4,17,18]. Ref. [19] analyzed the circumnavigation problem in a two-dimensional plane, assuming that each agent has a fixed velocity. Ref. [20] used the self-propelled particle system to analyze the circumnavigation problem, but it required the circumnavigation target to be a static target. Most of the initial research assumed that each agent can accurately know the position information of the target [17,21], but this is difficult to achieve in practice. In order to overcome this drawback, researchers have proposed the use of distance measurement to obtain the position of the target. In [22], by measuring the distance between the agent and the target, the backstepping control is used to analyze the circumnavigation problem. In [23], a sliding mode control method based on distance measurement is designed to enable the UAV to achieve circumnavigation without using GPS. In practice, achieving distance measurement requires high-accuracy and expensive sensors, but UAVs are small and have limited payload capacity, so this approach is not realistic for UAVs. In contrast, angle measurement is relatively easy to implement. UAVs only need to carry cameras or other small sensors to measure angles.

In [24], the kinematic model of the underwater vehicle is combined with the dynamic model, and a cascade-based distributed circumnavigation control system is proposed, which can realize the circumnavigation of static and dynamic targets. In [25], the navigation control is realized by using the angle information between adjacent UAVs and the angle information between the UAV and the target. According to the local information of the target, ref. [26] uses the estimation method to estimate the location information of the target, so as to achieve circumnavigation, and this method does not need to consider the initial state of the agent. Refs. [27,28] propose a wireless speed measurement method to keep the multi-agent formation in the desired formation. Ref. [29] proposes a swarm control strategy for fixed-wing UAVs using multi-objective reinforcement learning. Taking the underwater robot as the research object, ref. [30] uses fast non-singular terminal sliding technology to provide faster convergence for full-drive underwater robot trajectory tracking control, while using adaptive technology to remove the restrictions that require system parameters.

However, the above studies were all conducted in the two-dimensional plane, but the UAV is active in the complex three-dimensional space; it is obviously more challenging to analyze the circumnavigation problem in three-dimensional space. At the same time, it should be noted that the formations proposed in these works are generally circular formations, which limits their practical application. In this paper, we set the formation to be elliptical and allowed the UAVs to be deployed in different orbits. The main contributions of this paper are summarized as follows:

(1) A circumnavigation control law in three-dimensional space using only angle information is proposed, and an estimation method is used to obtain the position information of the target. In this way, the limitation of requiring knowledge of both target position information and angle information at the same time is eliminated.
(2) The circumnavigation trajectory is set as an ellipse instead of being limited to a circular trajectory, and the major and minor semi-axes of the ellipse can be arbitrarily set. At the same time, UAVs can be deployed on multiple orbits by setting different coefficients.
(3) Using the dynamic equation of the UAV, the three-dimensional position estimator and the adjustable elliptical orbit, the relative ideal velocity equation is designed, and by constructing the dynamic error between the ideal relative velocity and the actual velocity, the circumnavigation control problem is transformed into the tracking problem of relative velocity. At the same time, by adopting sliding mode control, the robustness of the system is greatly improved, and the stability of the system is proved by the Lyapunov method.

## 2. Problem Formulation

### 2.1. Define the Desired Angle

In the three-dimensional space, UAV $i$ moves with speed $v_i$, and a space coordinate system is established with the UAV $i$ as the origin, as shown in Figures 1 and 2. The target is denoted by $h$, and its projection point on the i-plane is denoted by $h'$. The distance between $i$ and $h'$ is denoted by $l'_i$, and $l_d$ is the desired radius, that is, the desired distance between the UAV emphi and the target projection point $h'$.

**Definition 1.** *$\delta_i$ is the angle between $ih$ and $ih'$. $\gamma_i$ is the angle formed by $ih'$ and the positive direction of the X-axis. $\theta_i$ is the pitch angle of $i$, and $\alpha_i$ is the heading angle of $i$. Next, make the following assumption:*

**Assumption 1.** *Assume that every UAV can know the angles in Definition 1, and the communication between UAVs is realized under a cyclic directed graph, i.e., UAV $i + 1$ can get the information of UAV $i$; at the same time, there is no interference in the communication between UAVs.*

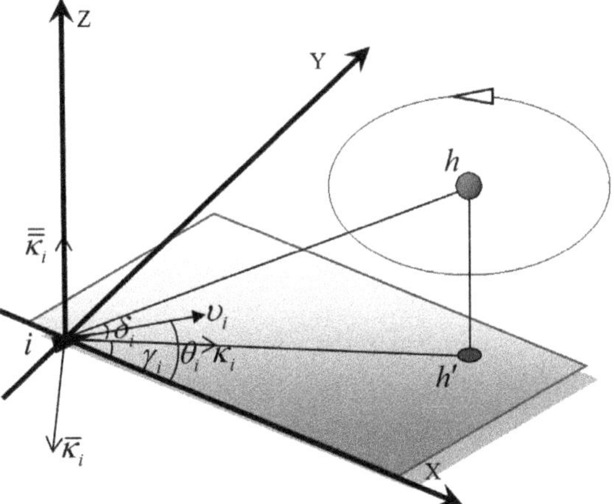

**Figure 1.** Schematic diagram of three-dimensional space circumnavigation.

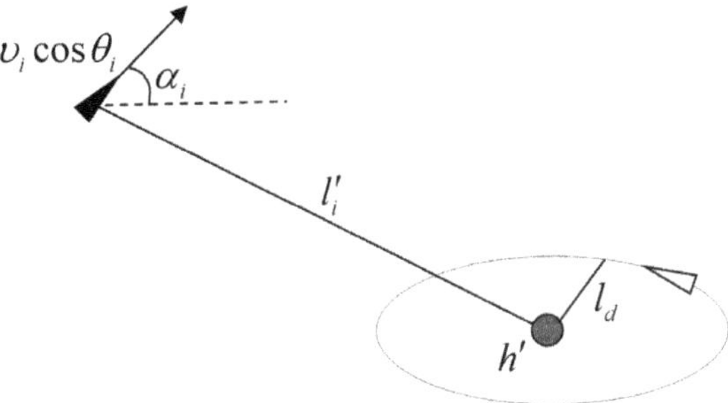

**Figure 2.** Top view of circumnavigation.

**Definition 2.** *Define a set of orthogonal vectors* $[\kappa_i, \ \bar{\kappa}_i, \ \bar{\bar{\kappa}}_i]$ *as shown in Figure 1, where* $\kappa_i$ *represents the unit vector pointed by the i to* $h'$*;* $\bar{\kappa}_i$ *represents the unit vector in the XOY plane, perpendicular to* $\kappa_i$*; and* $\bar{\bar{\kappa}}_i$ *represents the vector perpendicular to the XOY plane.*

$$\begin{cases} \kappa_i = \begin{bmatrix} \cos\gamma_i & \sin\gamma_i & 0 \end{bmatrix}^T, \\ \bar{\kappa}_i = \begin{bmatrix} \sin\gamma_i & -\cos\gamma_i & 0 \end{bmatrix}^T, \\ \bar{\bar{\kappa}}_i = \begin{bmatrix} 0 & 0 & \sin\delta_i \end{bmatrix}^T. \end{cases} \tag{1}$$

### 2.2. Multi-Orbit Circumnavigation

The multi-orbit circumnavigation is shown in Figure 3. For agent $i$, its required circumnavigation radius at time $t$ should be set as $l_i(t) - \tau_i l_d$. $l_i(t)$ represents the actual distance between the UAV $i$ and the target projection point. $\tau_i$ and $l_d$ are positive real numbers, and by adjusting different $\tau_i$, different circumnavigation radii can be obtained.

Therefore, in the multi-orbit circumnavigation, different orbits are realized according to the given basic circumnavigation radius $l_d$ and multiplied by the corresponding $\tau_i$. The relationship between different orbits is based on the multiple relationship of $l_d$.

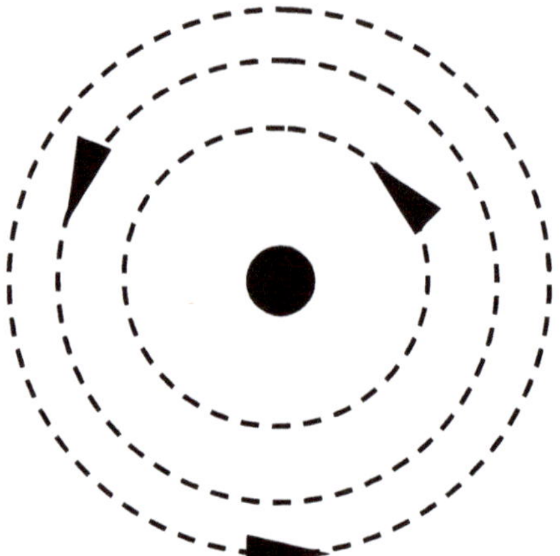

**Figure 3.** Multi-orbit circumnavigation.

*2.3. Dynamic Model of UAVs*

First, we make the following assumption:

**Assumption 2.** *In this paper, the quadrotor UAV is taken as the research object. Next, the quadrotor UAV dynamic model in the inertial coordinate system can be described as [31]*

$$\begin{cases} \ddot{x}_i = \frac{F_i}{m_i}(\cos \psi_i \cos \phi_i \sin \theta_i + \sin \psi_i \sin \phi_i) + d_{xi}, \\ \ddot{y}_i = \frac{F_i}{m_i}(\sin \psi_i \cos \phi_i \sin \theta_i + \cos \psi_i \sin \phi_i) + d_{yi}, \\ \ddot{z}_i = \frac{F_i}{m_i}(\cos \phi_i \cos \theta_i) - g + d_{zi}, \end{cases} \tag{2}$$

*where $m_i$ represents the mass of the UAV. Defining $\boldsymbol{p_i} = [x_i, y_i, z_i]^T$ indicates the position of the i-th UAV in the inertial coordinate system, and $\boldsymbol{B_i} = [\psi_i, \theta_i, \phi_i]^T$ indicates the roll, pitch and yaw of the UAV in the body coordinate system. The relationship between the inertial coordinate system and the body coordinate system is given in Figure 4. $F_i$ is the total thrust generated by the motors of the UAV, and g is the acceleration of gravity. $\boldsymbol{d_i} = [d_{xi}, d_{yi}, d_{zi}]^T$ indicates system external interference.*

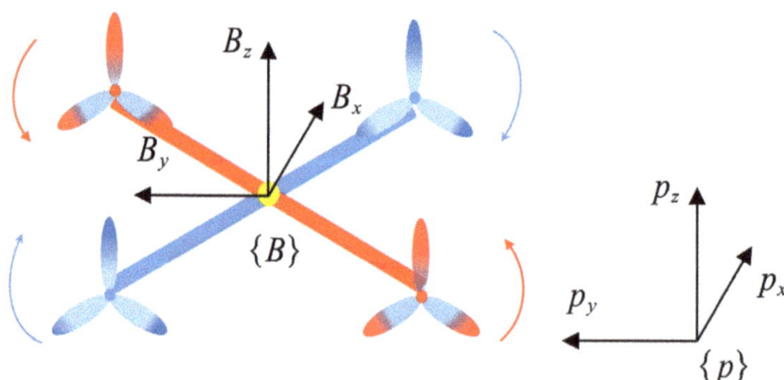


**Figure 4.** Quadcopter UAV coordinate system.

In order to convert the coordinates of the UAV from the body coordinate system to the inertial coordinate system, we need the following rotation matrix $\boldsymbol{\vartheta_i}$

$$\boldsymbol{\vartheta_i} = \begin{bmatrix} \cos\theta_i\cos\psi_i & \sin\theta_i\cos\psi_i\sin\phi_i - \sin\psi_i\cos\phi_i & \sin\theta_i\cos\psi_i\cos\phi_i + \sin\psi_i\sin\phi_i \\ \cos\theta_i\sin\psi_i & \sin\theta_i\sin\psi_i\sin\phi_i + \cos\psi_i\cos\phi_i & \sin\theta_i\sin\psi_i\cos\phi_i - \cos\psi_i\sin\phi_i \\ -\sin\theta_i & \cos\theta_i\sin\phi_i & \cos\theta_i\cos\phi_i \end{bmatrix}. \tag{3}$$

After some transformations, the dynamics model can be written as in [32]

$$m_i \begin{bmatrix} \ddot{p}_{xi} \\ \ddot{p}_{yi} \\ \ddot{p}_{zi} \end{bmatrix} = \vartheta_i F_i + \begin{bmatrix} 0 \\ 0 \\ -m_i g \end{bmatrix} + m_i \begin{bmatrix} d_{xi} \\ d_{yi} \\ d_{zi} \end{bmatrix}, \tag{4}$$

where $T_i$ is a continuous variable that represents the external input of UAV $i$, and can be written as

$$T_i = m_i \boldsymbol{\vartheta_i}^T [u_{xi}, u_{yi}, u_{zi} + g]^T, \tag{5}$$

where $u_i = [u_{xi}, u_{yi}, u_{zi}]^T$ represents the components of the control force of the UAV $i$ on the three axes in the inertial coordinate system.

Next, combining (2), (4) and (5), a new expression can be obtained

$$\begin{aligned} F_i =& m_i u_{xi}(\cos\psi_i\cos\phi_i\sin\theta_i + \sin\psi_i\sin\phi_i)^{-1} \\ &+ m_i u_{yi}(\sin\psi_i\cos\phi_i\sin\theta_i + \cos\psi_i\sin\phi_i)^{-1} \\ &+ m_i u_{zi}(\cos\phi_i\cos\theta_i)^{-1}. \end{aligned} \tag{6}$$

From Equation (6), it can be seen that by designing the control $u_i = [u_{xi}, u_{yi}, u_{zi}]^T$, the UAV can be controlled by applying the external input $T_i$ and the attitude angle $B_i = [\psi_i, \theta_i, \phi_i]^T$, so as to realize the circumnavigation control.

*2.4. Control Objective*

Given a group of $n$ UAVs and a moving target $h$, the purpose of this paper is to design a control law to make the UAV group around the given target in elliptical orbits. Ultimately, the UAV group needs to meet the following conditions:

(1) The UAV group should circumnavigate the target on multiple elliptical orbits, so the circumnavigation radius of each UAV is different.
(2) During the circumnavigation process, the circumnavigation angular velocity of the UAV group should be consistent.
(3) The angular spacing between adjacent UAVs should remain unchanged.

The above objective can be written in the following form:

$$\begin{cases} \lim\limits_{t\to+\infty} l_i(t) = \tau_i l_d, \\ \lim\limits_{t\to+\infty} \dot{\alpha}_i = \omega_d, \qquad \forall i, j \in N, \\ \lim\limits_{t\to+\infty} \alpha_i - \alpha_j = \beta_d, \end{cases} \tag{7}$$

where $N = \{1, \cdots, n\}$, $\omega_d$ represents the desired circumnavigation angular velocity; $\beta_d$ is the desired angular spacing between two UAVs, with a value of $2\pi/n$; and $n$ is the number of UAVs. This means that at time $t \to +\infty$, the angular spacing between adjacent UAVs will be a fixed value.

## 3. Circumnavigation Control

Due to the target coordinate position, an estimator is needed to estimate the position of the target. We use the following estimator

$$\dot{\boldsymbol{p}}_{ti} = k(\boldsymbol{I} - \boldsymbol{\varepsilon}(t)\boldsymbol{\varepsilon}(t)^T)(\boldsymbol{p}_i - \hat{\boldsymbol{p}}_{ti}),$$

$$\boldsymbol{\varepsilon}(t) = \frac{\boldsymbol{p}_t - \boldsymbol{p}_i}{\|\boldsymbol{p}_t - \boldsymbol{p}_i\|} = \begin{bmatrix} \cos\delta_i\cos\gamma_i \\ \cos\delta_i\sin\gamma_i \\ \sin\delta_i \end{bmatrix}, \tag{8}$$

where $k$ is a constant gain, $\boldsymbol{p}_t = [x_t, y_t, z_t]^T$ is the coordinates of the target, and $\hat{\boldsymbol{p}}_{ti} = [\hat{x}_{ti}, \hat{y}_{ti}, \hat{z}_{ti}]$ is the coordinate estimation of the target by the UAV $i$. $\boldsymbol{I}$ is the identity matrix. Through this three-dimensional position estimator, the position of the target can be estimated with only the angle information known. The desired elliptical circumnavigation radius is set as follows

$$l_d = \frac{ab}{\sqrt{a^2\sin^2\gamma_i + b^2\cos^2\gamma_i}}, \tag{9}$$

where $a$ and $b$ are the major and minor axes of the ellipse, respectively. By setting different values for $a$ and $b$, the desired elliptical orbit $l_d$ can be obtained. Using $l_d$ multiplied by constant $\tau_i$, we can obtain different elliptical orbits; then, we can cause each UAV circle to be in its own orbit. It is worth noting that when $a = b$, it is a circular orbit.

Next, we write the dynamic Equation in (4) in the following form:

$$\dot{\boldsymbol{p}}_i = s_i, \dot{s}_i = u_i + d_i. \tag{10}$$

At the same time, the angle in Definition 1 can be represented by known information and estimated values

$$\begin{cases} \delta_i = \arctan\dfrac{\hat{z}_{ti} - z_i}{\sqrt{(\hat{x}_{ti} - x_i)^2 + (\hat{y}_{ti} - y_i)^2}}, \\ \gamma_i = (-1)^\zeta \arccos\dfrac{\hat{x}_{ti} - x_i}{\sqrt{(\hat{x}_{ti} - x_i)^2 + (\hat{y}_{ti} - y_i)^2}} + 2\pi\zeta, \end{cases} \tag{11}$$

where

$$\zeta = \begin{cases} 0, \hat{y}_{ti} < y_i, \\ 1, \hat{y}_{ti} \geq y_i. \end{cases} \tag{12}$$

It can be seen that $\gamma_i$ is discontinuous, but the derivative of $\gamma_i$ with respect to time can be expressed in the following form

$$\dot{\gamma}_i = \frac{v_i\cos\theta_i\sin(\gamma_i - \alpha_i)}{l'_i(t)} - \omega_i, \tag{13}$$

where $l'_i(t)$ is the distance between the UAV and the target projection point.

**Proof.** Suppose the motion of UAV $i$ at time $t$ is as shown in Figure 5; at this time, the relationship between the angles $\gamma_i$ and $\alpha_i$ is also marked in Figure 5. The velocity of $i$ on the horizontal plane can be decomposed orthogonally into $v_{ia}$ and $v_{ib}$; then, $v_{ia}$ can be written as:

$$v_{ia} = v_i\cos\theta_i\sin(\gamma_i - \alpha_i) \tag{14}$$

The derivative of $\gamma_i + (2\pi - \alpha_i)$ can be written as:

$$\dot{\gamma}_i - \dot{\alpha}_i = \frac{v_{ia}}{l'_i(t)}. \tag{15}$$

Then, (14) and (15) can be combined to obtain the result in (13). The proof is finished. $\square$

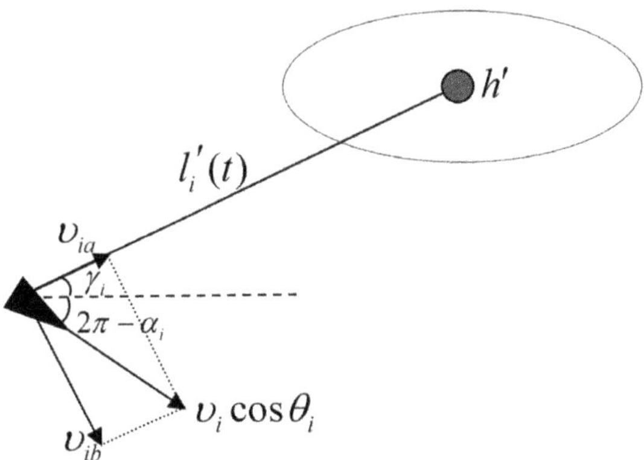



**Figure 5.** Top view of the UAV $i$ at time $t$.

Next, the ideal relative velocity is constructed from the orthogonal vector in (1)

$$v_{id} = c_1(l'_i(t) - \tau_i l_d)\kappa_i + \tau_i l_d \omega_d \left(1 - \frac{2}{\pi}\arctan\beta_i\right)\bar{\kappa}_i + c_2\bar{\bar{\kappa}}_i, \tag{16}$$

where $c_1 > 0, c_2 > 0$ , $\beta_i = \alpha_i - \alpha_j - (i-j)\frac{2\pi}{n}$.

$\sigma_{1i}$ is defined as the difference between the actual relative position and the ideal relative position of the UAV $i$, and $\sigma_{2i}$ as the difference between the actual relative speed and the ideal relative speed of the UAV $i$; thus, we can obtain

$$\begin{cases} \sigma_{1i} = p_i - p_t - \int v_{id}dt, \\ \dot{\sigma}_{1i} = \sigma_{2i} = \dot{p}_i - \dot{p}_t - v_{id}, \\ \dot{\sigma}_{2i} = u_i + d_i - \ddot{p}_t - \dot{v}_{id}. \end{cases} \tag{17}$$

In order to improve the robustness of the system, sliding mode control is adopted. The sliding surface is designed as

$$S = \chi\sigma_{1i} + \sigma_{2i}, \tag{18}$$

where $\chi > 0$, using the exponential reaching law as follows

$$\dot{S} = -c_3 S - c_4 \text{sgn}(S), \tag{19}$$

where $c_3 > 0, c_4 > 0$. Then, we obtain the following control law

$$u_i = -c_3 S - c_4 \text{sgn}(S) - \chi\sigma_{2i} + \ddot{p}_t + \dot{v}_{id} - d_i. \tag{20}$$

**Theorem 1.** *Considering a group of UAVs as a nonlinear system (4), if the three-dimensional position estimators are set as (8), the ellipse surround radius is set as (9), and the parameters $c_1$, $c_2$, $c_3$, $c_4$, and $\chi$ are selected to be greater than 0, then the controller (20) can make the group of UAVs circle the target on multiple elliptical orbits.*

**Proof.** Considering the Lyapunov candidate function $V = \frac{1}{2}S^2$, we take its derivative with respect to time as

$$\begin{aligned} \dot{V} &= S\dot{S} \\ &= S(\chi\sigma_{2i} + \dot{\sigma}_{2i}) \\ &= S(\chi\sigma_{2i} + (u_i + d_i - \ddot{p}_t - \dot{v}_{id})). \end{aligned} \tag{21}$$

Next, we substitute (20) into (21)

$$\dot{V} = S(\chi\sigma_{2i} - c_3 S - c_4\mathrm{sgn}(S) - \chi\sigma_{2i} + \ddot{p}_t + \dot{v}_{id} - d_i - \ddot{p}_t - \dot{v}_{id} + d_i)$$
$$= S(-c_3 S - c_4\mathrm{sgn}(S)) \tag{22}$$
$$= -c_3\|S\|^2 - c_4\|S\|,$$

where $c_3 > 0$, and $c_4 > 0$; therefore, $\dot{V} \leq 0$ satisfies the Lyapunov stability criterion. $\square$

## 4. Simulation Results

In this section, the performance of the estimator (8) and control law (20) is verified by considering a group of five UAVs circling a moving target on multiple trajectories, where $\beta_d$ is $2\pi/5$. We set the mass of each UAV to $m_i = 1$ kg, and the acceleration of gravity to $g = 9.8$ m/s$^2$. For the ideal relative velocity in (16), $c_1 = 1$ and $c_2 = 5$ were selected. Parameters in the controller (20) were selected as $c_3 = 5$, $c_4 = 10$, $\chi = 1$.

### 4.1. Case 1: Target Moves in a Straight Line

The speed of the moving target in three-dimensional space is selected as $\dot{p}_t = [2 \quad 2 \quad 4]^T$.

First, the parameters of the estimator are selected, $k = 1$, and the initial positions of the five UAVs are randomly generated. Figure 6 presents the simulation results of the convergence effect of the position estimator.

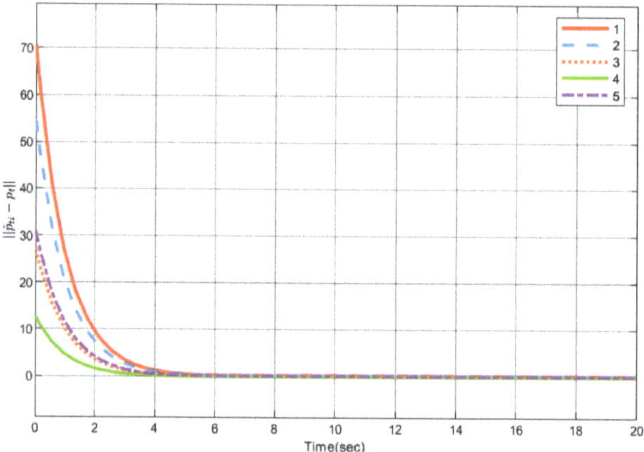


**Figure 6.** Position estimator error convergence $\|\hat{p}_{ti} - p_t\|$.

Next, when external interference is not considered, that is $d_i = [0,0,0]^T$, the parameters of the elliptical orbit are taken as $a = 25$ and $b = 15$, with $\tau_i$ selected as $\tau_i = [1, 1.5, 1.8, , 2.3, 2.8]$.

Figure 7 shows the angular spacing between two adjacent UAVs, Figure 8 gives the circumnavigation control error, and Figure 8a shows distance between each UAV and the target, that is $\|p_i - p_t\|$. Since the circumnavigation track is elliptical, the distance should change all the time. Figure 8b is $\|p_i - p_t\| - \tau_i l_d$, that is, the error between the actual distance and estimated distance from the position of UAV to target. The three-dimensional simulation diagram of the UAVs circumnavigation is given in Figure 9.

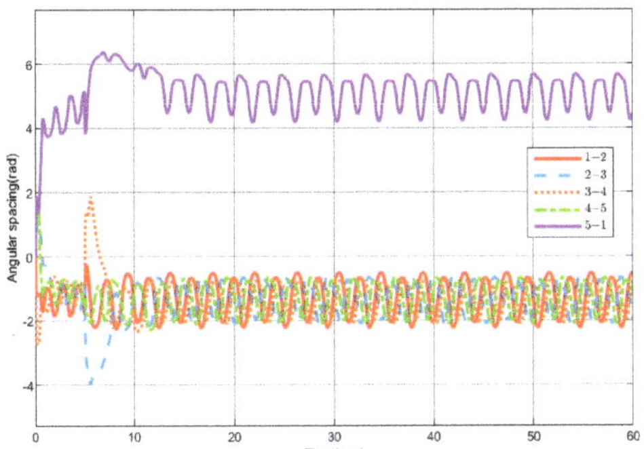

**Figure 7.** Angular spacing between adjacent UAVs.

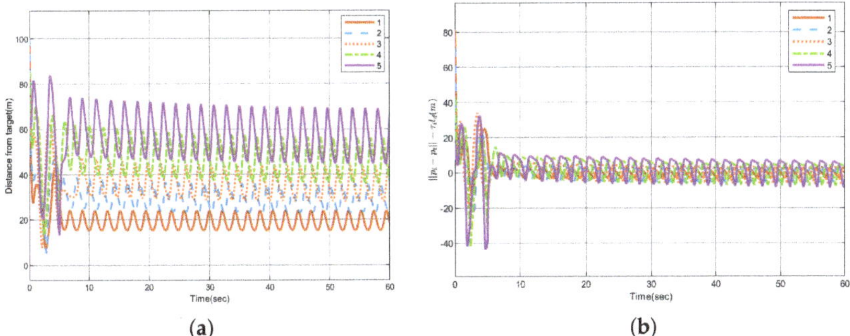

(a)

(b)

**Figure 8.** Circumnavigation control error. (**a**) $||p_i - p_t||$. (**b**) $||p_i - p_t|| - \tau_i l_d$.

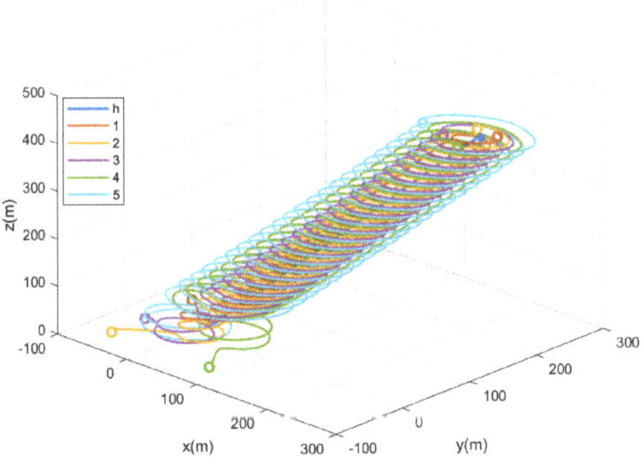

**Figure 9.** Three−dimensional diagram of circumnavigation control.

Since the UAV is easily interfered with by external factors during the operation, the robustness of the control system is tested. When $t = 70$ s, the external interference of the third UAV is set to $\boldsymbol{d_3} = [0, 30, 70]^T$. Figure 10 shows the change in the angular spacing between each UAV, and Figure 11 shows the circumnavigation control error, where Figure 11a is $||p_i - p_t||$, and Figure 11b is $||p_i - p_t|| - \tau_i l_d$. It can be seen from the above pictures that when there is interference, the sliding mode control can quickly eliminate the influence of the interference and maintain the formation of the UAV group.

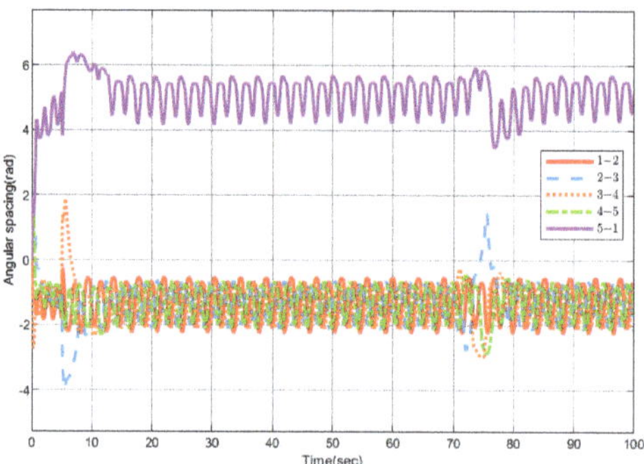

**Figure 10.** Variation in angular separation between adjacent UAVs when interference occurs.

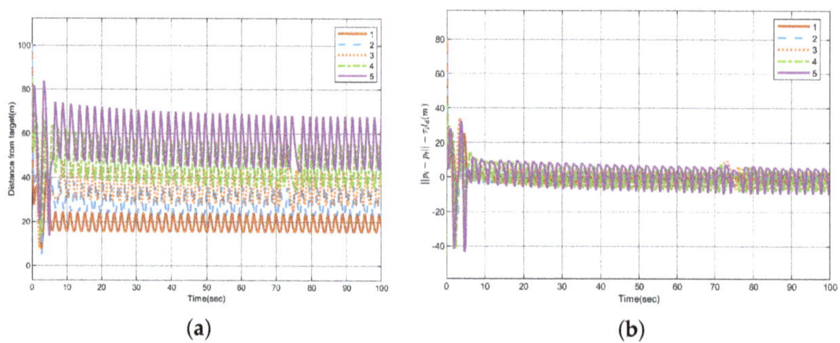

(a)                                  (b)

**Figure 11.** The circumnavigation control error when disturbance occurs. (a) $||p_i - p_t||$. (b) $||p_i - p_t|| - \tau_i l_d$.

### 4.2. Case 2: Target Moves in a Curve

The speed of the moving target in three-dimensional space is selected as $\dot{p}_t = [3\cos(0.5t) \quad 3\sin(0.5t) \quad t]^T$.

The parameters of the estimator are selected, $k = 1$, and the initial positions of the five UAVs are randomly generated. Figure 12 presents the simulation results of the convergence effect of the position estimator.

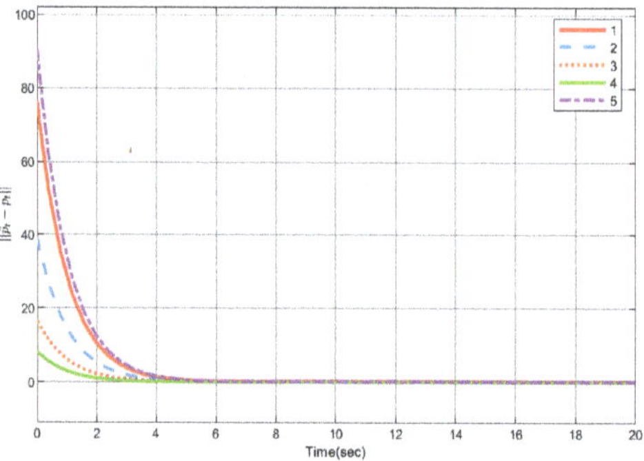

**Figure 12.** Position estimator error convergence $||\hat{p}_{ti} - p_t||$.

When $d_i = [0, 0, 0]^T$, we take the parameters of the elliptical orbit as $a = 30$ and $b = 20$, with $\tau_i$ selected as $\tau_i = [1, 1.2, 1.4, , 1.6, 1.8]$.

Figure 13 shows the angular spacing between two adjacent UAVs, Figure 14 gives the circumnavigation control error, and Figure 14a shows the distance between each UAV and the target, that is $||p_i - p_t||$. Figure 14b is $||p_i - p_t|| - \tau_i l_d$. The three-dimensional simulation diagram of the UAVs circumnavigation is given in Figure 15.

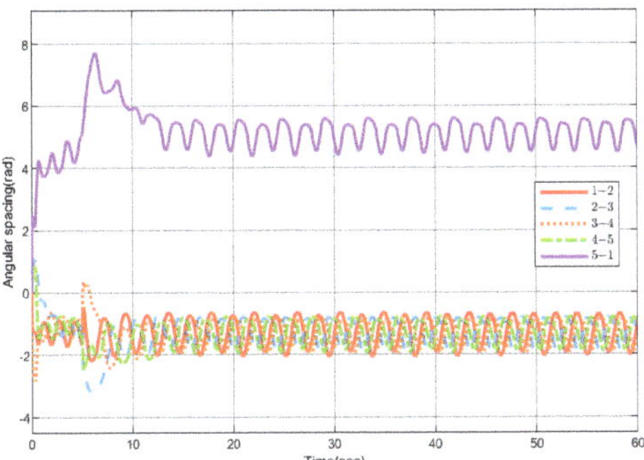

**Figure 13.** Angular spacing between adjacent UAVs.

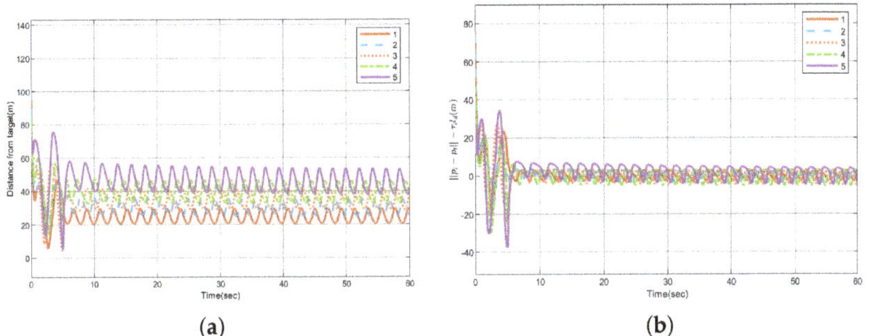

**Figure 14.** Circumnavigation control error. (**a**) $||p_i - p_t||$. (**b**) $||p_i - p_t|| - \tau_i l_d$.

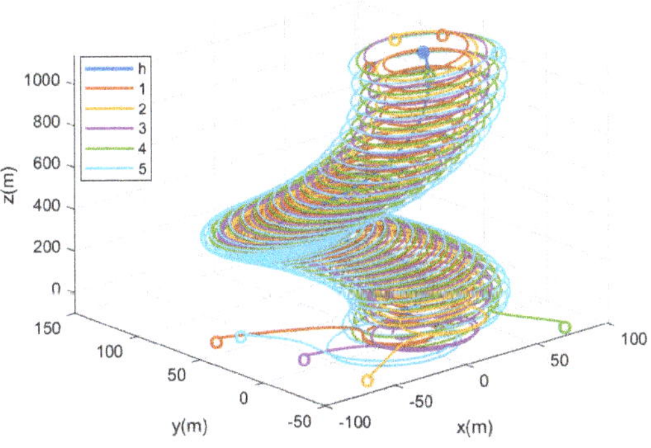

**Figure 15.** Three−dimensional diagram of circumnavigation control.

The robustness of the system is verified again. When $t = 70$ s, the external interference of the fifth UAV is set to $d_3 = [30, 10, 50]^T$. Figure 16 shows the change in the angular spacing between each UAV, and Figure 17 shows the circumnavigation control error, where Figure 17a is $||p_i - p_t||$, and Figure 17b is $||p_i - p_t|| - \tau_i l_d$. The above experiments once again prove that the sliding mode control adopted in this paper has strong robustness.

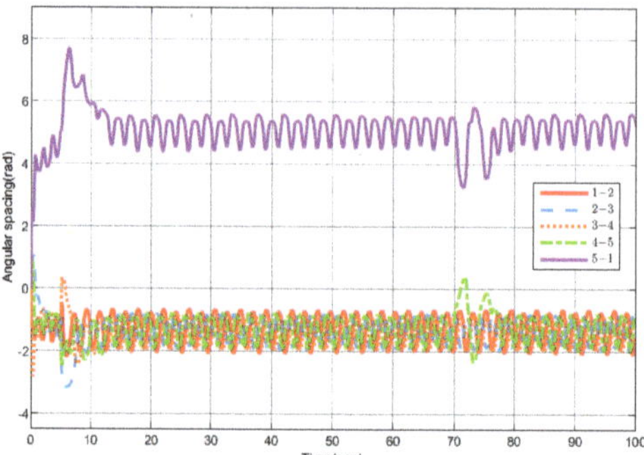

**Figure 16.** Variation in angular separation between adjacent UAVs when interference occurs.

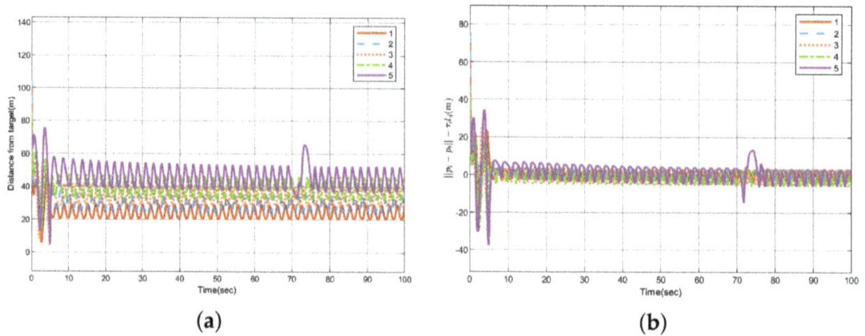

(a)  (b)

**Figure 17.** The circumnavigation control error when disturbance occurs. (a) $||p_i - p_t||$. (b) $||p_i - p_t - \tau_i l_d||$.

## 5. Conclusions

In this paper, the problem of circumnavigation control in three-dimensional space is studied, the group of UAVs is made to circle the target in elliptical formation and with multiple orbits. The target is made to move with curvilinear variable speed, and UAVs estimate the position of the target only from the angle information. The error dynamic equation is constructed using the ideal relative velocity and the actual relative velocity, and the circumnavigation control is transformed into a velocity tracking problem. In order to improve the robustness of the system, sliding mode control is used to design the control law. Finally, the effectiveness of the proposed control law is proved by simulation, and disturbance is added to the simulation process to verify the robustness of sliding mode control.

**Author Contributions:** Conceptualization, Y.L. and Z.W.; methodology, Z.W.; software, Z.W.; validation, Y.L. and Z.W.; formal analysis, Z.W.; investigation, Z.W.; resources, Y.L.; data curation, Z.W.; writing—original draft preparation, Z.W.; writing—review and editing, Y.L.; visualization, Z.W.; supervision, Y.L.; project administration, Y.L.; funding acquisition, Y.L. All authors have read and agreed to the published version of the manuscript.

**Funding:** This research was funded by the National Natural Science Foundation of China (62173082), Shenyang Science & Technology Innovation Program for Young and Middle-aged Talents (RC210503), and the Liaoning Joint Fund of the National Natural Science Foundation of China (U1908217).

**Institutional Review Board Statement:** Not applicable.

**Informed Consent Statement:** Not applicable.

**Data Availability Statement:** Not applicable.

**Conflicts of Interest:** The authors declare no conflict of interest.

# References


1. Luo, Y.; Yu, X.; Yang, D.; Zhou, B. A survey of intelligent transmission line inspection based on unmanned aerial vehicle. *Artif. Intell. Rev.* **2022**, 1–29. [CrossRef]
2. Israr, A.; Ali, Z.A.; Alkhammash, E.H.; Jussila, J.J. Optimization Methods Applied to Motion Planning of Unmanned Aerial Vehicles: A Review. *Drones* **2022**, *6*, 126. [CrossRef]
3. Ming, Z.; Huang, H. A 3d vision cone based method for collision free navigation of a quadcopter UAV among moving obstacles. *Drones* **2021**, *5*, 134. [CrossRef]
4. Lan, Y.; Lin, Z.; Cao, M.; Yan, G. A distributed reconfigurable control law for escorting and patrolling missions using teams of unicycles. In Proceedings of the 49th IEEE Conference on Decision and Control, Atlanta, GA, USA, 15–17 December 2010; pp. 5456–5461.
5. Zhang, Y.; Wen, Y.; Li, F.; Chen, Y. Distributed observer-based formation tracking control of multi-agent systems with multiple targets of unknown periodic inputs. *Unmanned Syst.* **2019**, *7*, 15–23. [CrossRef]
6. Zhang, M.; Jia, J.; Mei, J. A composite system theory-based guidance law for cooperative target circumnavigation of UAVs. *Aerosp. Sci. Technol.* **2021**, *118*, 107034. [CrossRef]
7. Le, W.; Xue, Z.; Chen, J.; Zhang, Z. Coverage Path Planning Based on the Optimization Strategy of Multiple Solar Powered Unmanned Aerial Vehicles. *Drones* **2022**, *6*, 203. [CrossRef]
8. Yan, J.; Yu, Y.; Wang, X. Distance-Based Formation Control for Fixed-Wing UAVs with Input Constraints: A Low Gain Method. *Drones* **2022**, *6*, 159. [CrossRef]
9. Leonard, N.E.; Paley, D.A.; Lekien, F.; Sepulchre, R.; Fratantoni, D.M.; Davis, R.E. Collective motion, sensor networks, and ocean sampling. *Proc. IEEE* **2007**, *95*, 48–74. [CrossRef]
10. Kothari, M.; Sharma, R.; Postlethwaite, I.; Beard, R.W.; Pack, D. Cooperative target-capturing with incomplete target information. *J. Intell. Robot. Syst.* **2013**, *72*, 373–384. [CrossRef]
11. Sepulchre, R.; Paley, D.A.; Leonard, N.E. Stabilization of planar collective motion: All-to-all communication. *IEEE Trans. Autom. Control* **2007**, *52*, 811–824. [CrossRef]
12. Sepulchre, R.; Paley, D.A.; Leonard, N.E. Stabilization of planar collective motion with limited communication. *IEEE Trans. Autom. Control* **2008**, *53*, 706–719. [CrossRef]
13. Marshall, J.A.; Broucke, M.E.; Francis, B.A. Formations of vehicles in cyclic pursuit. *IEEE Trans. Autom. Control* **2004**, *49*, 1963–1974. [CrossRef]
14. Deghat, M.; Shames, I.; Anderson, B.D.; Yu, C. Localization and circumnavigation of a slowly moving target using bearing measurements. *IEEE Trans. Autom. Control* **2014**, *59*, 2182–2188. [CrossRef]
15. Wang, J.; Ma, B.; Yan, K. Mobile Robot Circumnavigating an Unknown Target Using Only Range Rate Measurement. *IEEE Trans. Circ. Syst. I Express Briefs* **2021**, *69*, 2. [CrossRef]
16. Greiff, M.; Deghat, M.; Sun, Z.; Robertsson, A. Target Localization and Circumnavigation with Integral Action in $R^2$. *IEEE Control Syst. Lett.* **2021**, *6*, 1250–1255. [CrossRef]
17. Summers, T.H.; Akella, M.R.; Mears, M.J. Coordinated standoff tracking of moving targets: Control laws and information architectures. *J. Guid. Control Dyn.* **2009**, *32*, 56–69. [CrossRef]
18. Huo, M.; Duan, H.; Fan, Y. Pigeon-inspired circular formation control for multi-UAV system with limited target information. *Guid. Navig. Control* **2021**, *1*, 2150004. [CrossRef]
19. Seyboth, G.S.; Wu, J.; Qin, J.; Yu, C.; Allgöwer, F. Collective circular motion of unicycle type vehicles with nonidentical constant velocities. *IEEE Trans. Control Netw. Syst.* **2014**, *1*, 167–176. [CrossRef]
20. Hernandez, S.; Paley, D.A. Three-dimensional motion coordination in a spatiotemporal flowfield. *IEEE Trans. Autom. Control* **2010**, *55*, 2805–2810. [CrossRef]
21. Kim, T.H.; Hara, S.; Hori, Y. Cooperative control of multi-agent dynamical systems in target-enclosing operations using cyclic pursuit strategy. *Int. J. Control* **2010**, *83*, 2040–2052. [CrossRef]
22. Dong, F.; You, K.; Song, S. Target encirclement with any smooth pattern using range-based measurements. *Automatica* **2020**, *116*, 108932. [CrossRef]
23. Cao, Y. UAV circumnavigating an unknown target under a GPS-denied environment with range-only measurements. *Automatica* **2015**, *55*, 150–158. [CrossRef]
24. Shi, L.; Zheng, R.; Liu, M.; Zhang, S. Distributed circumnavigation control of autonomous underwater vehicles based on local information. *Syst. Control Lett.* **2021**, *148*, 104873. [CrossRef]
25. Yu, X.; Liu, L. Cooperative control for moving-target circular formation of nonholonomic vehicles. *IEEE Trans. Autom. Control* **2016**, *62*, 3448–3454. [CrossRef]
26. Sen, A.; Sahoo, S.R.; Kothari, M. Circumnavigation on multiple circles around a nonstationary target with desired angular spacing. *IEEE Trans. Cybern.* **2019**, *51*, 222–232. [CrossRef]
27. Mehrabian, A.R.; Tafazoli, S.; Khorasani, K. *Coordinated Attitude Control of Spacecraft Formation without Angular Velocity Feedback: A Decentralized Approach*; Aerospace Research Central: Lawrence, KS, USA, 2009; p. 6289.

28. Chen, Y.; Liang, J.; Miao, Z.; Wang, Y. Distributed Formation Control of Quadrotor UAVs Based on Rotation Matrices without Linear Velocity Feedback. *Int. J. Control Autom. Syst.* **2021**, *19*, 3464–3474. [CrossRef]

29. Speck, C.; Bucci, D.J. Distributed uav swarm formation control via object-focused, multi-objective sarsa. In Proceedings of the 2018 Annual American Control Conference, Milwaukee, WI, USA, 27–29 June 2018; pp. 6596–6601.

30. Qiao, L.; Zhang, W. Adaptive second-order fast nonsingular terminal sliding mode tracking control for fully actuated autonomous underwater vehicles. *IEEE J. Ocean. Eng.* **2018**, *44*, 363–385. [CrossRef]

31. Bai, A.; Luo, Y.; Zhang, H.; Li, Z. L2-gain robust trajectory tracking control for quadrotor UAV with unknown disturbance. *Asian J. Control* **2021**, 1–13. [CrossRef]

32. Li, D.; Cao, K.; Kong, L.; Yu, H. Fully Distributed Cooperative Circumnavigation of Networked Unmanned Aerial Vehicles. *IEEE/ASME Trans. Mechatron.* **2021**, *26*, 709–718. [CrossRef]

*Review*

# Comprehensive Review of UAV Detection, Security, and Communication Advancements to Prevent Threats

Ghulam E. Mustafa Abro [1,2,3,4,*], Saiful Azrin B. M. Zulkifli [1,3], Rana Javed Masood [5], Vijanth Sagayan Asirvadam [3] and Anis Laouti [2]

1   Center for Automotive Research and Electric Mobility (CAREM), Universiti Teknologi PETRONAS, Seri Iskandar 32610, Perak, Malaysia
2   Samovar, Telecom SudParis, CNRS, Institut Polytechnique de Paris, 9 Rue Charles Fourier, 91011 Paris, France
3   Electrical and Electronic Engineering Department, Universiti Teknologi PETRONAS, Seri Iskandar 32610, Perak, Malaysia
4   Condition Monitoring Systems (CMS) Lab, NCRA, Mehran University of Engineering and Technology (MUET), Jamshoro 67480, Sindh, Pakistan
5   Electronic Engineering Department, Usman Institute of Technology (U.I.T.), Karachi 75300, Sindh, Pakistan
*   Correspondence: ghulam_20000150@utp.edu.my or mustafa.abro@ieee.org

**Citation:** Abro, G.E.M.; Zulkifli, S.A.B.M.; Masood, R.J.; Asirvadam, V.S.; Laouti, A. Comprehensive Review of UAV Detection, Security, and Communication Advancements to Prevent Threats. *Drones* **2022**, 6, 284. https://doi.org/10.3390/drones6100284

Academic Editor: Diego González-Aguilera

Received: 22 August 2022
Accepted: 11 September 2022
Published: 1 October 2022

**Publisher's Note:** MDPI stays neutral with regard to jurisdictional claims in published maps and institutional affiliations.

**Abstract:** It has been observed that unmanned aerial vehicles (UAVs), also known as drones, have been used in a very different way over time. The advancements in key UAV areas include detection (including radio frequency and radar), classification (including micro, mini, close range, short range, medium range, medium-range endurance, low-altitude deep penetration, low-altitude long endurance, and medium-altitude long endurance), tracking (including lateral tracking, vertical tracking, moving aerial pan with moving target, and moving aerial tilt with moving target), and so forth. Even with all of these improvements and advantages, security and privacy can still be ensured by researching a number of key aspects of an unmanned aerial vehicle, such as through the jamming of the control signals of a UAV and redirecting them for any high-assault activity. This review article will examine the privacy issues related to drone standards and regulations. The manuscript will also provide a comprehensive answer to these limitations. In addition to updated information on current legislation and the many classes that can be used to establish communication between a ground control room and an unmanned aerial vehicle, this article provides a basic overview of unmanned aerial vehicles. After reading this review, readers will understand the shortcomings, the most recent advancements, and the strategies for addressing security issues, assaults, and limitations. The open research areas described in this manuscript can be utilized to create novel methods for strengthening the security and privacy of an unmanned aerial vehicle.

**Keywords:** unmanned aerial vehicle; advancement; classification; tracking and communication threats

## 1. Introduction

Technology has advanced, and as a result, the world of today has seen a number of ground-breaking developments. These outcomes have been demonstrated to be more trustworthy, approachable, and economical in our everyday lives. In addition, people now engage with one another in novel ways in their social circles. Additionally, unmanned aerial vehicles (UAVs) are employed for both commercial and private purposes in addition to being heavily utilized in military contexts. The market potential for medium-sized drones has been estimated by the China Unmanned Aerial Vehicle Industry (CUAVI) to reach CNY 80 billion by 2025 [1], whereas the Federal Aviation Administration (FAA) concluded that there are currently 3 million drones flying in the US sky and that number will increase by four times by the end of 2022 [2]. Drone use is increasing because of its value in a variety of jobs, including through the live broadcasting of events, aerial video shoots, the mobility to move packages from one location to another, and simple navigation as shown in Figure 1. These drones are commonly employed for transportation purposes due to their cheap maintenance requirements, ability to take-off and land vertically, ability to hover, and high degree of mobility. These drones are frequently outfitted with computer vision and internet of things (IoT)-like features, particularly for the swarming of drones [3,4], and they have proven to be an effective choice for surveillance and rescue-related missions [5]. There are, however, a few important elements that are connected to UAV security worries. This collection contains the story of the Iranian military jamming an American drone's control signals [6]; nonetheless, it is still difficult to create a security control module for UAVs that is completely foolproof. Additionally, the first drones were unmanned balloons loaded with explosives that were used to assault Venice in Italy [7]. Later, in 1915, the British troops employed these unmanned balloons for photographic-based surveillance during the renowned Battle of Neuve Chapelle [8]. During this time, cameras were not as advanced; hence, this strategy was suggested to improve visibility [8]. In order to find various terrorists, several of them were also used during the Afghan War [9,10]. Prior to today, these drones were usually utilized for military operations, but they are now also chosen for the majority of domestic applications, to the point that Amazon began using drones to carry packages in 2014 [11]. They have also been utilized in fields including agriculture [1,2], for checking building sites [3], and to greatly assist law enforcement organizations with emergency rescue operations. The United States of America started making pilotless aircrafts that could maneuver for roughly a kilometer in the early 1910s. During World War II, the US started developing advanced UAV programs, such as the N2C-2 drone and the OQ-2 communications plane [9], but these endeavors were both expensive and unreliable. In the late 1980s, the US started developing sophisticated drones, and they already have some top-notch micro unmanned aerial vehicles. Drones are also being used in the media business for aerial photography and filmmaking. Drone use is expanding quickly, and at the same time, security and privacy issues have grown more complicated and serious.

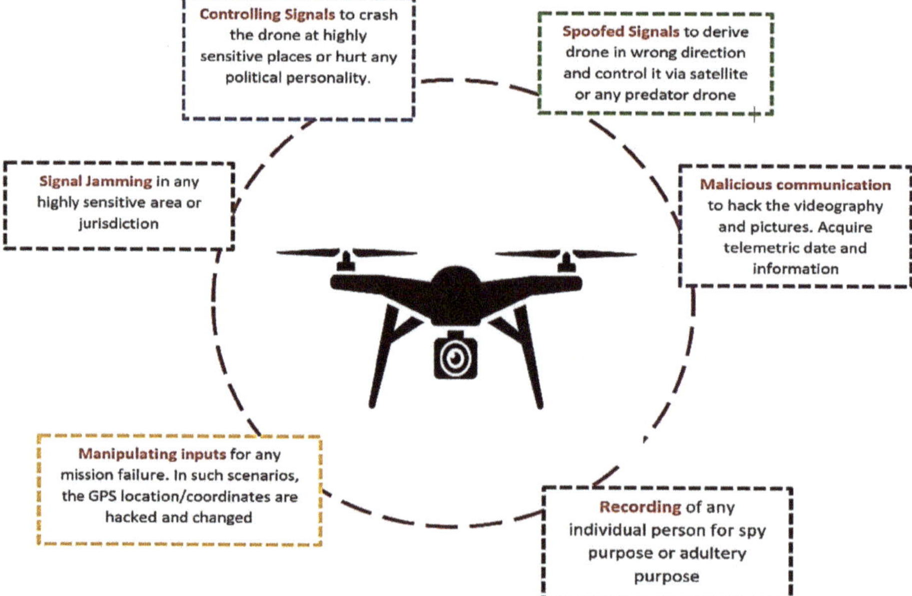


**Figure 1.** Security and privacy threats of UAVs.

The major goal of this review article is to give readers a comprehensive understanding of the new advancements that have led to the issues surrounding unmanned aerial vehicles (UAVs), including security threats, privacy concerns, and other limits that are important and cannot be disregarded. To give readers a thorough understanding of the subject, the manuscript has been organized into 10 sections. The major goal is to identify these issues and give all scholars access to a single resource that will allow them to fully understand the most recent trends and work to advance their field.

Section 2 contains the regulatory standards, whereas Section 3 describes the classification of various unmanned aerial vehicles (UAVs). Section 4 of the document discusses the structures and techniques of communication. In Section 5, it is specifically mentioned how and why drones are used. Section 6 covers the key security challenges and weaknesses, whereas Section 7 covers the present constraints. The most recent methods to address these restrictions are also discussed in Section 8, along with open research areas and recommendations in Section 9. The thorough conclusion to this work can be found last, but by no means least, in Section 10.

## 2. Study Related to Regulations

Many countries have been following the standard regulations to ensure the security and privacy implications of drones. Many of them have started to propose several step-by-step procedures to license their UAVs [12,13]. If these regulations are not followed, then unlicensed drones are taken under custody and proper legal action is taken against the pilot [12]. As per the media news broadcasted by the British Broadcasting Corporation (BBC), the CAA and FAA have declared some standard operating procedures (SOPs) to maneuver the UAVs at a low altitude [14], which are mentioned below:

- The users or operators of a registered UAV must carry the proof of license while operating the UAV.
- The maximum height at which the UAV can maneuver is 400 feet only.
- UAVs must be kept away from the airfields and, in case of necessity, one may acquire the written permission from relevant boards or authorities.
  - In the case of a UAV crash, legal action can be taken against any harmful actions or the damage that occurred from UAV failure.
  - UAVs with computer vision or camera surveillance are not allowed to maneuver within 50 m of people or any crowd.
  - UAVs will be summoned if they are not flown within the operator's line of sight.
  - UAVs will be summoned if they are flown at night without proper lighting.

The above standard rules and SOPs ensure the secure operation of drones.

It is also noted that with a dramatic rise in the drone industry, various countries have inducted their own rules as well [15]. Mainly, to operate a UAV, there are three fundamental components: the first is the ground control room (GCR); the second one is the communication method, for example, satellite, radio frequency, etc., as illustrated in Figure 2; and last, but certainly not the least, is the UAV itself. There are three different methods to communicate with a drone, i.e., satellite, radio signal, and internet, as shown in Figure 2 [15]. The essential license-exempted radio equipment, along with the frequencies, is mentioned in Table 1 [16].

The standard bandwidth through which a communication is established between a UAV and the ground control room (GCR) is mentioned in Table 1, whereas the other standards are still in progress for the safe operations of a UAV in any vicinity [17].

**Figure 2.** Communication channels mostly used to control UAVs.

**Table 1.** Frequencies through which GCR communicates with UAVs.

| Sr. No. | Bandwidth | Description |
|---|---|---|
| 1 | 2.4 KHz to 2483.500 MHz | The appropriate standard is EN 300 328, which is digital wideband data transmission equipment, and sometimes, the standard used is EN 300 440, which is general short-range devices. **Purpose:** Mostly used for short-range surveillance or short-range maneuvering missions. |
| 2 | 5.47 KHz to 57250 MHz | Operational power is less or equal to 1 watt, whereas the power spectral density is less the 50mW/1 MHz frequency. The standard is EN 201 893, which is known as RLAN equipment. **Purpose:** Long stay in sky operations, used mainly for aerial photography. |
| 3 | 5.725 KHz to 5875 MHz | Its operational power rating is less than 25 mW and standard is EN 300 440, which is general short-range devices. **Purpose:** Used for short-range surveillance with fast maneuvering and manipulating tasks. |
| 4 | 5030 to 5091 MHz | This is the frequency used only for the International Telecommunication Union (ITU) and, therefore, cannot be used for communication with drones. **Purpose:** Used in such operations where data sharing is important with ground control room (GCR). |

### 3. Classification of UAVs

One must understand the real sense of calling any drone a UAV. Not all drones can be classified as UAVs. A UAV can be controlled autonomously without a pilot and can be controlled remotely [17].

*3.1. Classification of Drones*

Drone is a very generic term and can refer to intelligent or autonomous vehicles such that there are unmanned aerial vehicles of different types. This can be hexarotors, quadrotors, multirotors or wing-based air vehicles. Mainly discussing the flying drones, they can be classified into three main categories as follows:

- Rotary-wing drones;
- Fixed-wing drones;
- Hybrid-wing drones;
- Flapping-wing drones.

The drones with a vertical take-off and landing (VTOL) feature and that can hover at a high rate are known as rotary-wing drones. The most common example is a quadrotor unmanned aerial vehicle that has four brushless DC motors. Drones with the capability to fly aggressively and glide even with heavy payloads are known as fixed-wing drones. They perform a horizontal take-off and landing (HTOL). Lastly, the drones that have both fixed and rotary wings are known as hybrid-wing drones. They are designed to have both features of rotary- and fixed-wing drones so that they can perform HTOL and VTOL along with high-rate hovering. Some of the robots are designed to exhibit the flying motion of a fly [18], and here they proposed a 5% more power-efficient wing by changing the wing stiffener pattern parametrically. An experimental aerodynamic analysis found that this could relate to increased wing stiffness, as well as indications of vortex generation during the flap cycle. The experiments reported an improved generated lift, allowing the DelFly to be outfitted with a yaw-rate gyro, pressure sensor, and microprocessor. These flapping-wings were later scaled to the micro level and are known as flapping-wing micro air vehicles (FWMAVs), due to inspiration taken from microscopic insects. These FWMAVs have the ability to perform activities in urban and interior environments. However, there are many hurdles for the successful flight of these vehicles that are replicating insect flight, including their design, manufacture, control, and propulsion [19,20].

### 3.2. Classification of UAVs Based on Ground Command and Control

It is already shown in Figure 2 that any UAV can be controlled remotely using a ground command and control mechanism, either by mobile phone, radio channel frequency, or the internet of things [21]. Therefore, these UAVs are classified based on their ability to fly over long distances without any intervention. These types are mentioned below as:

- Fully autonomous controlled UAVs: These are the UAVs that can perform different tasks without any intervention from human beings and are fully automated.
- Remotely operated UAVs: These UAVs are designed to execute the task as directed by a human being. Thus, they have a human as their main operator.
- Remotely pilot-controlled UAVs: Drones where all tasks and maneuvers are performed by the human-based remote control from the GCR.

The above-referred classification is summarized in Table 2 [22] along with the pros and cons of the UAVs.

**Table 2.** Classification of UAVs based on wing type and altitude.

| Factors | Based on Wing Type | | | Based on Altitude | |
|---|---|---|---|---|---|
| | Fixed-Wing Type | Rotary-Wing Type | Hybrid-Wing Type | Low Altitude Below 400 ft | High Altitude Above 400 ft |
| Hovering | No | Yes | Yes | Yes | Yes |
| Small Size | No | Yes | Yes | Yes | No |
| Transport goods | Yes | Low weight | Yes | Yes | No |
| Battery time (in hour) | >1 h | 1 h | >1 h | >1 h | >1 h |
| Maneuver speed | High Speed | Low speed | High speed | High Speed | High Speed |
| Flexible deployment of communication | No | No | No | Yes | No |
| Cost effective | Expensive | Cheap | Expensive | Cheap | Cheap |
| Endurance | High | Low | Medium | Low | High |

In the above, the reader might be confused with the term of flexible deployment of communication. This states the proper coverage at which the drone can be controlled and stabilized for any sort of task. There are some research contributions that have classified the drones based on their altitude, as demonstrated in Figure 3 along with their examples [23,24].

- **Tier (Netra, Batmav):** Small/Micro UAV
- **Tier I (Rq-2, Pioneer):** Low Altitude, Long endurance UAVs.
- **Tier II (MQ/1 Predator):** Medium Altitude, Long Endurance UAV
- **Tier III (Global Hawk, PA-B):** High Altitude with long endurance UAV
- **Tier IV (Rq-17 Sentinel):** High Altitude, Long endurance and low observation

**Figure 3.** Classification of UAVs in terms of altitude.

## 4. Communication Methods and Architecture

Today, one may see the variety of several drones being opted for commercial and domestic use. This is because they are cost effective and are controlled remotely from anywhere. In military operations, mostly the micro or miniature-type UAVs are used, but there are some major limitations in terms of size and weight. In Figure 3, tier II and III UAVs have several requirements such as being able to deploy sensors, and having a global positioning system (GPS), communication module, and efficient batteries. This is illustrated further in Figure 4. Although there are huge advancements being noticed in the field of drones or unmanned aerial vehicles (UAVs), at the same time, there are some limitations associated with the software and hardware support.

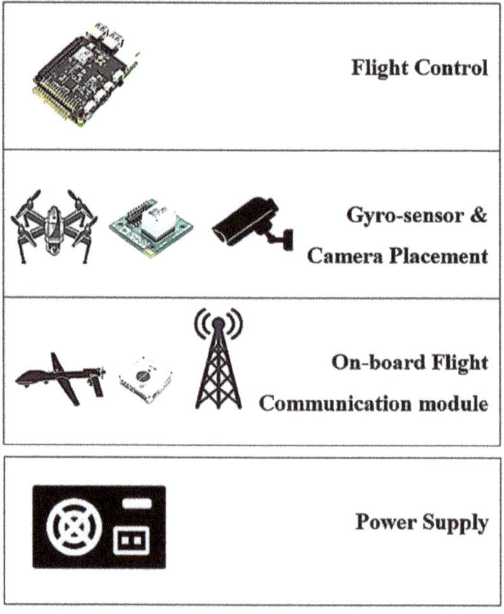


**Figure 4.** Components of an unmanned aerial vehicle.

The UAVs which are proposed for military operations have some advanced sensors which are not accessible to ordinary people. These sensors enable drones to carry additional payloads. After studying different UAVs, one may subdivide UAVs into their five major components as follows:

- Drone airframe;
- Onboard controller;
- Payload capability;
- Communication system;
- Efficient batteries.

Discussing the airframe of a UAV, one must consider aspects such as aerodynamics, a lightweight structure, and stability. These can be some of the constraints to designing the UAV airframe. Moreover, the onboard controller is the main thing that maneuvers the drone. Therefore, it must be equipped with all essential sensors such as the accelerometer, gyro sensor, pressure sensor, GPS, and camera. While designing the drone, one should consider the factor of payload variation. In this way, the drone may carry some nominal amount of weight from one place to another [25]. Another important component is the communication system where the drone requires some communication equipment such as a sitcom, modem, or radio channel-based equipment. This will ensure the communication and control between a UAV and the ground control room (GCR). Lastly, the UAV must have a reliable power source that can help it to fly for a specific time to fulfil the task. Mostly, lithium batteries are considered as the main power source for these UAVs [26].

*Communication Methods*

When one discusses the communication aspect of UAVs, one is directed toward several integral subcomponents such as the communication protocols, the network type, and the UAV model itself. This means that with the change of communication method, one may induct the number of components and this will change the architecture of the system [27]. Many researchers have suggested several topologies and designs. This has been illustrated in Figure 4 along with the altitude range. One may opt for a different communication module and protocol as per the mission and the nature of task type [28]. Moreover, with the advent of 5G networks, several constraints such as data rate, latency, and coverage have been resolved. This advancement in communication technology not only improved these areas, but also helps to improvise the positioning and control of drones in several critical rescue and surveillance missions; for that purpose, people have used advanced flight controls and multiple sensors as well, along with camera placement, controlled and communicated in different ways as per their altitude as shown in Figure 5. These technologies are summarized later in Table 3 [29].

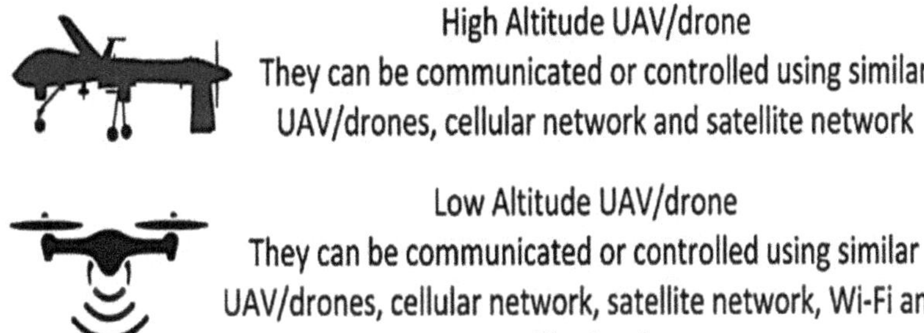


**Figure 5.** Communication methods for high- and low-altitude levels.

**Table 3.** Classification of UAVs based on communication channels.

| Technique | Channel Width | Band | Bit Rate | Range | Latency | Mobility Support |
|---|---|---|---|---|---|---|
| Wi-Fi | 20 MHz | 2.4 GHz to 5.2 GHz | 6–54 Mbps | 100 m | 10 ms | Low |
| GPS | 2 MHz | 1176 to 1576 MHz | 50 bps | - | 10 ms | Higher |
| UMTS | 5 MHz | 700 to 2600 MHz | 2 Mbps | 10 Km | 20–70 ms | High |
| 5G | 2.16 GHz | 57 to 64 GHz | Up to 4 Gbps | 50 m | - | Ultra-High |
| LTE | 20 MHz | 700 to 2690 MHz | Up to 300 Mbps | 30 Km | 10 ms | Very High |
| LTE-A | Up to 100 MHz | 450 MHz to 4.99 GHz | Up to 1 Gbps | | - | Very High |

Due to high security concerns, the modern UAVs are controlled using the line-of-sight method at a low altitude, whereas for high altitudes, researchers have given preference to GPS and the beyond-line-of-sight (BLoS) technique. Table 4 describes these techniques in brief [30].

**Table 4.** Communication based on satellite type.

| Type of Communication | Elevation in Km | Number of Satellites | Satellite Life | Handoff Frequency | Doppler | Gateway Cost | Propagation Path Loss |
|---|---|---|---|---|---|---|---|
| Geostationary Earth orbit (GEO) | Up to 36,000 | 3, no polar coverage | 15+ | NA | Low | Very expensive | Highest |
| Medium Earth orbit (MEO) | 5000–15,000 | 8–20 global | 10–15 | Low | Medium | Expensive | High |
| Low Earth orbit (LEO) | 500–1500 | 40–800 global | 3–7 | High | High | Cheap | Least |

Tables 3 and 4 are very important for the readers to understand the significance of the several channels based on different bandwidths and satellites, respectively. Discussing Table 3, it communicates the different wireless communication methods, but at the same time, it shares that the communication will have a latency rate as well. The table also shares the channel width, band interval, and most importantly, the mobility support for the readers to design their drones accordingly.

Discussing Table 4, it communicates the type of communication based on satellite type. This will help the reader to see the elevation in kilometers, number of satellites, satellite life, handoff frequency doppler gateway cost, and most importantly, propagation path loss of each satellite communication [31].

## 5. Utilization of UAVs in Different Domains

The potential of UAVs and drones has been proved already, and this domain covers every type of utilization from personal usage to military purpose, as illustrated in Figure 6.

These UAVs can be more efficient in performing several missions if they are equipped with a camera, smart sensors, and processors. With these essential components, one may see 100 plus applications of drones mentioned by several researchers, such as in [32].

Please note that Figure 6 shows the areas where UAVs are utilized mostly in general, whereas Figure 7 shows the benevolent usage where UAVs are commonly used. The term malevolent usage shows the areas and specific domains where people have witnessed an incremental increase in utilizing the drones over the last few decades. Factors such as diligence, cost, mobility in the areas where humans are unable to reach, payload options, and risk compel everyone to use drones/UAVs. Now, depending on the type of drone, they may be used in a better way. Commonly, it is seen that the design of drones is dependent on the type of mission they perform in the field [33]. Thus, categorizing them all with respect to their domains may lead to understanding their architecture in a better way. This has been illustrated in Figure 7, where every domain has its own privacy and security needs [33,34].

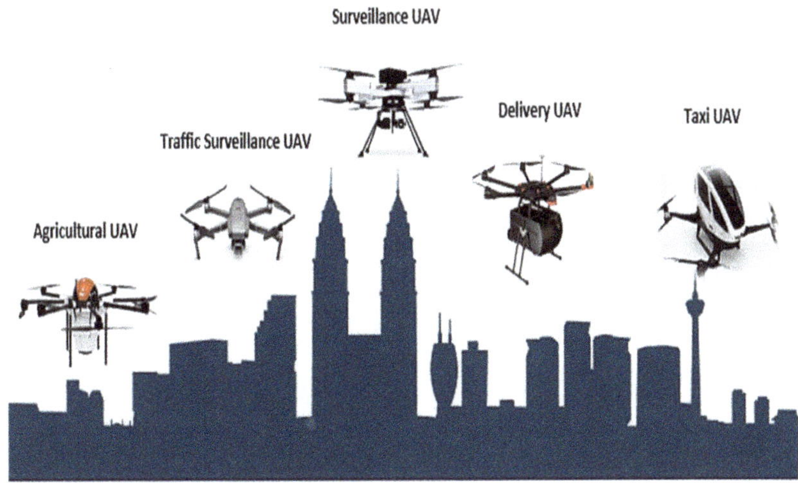


**Figure 6.** Utilization of drones in several domains.

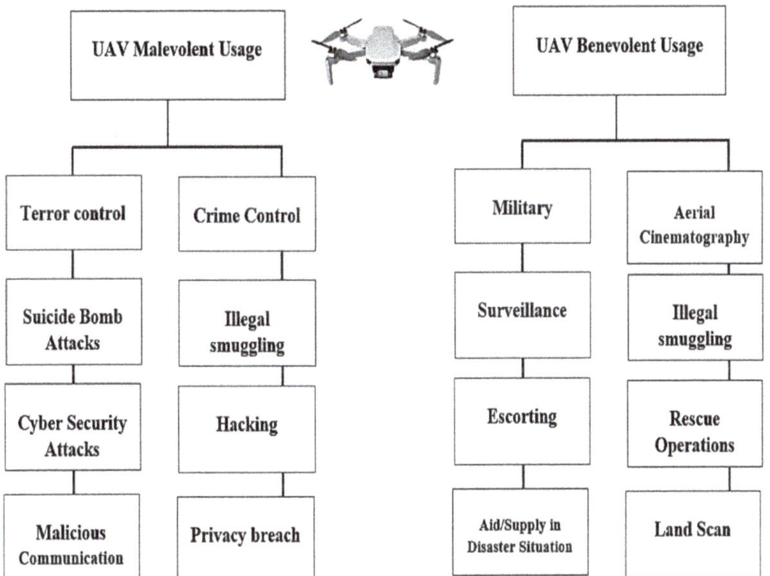


**Figure 7.** Malevolent and benevolent usages of a drone.

## 6. Security Threats Related to UAVs

UAVs offer several perks as the technology advances, but still, there are some constraints associated with privacy, security, and safety concerns [35,36]. Regularization and some important measures to license the utilization of drones is a very significant aspect. This limits unnecessary aerial photography. Most of the authorities in the world ensure this aspect and provide strict policies over uninformed aerial photography. If one discusses the network security point and the risk analysis, it is an admitted fact that the coverage is quite different as compared to any sort of wireless sensor network (WSN) or any mobile ad hoc networks (MANETs) [37]. The reason is because of the resource constraints, as UAV-related coverage is broader and wider than WSNs or MANETs.

The framework that sets the rules to operate drones in any vicinity is known as authentication, authorization, and accounting (AAA), which states several privileges to the controller of a drone to operate as per the mentioned administrative rights, whereas it also shares some of the rigid authentication procedures for drones to protect the control of a drone so that it may not be diverted to any other unknown entity. Moreover, in case of any uncertainty or illegal activity by drone, one may easily track down the operator. This is done to limit illegal surveillance, cyberattacks, and privacy threats. Thus, several mechatronic engineering solutions have been presented to overcome these malicious activities [38].

These drones are low cost and easily available in markets nowadays, and therefore, they are easy to use for any sort of criminal activity. Their ability to carry a wide range of external payloads make them more dangerous as it could lead to drones carrying any harmful chemical or explosive thing. Moreover, their ability to reach places where normal human beings cannot makes them more lethal because they can deliver anything without coming under anyone's notice [39]. It should be noted that security is not the only concern, but one may also see a safety concern if drones are flying over any populous place and, due to any number of faults, may crash, which can lead to several types of tragedies [40]. These sorts of incidents have been reported often. One of the examples is when a UAV faced a collision with British Airways BA727, which was a passenger aircraft in April 2016. After looking over these incidents and issues, one may ensure below the mentioned public safety measures:

- It is a high probability that a drone can be hacked or may deviate from its path due to heavy wind disturbance. Thus, there should be a reset option available which may turn the drone to a hovering condition only and help to gain the control back.
- There are certain areas where drones may face signal jammers and, later, can be controlled for a cyberattack. Thus, drones must have some sort of filter that may detect if there is any signal jammer nearby.

The third safety measure is related to its design, as most drones have open propellers as shown in Figure 8. In case of uncertainty, these propellers may go off and may harm anyone nearby; thus, the safety design as shown in Figure 9 is necessary to avoid any harm during a crash.

**Figure 8.** UAVs with open propellers.

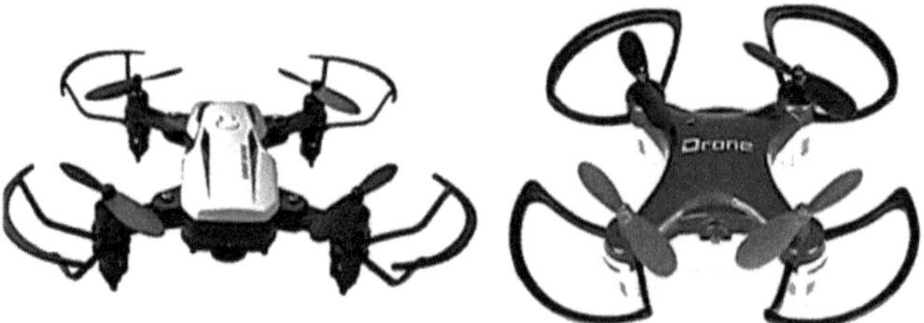

**Figure 9.** UAVs with closed and safety propellers.

Lastly, there are some serious privacy concerns as well. Since UAVs on the market can easily be procured with high-definition cameras, this may lead to the recording of any private property without permission. Due to this reason, Canadian Public Safety (CPS) stated that these UAVs are prohibited from flying over any property without mutually agreed permission [41].

## 7. Current Vulnerability Issues of UAVs

For these UAVs, unfortunately, there is neither a standardization of policies nor the availability of wireless security [42–44]. This leads to several threats, as highlighted in Table 5. There are researchers who have addressed different types of cyber-attacks associated with the several types of UAVs in a pre-controlled environment [45–50]. Such practical validations include the crashing of drones with many parallel requests and modifying the request packet known as the buffer-overflow attack, whereas some researchers went for the cache-poisoning approach that leads to the shutdown of communication between the drone and GCR. In all conditions, most attacks occur to target the operating system or, in other words, the microcontroller of the drone [51]. Since there are huge advancements in the technology, UAVs have a high probability of experiencing such attacks, as shown in Figure 9 [52–58]. From these attacks, the most common attack is GPS spoofing, such as signal jamming, de-authentication. and zero-day attacks.

**Table 5.** Summary of all current vulnerability issues in UAVs.

| Vulnerability Type | Description |
| --- | --- |
| Malware issue | In various cases, it has been observed that these UAVs are generally connected and controlled via cell phone or any sort of remote control. These techniques are, thus far, not safe [43] and, therefore, the UAVs are easy to be hacked using a reverse-shell TCP payload that can be injected into UAV memory. Furthermore, this leads to installation of malware over UAVs automatically. |
| Spoofing | These are the issues related to the communication method, usually with serial port connections that are not encrypted properly [44]. Due to this spoofing issue, the information associated with GPS can be taken and altered. |
| Manipulation and other common concerns | The flying paths which UAVs must track are pre-programmed before; therefore, these paths can be altered [45], whereas the common issues are related to wind, overheating, or any predator bird harming the lightweight drone easily [46]. |
| Physical design and control system constraints | There are various challenges with unmanned aerial vehicle control system design, such as the sluggish convergence rate, which prevents the drone from performing fast or aggressive maneuvers, and one may notice faults in the flight or divergence from the target trajectory [47,59,60]. This slow convergence rate and glitches are caused by the physical architecture of drones or the planned control system, which is primarily intended to stabilize the drone in uncertain conditions. |
| Sensorization issue | Since these UAVs depend on sensors, thus, it is also proved that the ultrasonic waves may attack the MEMS gyro sensors [47]. |
| Wi-Fi constraints | Operating drones using a Wi-Fi facility may be risky. This is proved in [48] where the connection was disrupted with the help of software and changing the control of the UAV. |
| GPS issue | Automatic Dependent Surveillance–Broadcast depends on the GPS module, which is not encrypted sometimes and may lead to spoofing [49]. |
| Firmware issue | The bugs available in the first prototype and first algorithm which come to the front after usage [50]. |
| Sky Jack-based attacks | Sky Jack is one software used to conduct the attacks related to de-authentication of targets during control [51]. |
| Controller issues | These issues are related to the operation control unit and may puzzle the controller by changing the live feed to some other video [52]. |

## 8. Current State-of-the-Art Solutions

The very first and significant thing to resolve the threats is to identify them first. Thus, Table 5 and Figure 10 classify these attacks. Moreover, there are several contributions which address these sorts of attacks along with the suitable measures [58]. In recent times, researchers have utilized machine learning approaches to demonstrate the intrusion detection system (IDS). Thus, machine learning (ML)-based IDS is one of the areas where researchers are still working to improve the results [59,60]. Blockchain is also among the most effective approaches for UAV/drone security and privacy [12]. This ML-based IDS technique is from the robust technique, and it is categorized into three kinds as mentioned below:

- Rule-based IDS;
- Signature-based IDS;
- Anomaly-based IDS.

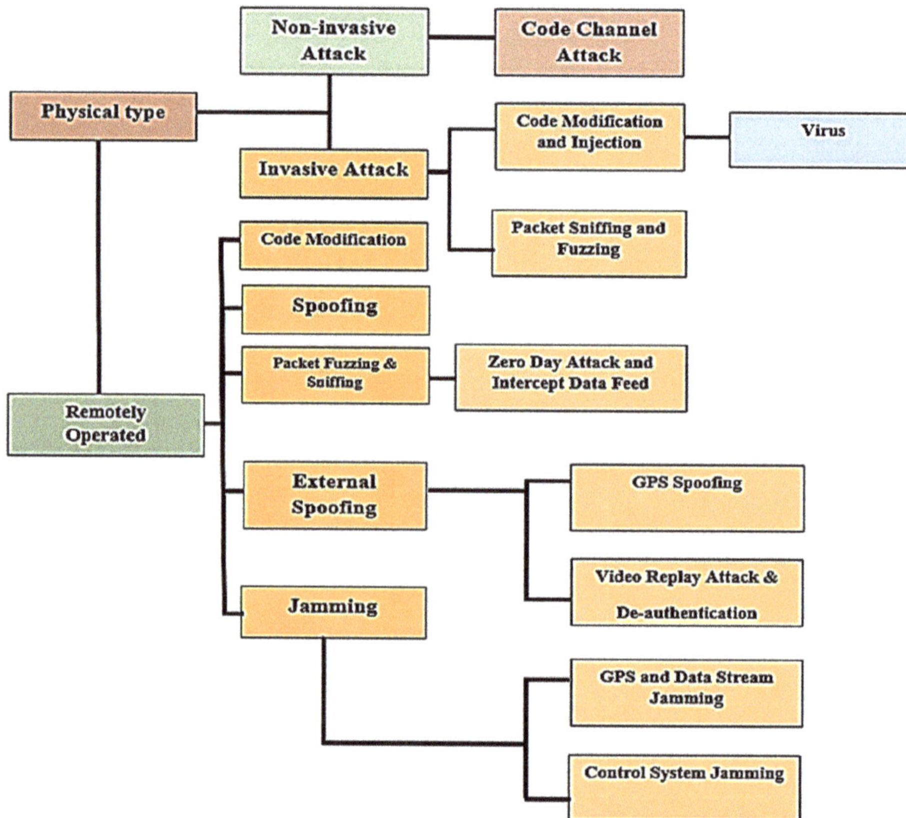


**Figure 10.** UAV attack vector with reported incidents.

Above are the major approaches for detecting threats that intrusion detection systems utilize to inform the operator in the ground control room (GCR). Rule-based threat detection is a new approach enabled by artificial intelligence (AI) [61]. In comparison to others, it is more reliant on technology and less on manual interaction. Signature-based detection works well for recognizing known threats. It uses a pre-programmed list of known threats and their indicators of compromise to operate (IOCs). An indicator of compromise (IOC) could be a distinctive behavior that typically precedes a malicious network attack, such as file hashes, malicious sites, known byte sequences, or even the content of email subject headings. A signature-based IDS examines network packets and compares them to a database of known IOCs or attack signatures to detect any suspicious behavior. Anomaly-based intrusion detection systems, on the other hand, can alert you to unusual behavior. An anomaly-based detection system uses machine learning to educate the detection system to recognize a normalized baseline rather than searching for known threats. All network activity is compared to the baseline, which represents how the system ordinarily performs. Rather than looking for known IOCs, anomaly-based IDS simply detects any unusual behavior to generate alarms.

To identify the false data injection attacks, one may use the rule-based approach. This is used to target the signal strength in between the UAV and ground control room (GCR) and can be useful for any sort of known attack, pattern, or technique only. Some research papers have proposed a signature-based IDS over drones [62] where they addressed bio-inspired cyberattacks associated with air-born networks. Last, but certainly not least, is the anomaly-based IDS scheme which is used against jamming attacks [62]. The only limitation of anomaly IDS is the huge resource requirement.

Similarly, there are some researchers who have suggested some algorithm schemes with forensic methods to address the advanced and complex attacks. They are complex and difficult to identify [63–69]. With the help of forensics, both perpetrator and method of attack can be identified. With the identification of the attack type, appropriate countermeasures can be implemented to avoid any future incident. As per the survey of [70], between 2014 and 2017 incidents among airplanes and drones amplified from 6 to 93, which makes it very important for the authorities to address security and privacy issues for UAVs. Due to an increase in cyberattacks on drones/UAVs, the government needs to introduce strict policies and standards to minimize these concerns. With the popularity of UAVs among the civilian population, attacks and the illegal use of UAVs will likely proliferate. Civilian or domestic UAV countermeasures are divided into physical and local countermeasures, which are already proposed but still can be improved.

There are several survey papers that address the latest integration of UAVs into cellular networks and discuss the inference issues [71], as well as those, like this paper, that address the significant concerns related to the standardization and regulation of drones and their privacy. In addition to this, the manuscript focuses on the issues related to addressing these limitations while communicating from the drone to the ground control room (GCR). Some of the researchers proposed survey papers also on the quantity and quality of service requirements and discussed network-relevant mission parameters [72], which is unlike this paper that discusses the safety, privacy, and adaptability features of drones.

## 9. Open Research Areas and Recommendations

After studying the previous sections, it is noted that there is still a need to improve some of the areas associated with unmanned aerial vehicles (UAVs). These areas are very significant and one may address these concerns to enhance the utilization of drones [73,74]. One of the important areas is path loss, where one needs to propose the channel model to hold on to the communication at higher carrier frequencies, and even in the presence of tall concrete buildings. This area is in regard to the latency in the communication, which should be less than 1 millisecond and remains as an area of concern [75,76].

In addition to these areas, one may work over the reliability aspect, where one may improve the drones with ultra-reliable communication so that even with the increase in UAVs in the sky, the communication can be performed easily. It is noted that these drones have not been operated in the sky for a long time, which is because of the battery life. Hence, battery optimization for drones is also one of the areas where researchers may engage themselves to increase the flight time.

Last, but certainly not least, is the amalgamation of artificial intelligence and computer vision algorithms in a drone to improve the mobility of the drone without any collision. This will protect the drone in terms of data logging and security [77–84]. This manuscript also suggests some of the recommendations that may improve the privacy and security aspects, such as the registration of drone licenses. This will ensure the authorities identify the specific drone that has created an inconvenience in the jurisdiction [85–91]. Moreover, there must be flying permits allotted after necessary training to limit the illegal utilization of drones in any activity.

Another recommendation is to educate the public about the legal and illegal usage of such autonomous unmanned aerial vehicles and the laws related to it so that if they witness anything around, they may easily report it. In any vicinity, there are some restricted zones; thus, the market drones must be operated based on a built-in map [92–95] as per the local regions. In this way, when any drone is forced to enter into any barred jurisdiction, it will automatically revert to the ground using the vertical take-off and landing (VTOL) mode. In terms of security tools, this paper proposes the machine learning-based IDS system [96–98] to improve the security infrastructure of UAVs, and lastly, there should be rigid multi-factor authentication methods that tackle the security threats easily.

There are several future recommendations to increase the standards for the security and privacy aspects of UAVs. These aspects are improved by proposing an approach which is based on a pairing certificate so that other strange entities may not easily connect or communicate with our UAV. Some of them are based on identification/authentication protocols [99–103]. To secure the drone more and make it less vulnerable, researchers have also used a symmetric searchable encryption method (SSE) [104] as well. Some researchers have proposed an internet of things (IoT) feature also for the same purpose [105–108]. In terms of identifying an unknown input observance, one may see an intelligent control algorithm that stabilizes the UAV in the presence of an unknown input [108–110] and devise it in trajectory and altitude levels to identify unknown system dynamics online by utilizing filtering manipulations that possess a concise structure, low calculation consumption, and asymptotic error convergence. In addition to this, the manuscript highlights the major domains and compares them with some of the latest review papers on UAVs for contrast. This is the significance of this article, which is seen in Table 6.

**Table 6.** Summary of all major domains along with reference list.

| Area/Domain | [82] | [75,76] | [57] | [47] | [20,32] | [4] |
|---|---|---|---|---|---|---|
| Regulations and classification | | | | | • | |
| Communication methods | | | | • | • | • |
| Applications | • | • | | • | | • |
| Security issues and solutions | • | • | | | • | |
| Physical and logical attacks | | | | | • | |
| Open research area | | | | • | • | • |
| Recommendations | | | • | • | • | |

One can observe from the above chart that the first column lists the issues that have already been covered in the paper, whereas the first row lists the number of review manuscripts that state or debate the same theme. After reading through Table 6, one can find this article to be more thorough in determining the future answer quickly and effectively. A black dot in the table above indicates the articles in which the topics were directly mentioned.

A statement in support of integrating UAVs with contemporary trends of communication and a control system is developed by evaluating various research contributions linked to UAVs and communication aspects to identify the limits [104,111]. To get over the limitations, more research is still needed to examine the subtopics below:

- There is a need to address the area of high-speed mobility, as there are huge chances to hack the communication links through the ground control room or with neighboring UAVs.

- In some of the integrated solutions, i.e., the space-air-ground network, one may see a frequent issue of synchronization, and thus, it is desirable to re-design some cooperation incentives for using cross-layered protocols with linked reliability. In this way, there will be less chances of any security attacks.
- One more aspect is to recommend a lightweight mechanism for UAVs to prevent attacks, such as eavesdropping, a man-in-the-middle attack [112,113], and so on. There are a number of artificial intelligence solutions which are recommended in [28] for addressing the security in cellular network-based controlled UAVs for delivering packages.
- Integrating UAVs with the IoT can result in endurance and reliability, but at the same time, it consumes the maximum battery capacity which is generally small; thus, this may lead UAVs toward possible collisions and can be a high-risk threat.
- Lastly, proposing a big data deep reinforcement learning approach to enable the dynamic arrangement of networking, caching, and computing resources for improving the performance of UAVs with secure operations in smart cities.

Thus, with all the above recommendations and in contrast to these topics, our manuscript provides a detailed direction for future work.

## 10. Conclusions

The use of UAVs has increased dramatically, ushering in an era of autonomous systems and vehicles. These drones are quite important because they have many benefits for both civil and military concerns. However, with this rise in usage, severe privacy and security concerns are also evident. The most frequent reason why these drones are chosen in any sneaky assault is because they are readily available and inexpensive to obtain.

There have been numerous scientific contributions that have already addressed the countermeasures to these worries; however, there are still some areas that have not been addressed and can, therefore, still be exploited for any negative purposes, such as privacy and security issues. In this current, technological age, these two challenges cannot be disregarded. As a result, this review paper offers a thorough examination of these two pressing issues by providing a quick summary of the causes of each worry, as well as potential solutions. The existing solutions and a number of recommendations are presented in this study, which claims that the UAV drones can be enhanced if proper data integration, authentication, and accessibility factors are treated seriously.

**Author Contributions:** Conceptualization, G.E.M.A., A.L., R.J.M., V.S.A. and S.A.B.M.Z.; methodology, A.L.; software, G.E.M.A. and A.L.; validation, V.S.A., R.J.M. and S.A.B.M.Z.; formal analysis, A.L.; investigation, V.S.A. and A.L.; resources, S.A.B.M.Z. and G.E.M.A.; data curation, S.A.B.M.Z., V.S.A. and A.L.; writing—original draft preparation, G.E.M.A.; writing—review and editing, G.E.M.A. and S.A.B.M.Z.; visualization, V.S.A. and S.A.B.M.Z.; supervision, A.L., S.A.B.M.Z., R.J.M. and V.S.A.; project administration, G.E.M.A.; funding acquisition, S.A.B.M.Z. and G.E.M.A. All authors have read and agreed to the published version of the manuscript.

**Funding:** The article processing charges were financed by the Research Management Center, Universiti Teknologi PETRONAS, Malaysia, under the Research Fund, which was supported by Yayasan Universiti Teknologi Petronas (YUTP), award number 015LC0-316.

**Data Availability Statement:** Not applicable.

**Acknowledgments:** The authors would like to express their gratitude to the Centre of Graduate Studies (CGS), Universiti Teknologi, PETRONAS, Malaysia, and the Erasmus+ Program for giving them the opportunity to study abroad as exchange students at Telecom SudParis, France and conduct this research using the cutting-edge resources of that country's university. Finally, I would like to express my gratitude to the Department of Electrical and Electronic Engineering for offering me a position to pursue PhD studies through YUTP Funding, grant number 015CL0-316.

**Conflicts of Interest:** The authors declare no conflict of interest.

## References

1. Bombe, M.K. Unmanned Aerial Vehicle (UAV) Market Worth $21.8 billion by 2027- Pre and Post COVID-19 Market Analysis Report by Meticulous Research. 11 June 2020. Available online: https://www.meticulousresearch.com/download-sample-report/cp_id=5086 (accessed on 18 August 2022).
2. Kumar, R.; Kumar, P.; Tripathi, R.; Gupta, G.P.; Gadekallu, T.R.; Srivastava, G. SP2F: A secured privacy-preserving framework for smart agricultural Unmanned Aerial Vehicles. *Comput. Networks* **2021**, *187*, 107819. [CrossRef]
3. Israr, A.; Abro, G.E.M.; Sadiq Ali Khan, M.; Farhan, M.; Zulkifli, B.M.; ul Azrin, S. Internet of things (IoT)-Enabled unmanned aerial vehicles for the inspection of construction sites: A vision and future directions. *Math. Problems Eng.* **2021**. [CrossRef]
4. Chen, R.; Yang, B.; Zhang, W. Distributed and Collaborative Localization for Swarming UAVs. *IEEE Internet Things J.* **2020**, *8*, 5062–5074. [CrossRef]
5. Hayat, S.; Yanmaz, E.; Muzaffar, R. Survey on Unmanned Aerial Vehicle Networks for Civil Applications: A Communications Viewpoint. *IEEE Commun. Surv. Tutorials* **2016**, *18*, 2624–2661. [CrossRef]
6. Chan, K.W.; Nirmal, U.; Cheaw, W.G. Progress on drone technology and their applications: A comprehensive review. *AIP Conf. Proc.* **2018**, *2030*, 020308. [CrossRef]
7. Hartmann, K.S.C. The vulnerability of UAVs to cyber, in Cyber Conflict (CyCon). In Proceedings of the 2013 5th International Conference, Tallinn, Estonia, 4–7 June 2013.
8. Bowden, M. How the Predator Drone Changed the Character of War, Smithsonian Magazine. November 2013. Available online: https://www.smithsonianmag.com/history/how-the-predatordrone-changed-the-character-of-war-3794671/ (accessed on 18 August 2022).
9. Ekramul, D. First Successful Air-Raid in History. 22 August 2019. Available online: https://www.daily-bangladesh.com/english/First-successful-Air-raid-in-history/27424 (accessed on 18 August 2022).
10. O'Donnell, S. Consortiq. Available online: https://consortiq.com/short-history-unmanned-aerialvehicles-uavs/ (accessed on 18 August 2022).
11. Berg, T.R. Air Space Mag. 10 January 2020. Available online: https://www.airspacemag.com/daily-planet/first-map-compiled-aerial-photographs-180973929/ (accessed on 18 August 2022).
12. Abdullah, Q.A. Introduction to the Unmanned Aircraft Systems. Available online: https://www.eeducation.psu.edu/geog892/node/643 (accessed on 18 August 2022).
13. Miah, A. *Drones: The Brilliant, the Bad and the Beautiful*; Emerald Group Publishing: Bently, UK, 2020. [CrossRef]
14. Ali, B.S.; Saji, S.; Su, M.T. An assessment of frameworks for heterogeneous aircraft operations in low-altitude airspace. *Int. J. Crit. Infrastruct. Prot.* **2022**, *37*, 100528. [CrossRef]
15. Wright, S. Ethical and safety implications of the growing use of civilian drone. UK Parliam. *Website Sci. Technol. Commit.* **2019**.
16. Coach, U. Master List of Drone Laws (Organized by State & Country). Available online: https://uavcoach.com/drone-laws/ (accessed on 18 August 2022).
17. Aljehani, M.; Inoue, M.; Watanbe, A.; Yokemura, T.; Ogyu, F.; Iida, H. UAV communication system integrated into network traversal with mobility. *SN Appl. Sci.* **2020**, *2*, 2749. [CrossRef]
18. de Croon, G.C.H.E.; Groen, M.A.; De Wagter, C.; Remes, B.; Ruijsink, R.; van Oudheusden, B.W. Design, aerodynamics and autonomy of the DelFly. *Bioinspir. Biomim.* **2012**, *7*, 025003. [CrossRef]
19. Cheaw, B.H.; Ho, H.W.; Abu Bakar, E. Wing Design, Fabrication, and Analysis for an X-Wing Flapping-Wing Micro Air Vehicle. *Drones* **2019**, *3*, 65. [CrossRef]
20. Teoh, Z.E.; Fuller, S.B.; Chirarattananon, P.; Prez-Arancibia, N.O.; Greenberg, J.D.; Wood, R.J. A hovering flapping-wing microrobot with altitude control and passive upright stability. In Proceedings of the 2012 IEEE/RSJ International Conference on Intelligent Robots and Systems, Vilamoura-Algarve, Portugal, 7–12 October 2012; pp. 3209–3216. [CrossRef]
21. Professionals, Drones and Remotely Piloted Aircraft (UAS/RPAS)-Frequencies and Radio Licenses, Traficom. 17 July 2021. Available online: https://www.traficom.fi/en/transport/aviation/drones-and-remotely-piloted-aircraft-uasrpasfrequencies-and-radio-licences (accessed on 18 August 2022).
22. Carnahan, ISO/TC 20/SC 16 Unmanned Aircraft Systems. 2014. Available online: https://www.iso.org/committee/5336224.html (accessed on 18 August 2022).
23. Irizarry, M.J.; Gheisari, B. Walker, Usability Assessment of Drone Technology as Safety Inspection Tools. *Electron. J. Inf. Technol. Constr.* **2012**, *17*, 194–212.
24. Mozaffari, M.; Saad, W.; Bennis, M.; Nam, Y.-H.; Debbah, M. A tutorial on UAVs for wireless networks: Applications, challenges, and open problems. *IEEE Commun. Surv. Tut.* **2019**, *21*, 2334–2360. [CrossRef]
25. Lowbridge, C. Are Drones Dangerous or Harmless Fun? *BBC News*. 5 October 2015. Available online: https://www.bbc.com/news/uk-england-34269585 (accessed on 18 August 2022).
26. Federal Aviation Authorities, Recreational Flyers & Modeler Community-Based Organizations. 18 February 2020. Available online: https://www.faa.gov/uas/recreational_fliers/ (accessed on 18 August 2022).
27. Pilot, What's the Difference Between Drones, UAV, and UAS? Definitions and Terms. Pilot Institute. 22 March 2020. Available online: https://pilotinstitute.com/drones-vs-uav-vs-uas/ (accessed on 18 August 2022).

28. Fotouhi, A.; Qiang, H.; Ding, M.; Hassan, M.; Giordano, L.G.; Garcia-Rodriguez, A.; Yuan, J. Survey on UAV Cellular Communications: Practical Aspects, Standardization Advancements, Regulation, and Security Challenges. *IEEE Commun. Surv. Tutorials* **2019**, *21*, 3417–3442. [CrossRef]
29. Nagpal, K. Unmanned Aerial Vehicles (UAV) Market, Q Tech Synergy. 24 December 2016. Available online: https://defproac.com/?p=2041 (accessed on 18 August 2022).
30. Pastor, E.; Lopez, J.; Royo, P. A Hardware/Software Architecture for UAV Payload and Mission Control. In Proceedings of the 2006 IEEE/AIAA 25TH Digital Avionics Systems Conference, Portland, OR, USA, 15–18 October 2006; pp. 1–8. [CrossRef]
31. VanZwol, J. Design Essentials: For UAVs and Drones, Batteries are Included, Machine Design. 4 April 2017. Available online: https://www.machinedesign.com/mechanical-motionsystems/article/21835356/design-essentials-for-uavs-and-drones-batteries-are-included (accessed on 18 August 2022).
32. Sharma, A.; Vanjani, P.; Paliwal, N.; Basnayaka, C.M.; Jayakody, D.N.K.; Wang, H.-C.; Muthuchidambaranathan, P. Communication and networking technologies for UAVs: A survey. *J. Netw. Comput. Appl.* **2020**, *168*, 102739. [CrossRef]
33. Ullah, H.; Nair, N.G.; Moore, A.; Nugent, C.; Muschamp, P.; Cuevas, M. 5G Communication: An Overview of Vehicle-to-Everything, Drones, and Healthcare Use-Cases. *IEEE Access* **2019**, *7*, 37251–37268. [CrossRef]
34. Luo, C.; Miao, W.; Ullah, H.; McClean, S.; Parr, G.; Min, G. Unmanned aerial vehicles for disaster management. In *Geological Disaster Monitoring Based on Sensor Networks*; Springer: Singapore, 2019; pp. 83–107.
35. Hosseini, N.; Jamal, H.; Haque, J.; Magesacher, T.; Matolak, D.W. UAV Command and Control, Navigation and Surveillance: A Review of Potential 5G and Satellite Systems. In Proceedings of the 2019 IEEE Aerospace Conference, Big Sky, MT, USA, 2–9 March 2019; pp. 1–10. [CrossRef]
36. IvyPanda, Unmanned Aerial Vehicles Essay. 13 January 2020. Available online: https://ivypanda.com/essays/unmanned-aerial-vehicles-essay/ (accessed on 20 December 2020).
37. Valavanis, K.P.; Vachtsevanos, G.J. UAV Applications: Introduction. In *Handbook of Unmanned Aerial Vehicles*; Springer: Dordrecht, The Netherlands, 2015; pp. 2639–2641. [CrossRef]
38. Shakhatreh, H.; Sawalmeh, A.H.; Al-Fuqaha, A.; Dou, Z.; Almaita, E.; Khalil, I.; Othman, N.S.; Khreishah, A.; Guizani, M. Unmanned Aerial Vehicles (UAVs): A Survey on Civil Applications and Key Research Challenges. *IEEE Access* **2019**, *7*, 48572–48634. [CrossRef]
39. Cook, K.L.B. The Silent Force Multiplier: The History and Role of UAVs in Warfare. In Proceedings of the 2007 IEEE Aerospace Conference, Big Sky, MT, USA, 3–10 March 2007; pp. 1–7. [CrossRef]
40. Siddiqi, M.A.; Khoso, A.M. Aziz, Analysis on Security Methods of Wireless Sensor Network (WSN). In Proceedings of the SJCMS 2018, Sukkur, Pakistan, 10 December 2018.
41. Cavoukian, A. *Privacy and Drones: Unmanned Aerial Vehicle*; Information and Privacy Commissioner: Toronto, ON, Canada, 2012.
42. Kafi, M.A.; Challal, Y.; Djenouri, D.; Doudou, M.; Bouabdallah, A.; Badache, N. A Study of Wireless Sensor Networks for Urban Traffic Monitoring: Applications and Architectures. *Procedia Comput. Sci.* **2013**, *19*, 617–626. [CrossRef]
43. Mansfield, K.; Eveleigh, T.; Holzer, T.H.; Sarkani, S. Unmanned aerial vehicle smart device ground control station cyber security threat model. In Proceedings of the 2013 IEEE International Conference Technology Homel Security (HST), Walthan, MA, USA, 12–14 November 2013; pp. 722–728. [CrossRef]
44. Smith, K.W. Drone Technology: Benefits, Risks, and Legal Considerations. *Seattle J. Environ. Law (SJEL)* **2015**, *5*, 291–302.
45. Eyerman, J.; Hinkle, K.; Letterman, C.; Schanzer, D.; Pitts, W.; Ladd, K. *Unmanned Aircraft and the Human Element: Public Perceptions and First Responder Concerns*; Institute of Homeland Security and Solutions: Washington, DC, USA, 2013.
46. Syed, N.; Berry, M. *Journo-Drones: A Flight over the Legal Landscape*; American Bar Association: Chicago, IL, USA, 2014.
47. Rahman, M.F.B.A. *Smart CCTVS for Secure Cities: Potentials and Challenges*; Rajaratnam School of International Studies (RSIS): Singapore, 2017.
48. Kim, A.; Wampler, B.; Goppert, J.; Hwang, I.; Aldridge, H. Cyber Attack Vulnerabilities Analysis for Unmanned Aerial Vehicles. *Aerospace Res. Cent.* **2012**, 2438. [CrossRef]
49. Zeng, Y.; Zhang, R.; Lim, T.J. Wireless communications with unmanned aerial vehicles: Opportunities and challenges. *IEEE Commun. Mag.* **2016**, *54*, 36–42. [CrossRef]
50. Soria, P.R.; Bevec, R.; Arrue, B.C.; Ude, A.; Ollero, A. Extracting Objects for Aerial Manipulation on UAVs Using Low Cost Stereo Sensors. *Sensors* **2016**, *16*, 700. [CrossRef]
51. Erdelj, M.; Natalizio, E. Drones, Smartphones and Sensors to Face Natural Disasters. In Proceedings of the 4th ACM Workshop on Micro Aerial Vehicle Networks, Systems, and Applications, Paris, France, 10–15 June 2018; pp. 75–86. [CrossRef]
52. Son, Y.; Shin, H.; Kim, D.; Park, Y.; Noh, J.; Choi, K. Rocking Drones with Intentional Sound Noise on Gyroscopic Sensors. In Proceedings of the 24th USENIX Security Symposium, Washington, DC, USA, 12–14 August 2015.
53. Zhi, Y.; Fu, Z.; Sun, X.; Yu, J. Security and Privacy Issues of UAV: A Survey. *Mob. Netw. Appl.* **2019**, *25*, 95–101. [CrossRef]
54. Strohmeier, M.; Schafer, M.; Lenders, V.; Martinovic, I. Realities and challenges of nextgen air traffic management: The case of ADS-B. *IEEE Commun. Mag.* **2014**, *52*, 111–118. [CrossRef]
55. Hooper, M.; Tian, Y.; Zhou, R.; Cao, B.; Lauf, A.P.; Watkins, L.; Robinson, W.H.; Alexis, W. Securing commercial WiFi-based UAVs from common security attacks. In Proceedings of the MILCOM 2016-2016 IEEE Military Communications Conference, Baltimore, MD, USA, 1–3 November 2016; pp. 1213–1218. [CrossRef]

56. Hartmann, K.; Giles, K. UAV exploitation: A new domain for cyber power. In Proceedings of the 2016 8th International Conference Cyber Conflict, Tallinn, Estonia, 31 May–3 June 2016; pp. 205–221. [CrossRef]

57. Rivera, E.; Baykov, R.; Gu, G. A Study on Unmanned Vehicles and Cyber Security. In Proceedings of the Rivera 2014 ASO, Austin, TX, USA, 2014.

58. Junejo, I.N.; Foroosh, H. GPS coordinates estimation and camera calibration from solar shadows. *Comput. Vis. Image Underst.* **2010**, *114*, 991–1003. [CrossRef]

59. Wang, W.; Huang, H.; Zhang, L.; Su, C. Secure and efficient mutual authentication protocol for smart grid under blockchain. *Peer-to-Peer Netw. Appl.* **2020**, *14*, 2681–2693. [CrossRef]

60. Zhang, L.; Zhang, Z.; Wang, W.; Jin, Z.; Su, Y.; Chen, H. Research on a Covert Communication Model Realized by Using Smart Contracts in Blockchain Environment. *IEEE Syst. J.* **2021**, *16*, 2822–2833. [CrossRef]

61. Currier, C.; Moltke, H. *Spies in the Sky*; The Intercept: New York, NY, USA, 2016.

62. Yağdereli, E.; Gemci, C.; Aktaş, A.Z. A study on cyber-security of autonomous and unmanned vehicles. *J. Déf. Model. Simulation: Appl. Methodol. Technol.* **2015**, *12*, 369–381. [CrossRef]

63. Lee, Y.S.; Dongseo University; Kang, Y.-J.; Lee, S.-G.; Lee, H.; Ryu, Y. An Overview of Unmanned Aerial Vehicle: Cyber Security Perspective. *IT Converg. Technol.* **2016**, *4*, 30. [CrossRef]

64. Wu, L.; Cao, X.; Foroosh, H. Camera calibration and geo-location estimation from two shadow trajectories. *Comput. Vis. Image Underst.* **2010**, *114*, 915–927. [CrossRef]

65. Krishna, C.G.L.; Murphy, R.R. A review on cybersecurity vulnerabilities for unmanned aerial vehicles. In Proceedings of the 2017 IEEE International Symposium on Safety, Security and Rescue Robotics (SSRR), Shanghai, China, 11–13 October 2017; pp. 194–199. [CrossRef]

66. Siddiqi, M.A.; Pak, W. Optimizing Filter-Based Feature Selection Method Flow for Intrusion Detection System. *Electronics* **2020**, *9*, 2114. [CrossRef]

67. Strohmeier, M.; Lenders, V.; Martinovic, I. Intrusion Detection for Airborne Communication Using PHY-Layer Information. In *Proceedings of the International Conference Detection of Intrusions and Malware, and Vulnerability Assessment*; Springer: Cham, Switerland, 2015; pp. 67–77. [CrossRef]

68. Gil Casals, S.; Owezarski, P.; Descargues, G. Generic and autonomous system for airborne networks cyber-threat detection. In Proceedings of the 2013 IEEE/AIAA 32nd Digital Avionics Systems Conference (DASC), IEEE, New York, NY, USA, 5–10 October 2013; pp. 4A4-1–4A4-14. [CrossRef]

69. Rani, C.; Modares, H.; Sriram, R.; Mikulski, D.; Lewis, F.L. Security of unmanned aerial vehicle systems against cyber-physical attacks. *J. Déf. Model. Simul. Appl. Methodol. Technol.* **2015**, *13*, 331–342. [CrossRef]

70. Zhang, G.; Wu, Q.; Cui, M.; Zhang, R. Securing UAV Communications via Joint Trajectory and Power Control. *IEEE Trans. Wirel. Commun.* **2019**, *18*, 1376–1389. [CrossRef]

71. Shao, X.; Wang, L.; Li, J.; Liu, J. High-order ESO based output feedback dynamic surface control for quadrotors under position constraints and uncertainties. *Aerosp. Sci. Technol.* **2019**, *89*, 288–298. [CrossRef]

72. Li, B.; Fei, Z.; Zhang, Y. UAV Communications for 5G and Beyond: Recent Advances and Future Trends. *IEEE Internet Things J.* **2018**, *6*, 2241–2263. [CrossRef]

73. Lee, Y.-S.; Kim, E.; Kim, Y.-S.; Seol, D.-C. Effective Message Authentication Method for Performing a Swarm Flight of Drones. *Emergency* **2015**, *3*, 95–97. [CrossRef]

74. Pilli, E.S.; Joshi, R.; Niyogi, R. A Generic Framework for Network Forensics. *Int. J. Comput. Appl.* **2010**, *1*, 251–408. [CrossRef]

75. Beebe, N.L.; Clark, J.G. A hierarchical, objectives-based framework for the digital investigations process. *Digit. Investig.* **2005**, *2*, 147–167. [CrossRef]

76. Jain, U.; Rogers, M.; Matson, E.T. Drone forensic framework: Sensor and data identification and verification. In Proceedings of the 2017 IEEE Sensors Application Symposium (SAS), Glasgow, UK, 31 October 2017; pp. 1–6. [CrossRef]

77. Roder, K.; Choo, N.K.R.A. Le-Khac, Unmanned Aerial Vehicle Forensic Investigation Process: Dji Phantom 3 Drone As A Case Study. *arXiv Prepr.* **2018**, arXiv:1804.08649.

78. Siddiqi, M.A.; Ghani, N. Critical Analysis on Advanced Persistent Threats. *Int. J. Comput. Appl.* **2016**, *141*, 46–50. [CrossRef]

79. Siddiqi, M.A.; Yu, H.; Joung, J. 5G Ultra-Reliable Low-Latency Communication Implementation Challenges and Operational Issues with IoT Devices. *Electronics* **2019**, *8*, 981. [CrossRef]

80. Khan, N.A.; Jhanjhi, N.Z.; Brohi, S.N.; Almazroi, A.A.; Almazroi, A.A. A secure communication protocol for unmanned aerial vehicles. *CMC-COMPUTERS MATERIALS CONTINUA* **2022**, *70*, 601–618. [CrossRef]

81. Wild, G.; Murray, J.; Baxter, G. Exploring Civil Drone Accidents and Incidents to Help Prevent Potential Air Disasters. *Aerospace* **2016**, *3*, 22. [CrossRef]

82. Goodrich, M. Drone Catcher: "Robotic Falcon" can Capture, Retrieve Renegade Drones, Michigan Tech. 7 January 2016. Available online: https://www.mtu.edu/news/stories/2016/january/drone-catcher-robotic-falcon-can-capture-retrieve-renegade-drones.html (accessed on 18 August 2022).

83. McNabb, M. Dedrone Acquires the Anti Drone Shoulder Rifle, Batelle's Drone Defender, Drone Life. 9 October 2019. Available online: https://dronelife.com/2019/10/09/dedrone-acquires-theanti-drone-shoulder-rifle-batelles-drone-defender/ (accessed on 18 August 2022).

84. Capello, E.; Dentis, M.; Mascarello, L.N.; Primatesta, S. Regulation analysis and new concept for a cloud-based UAV supervision system in urban environment. In Proceedings of the Workshop on Research, Education and Development of Unmanned Aerial Systems (RED-UAS), IEEE, (2017), Carnfiled, UK, 25–27 November 2017; pp. 90–95. [CrossRef]

85. Joglekar, R. 4 Strategies for Stopping 'Rogue' Drones from Flying in Illegal Airspace. *ABC News*. 23 December 2018. Available online: https://abcnews.go.com/Technology/strategies-stoppingrogue-drones-flying-illegal-airspace/story?id=59973853 (accessed on 18 August 2022).

86. Friedberg, S. A Primer on Jamming, Spoofing, and Electronic Interruption of a Drone, Dedrone. 19 April 2018. Available online: https://blog.dedrone.com/en/primer-jamming-spoofing-andelectronic-interruption-of-a-drone (accessed on 18 August 2022).

87. Cyber, T.E.O. How To Crack WPA/WPA2 Wi-Fi Passwords Using Aircrack-ng, Medium. 5 November 2019. Available online: https://medium.com/@TheEyeOfCyberBuckeyeSecurity/howto-crack-wpa-wpa2-wi-fi-passwords-using-aircrack-ng-8cb7161abcf9 (accessed on 18 August 2022).

88. Yaacoub, J.-P.; Noura, H.; Salman, O.; Chehab, A. Security analysis of drones systems: Attacks, limitations, and recommendations. *Internet Things* 2020, 11, 100218. [CrossRef]

89. Bonilla, C.A.T.; Parra, O.J.S.; Forero, J.H.D. Common Security Attacks on Drones. *Int. J. Appl. Eng. Res.* **2018**, *13*, 4982–4988.

90. Gaspar, J.; Ferreira, R.; Sebastião, P.; Souto, N. Capture of UAVs Through GPS Spoofing Using Low-Cost SDR Platforms. *Wirel. Pers. Commun.* **2020**, *115*, 2729–2754. [CrossRef]

91. Ezuma, M.; Erden, F.; Anjinappa, C.K.; Ozdemir, O.; Guvenc, I. Micro-UAV Detection and Classification from RF Fingerprints Using Machine Learning Techniques. In Proceedings of the 2019 IEEE Aerospace Conference; IEEE: Piscataway, NJ, USA, 2019; pp. 1–13. [CrossRef]

92. Digulescu, A.; Despina-Stoian, C.; Stănescu, D.; Popescu, F.; Enache, F.; Ioana, C.; Rădoi, E.; Rîncu, I.; Șerbănescu, A. New Approach of UAV Movement Detection and Characterization Using Advanced Signal Processing Methods Based on UWB Sensing. *Sensors* **2020**, *20*, 5904. [CrossRef]

93. Bisio, I.; Garibotto, C.; Lavagetto, F.; Sciarrone, A.; Zappatore, S. Unauthorized Amateur UAV Detection Based on WiFi Statistical Fingerprint Analysis. *IEEE Commun. Mag.* **2018**, *56*, 106–111. [CrossRef]

94. Choudhary, G.; Sharma, V.; Gupta, T.; Kim, J.; You, U. Internet of Drones (IoD): Threats, Vulnerability, and Security Perspectives. In Proceedings of the MobiSec 2018: The 3rd International Symposium on Mobile Internet Security, Cebu, Philippines, 29 Auguest–1 September 2018.

95. Noura, H.N.; Salman, O.; Chehab, A.; Couturier, R. DistLog: A distributed logging scheme for IoT forensics. *Ad Hoc Networks* **2019**, *98*, 102061. [CrossRef]

96. Cohen, R.S. The Drone Zappers, Air Force Magazine. 22 March 2019. Available online: https://www.airforcemag.com/article/the-drone-zappers/ (accessed on 18 August 2022).

97. Mizokami. Air Force Downs Several Drones with New ATHENA Laser Weapon System, Popular Mechanics. 8 November 2019. Available online: https://www.popularmechanics.com/military/research/a29727696/athena-laser-weapon/ (accessed on 18 August 2022).

98. Federal Aviation Administration, Become a Drone Pilot, Become a Pilot. 19 May 2021. Available online: https://www.faa.gov/uas/commercial_operators/become_a_drone_pilot/ (accessed on 18 August 2022).

99. Transport Canada, Where to Fly your Drone, Drone Safety. 19 February 2021. Available online: https://tc.canada.ca/en/aviation/drone-safety/where-fly-your-drone (accessed on 18 August 2022).

100. Ch, R.; Srivastava, G.; Gadekallu, T.R.; Maddikunta, P.K.R.; Bhattacharya, S. Security and privacy of UAV data using blockchain technology. *J. Inf. Secur. Appl.* **2020**, *55*, 102670. [CrossRef]

101. Sinha Satyajit, Securing IoT with Blockchain, Counterpoint. 11 May 2018. Available online: https://www.counterpointresearch.com/securing-iot-blockchain/ (accessed on 18 August 2022).

102. Wang, W.; Xu, H.; Alazab, M.; Gadekallu, T.R.; Han, Z.; Su, C. Blockchain-Based Reliable and Efficient Certificateless Signature for IIoT Devices. *IEEE Trans. Ind. Inform.* **2021**, *18*, 7059–7067. [CrossRef]

103. Zhang, L.; Zou, Y.; Wang, W.; Jin, Z.; Su, Y.; Chen, H. Resource allocation and trust computing for blockchain-enabled edge computing system. *Comput. Secur.* **2021**, *105*, 102249. [CrossRef]

104. Zhang, L.; Peng, M.; Wang, W.; Jin, Z.; Su, Y.; Chen, H. Secure and efficient data storage and sharing scheme for blockchain-based mobile-edge computing. *Trans. Emerg. Telecommun. Technol.* **2021**, *32*, e4315. [CrossRef]

105. Shao, X.; Yue, X.; Liu, J. Distributed adaptive formation control for underactuated quadrotors with guaranteed performances. *Nonlinear Dyn.* **2021**, *105*, 3167–3189. [CrossRef]

106. Li, M.; Guo, C.; Yu, H.; Yuan, Y. Event-triggered containment control of networked underactuated unmanned surface vehicles with finite-time convergence. *Ocean Eng.* **2022**, *246*, 110548. [CrossRef]

107. Rahman, Z.; Yi, X.; Khalil, I. Blockchain based AI-enabled Industry 4.0 CPS Protection against Advanced Persistent Threat. *IEEE Internet Things J.* **2022**. [CrossRef]

108. Yu, S.; Das, A.K.; Park, Y.; Lorenz, P. SLAP-IoD: Secure and Lightweight Authentication Protocol Using Physical Unclonable Functions for Internet of Drones in Smart City Environments. *IEEE Trans. Veh. Technol.* **2022**. [CrossRef]

109. Gaurav, B.; Kumar, D.; Vidyarthi, D.P. BARA: A blockchain-aided auction-based resource allocation in edge computing enabled industrial internet of things. *Future Gener. Comput. Syst.* **2022**.

110. Yuan, L.; Zhang, Y.; Wang, J.; Xiang, W.; Xiao, S.; Chang, L.; Tang, W. Performance analysis for covert communications under faster-than-Nyquist signaling. *IEEE Commun. Lett.* **2022**.
111. Zhang, W.; Shao, X.; Zhang, W.; Qi, J.; Li, H. Unknown input observer-based appointed-time funnel control for quadrotors. *Aerosp. Sci. Technol.* **2022**, *126*, 107351. [CrossRef]
112. Challita, U.; Ferdowsi, A.; Chen, M.; Saad, W. Machine Learning for Wireless Connectivity and Security of Cellular-Connected UAVs. *IEEE Wirel. Commun.* **2019**, *26*, 28–35. [CrossRef]
113. Sanjab, A.; Saad, W.; Basar, T. Prospect theory for enhanced cyber-physical security of drone delivery systems: A network interdiction game. In Proceedings of the 2017 IEEE International Conference on Communications (ICC), Paris, France, 21–25 May 2017; pp. 1–6. [CrossRef]

*Article*

# Detection of Micro-Doppler Signals of Drones Using Radar Systems with Different Radar Dwell Times

Jiangkun Gong [1], Jun Yan [1], Deren Li [1] and Deyong Kong [2,*]

[1] State Key Laboratory of Information Engineering in Surveying, Mapping and Remote Sensing, Wuhan University, Wuhan 430072, China
[2] School of Information Engineering, Hubei University of Economics, Wuhan 430205, China
* Correspondence: kdykong@hbue.edu.cn; Tel.: +86-027-68778527

**Abstract:** Not any radar dwell time of a drone radar is suitable for detecting micro-Doppler (or jet engine modulation, JEM) produced by the rotating blades in radar signals of drones. Theoretically, any X-band drone radar system should detect micro-Doppler of blades because of the micro-Doppler effect and partial resonance effect. Yet, we analyzed radar data detected by three radar systems with different radar dwell times but similar frequency and velocity resolution, including Radar$-\alpha$, Radar$-\beta$, and Radar$-\gamma$ with radar dwell times of 2.7 ms, 20 ms, and 89 ms, respectively. The results indicate that Radar$-\beta$ is the best radar for detecting micro-Doppler (i.e., JEM signals) produced by the rotating blades of a quadrotor drone, DJI Phantom 4, because the detection probability of JEM signals is almost 100%, with approximately 2 peaks, whose magnitudes are similar to that of the body Doppler. In contrast, Radar$-\alpha$ can barely detect any micro-Doppler, and Radar$-\gamma$ detects weak micro-Doppler signals, whose magnitude is only 10% of the body Doppler's. Proper radar dwell time is the key to micro-Doppler detection. This research provides an idea for designing a cognitive micro-Doppler radar by changing radar dwell time for detecting and tracking micro-Doppler signals of drones.

**Keywords:** cognitive micro-Doppler radar; drone detection; Doppler resolution; JEM signals; radar dwell time

**Citation:** Gong, J.; Yan, J.; Li, D.; Kong, D. Detection of Micro-Doppler Signals of Drones Using Radar Systems with Different Radar Dwell Times. *Drones* **2022**, *6*, 262. https://doi.org/10.3390/drones6090262

Academic Editors: Daobo Wang and Zain Anwar Ali

Received: 29 August 2022
Accepted: 17 September 2022
Published: 19 September 2022

**Publisher's Note:** MDPI stays neutral with regard to jurisdictional claims in published maps and institutional affiliations.

## 1. Introduction

Recently, researching topics about using micro-Doppler to detect, classify, and track radar echoes of drones have been hot spots. The most common drones in these studies are drones with rotating blades, such as single-rotor drones, quadrotor drones, six-rotor drones, and even hybrid vertical take-off and landing (VTOL) drones. They are small in size, fly at a slow speed, and are mainly active at low-altitude airspace [1,2]. The rotating movement of rotating blades can modulate the incident radar wave and produce an additional micro-Doppler on the base of the body Doppler contributed by the flying motion of the drone body. Micro-Doppler signals are thought to be useful signatures for radar applications.

Not any one radar system is suitable for detecting and classifying micro-Doppler in radar signals of drones. Currently, both academia and industry have revealed many drone detection radar solutions. They are a wide and diverse variety of types, such as pulse-Doppler marine radar [3], FMCW (frequency-modulated continuous wave) radar [4–7], millimeter-wave radar [8], CW (continuous wave) radar [9], staring radar [10,11], airborne weather radar [12], multistatic radar [13], wide/ultrawideband radar systems [14,15], and even radar networks [16]. Radar vendors also launch commercial off-the-shelf (COTS) counter-drone radars from generation to generation; for example, there are some commercial drone detection radar systems listed in Table 1. Their radar dwell times can be estimated roughly using the rotating rate of the antennas. Generally, with a faster rotating rate comes a shorter radar dwell time. No matter what drone radar is configured, radar dwell time is one of many factors to extracting both body Doppler signals and micro-Doppler signals of drones in background clutter.

**Table 1.** Some drone detection radars [1].

| Model<br>(Vendor; Country) | Radar Band | Update Rate<br>(Hz) [2] | Range<br>(km) [3] | Identification Strategy [4] |
|---|---|---|---|---|
| Retinar FAR-AD (Meteksan; Turkey) | Ku | 4/15 | 4.4 | Micro-Doppler |
| Gamekeeper 16U<br>(AVEILLANT; UK) | L | 4 | 5 | Micro-Doppler, tracking data. |
| A800(Blighter; UK) | Ku | 1/4 | 3 | Micro-Doppler |
| XENTA-M1<br>(Weibel; Danish) | X | 1 | 10 | Range-Doppler,<br>micro-Doppler. |
| ReGUARD<br>(Retia; Czech Republic) | X | 1/4 | 6 | Rada cross section (RCS) |
| ELM/2026BF<br>(IAI; Israel) | X | | 5.2 | Tracking data |
| Spyglass™<br>(Numerica; USA) | Ku | | | Tracking data |
| Gryphon<br>R1400/R1410<br>(SRC; USA) | X | | 8.5 | Tracking data |
| ELVIRA<br>(Robin; Netherlands) | X | 2/3 | 2.7 | AI, micro-Doppler |
| Giraffe 1X<br>(SAAB; Sweden) | X | 1 | 13 | AI, kinematic, RCS<br>micro-Doppler, etc. |
| GO20 MM<br>(Thales; France) | X | 1/6 | 4 | AI, micro-Doppler |

[1] These data can be found on their official websites. [2] The update rate is the typical value. Some of them can be selectable. [3] The detection range is for drones with RCS of ~0.01 m$^2$, such as DJI Phantom-4. The classification range is normally shorter than the detection range. [4] The specific identification signatures are not available, and those terms are reported in their official brochures.

Generally, the longer the radar dwell time means the better Doppler resolution. Yet, there is still an upper limitation of radar dwell time for a practical radar sensor. First of all, a radar needs a rotating motion with a rotating rate to achieve the 360° cycle-scanning ability. Then, there is a conflict in radar parameter design, in that a rapid rotating update rate and fast beam scanning will result in a short dwell time, and then poor Doppler resolution. The result of cycle-scanning is that the radar dwell time in one scanning cannot be infinitely long. Second, even if a radar can stare in some direction and obtain a long radar dwell time, the micro-Doppler could migrate between different space resolution cells during the long radar dwell time, and then there is an overlap of micro-Doppler either in the range-Doppler cell or time-Doppler cell. Moreover, Radar dwell time not only affects micro-Doppler but also body Doppler. Since micro-Doppler is the additional Doppler around the body Doppler, the ratio of micro-Doppler to body Doppler is also the factor related to the signal extraction. Although some typical high-resolution algorithms have also been investigated for improving Doppler resolution, such as compressed sensing (CS) [17], minimum variance distortionless response (MVDR) [18], multiple signal classification (MUSIC) [19], iterative adaptive algorithm (IAA) [20], and machine learning (ML) technology [21], radar dwell time is still the base factor related to the micro-Doppler. A balance of a suitable radar dwell time is required for extracting and observing micro-Doppler in radar signals of drones.

In this paper, we investigate the proper radar dwell time for detecting the micro-Doppler signals (i.e., jet engine modulation, JEM) modulated by the rotating blades of drones. In Section 2, we discuss the relationship between radar dwell time and the micro-Doppler produced by the rotating blades theoretically and then introduce the three typical radar dwell times of our three radar systems, i.e., Radar$-\alpha$, Radar$-\beta$, and Radar$-\gamma$. In Section 3 and Section 4, we analyze the detection performance of micro-Doppler detected by the three radars and then propose our explanation of the detection results. Finally, we conclude our point in Section 5. The objectives of this paper include three points. (1) We argue that not any radar dwell time is suitable for detecting micro-Doppler produced by the rotating blades in radar signals of drones, and the proper radar dwell time depends on the rotating period of the blades of drones. (2) We propose that two parameters can be used for evaluating the detection performance of micro-Doppler modulated by drones' blades, including the JEM number and the ratio of the first blade's magnitude to that of the body. (3) We suggest that a cognitive radar can be designed by adjusting the radar dwell time to detect micro-Doppler signals of drones.

## 2. Materials and Methods

### 2.1. Micro-Doppler of Rotating Blades of Drones

Micro-Doppler is the additional Doppler related to the micromotion in addition to the body Doppler. Figure 1a demonstrates the geometry of a radar and a rotating rotor blade of a quadrotor drone. Assume that the azimuth angle of $\alpha$ and the elevation angle of $\beta$ is zero. The shape of a blade is not, and nor can it be seen as, a thin rectangular bar with rotating movement. According to micro-Doppler theory in [22], the received signals of the $k$th blade and its micro-Doppler frequency are given by

$$\left| S_{k,mD}(t) \right| = \sum_{k=0}^{N-1} Lsinc\left( \frac{2\pi L}{\lambda} sin\left( \Omega t + \theta_{0,k} + \frac{k2\pi}{N} \right) \right) \tag{1}$$

$$f_{k,mD}(t) = \sum_{k=0}^{N-1} \frac{2\pi L}{\lambda} \Omega \left[ -sin\left( \theta_{0,k} + \frac{k2\pi}{N} \right) sin(\Omega t) + cos\left( \theta_{0,k} + \frac{k2\pi}{N} \right) cos(\Omega t) \right] \tag{2}$$

where $N$ is the number of blades, $L$ is the blade length, $\lambda$ is the wavelength, $\Omega$ is the rotation rate, and $\theta_{0,k}$ is the initial rotation angle. Equations (1) and (2) indicate that such micro-Doppler signals are modulated by the rotation rate $\Omega$ through two sinusoidal functions. The maximum values of both $\left| S_{k,mD}(t) \right|$ and $f_{k,mD}(t)$ appear when the direction of the incident wave is perpendicular to the long face of the blade, which means that the rotation angle is 90°, and then we can obtain the "blade flash" signals in the time series and the JEM-like peaks in the spectrum [22–25].

The ideal micro-Doppler or JEM could be simulated when the radar dwell time and frequency resolution are enough. Figure 1b shows the simulated micro-Doppler of rotating blades. The number of blades is 1, and the rotating rate is 100 Hz. The length of a single blade is 0.2 m. The elevation angle is 15°, and the detection range is 20 km. The test band is the X-band, working on 10 GHz. The simulated radar time is 100 ms. Figure 1b shows the flash signals produced by the blade, which are modulated by the rotating rate on the radar images processed by short-time Fourier transform (STFT) algorithm. Figure 1c demonstrates the "blade flash" signals in the time domain with a modulation period of 5 ms and the JEM signals with a frequency interval of about 100 Hz.

Radar dwell time is the key to observing such micro-Doppler. According to Nyquist Theorem, the radar dwell time must be at least longer than twice the rotating period of the rotating blades Given that the rotating rate of the blades of the drone is $\Omega$, the minimum radar dwell time for obtaining such sufficient micro-Doppler is given by

$$T_s = \frac{2}{\Omega} \tag{3}$$

where $T_s$ is the minimum radar dwell time. If the general rotation rate of blades of drones is 100 Hz, the minimum dwell time is approximately 20 ms. Different radar systems have different radar dwell times. If radar dwell time of radar is much shorter than the required one, $T_s$, then we may observe the insufficient micro-Doppler with weaker magnitude and a smaller number of JEM peaks. Furthermore, what happens when the radar dwell time of radar is much longer than the required one? Can we obtain a much clearer micro-Doppler in the radar signals of drones? What is the best radar dwell time for observing such micro-Doppler?

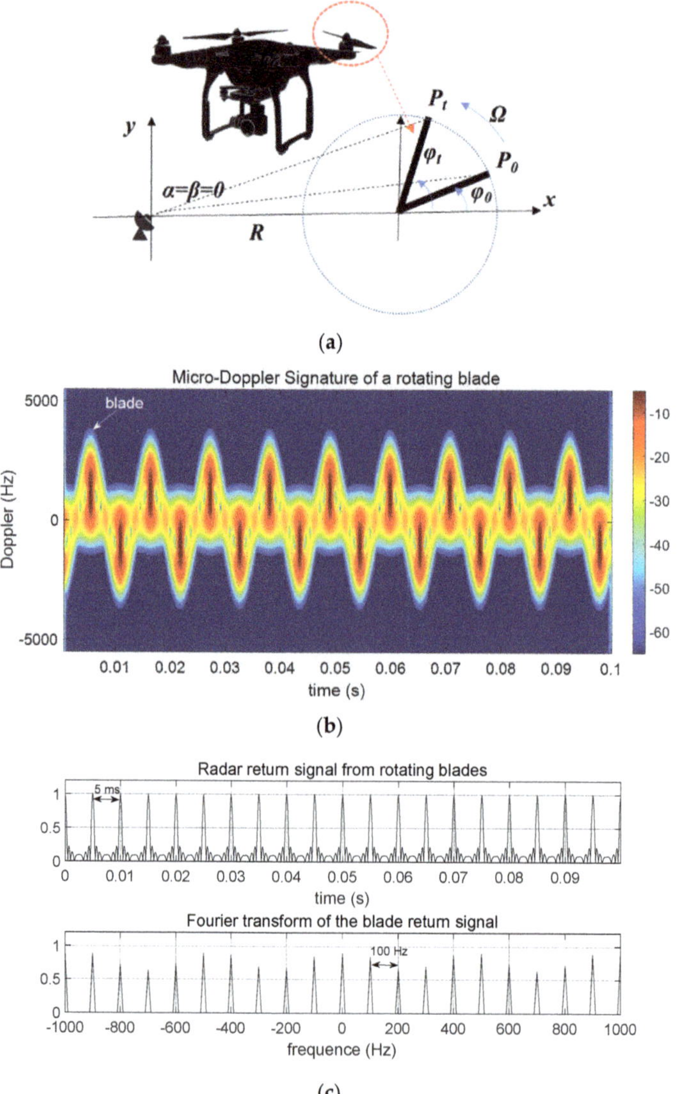

**(a)**

**(b)**

**(c)**

**Figure 1.** Simulated micro-Doppler of rotating blades within X-band data. (**a**) The geometry of the radar and the rotating rotor blades, (**b**) micro-Doppler on the STFT image, (**c**) blade flash and JEM signals.

To evaluate the detection performance of micro-Doppler (i.e., JEM) of rotating blades of drones, we select some parameters about JEM signatures. They are (1) the number of JEM peaks, and (2) the ratio of the first blade's magnitude to the body's magnitude. They can be given, respectively, by

$$N_{mD} = N - 1 \tag{4}$$

where $N_{mD}$ is the number of JEM peaks, $N$ is the number of Doppler peaks in the spectrum, and the 1 represents the body Doppler (i.e., bulk Doppler); and

$$r_{mb} = \frac{A_{mD}}{A_{bD}} \tag{5}$$

where $r_{mb}$ is the ratio of blade's magnitude to body's magnitude, $A_{mD}$ is the magnitude of the first neighboring blade Doppler, and $A_{bD}$ is the magnitude of the body Doppler. Theoretically, the better detection performance of such micro-Doppler (i.e., JEM) means a bigger $N_{mD}$ and a higher $r_{mb}$.

### 2.2. Experimental Conditions

To investigate the detection performance of such micro-Doppler, a software-defined radar platform that can be changed with radar dwell times as well as other parameters should be used for collecting radar data of drones and then seeking the best radar dwell time for detecting micro-Doppler after investigating the relationship between the radar dwell time and micro-Doppler. Unfortunately, we do not have such resources to conduct this research, and we can only use some radar data of drones detected by three radar systems with typical radar dwell times to explore the first step of such a topic. Instead, we obtained some drone detection radar data detected by three radar systems with typical radar dwell times (i.e., insufficient, moderate, and sufficient). Table 2 lists some parameters of the three radar systems. They are pulse-Doppler radar systems equipped with phased-array antennas. The pulse repetition frequency (PRF) and Doppler resolutions of three radar sensors are similar, but Radar$-\alpha$ (the insufficient one) has the shortest radar dwell time in one coherent pulse interval (CPI) of 2.7 ms, Radar$-\beta$ (the moderate one) has a moderate time of 20 ms, and Radar$-\gamma$ (the sufficient one) has the longest time of 89 ms. Thereby, given the general rotating period (i.e., 10 ms) of the blade, Radar$-\alpha$ can detect insufficient micro-Doppler because the radar dwell time is only 27% of the rotating period, Radar$-\gamma$ can detect sufficient micro-Doppler because the radar dwell time is about 4 times than the rotating period, and Radar$-\beta$ may detect either sufficient or insufficient micro-Doppler based on the rotating rate of the blades in actual cases. The quadcopter drone is a DJI Phantom 4, fabricated by DJI Inc., China. It is a small drone with a flight weight of 1.38 kg. Both its body and propellers are mainly composed of plastic. Its wheelbase is approximately 0.35 m. There are four rotor blades with a length of 0.2 m. The cruise speed is approximately 15 m/s. The maximum flight time is approximately 28 min, with a maximum flight height lower than 500 m. The rotating rate of blades is from 5000 RPM to 7000 RPM (revolutions per minute).

**Table 2.** Parameters of the three drone detection radars.

| Parameters | Radar$-\alpha$ | Radar$-\beta$ | Radar$-\gamma$ |
|---|---|---|---|
| Radar band | X | X | X |
| CPI (ms) | 2.7 | 20 | 89 |
| PRF (kHz) | 33.3 | 5 | 2.8 |
| Sampling points after zero padding | 2048 | 256 | 256 |
| Frequency resolution (Hz) | 16 | 19 | 11 |
| Doppler resolution (m/s) | 0.163 | 0.285 | 0.165 |
| Range resolution (m) | 3.75 | 12 | 10 |
| Beamwidth | 0.97° | 0.72° | 2° |
| Detection range (m) | 3000 | 10,000 | 6000 |
| Width of the wavefront (m) | 50.7 | 125.6 | 209.4 |
| Space resolution (m$^2$) | 190.1 | 1507.2 | 2094 |
| Radar dwell time per square meter (ms/m$^2$) | 0.014 | 0.013 | 0.042 |

The radars collected these data at three areas. The detection background is mainly ground clutter, but the detection ranges were different. The range of the drones from

Radar$-\alpha$ was about 3 km, the one from Radar$-\beta$ was about 10 km, and that from Radar$-\gamma$ was about 6 km. The drones were flying in the radar beams, and the radars worked in a tracking mode. Thereby, we collected tracking data of drones. The drones were flying in a range widow with a size of 1 km, at an altitude below 300 m. The width of the wavefront at the radar range of a target is calculated by the beamwidth and the detection range, which is

$$W = \theta_{re}R \tag{6}$$

where $W$ is the width of the wavefront, $\theta_{re}$ is the beamwidth, and $R$ is the detection range of the target. Thereby, the spatial resolution of the sector where the target is in can be given approximately by

$$Sp_r = WR_{re} \tag{7}$$

where $Sp_r$ is the spatial resolution and $R_{re}$ is the range resolution. Table 2 also lists the range resolutions, detection ranges, and space resolutions in the three cases. If we divide the spatial resolution by the CPI time, we can obtain the radar dwell time per square meter values of the three radars, which are 0.014 ms/m$^2$, 0.013 ms/m$^2$, and 0.042 ms/m$^2$, respectively. These numbers are similar to each other. Figure 2 shows an example of detecting and tracking the DJI Phantom 4, using Radar$-\gamma$ working in a tracking mode. The blue solid line was the radar beam, and the blue dotted lines were the tracking trace of the drone when flying towards the radar location. We selected some radar data collected in this area and used them in this paper. Similarly, we also collected some data using Radar$-\alpha$ and Radar$-\beta$ in some other areas.

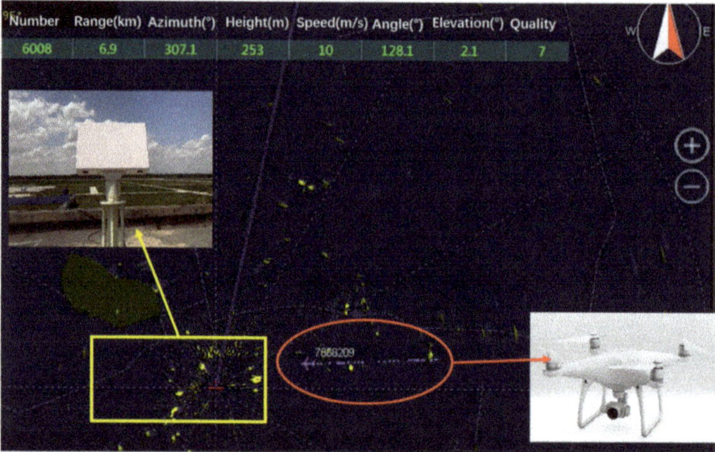


**Figure 2.** Example of tracking trace of the drone on the radar (i.e., Radar$-\gamma$) screenshot.

Although our radar data were collected in a different area, they could still be used for this research for the following reasons. First, we investigated the detection performance of micro-Doppler signals of drones over the radar signals of drones. This means that the radar data of drones were already detected and tracked in the background clutter. Second, we compared the micro-Doppler with the body Doppler; thus, it involves no background clutter. We can also use the normalized magnitudes to remove the interference from the background environment. Thereby, these radar data collected from different ranges in different areas can be used for research to investigate micro-Doppler signals of drones.

## 3. Results

Radar detection is always accompanied by interference from background clutter, and different radar systems with different dwell times can have different performances.

Generally, the longer the radar dwell time, the higher the SNR value of the target. Yet, the noise level of clutter also increases along with the increased radar dwell time. Figure 3 demonstrates radar signals of drones in a similar range window. The black frames in the range-Doppler images mark the range bin that the drone was in. The length values of range windows are 78.75 m, 84 m, and 70 m, respectively. Generally, the ground clutter is mainly around 0 Hz, with different spectral widths, which are 728 Hz, 80 Hz, and 106 Hz when using Radar$-\alpha$, Radar$-\beta$, and Radar$-\gamma$. Although the SNR values of the drone (65.68 dB, 11.43 dB, 11.55 dB) are different in the three cases, its radar echoes can be detected in the clutter. Yet, Radar$-\beta$ can detect the most legible radar signals of the drone because the micro-Doppler signals produced by the rotating blades can also be detected and identified along with the body Doppler (three pots in the black frame in Figure 3b). Other radars seem to detect only body Dopplers of drones on their range-Doppler images (Figure 3a,c). In total, Radar$-\beta$ enjoys a better detection performance for detecting drones than Radar$-\alpha$ and Radar$-\gamma$.

Micro-Doppler signatures of drones produced by the rotating blades are related directly to the radar dwell time. Figure 4 compares the raw radar signals and the spectrums of drones using the three radar systems. The data are extracted from the range windows in Figure 3. The blues words register the Doppler peaks corresponding to either blades or the body of drones. When the drone is flying away from the radar, its Doppler velocity is negative, and when it is flying approaching the radar, the velocity is positive. Each subfigure shows two cases in which the drone flew in a different direction, relative to the radars. According to Table 2, both the frequency resolution and the velocity resolution of the three radars are similar to each other, but the radar dwell times (i.e., 2.7 ms, 20 ms, and 89 ms) are different. The rotating rate of the blades ranges from 5000–7000 RPM (revolutions per minute). According to Formula (3), the minimum radar dwell time to obtain the sufficient micro-Doppler of the rotating blades is 17–24 ms. Therefore, Radar$-\alpha$ can only detect insufficient micro-Doppler signals of the drone, and Radar$-\beta$ observes either sufficient micro-Doppler signals or insufficient micro-Doppler signals based on the rotating rate of the blades, but Radar$-\gamma$ detected only sufficient micro-Doppler signals.

There seems to be only one bulk Doppler (i.e., body Doppler) detected by Radar$-\alpha$, in Figure 4a, which is −9.4 m/s (or +8.6 m/s). In contrast, several Doppler peaks including one body Doppler and two or three micro-Dopplers appear in the spectrums detected by Radar$-\beta$ and Radar$-\gamma$, which are −13.2 m/s, −8.1 m/s, −6 m/s, −3.3 m/s (or 2.1 m/s, 4.2 m/s, 6.9 m/s) in Figure 4b, and −20.2 m/s, −17.0 m/s, −13.8 m/s, −12.0 m/s, −10.8 m/s (or 6.6 m/s, 9.2 m/s, 11.7 m/s, 14.2 m/s) in Figure 4c. The number of micro-Dopplers (i.e., $N_{mD}$) detected by Radar$-\gamma$ in Figure 4c is about four, and the number detected by Radar$-\beta$ in Figure 4b is about three. Moreover, the ratio of strongest micro-Doppler magnitude to that of body-Doppler (i.e., $r_{mb}$) in Figure 4b is about 1, but this number decreases below 0.2 in Figure 4c. Thereby, Radar$-\gamma$ seems to be able to detect more micro-Dopplers than Radar$-\beta$, but Radar$-\beta$ can detect much stronger micro-Doppler than Radar$-\gamma$. Yet, Radar$-\alpha$ detected the poorest micro-Doppler. This means that a moderate dwell time is better for the micro-Doppler of drones.

It is not proper that the longest radar dwell times come with the best detection performance of micro-Doppler signals. Figure 5 presents the tracking Doppler signals of drones using the three radar systems. The tracking intervals are different from each other. The black dotted curves in Figure 5 describe the changing body Doppler of drones in the three cases. First, similar to the range-Doppler images in Figure 3, there is always clutter when detecting drones, and the Doppler of background clutter mainly stays around 0 m/s. Doppler detection can separate the radar signals of drones from the clutter with small velocities. Second, Radar-$\beta$ can track more enriched micro-Doppler signatures than Radar$-\alpha$ and Radar$-\gamma$. There are always attendant spots around the body Doppler in Figure 5b, which represent the distributed pattern of micro-Doppler modulated by the rotating blades of drones. Yet, these micro-Doppler spots seem to disappear on the images in Figure 5a,c. As we stated in Figure 3, Radar$-\alpha$ cannot detect insufficient micro-Doppler, and then the micro-Doppler spots disappear in the tracking results. However, Radar$-\gamma$ can detect very strong body Doppler, which is much stronger than micro-Dopplers, and then the micro-Dopplers are suppressed by the body Dopplers and hidden in the images of Figure 5c.

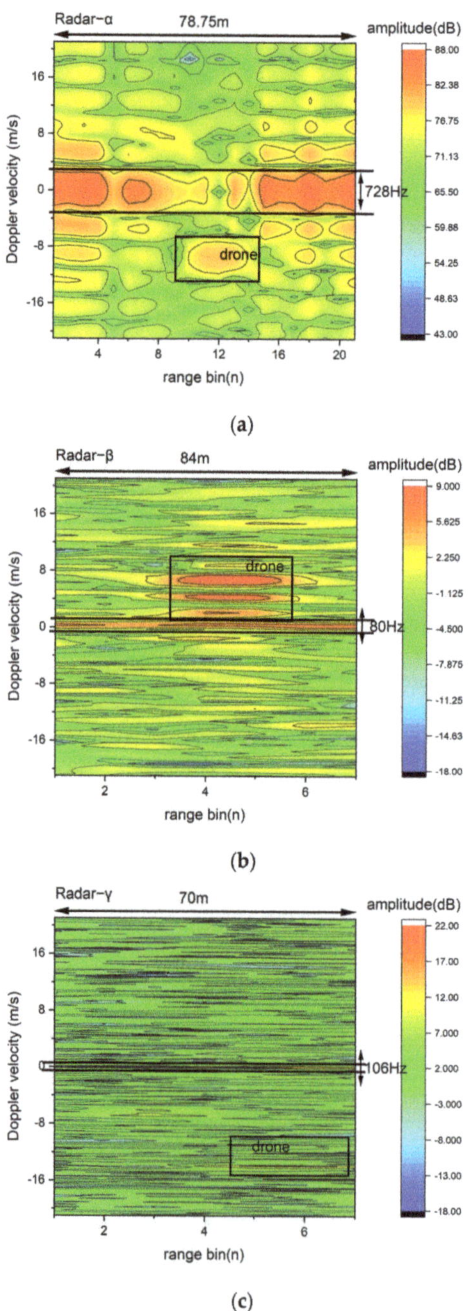



**Figure 3.** Range-Doppler data of drones. (**a**) Radar$-\alpha$, (**b**) Radar$-\beta$, (**c**) Radar$-\gamma$.

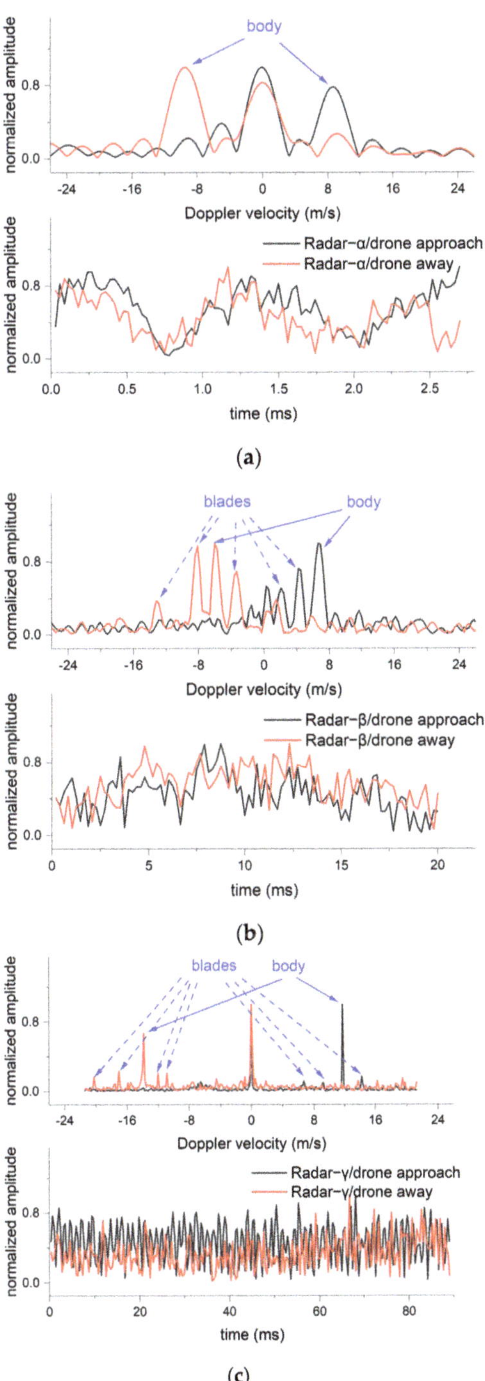


**Figure 4.** Radar signals and spectrums of drones. (**a**) Radar−α, (**b**) Radar−β, (**c**) Radar−γ.

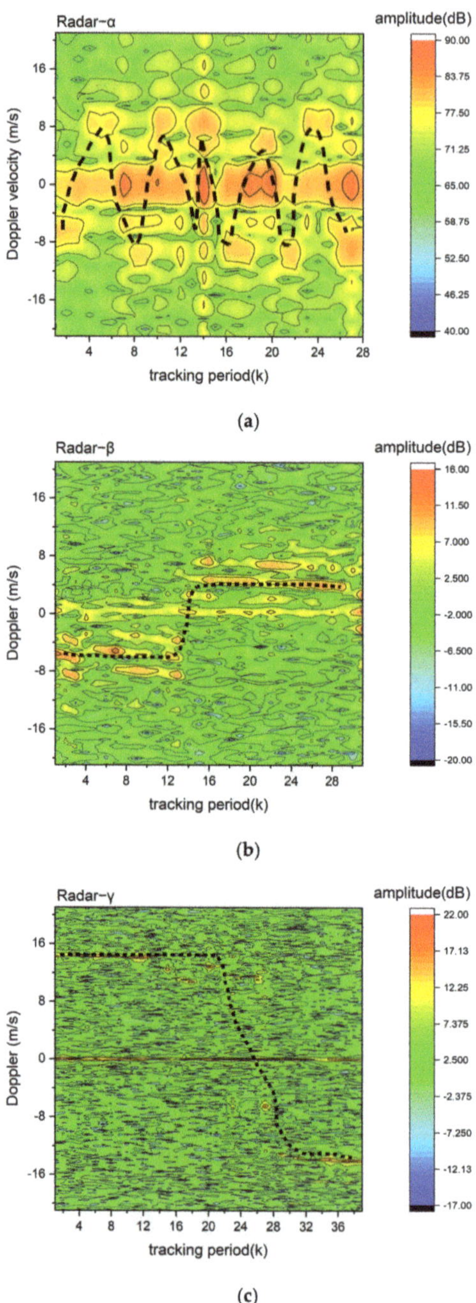

**Figure 5.** Tracking Doppler data of drones. (**a**) Radar−α, (**b**) Radar−β, (**c**) Radar−γ.

The quantification analysis of tracking results indicates that compared to Radar−α and Radar−γ, Radar−β with moderate radar dwell time is a better solution for detecting and tracking micro-Doppler signals of drones among the three radars. Figure 6 demonstrates the tracking parameters of three cases in Figure 5. SNR means signal-to-noise ratio, which describes the scattering power of a target, and SCR means signal-to-clutter ratio, which presents the scattering superiority of a target to the clutter. The JEM number is the number of micro-Doppler peaks in the spectrum, and blade/body means the ratio of the micro-Doppler's magnitude to the body Doppler's.

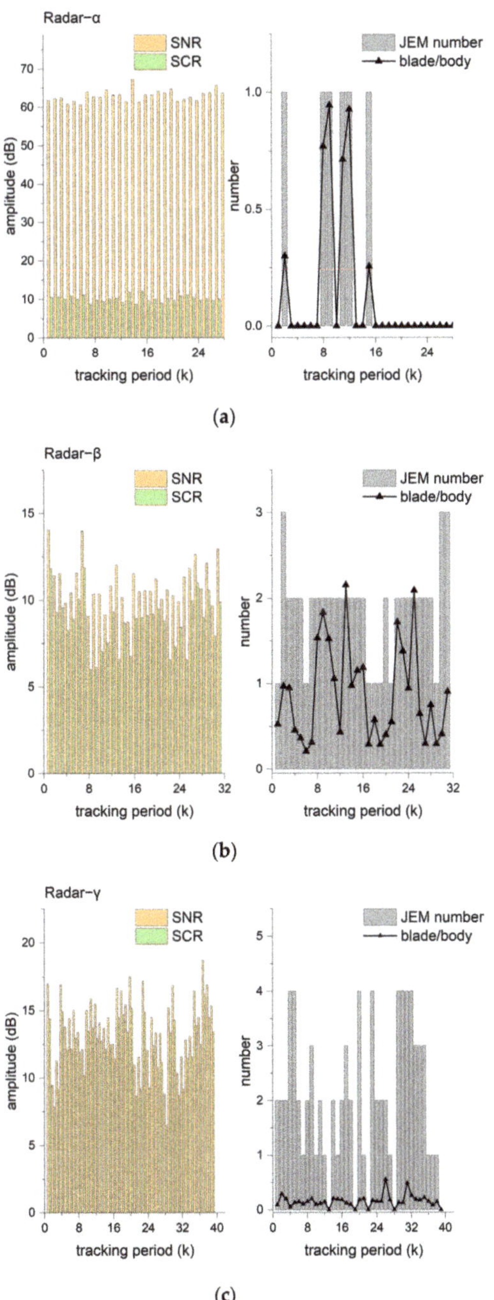

**Figure 6.** Detection results of drones. (**a**) Radar$-\alpha$, (**b**) Radar$-\beta$, (**c**) Radar$-\gamma$.

Table 3 shows the statical data of results in Figure 6. First, both SNR and SCR values are fluctuating in three cases. Since different radar systems have different transmitted power levels and noise levels, the SNR values of drones can fluctuate considerably. For example, the mean SNR of drones detected by Radar$-\alpha$ is 63.01 dB, about six times that of the 10.98 dB detected by Radar$-\beta$. Yet, SCR seems to be much more stable, with only a range of less than 4 dB. To this degree, SCR is a better value for detecting radar signals of targets. Second, Radar$-\alpha$ can detect possible micro-Doppler in some cases (e.g., #2, #8, #9, etc.) and Radar$-\gamma$ sometimes detects no micro-Doppler (e.g., #13, #19, #2, etc.), but Radar$-\beta$ can always detect micro-Doppler in all sampling cases. Thereby, the detection probabilities of JEM detected by the three radars are 21.42%, 100%, and 12.82%, respectively. Third, the numbers of micro-Doppler detected by Radar$-\beta$ and Radar$-\gamma$ are similar at 2, much bigger than the 0.18 detected by Radar$-\alpha$. It means that as long as the micro-Doppler is detected by the radar, there are at least two JEM peaks. Fourth, Radar$-\beta$ can detect the strongest micro-Dopplers with the ratio of blade's signal to body's signal of 0.88, but the number (i.e., 0.16) is very small in the cases detected by Radar$-\gamma$. In some words, even if the micro-Doppler is in the radar signals, it can be neglected by an extraction algorithm. Although the number of cases containing micro-Doppler detected by Radar$-\alpha$ is only 6 among the whole 28, the ratio of blade's signal to body's signal is still about 0.65. Fifth, the frequency offsets between the body's Doppler and the first neighboring blade's Doppler are 469 Hz, 165 Hz, and 158 Hz, respectively.

**Table 3.** Comparison of radar detection of a drone1.

| Contents | Radar$-\alpha$ | Radar$-\beta$ | Radar$-\gamma$ |
|---|---|---|---|
| Detection range (km) | ~3 km | ~10 km | ~6 km |
| Doppler velocity (m/s) | 8.29 | 4.70 | 13.00 |
| Mean SNR (dB) | 63.01 | 10.98 | 14.22 |
| Mean SCR (dB) | 10.23 | 8.71 | 12.17 |
| Probability of JEM signals | 21.42% | 100% | 12.82% |
| Number of JEM peaks | 0.18 | 1.87 | 2.05 |
| The ratio of the blade's magnitude to that of the body | 0.65 | 0.88 | 0.16 |
| Frequency offset between blade and body (Hz) | 469 | 165 | 158 |

## 4. Discussion

Why can Radar$-\beta$ with moderate radar dwell time detect the best micro-Doppler among the three radar systems? Radar dwell time and radar wavelength are the key factors. Micro-Doppler is the additional Doppler related to the micromotion of the microcomponent on the body of a target. For the drone, the blades are the microcomponent, and the rotating motion is the source of the additional Doppler, as shown in Figure 7. According to Equations (1) and (2), the maximum values of $\left|S_{k,mD}(t)\right|$ and $f_{k,mD}(t)$ appear when the direction of the incident wave is perpendicular to the long face of the blade, which means that the rotation angle is 90°, and then "blade flash" signals appear in the time series, and JEM-like peaks occur in the spectrum. Furthermore, the partial resonance effect will also contribute to the scattering power of blades. It is known that the resonance effect occurs when the sizes of a target (e.g., drones) are comparable with the radar wavelengths, so their scattering properties are calculated via Mie theory [26,27]. As such, the scattering power of the target is an oscillating function of the size, the materials contents, and the wavelength so that the radar reflectivity values can be amplified at another wavelength. Since the wavelength of the X-band is similar to the width of the blade of the drone, and when the transmitted wave is shot directly onto the blade in the direction of perpendicular to the transmission, the partial resonance effect will amplify only the scattering power of the blades. The micro-Doppler effect and the partial resonance effect together cause the strong JEM in the spectrum.

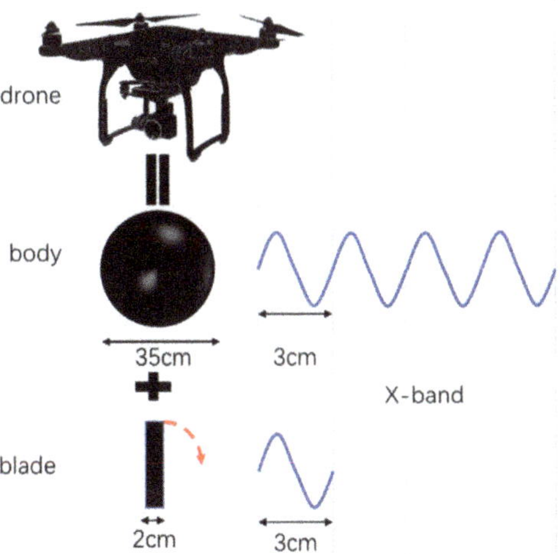



**Figure 7.** Diagram showing radar wave intervals with different structures of a quadrotor drone.

The different coherent integration times (i.e., radar dwell time) on the drone's body and blades also affect the scattering power from the body and the blades. It is known that if the coherent integration is performed and the magnitudes of the returns from all $N$ pulses are added, the SNR of a target increases as $N$. Note that due to the rotating motion of the blades, the real radar dwell times on the blades are always shorter than the body. In our cases, the three radar systems (i.e., Radar$-\alpha$, Radar$-\beta$, Radar$-\gamma$) have radar dwell times of 2.7 ms, 20 ms, and 89 ms, respectively, and the rotating period of the blades of drones is about 10 ms. Only the radar dwell time of 20 ms of Radar$-\beta$ is similar to the two-rotation period of the blades, and then the magnitude of the blade Doppler is similar to that of the body Doppler because of the partial resonance effect. In contrast, the radar dwell time of 2.7 ms of Radar$-\alpha$ is too short, so there is only a 10% probability of "blade flash" signals (or JEM signals) contributed by the micro-Doppler effect and the partial resonance effect. Furthermore, the radar dwell time of 89 ms of Radar$-\gamma$ is about four times the required one. During this period, the blades have eight rotating periods, which means that the magnitude of the blade Doppler is smaller than 1/4 of the body Doppler's magnitude. Thereby, due to improper radar dwell time, Radar$-\alpha$ and Radar$-\gamma$ are not suitable for detecting micro-Doppler of drones.

The promising application of this finding that proper radar dwell time is the key to detecting micro-Doppler of targets is to design the cognitive radar systems detecting micro-Doppler by adjusting the radar dwell time. Cognitive radar systems use adaption between the information extracted from the sensor and the transmission of subsequent illuminating waveforms. A practicable cognitive radar is to use the micro-Doppler information and then enjoy the best detection performance. As shown in Figure 8, a cognitive micro-Doppler drone detection radar can change its radar dwell time and PRF to obtain the best performance in detecting micro-Doppler signals of drones. The radar can learn the transmitted parameters of Radar$-\beta$ and adjust the transmitted parameters by using the JEM signatures including the number of JEM Dopplers, and the ratio of the blade's signal to the body's signal. We believe that this new drone detection radar will work well, like Radar$-\beta$ in this paper. In the future, we will continue to conduct related research, design the cognitive micro-Doppler radar for detecting drones using a software-defined radar platform, and evaluate its performance.

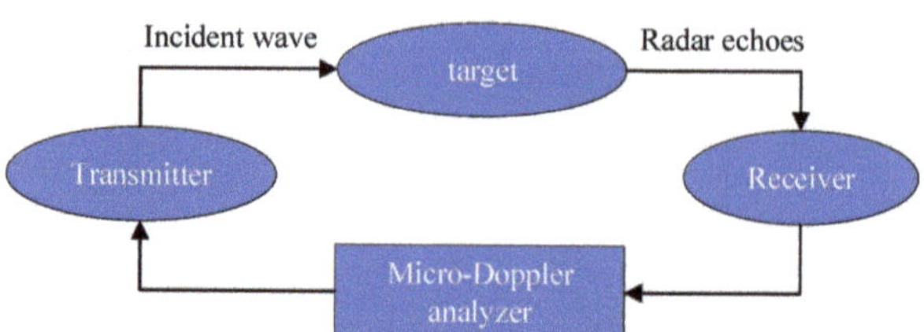


**Figure 8.** Block diagram of a cognitive micro-Doppler radar system.

## 5. Conclusions

The radar dwell time of a drone detection radar is the key to detecting micro-Doppler in radar signals of drones. Theoretically, any X-band drone radar system should detect micro-Doppler signals because of the micro-Doppler effect and partial resonance effect. In this paper, we analyze radar data detected by three radar systems with different radar dwell times and similar frequency and velocity resolution. Radar$-\alpha$, Radar$-\beta$, and Radar$-\gamma$ have radar dwell times of 2.7 ms, 20 ms, and 89 ms, respectively. We use two parameters to evaluate the detection performance of micro-Doppler using the three radar systems, including the number of JEM peaks (See Equation (4)), and the ratio of the first blade's magnitude to the body's magnitude (See Equation (5)). The detection results indicate that Radar$-\beta$ is the best radar for detecting micro-Doppler (i.e., JEM signals) produced by the rotating blades of a quadrotor drone, DJI Phantom 4, because the probability of the JEM signals is almost 100%, with approximately 2 peaks, whose magnitudes are similar to that of the body Doppler. In contrast, Radar$-\alpha$ can barely detect any micro-Doppler, and Radar$-\gamma$ detects weak micro-Doppler signals, whose magnitude is only 10% of the body Doppler's. Furthermore, the best radar dwell time for detecting such micro-Doppler is similar to the two-rotation period of the blades. Our findings demonstrate that micro-Doppler signals could be used for designing a cognitive radar for detecting and tracking micro-Doppler signals of drones.

**Author Contributions:** Conceptualization, J.Y.; methodology, J.G.; software, D.K.; validation, J.Y.; formal analysis, J.G.; investigation, J.Y.; resources, D.L.; data curation, J.Y.; writing—original draft preparation, J.G.; writing—review and editing, D.K..; visualization, D.K.; supervision, J.Y.; project administration, D.L.; funding acquisition, D.K. All authors have read and agreed to the published version of the manuscript.

**Funding:** This research received some support from the Natural Science Foundation of Hubei Providence (General Program: 2021CFB309).

**Institutional Review Board Statement:** Not applicable.

**Informed Consent Statement:** Not applicable.

**Data Availability Statement:** Some of the data presented in this study may be available on request from the corresponding author. The data are not publicly available due to the internal restriction of the research group.

**Acknowledgments:** We appreciate both the testers during the collection of the data, and we also want to thank the authors whose photographs are reproduced in this study. Furthermore, we would like to thank Huiping Hu for her help with processing the figures in this paper.

**Conflicts of Interest:** The authors declare that they have no conflict of interest.

## References


1. Musa, S.A.; Abdullah, R.S.A.R.; Sali, A.; Ismail, A.; Rashid, N.E.A.; Ibrahim, I.P.; Salah, A.A. A review of copter drone detection using radar systems. *Def. S&T Tech. Bull.* **2019**, *12*, 16–38.
2. Wellig, P.; Speirs, P.; Schuepbach, C.; Oechslin, R.; Renker, M.; Boeniger, U.; Pratisto, H. Radar systems and challenges for C-UAV. In Proceedings of the 2018 19th International Radar Symposium (IRS), Bonn, Germany, 20–22 June 2018; pp. 1–8.

3.  Galati, G.; Pavan, G. Calibration of an X-band commercial radar and reflectivity measurements in suburban areas. *IEEE Aerosp. Electron. Syst. Mag.* **2019**, *34*, 4–11. [CrossRef]
4.  Roldan, I.; del-Blanco, C.R.; Duque de Quevedo, Á.; Ibañez Urzaiz, F.; Gismero Menoyo, J.; Asensio López, A.; Berjón, D.; Jaureguizar, F.; García, N. DopplerNet: A convolutional neural network for recognising targets in real scenarios using a persistent range–Doppler radar. *IET Radar Sonar Navig.* **2020**, *14*, 593–600. [CrossRef]
5.  De Wit, J.J.M.; Gusland, D.; Trommel, R.P. Radar Measurements for the Assessment of Features for Drone Characterization. In Proceedings of the 2020 17th European Radar Conference (EuRAD), Utrecht, The Netherlands, 10–15 January 2021; pp. 38–41. [CrossRef]
6.  Park, J.; Jung, D.H.; Bae, K.B.; Park, S.O. Range-Doppler Map Improvement in FMCW Radar for Small Moving Drone Detection Using the Stationary Point Concentration Technique. *IEEE Trans. Microw. Theory Tech.* **2020**, *68*, 1858–1871. [CrossRef]
7.  Zulkifli, S.; Balleri, A. Design and Development of K-Band FMCW Radar for Nano-Drone Detection. In Proceedings of the 2020 IEEE Radar Conference (RadarConf20), Florence, Italy, 21–25 September 2020; pp. 1–5.
8.  Balal, N.; Richter, Y.; Pinhasi, Y. Identifying low-RCS targets using micro-Doppler high-resolution radar in the millimeter waves. In Proceedings of the 14th European Conference on Antennas and Propagation, EuCAP 2020, Copenhagen, Denmark, 15–20 March 2020; pp. 1–5.
9.  Bjorklund, S.; Wadstromer, N. Target Detection and Classification of Small Drones by Deep Learning on Radar Micro-Doppler. In Proceedings of the 2019 International Radar Conference, RADAR 2019, Toulon, France, 23–27 September 2019; pp. 1–6.
10. Beasley, P.; Ritchie, M.; Griffiths, H.; Miceli, W.; Inggs, M.; Lewis, S.; Kahn, B. Multistatic Radar Measurements of UAVs at X-band and L-band. In Proceedings of the 2020 IEEE Radar Conference (RadarConf20), Florence, Italy, 21–25 September 2020; pp. 1–6.
11. Jahangtr, M.; Atkinson, G.M.; Antoniou, M.; Baker, C.J.; Sadler, J.P.; Reynolds, S.J. Measurements of Birds and Drones with L-Band Staring Radar. In Proceedings of the 2021 21st International Radar Symposium (IRS), Berlin, Germany, 21–22 June 2021; pp. 1–10.
12. Blake, W.; Burger, I. Small Drone Detection Using Airborne Weather Radar. In Proceedings of the 2021 IEEE Radar Conference (RadarConf21), Atlanta, GA, USA, 7–14 May 2021; pp. 1–4.
13. Palamà, R.; Fioranelli, F.; Ritchie, M.; Inggs, M.; Lewis, S.; Griffiths, H. Measurements and discrimination of drones and birds with a multi-frequency multistatic radar system. *IET Radar Sonar Navig.* **2021**, *15*, 841–852. [CrossRef]
14. Mizushima, T.; Nakamura, R.; Hadama, H. Reflection characteristics of ultra-wideband radar echoes from various drones in flight. In Proceedings of the 2020 IEEE Topical Conference on Wireless Sensors and Sensor Networks, WiSNeT 2020, Antonio, TX, USA, 26–29 January 2020; pp. 30–33.
15. Huang, A.; Sévigny, P.; Balaji, B.; Rajan, S. Fundamental Frequency Estimation of HERM Lines of Drones. In Proceedings of the 2020 IEEE International Radar Conference (RADAR), Washington, DC, USA, 28–30 April 2020; pp. 1013–1018.
16. Coluccia, A.; Parisi, G.; Fascista, A. Detection and classification of multirotor drones in radar sensor networks: A review. *Sensors* **2020**, *20*, 4172. [CrossRef] [PubMed]
17. Yang, W.Y.; Kim, H.J.; Lee, J.H.; Yang, S.J.; Myung, N.H. Automatic feature extraction from insufficient JEM signals based on compressed sensing method. In Proceedings of the 2015 Asia-Pacific Microwave Conference (APMC), Nanjing, China, 6–9 December 2016; Volume 2, pp. 1–3.
18. Wölfel, M.; McDonough, J. Minimum variance distortionless response spectral estimation. *IEEE Signal Process. Mag.* **2005**, *22*, 117–126. [CrossRef]
19. Elbir, A.M. DeepMUSIC: Multiple Signal Classification via Deep Learning. *IEEE Sens. Lett.* **2020**, *4*, 7001004. [CrossRef]
20. Sun, H.; Oh, B.S.; Guo, X.; Lin, Z. Improving the Doppler Resolution of Ground-Based Surveillance Radar for Drone Detection. *IEEE Trans. Aerosp. Electron. Syst.* **2019**, *55*, 3667–3673. [CrossRef]
21. Ritchie, M.; Capraru, R.; Fioranelli, F. Dop-net: A micro-Doppler radar data challenge. *Electron. Lett.* **2020**, *56*, 568–570. [CrossRef]
22. Chen, V.C. *The Micro-Doppler Effect in Radar*; Artech House: Norwood, MA, USA, 2011; ISBN 9781608070572/1608070573.
23. Kim, B.K.; Kang, H.S.; Park, S.O. Experimental Analysis of Small Drone Polarimetry Based on Micro-Doppler Signature. *IEEE Geosci. Remote Sens. Lett.* **2017**, *14*, 1670–1674. [CrossRef]
24. Molchanov, P.; Harmanny, R.I.A.; De Wit, J.J.M.; Egiazarian, K.; Astola, J. Classification of small UAVs and birds by micro-Doppler signatures. *Int. J. Microw. Wirel. Technol.* **2014**, *6*, 435–444. [CrossRef]
25. Gong, J.; Yan, J.; Li, D.; Chen, R.; Tian, F.; Yan, Z. Theoretical and experimental analysis of radar micro-doppler signature modulated by rotating blades of drones. *IEEE Antennas Wirel. Propag. Lett.* **2020**, *19*, 1659–1663. [CrossRef]
26. Skolnik, M. *Radar Handbook*, 3rd ed.; McGraw-Hill Education: New York, NY, USA, 2008.
27. Tait, P. *Introduction to Radar Target Recognition*; Institution of Electrical Engineers: London, UK, 2006; ISBN 9781849190831.

*Article*

# ARSD: An Adaptive Region Selection Object Detection Framework for UAV Images

Yuzhuang Wan [1], Yi Zhong [1], Yan Huang [2], Yi Han [1,*], Yongqiang Cui [3], Qi Yang [4], Zhuo Li [5], Zhenhui Yuan [6] and Qing Li [7]

[1] School of Information Engineering, Wuhan University of Technology, Wuhan 430070, China
[2] Zmvision Technology, Wuhan 430070, China
[3] College of Electronics and Information, South-Central Minzu University, Wuhan 430074, China
[4] School of Mathematics & Statistics, South-Central Minzu University, Wuhan 430074, China
[5] SAIC GM Wuling Automobile Co., Ltd., Liuzhou 545007, China
[6] Department of Computer and Information Science, Northumbria University, Newcastle upon Tyne NE1 8ST, UK
[7] Peng Cheng Laboratory, Shenzhen 518066, China
* Correspondence: hanyi@whut.edu.cn; Tel.: +86-27-87858005

**Abstract:** Due to the rapid development of deep learning, the performance of object detection has greatly improved. However, object detection in high-resolution Unmanned Aerial Vehicles images remains a challenging problem for three main reasons: (1) the objects in aerial images have different scales and are usually small; (2) the images are high-resolution but state-of-the-art object detection networks are of a fixed size; (3) the objects are not evenly distributed in aerial images. To this end, we propose a two-stage Adaptive Region Selection Detection framework in this paper. An Overall Region Detection Network is first applied to coarsely localize the object. A fixed points density-based targets clustering algorithm and an adaptive selection algorithm are then designed to select object-dense sub-regions. The object-dense sub-regions are sent to a Key Regions Detection Network where results are fused with the results at the first stage. Extensive experiments and comprehensive evaluations on the VisDrone2021-DET benchmark datasets demonstrate the effectiveness and adaptiveness of the proposed framework. Experimental results show that the proposed framework outperforms, in terms of mean average precision (mAP), the existing baseline methods by 2.1% without additional time consumption.

**Keywords:** UAV; object detection; deep learning; adaptive cluster

**Citation:** Wan, Y.; Zhong, Y.; Huang, Y.; Han, Y.; Cui, Y.; Yang, Q.; Li, Z.; Yuan, Z.; Li, Q. ARSD: An Adaptive Region Selection Object Detection Framework for UAV Images. *Drones* **2022**, *6*, 228. https://doi.org/10.3390/drones6090228

Academic Editors: Daobo Wang and Zain Anwar Ali

Received: 30 July 2022
Accepted: 28 August 2022
Published: 31 August 2022

**Publisher's Note:** MDPI stays neutral with regard to jurisdictional claims in published maps and institutional affiliations.

## 1. Introduction

Nowadays, as a fast-growing number of Unmanned Aerial Vehicles (UAVs) start carrying high-definition cameras, object detection technology in aerial images has been widely used in various practical applications, including agricultural planting [1,2], pedestrian tracking [3], urban security [4,5], inspecting buildings [6], search and rescue [7], and rare plant monitoring [8]. These applications all require accurate object detection in visible or infrared images taken by onboard cameras. However, detecting objects in UAV images is nontrivial. Varying purposes in different applications and the limited computing power of UAVs have brought challenges to this work. To solve these problems, object detection based on a Convolutional Neural Network (CNN) is gradually applied in UAV detection tasks.

The methods of object detection commonly used today are YOLO series [9–12] and Faster-RCNN [13]. They have achieved a good performance on large-scale datasets such as MS COCO [14], ImageNet [15], and VOC2007/2012 [16]. However, compared to these datasets, UAV images have the following features:

(1) UAV image datasets often provide higher resolution images, but the objects in these images are always in low resolution. For example, the image size in general image datasets VOC2007/2012 and MS COCO is approximately $500 \times 400$ and $600 \times 400$,

respectively. However, in the UAV image dataset VisDrone2021-DET [17], the image size is 2000 × 1500 while the object size is only about 50 × 50 pixels.

(2) The size of the objects depends on the altitude at which the drone takes the image. The higher the drone is, the smaller the object is in the images [18].

(3) The targets are not evenly distributed. Some regions in an image are plain backgrounds, while other regions are mostly occupied by objects.

To solve these issues, many researchers have attempted to change the structure of the object detection network. Extended from YOLOv5 [19], an improved network named YOLOv5-TPH [20] added a transformer model with attention mechanisms on the detection head of YOLOv5. It trained a 1536 × 1536 high-resolution network on VisDrone2021-DET and achieved 35.74% mAP. However, the high-resolution network and transformer model cost huge computing resources.

Another common solution is to partition a UAV image into several uniform subregions and then detect each of them. However, it cannot guarantee effective improvement by directly conducting uniform cropping or random cropping as this cannot locate key subregions. Based on the VisDrone2021-DET dataset, the system can achieve 43.5% mAP50 and 30.3% mAP when the sub-regions are obtained via sliding window search [21] and CNN-based clustered sub-network [22], respectively. However, they are either time consuming or training based. Although these detectors could achieve a better performance, they are inefficient at performing detection in every region. Because some regions only have large-scale objects, detecting these regions does not improve the overall accuracy and efficiency. The object-dense sub-regions need to be located first.

This paper proposes the Adaptive Region Selection Detection framework (ARSD), a novel two-stage detection model that combines the Region Detection Network and the Self-adaptive Intensive Region Selecting Algorithm. ARSD aims to significantly reduce computation resource consumption while maintaining high object detection accuracy. The first stage of ARSD uses an Overall Region Detection Network, which can coarsely locate where the target is. The model then applies a Self-adaptive Intensive Region Selecting Algorithm to generate object-dense sub-regions by cluster objects detected in the first stage and sends them to the next stage for further exquisite detection. The last stage is the Key Region Detection Network, which detects the object-dense sub-regions. Based on the first stage, this stage is extended with an additional small detection head based on the original detection heads.

To sum up, the novelty of this paper is as follows:

(1) An effective and efficient object detection framework is proposed to adaptively crop high-resolution UAV images according to object density based on clustering algorithms. This can significantly reduce the training and processing time of the UAV images.

(2) This paper proposes the Self-adaptive Intensive Region Selecting Algorithm to select the object-dense region in UAV images. It reduces the number of sub-regions for further object detection. This enables the framework to be more suitable for the limited UAV hardware computing power.

(3) This paper also proposes that an additional detection head is added to deal with the varying object sizes in UAV images. This helps the framework detect small objects more easily and increases detection accuracy.

In this way, the proposed framework can gradually reduce the computational complexity while maintaining high object detection accuracy.

The rest of the paper is organized as follows: Section 2 provides a comprehensive overview of the components of ARSD. The specific experimental details are given in Section 3, which demonstrates the performance of the proposed ARSD in various aspects, as well as comparisons with other works. Section 4 summarizes the experimental results. Finally, Section 5 concludes the paper and lists a collection of ongoing research and future work directions.

## 2. Materials and Methods

This section first introduces the overall framework of the proposed ARSD framework and then describes each module in detail. As shown in Figure 1, the ARSD framework consists of three parts. The first part is the Overall Region Detection Network (ORDN), which is used to roughly locate the objects. The Self-adaptive Intensive Region Selecting Algorithm (SIRSA), which consists of the Fixed Points Density-based Clustering Algorithm (FPDCA) and Adaptive Sub-regions Selection Algorithm (ASSA), is then adopted to properly select the object-dense sub-regions. Finally, the Key Region Detection Network (KRDN) is responsible for detecting the objects in sub-regions selected by SIRSA. The detected objects are combined with the results of ORDN. This framework has better detection accuracy for small targets in UAV images and reduces computing resources due to the filter of the sub-regions.

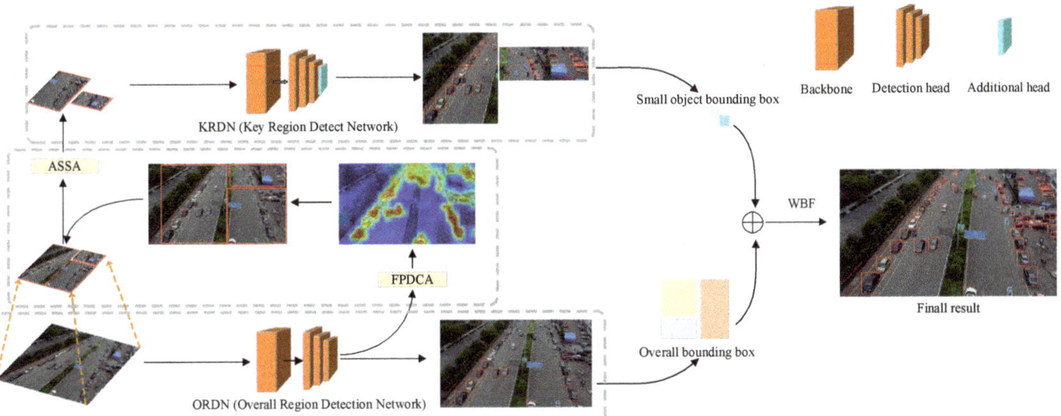


**Figure 1.** The pipeline of ARSD. The ORDN predicts overall bounding boxes from the original image, which is used for the subsequent region selection. FPDCA can cluster the center point of the overall bounding boxes to obtain candidate sub-regions. ASSA then filters sub-regions that needed to be detected in KRDN. The light blue box represents the additional head of KRDN. Finally, the small object bounding boxes obtained in KRDN are merged with the overall bounding boxes by Weighted Boxes Fusion (WBF) [23] and generate the final result.

### 2.1. Overall Region Detection Network (ORDN) and Key Region Detection Network (KRDN)

The ORDN module predicts object bounding boxes on the whole images, while KRDN predicts the object based on cropped object-dense sub-regions. After cropping sub-regions with SIRSA, KRDN can predict more accurate results in a high-resolution region. The results of ORDN and KRDN are merged by WBF and produce the final results.

It should be mentioned that although ORDN and KRDN are based on the same backbone structure, they differ in width and depth. Therefore, we can easily and inexpensively implement two detection networks with different precision and different time consumption. ORDN only needs to have an accurate recall rate, so it is designed to be more lightweight to save computing power. On the contrary, KRDN is wider and deeper than ORDN. This enables KRDN to have greater accuracy than ORDN without changing the network structure.

For an anchor-based object detection network, anchor size is an important factor affecting the accuracy. The targets in UAV images are on different scales and most are small-scale targets. An additional detection head is added to KRDN based on the original detection head in ORDN. Combined with the original detection head, KRDN can easily locate small objects and reduce the adverse influence caused by different object scales. The added detection head is generated from the low-level feature map of the backbone and the high-resolution feature map obtained by up-sampling in FPN [24]. Generally, feature information will be reduced as the size of the feature map decreases during the processing of CNN in the backbone. Since the image feature information does not decline gradually in the large-size feature map and more features of small targets are retained, this detection head can help to effectively detect small targets.

The object detection is performed by KRDN after object-dense sub-regions are clustered and selected by SIRSA. KRDN detects the object-dense region and obtains more details about the objects. This paper combines the results from two stages using WBF [23]. $B_o$ and $B_k$ represent the coordinates of the bounding boxes obtained by ORDN and KRDN, respectively. The final result of the bounding box $B$ is calculated by (1). The final confidence $C$ is obtained by (2), where $C_o$ and $C_k$ represent the confidence values of the bounding boxes obtained by ORDN and KRDN, respectively.

$$B_{(x,y)} = \frac{B_{o(x,y)} * C_o + B_{k(x,y)} * C_k}{2} \tag{1}$$

$$C = \frac{C_o + C_k}{2} \tag{2}$$

*2.2. Self-Adaptive Intensive Region Selecting Algorithm (SIRSA)*

SIRSA is composed of two parts: (1) the Fixed Points Density-based Clustering Algorithm (FPDCA) is used to cluster the center point of the object predicted in ORDN to obtain candidate sub-regions; (2) the adaptive region selection algorithm selects object-dense sub-regions.

### 2.2.1. Fixed Points Density-Based Clustering Algorithm (FPDCA)

This paper proposes a new clustering method named FPDCA, combining the advantages of two clustering algorithms: K-means [25] and Mean-Shift [26]. K-means is competent at generating fixed clustering centers, but it cannot use density information. Mean-Shift cannot confirm the number of clustering regions, which tends to generate a large number of small-size sub-regions or only one large sub-region. It consumes intensive computation power or cannot be applied to higher resolution images. However, Mean-Shift can perform region clustering based on object-dense information, which is neglected by K-means. Combining K-means and Mean-Shift can take both the density information and generating the fixed number of sub-regions into account.

As shown in Algorithm 1, the set of centers of the bounding box is denoted as $Q$, with a point $q$ in $Q$ randomly chosen as the initial cluster center. Point $q$ is updated by calculating the motion vector based on the points in the surrounding circle of radius $r$. The algorithm moves to the next Point $q$ and repeats the above step until all points in $Q$ have been processed. If the number of cluster centers is greater than the required number of sub-regions $N$, K-Means is used to regroup these clusters into $N$ clusters and then output the $N$ final candidate sub-regions. If the number of clusters does not exceed $N$, these clusters are discarded and K-Means are used directly based on the original $Q$ points to regroup to $N$ clusters and the corresponding $N$ final sub-regions.

### 2.2.2. Adaptive Sub-Regions Selection Algorithm (ASSA)

A large number of candidate sub-regions are produced by FPDCA. To save computing resources on hardware platforms, this paper proposes the Adaptive Sub-regions Selection Algorithm (ASSA) to filter sub-regions that needed to be detected in KRDN. Sub-regions with higher ASSA scores will be selected for further detection by KRDN, while the other sub-regions will be discarded.

Four criteria are defined in ASSA to evaluate whether a candidate sub-region $p$ requires further detection by KRDN, including regional target density, average confidence, the ratio of bounding boxes' total areas to the sub-region area, and average area of all bounding boxes in the sub-region. Regional target density is considered in [27], as defined in (3), where $L$ denotes the number of the predicted boxes in $p$ and $A$ is the area of $p$. However, this definition may not necessarily be accurate in that the size of the original image should also be considered. As shown in Figure 2, when reducing the image size from $200 \times 200$

(left image) to $100 \times 100$ (right image), the regional target density should remain the same. However, the result does not conform to this using (3). This paper extends (3) by considering the size of the original image $S$ in (4).

$$D = \frac{L^2}{A} \tag{3}$$

$$D = \frac{L^2 * S}{A} \tag{4}$$

---

**Algorithm 1:** Fixed Points Density-based Clustering Algorithm

---

**Input:**  $N$: number of sub-regions,
$Q$: the set of bounding box centers,
$r$: distance of algorithm,
$\xi$: threshold of the distance of vector.
**Output:** the set of sub-regions $S$
1: **for** $q(x, y)$ *in* $Q$ *but not in* $M$
2:    $M = M \cup q(x, y)$
3:    **for** $q(x_i, y_i)$ *in* $Q$ *but not in* $M$
4:       $S_k(q) = \{y : (x - x_i)^2 + (y - y_i)^2 < r^2\}$
5:       $M = M \cup S_k(q)$
6:       $C_i = C_i \cup S_k(q)$
7:       $V_{shift} = \frac{1}{k} \sum_{q \in S_k} (q_i - q)$
8:       $q = q + V_{shift}$
9:          **if** $\left| V_{shiftnew} - V_{shiftold} \right| < \xi$
*put $C_i$ in C*
10:                **break from line 3;**
11:             **end if**
12:       **end for**
13: **end for**
14: **if** $length(C) > N$
15:    $S = Kmeans(N, C)$
16: **else**
17:    $S = Kmeans(N, Q)$ *as random center*
18: **end if**

---

**Figure 2.** The left and right images should have the same target density.

The second criterion is the average confidence $M$ as defined in (5), where $score_i$ represents the confidence of bounding box $i$. It is considered that the sub-region with low average confidence is due to the classification inaccuracy caused by the lightweight network of ORDN. It is believed that the smaller the value of $M$, the greater the accuracy gain.

$$M = \frac{\sum_{i=1}^{L} score_i}{L} \tag{5}$$

ASSA then considers the ratio of the sum area of all bounding boxes to the sub-region area $R$ as defined in (6). The larger the value of $R$, the less background in this sub-region and the larger the area that contains objects. In addition, it can also reflect the overlapping degree of the objects to a certain extent, e.g., $R > 1$ means that there are overlaps of the object's bounding boxes in this sub-region. Larger $R$ values indicate that the detected object is big or that many objects are detected within the sub-region.

$$R = \frac{\sum_{i=1}^{L} area_i}{S} \tag{6}$$

Finally, the ratio of the sum area of all bounding boxes to the number of bounding boxes is defined in (7). It reflects the average size of the objects in the sub-region.

$$E = \frac{\sum_{i=1}^{L} area_i}{L} \tag{7}$$

As defined in (8), a final score $s_i$ is computed for each sub-region by adding the above four indicators, each with a corresponding weight calculated using information entropy; $w_j$ is the weight of each indicator and $p_{ij}$ is the indicator $j$ of sub-region $i$. The sub-regions with high scores are identified as object-dense sub-regions and sent to the KRDN stage.

$$s_i = \sum_{j=1}^{m} w_j \cdot p_{ij} \tag{8}$$

The weight of each indicator $w_j$ is calculated by information entropy $e_j$ as defined in (9); $w_j$ is then computed by (10) and (11).

$$e_j = -k \sum_{i=1}^{n} p_{ij} \ln(p_{ij}), \ k = 1/\ln(n) > 0 \ e_j \geq 0 \tag{9}$$

$$d_j = 1 - e_j \tag{10}$$

$$w_j = \frac{d_j}{\sum_{j=1}^{m} d_j} \tag{11}$$

Each final indicator $j$ of sub-region $i$, denoted as $p_{ij}$, is calculated by (14) after normalizing the origin indicators using (12) and (13). The higher the regional target density, the more likely that it should be sent to KRDN and the higher the final score. These indicators are defined as extremely large indicators. On the contrary, the smaller the average confidence, the more likely it should be sent to KRDN. These indicators are defined as extremely small indicators. To standardize and normalize these indicators, extremely small indicators are converted by (12) to extremely large indicators and extremely large indicators are normalized by (13).

$$x'_{ij} = \frac{x_{ij} - \min\{x_{ij}, \ldots, x_{nj}\}}{\max\{x_{ij}, \ldots, x_{nj}\} - \min\{x_{ij}, \ldots, x_{nj}\}} \tag{12}$$

$$x'_{ij} = \frac{\max\{x_{ij}, \ldots, x_{nj}\} - x_{ij}}{\max\{x_{ij}, \ldots, x_{nj}\} - \min\{x_{ij}, \ldots, x_{nj}\}} \tag{13}$$

$$p_{ij} = \frac{x_{ij}}{\sum_{i=1}^{n} x_{ij}}, \ i = 1, \ldots, n, j = 1, \ldots, m \tag{14}$$

The results of SIRSA are shown in Figure 3. A white rectangle indicates candidate sub-regions, while the transparency of sub-regions indicates their ASSA score. The clearer the sub-region, the higher its final ASSA score and the more it should be sent to KRDN.

**Figure 3.** Points of the same color indicate the center of the bounding box which belongs to the same sub-region. The obscurer the sub-regions, the lower the ASSA score.

## 3. Results

### 3.1. Datasets and Evaluation Metrics

***Datasets.*** The proposed approach is evaluated on the VisDrone2021-DET dataset. The VisDrone dataset is collected by drones in 14 different cities of China, at different heights and in different weather/light conditions. It contains a total of 10,209 images, which consists of 6471 training images, 548 validation images, and 3190 testing images. The objects in this dataset are mostly small and often clustered together. However, the dataset provides a high image resolution of approximately 2000 × 1500 pixels. Images are annotated with bounding boxes, including 10 predefined categories (pedestrian, person, car, van, bus, truck, motor, bicycle, awning-tricycle, and tricycle).

***Evaluation Metric.*** The proposed method is evaluated using the same evaluation protocol as described in MS-COCO, which has 12 evaluation indicators for object detection. AP, AP50, and AP75 are selected as the criteria for state-of-the-art comparison. AP is used to calculate the average index of all categories, and it generally defaulted to mAP. AP50 and AP75 represent indicators with threshold IOUs of 0.5 and 0.75, respectively. This paper also adopted the widely used precision–recall (PR) curves as our evaluation metric. The method of [28] is used to calculate the criteria proposed above.

***Implementation Details.*** The proposed methods are implemented using PyTorch. The backbone is pretrained on MS COCO. The following training of the model is completed on a server with one NVIDIA GeForce GTX 3080Ti GPU. The experiment uses Adam [29] to train ORDN and KRDN for 150 epochs, with the learning rate starting at 0.001.

### 3.2. Model Scaling Scheme for Two-Stage Framework

YOLOv5 has five networks of different sizes, namely YOLOv5x, YOLOv5l, YOLOv5m, YOLOv5s, and up-to-date YOLOv5n. Since it is the most notable and convenient one-stage detector with a flexible structure, the experiment in this paper chooses YOLOv5 as the basic model for ORDN and KRDN.

We test YOLOv5n and YOLOv5m in different sizes in Section 3.5. The experiment demonstrates that the large-size YOLOv5n is more accurate than the small-size YOLOv5m within the same execution time. Because ORDN is required to locate the bounding box in time, we balance time consumption and accuracy, properly setting YOLOv5n as the basic model of ORDN. Considering the limited performance of the UAV hardware platform, YOLOv5X is unsuitable at this stage due to its high time consumption, despite having the highest performance. YOLOv5m is balanced in its performance and computational requirement, and is therefore adopted as the KRDN base model. Finally, the input size of ORDN is set to 736 × 736 and the sub-regions are resized to 384 × 384 before being fed into KRDN.

### 3.3. Qualitative Result

To demonstrate the effectiveness of the proposed model ARSD, extensive experiments have been conducted to compare its results with the baseline method (YOLOv5) on the VisDrone2021 dataset, as shown in Figure 4. It can be noted that ARSD performs better in small-scale object detection, especially in object-dense sub-regions. In contrast, YOLOv5 misses many small-scale objects in the object-dense region. According to the partially enlarged image, there is almost no missing small-scale objects such as pedestrians. This is mainly because the object-dense sub-regions are on a small scale and do not need to resize dramatically before being sent to KRDN; thus, more features remain for more accurate detection.

**Figure 4.** *Cont.*

**Figure 4.** The qualitative comparisons among the detection results from baseline (**b,d**) and our method (**a,c**) on the validation set of the VisDrone2021-DET dataset.

Figure 5 shows the cluster results of the proposed FPDCA and Mean-Shift. The clustering number of Mean-Shift is dynamic; sometimes it has only one cluster. If this one sub-region is sent to KRDN, it cannot achieve any improvement in accuracy. In addition, it sometimes has too many clusters which will result in cost-intensive computing resources.

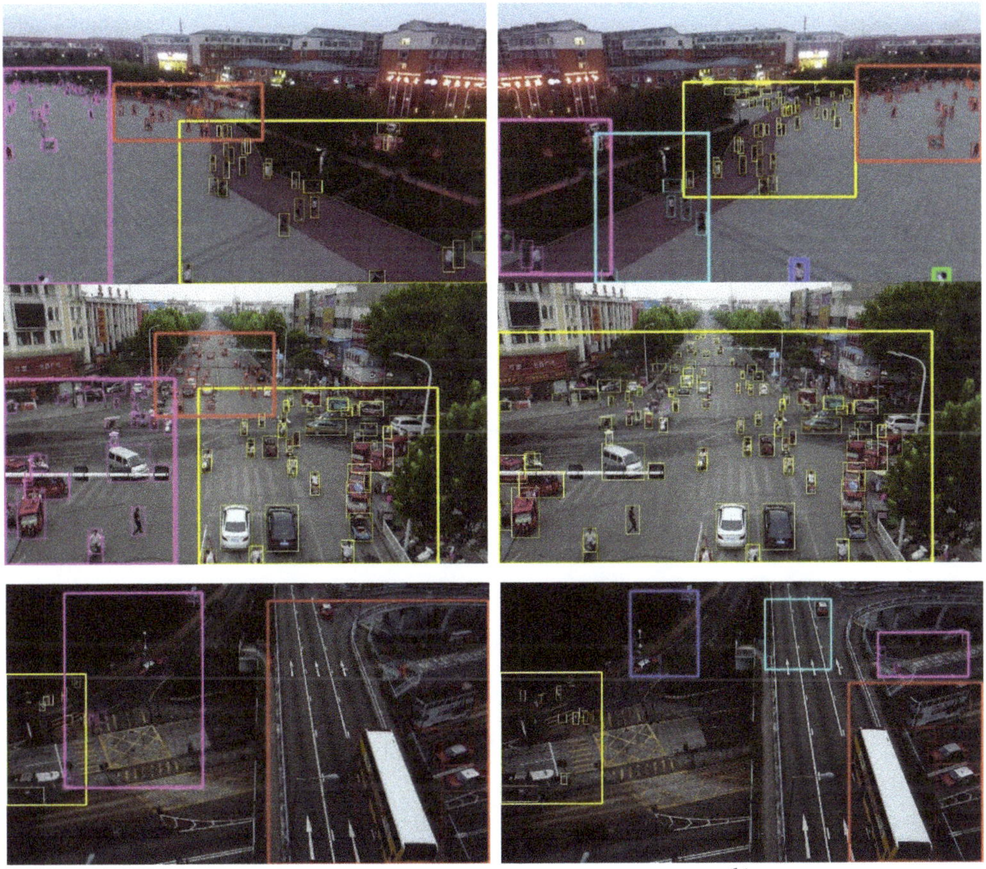

(a) (b)

**Figure 5.** The qualitative comparisons among the clusters result from different cluster methods on the test set of the VisDrone2021-DET dataset. For each testing image, we show the cluster results of the proposed method (**a**) and Mean-Shift (**b**).

### 3.4. Quantitative Evaluation

Since the traditional one-stage methods are easier to reproduce, we compare our method with them in each category; however, the comparison with state-of-the-art methods comes from their literature. Quantitative comparisons are conducted in two aspects using the VisDrone2021-DET dataset: (1) the detection accuracy of each category comparisons against different one-stage methods; (2) the detection accuracy and accuracy improvement comparisons against state-of-the-art two-stage framework methods.

Table 1 lists AP50 values for ARSD and other one-stage methods in the detection of 10 different categories of objects. The results of the first five methods are derived from [30], with the remaining results obtained from our conducted experiments. The proposed method, ARSD, consists of YOLOv5n + YOLOv5mAH and performs much better than the other methods in each category, especially in small-scale categories such as Pedestrian and People.

**Table 1.** AP50 comparison of each class among ARSD and other one-stage methods.

| Method | Pedestrian | People | Bicycle | Car | Van | Truck | Tricycle | Awning-Tricycle | Bus | Motor | Average AP50 |
|---|---|---|---|---|---|---|---|---|---|---|---|
| YOLOV3 | 18.1 | 9.9 | 2 | 56.6 | 17.5 | 17.6 | 6.7 | 2.9 | 32.4 | 17 | 17.1 |
| SlimYOLOv3 [31] | 17.4 | 9.3 | 2.4 | 55.7 | 18.3 | 16.9 | 9.1 | 3 | 26.9 | 17 | 17.6 |
| Faster-RCNN | 21.7 | 12.7 | 11.5 | 63.2 | 37.8 | 29.9 | 22.5 | 12.3 | 50.6 | 28.4 | 29.1 |
| FPN | 33 | 25.8 | 13.9 | 69.4 | 40 | 34.3 | 27.4 | 13.4 | 49.1 | 37.6 | 35.6 |
| YOLOv5m | 45.2 | 35.7 | 13.7 | 77.8 | 41 | 37.9 | 20.7 | 10.8 | 50.9 | 30.1 | 36.8 |
| YOLOv5l-TPH | 53.5 | 29.7 | 25.9 | 87 | 55.3 | 61.5 | 34.9 | 31.2 | 73.5 | 50.6 | 50.3 |
| **ARSD** | **68.8** | **56.8** | **40.68** | **88.17** | **61.53** | **53.74** | **49** | **26.19** | **72.78** | **61.3** | **57.9** |

As shown in Table 2, AP $\nearrow$, AP50 $\nearrow$, and AP75 $\nearrow$ highlight the outperformance of ARSD over other previous state-of-the-art detectors. Compared with GLASN and UCGNet, ARSD improves by 2.1% and 4.8% (AP50).

**Table 2.** Comparison with state-of-the-art object detection methods in the VisDrone2021-DET validation set.

| Method | AP | AP50 | AP75 | AP $\nearrow$ | AP50 $\nearrow$ | AP75 $\nearrow$ |
|---|---|---|---|---|---|---|
| ClusDet [32] | 26.7 | 50.6 | 24.7 | 8.34 | 7.3 | 11.91 |
| DMNet [33] | 28.2 | 47.6 | 28.9 | 6.84 | 10.3 | 7.71 |
| UCGNet [34] | 32.8 | 53.1 | 33.9 | 2.24 | 4.8 | 2.71 |
| GLASN [31] | 32.5 | 55.8 | 33 | 2.54 | 2.1 | 3.61 |
| **ARSD** | **35.04** | **57.9** | **36.61** | - | - | - |

### 3.5. Ablation Study

To validate the effectiveness of the additional detection head, different cluster methods, and a different number of sub-regions in detection tasks, this paper conducted extensive experiments on VisDrone2021-DET test-dev.

(1) **Effect of the large-scale and lightweight network.** The results are from different scales YOLOv5n and YOLOv5m, trained on VisDrone2021-DET, as shown in Figure 6. Within the same computation time, YOLOv5n performs better than YOLOv5m. Therefore, we choose the large-scale YOLOv5n as the base model of ORDN.

(2) **Effect of additional prediction head.** Though experiments show that adding a detection head for small-scale objects makes the GFLOPs increase from 48.1 to 59.1, the performance of an additional detection head is prominent. The experiment is to increase the size of the network to compare different accuracy under different time consumption. As shown in Figure 6, YOLOv5m-AH is YOLOv5m with an additional detection head. The mAP of the net with an additional detection head (blue line, YOLOv5m-AH) is 1.5% higher than without an additional head (green line, YOLOv5m) in the same processing time (3.0 ms). It not only saves the computing power of the hardware, but also improves the mAP considerably in each category.

(3)  **Effect of two-stage structure**. To demonstrate the effect of an additional detection head and two-stage framework, this paper chooses three networks with the same computation time: $768 \times 768$ YOLOv5m, $640 \times 640$ YOLOv5m-AH, and our two-stage framework ($768 \times 768$ YOLOv5n + $384 \times 384$ YOLOv5m-AH). Based on the results shown in Figure 7, our two-stage framework is more accurate in small-scale categories such as Pedestrian, People, and Bicycle. However, results in large-scale categories such as bus and truck are not as good as the one-stage structure. The reason for this phenomenon is that the increase is caused by the additional small object true positives predicted from sub-regions and the decrease is caused by the false positives predicted from sub-regions that match large ground truth boxes.

(4)  **Effect of FPDCA**. The proposed method performs well on clustering, as indicated by Figure 5. In addition, as shown by lines 1, 2, and 6 in Table 3, when using FPDCA as the SIRSA basic cluster method, the result outperforms K-means and Mean-Shift by 0.7% and 1.2%, respectively, in AP50. This shows that it is useful to consider density information when obtaining clusters. The object detection accuracy indicated by AP and AP50 increases as the number of clusters increases from 2 to 4. However, the gain is subtle while the computational complexity also increases. For example, lines 6 and 9 in Table 3 show that AP and AP50 only improves by 0.97% and 0.6%, respectively, when the number of clusters increases from 3 to 4.

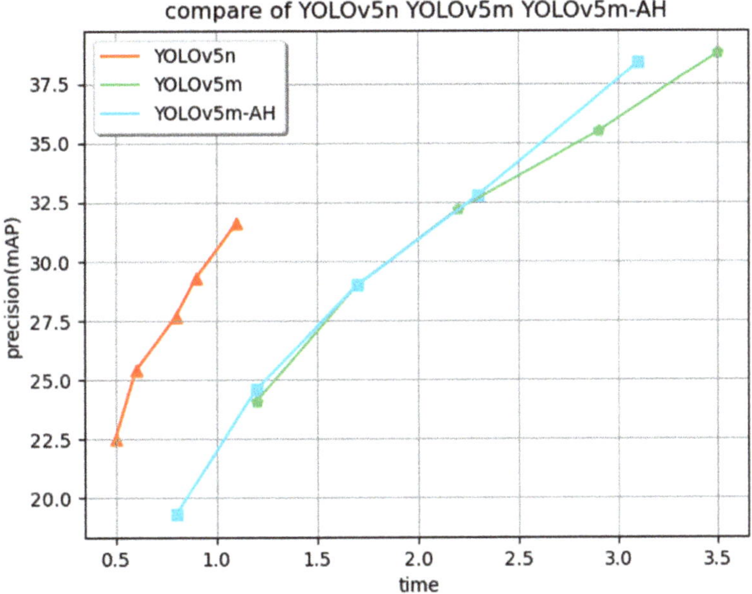


**Figure 6.** The comparison between the mAP of YOLOv5n, YOLOv5m, and YOLOv5m-AH for the same time consumption.

To improve processing speed in terms of FPS (frames per second), ASSA is used to discard the sub-regions by 1/3, 1/2, and 0 from the original set of sub-regions. Considering the balance of computing resources and accuracy, ASSA chooses to discard 1/3 of the candidate sub-regions in the inference phase. The processing time is reduced by about a third after discarding 1/3 of sub-regions.

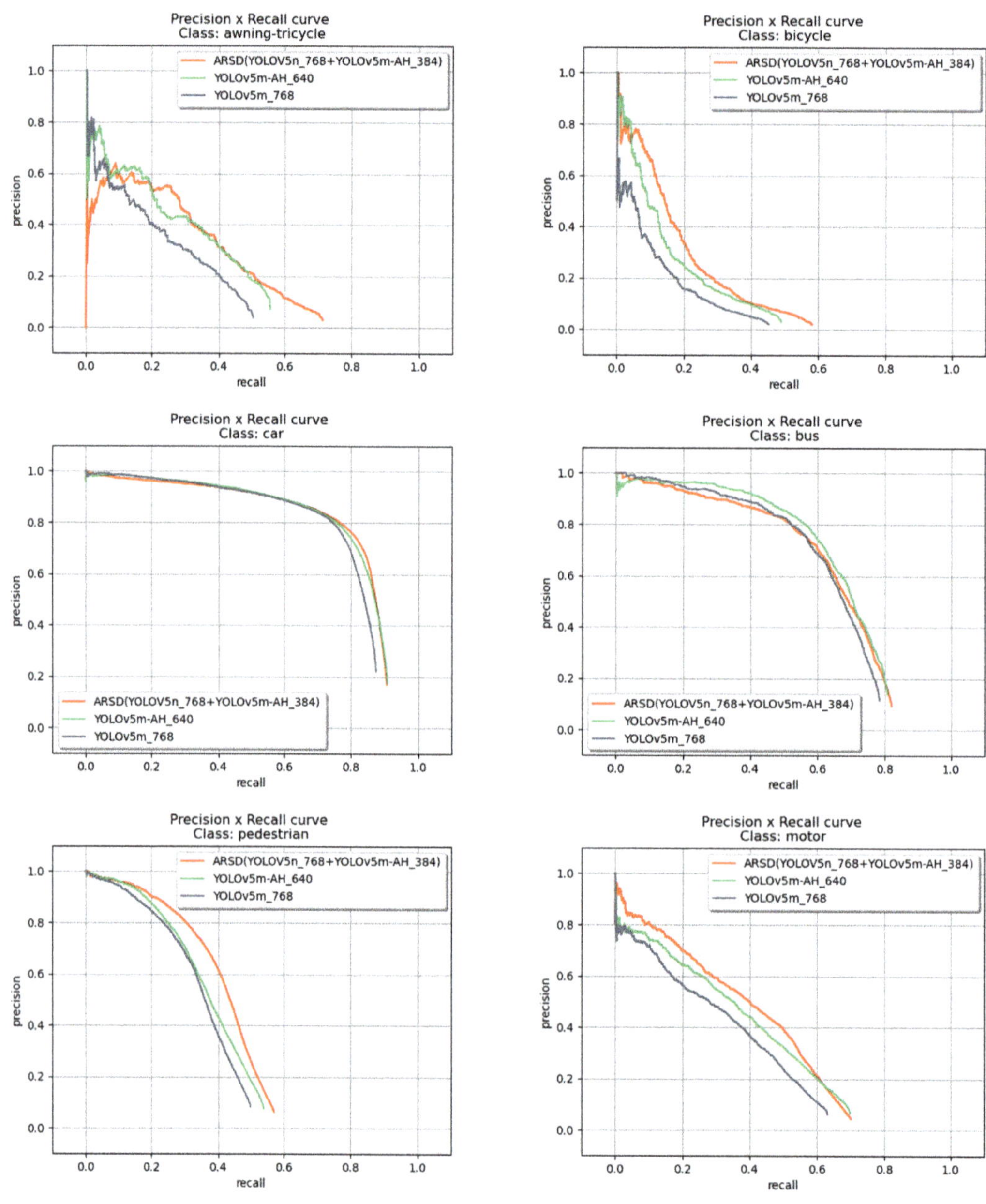

**Figure 7.** *Cont.*

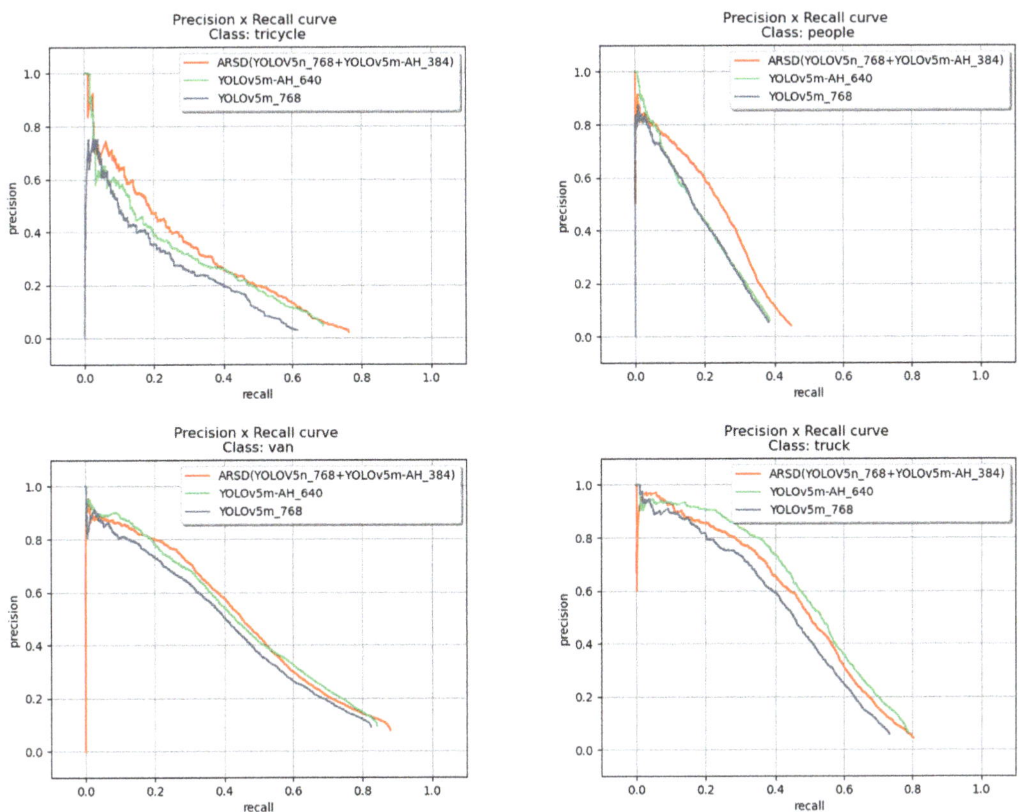

**Figure 7.** The comparison of 768 × 768 YOLOv5m, 640 × 640 YOLOv5m-AH, and our method for each category.

**Table 3.** Results with different cluster methods, clusters, and remaining sub-regions. Note: (x) indicates the ratio of sub-regions being discarded.

| | Methods | Cluster Methods | Number of Clusters | Candidate Sub-Regions | Remaining Sub-Regions | AP | AP50 |
|---|---|---|---|---|---|---|---|
| 1 | K-means | K-means | 3 | 4830 | 4830 | 22.64 | 40.6 |
| 2 | DNSCAN | Mean-Shift | - | 5559 | 5559 | 22.31 | 40.1 |
| 3 | ARSD | FPDCA | 2 | 3220 | 3220 (0) | 20.59 | 37.76 |
| 4 | ARSD | FPDCA | 2 | 3220 | 2146 (1/3) | 19.16 | 35.63 |
| 5 | ARSD | FPDCA | 2 | 3220 | 1610 (1/2) | 18.33 | 33.4 |
| 6 | ARSD | FPDCA | 3 | 4830 | 4830 (0) | 22.93 | 41.31 |
| 7 | ARSD | FPDCA | 3 | 4830 | 3220 (1/3) | 21.6 | 39.6 |
| 8 | ARSD | FPDCA | 3 | 4830 | 2415 (1/2) | 20.05 | 37.1 |
| 9 | ARSD | FPDCA | 4 | 6440 | 6440 (0) | 23.9 | 41.92 |
| 10 | ARSD | FPDCA | 4 | 6440 | 4293 (1/3) | 22.32 | 40.15 |
| 11 | ARSD | FPDCA | 4 | 6440 | 3220 (1/2) | 21.12 | 37.86 |

## 4. Discussion

The aim is to improve the object detection accuracy for UAV images in high resolution with a lot of small targets in the images. This paper proposes a two-stage object detection framework for UAV images, with extensive performance evaluations conducted. The qualitative results in Section 3.3 and quantitative evaluations in Section 3.4 directly show that the proposed framework performs better than the state-of-the-art detection methods. Compared with GLASN, it improves AP by 2.54% and AP50 by 2.1%. The ablation study includes four experiments. The first experiment performs tests on different network scales and the results demonstrate that YOLOv5n is the most suitable for ORDN. The second experiment shows that the network with an additional head performs better than others within the same processing time. The third experiment proves that this two-stage framework is more accurate in small-scale categories such as Pedestrians, People, and Bicycle. The final experiment analyses the influence of clustering algorithms and the number of clusters. The novelty mentioned in Section 1 has been verified by the above four experiments. The framework can improve the accuracy of object detection within the same processing time, especially in small-scale objects.

This method should balance the number of clusters and the ratio of the remaining sub-regions among all candidate sub-regions. If a higher number of clusters is set, the calculation time will undoubtedly increase, but the accuracy will also be improved. Our future work includes the study of selecting these parameters more precisely.

### 5. Conclusions

This paper proposes ARSD, an adaptive region selection detection framework, that is more efficient and more effective for UAV image object detection. The main idea is to locate object-dense sub-regions in high-resolution UAV images and feed these sub-regions into the second-stage detector. Finally, the results are merged by WBF. In addition, we developed an adaptive region selection algorithm which consists of the Fixed Points Density-based Clustering Algorithm and the Adaptive Sub-regions Selection Algorithm. FPDCA can generate a fixed number of sub-regions combined with density information, while ASSA can score each sub-region objectively. Additionally, a new detection head is added based on the original detection heads for better small object detection. The evaluations on VisDrone2021-DET datasets have demonstrated the effectiveness and adaptiveness of ARSD. We will extend this work to apply the proposed method to small-scale object detection in remote sensing and search and rescue.

**Author Contributions:** Conceptualization, Y.W., Y.H. (Yan Huang) and Y.H. (Yi Han); methodology, Y.W. and Y.Z.; software, Y.W., Y.H. (Yan Huang) and Q.Y.; validation, Y.W. and Y.H. (Yan Huang); formal analysis, Y.W., Y.Z. and Y.H. (Yan Huang); investigation, Y.W., Y.C., Z.L., Z.Y. and Q.L.; resources, Y.W. and Y.H. (Yan Huang); data curation, Y.W., Y.H. (Yi Han) and Z.L.; writing—original draft preparation, Y.W. and Y.H. (Yi Han); writing—review and editing, Y.W., Y.H. (Yi Han), Y.C., Z.L., Z.Y. and Q.L.; visualization, Y.W. and Q.Y.; supervision, Y.H. (Yi Han) and Z.Y.; project administration, Y.Z.; funding acquisition, Y.Z. All authors have read and agreed to the published version of the manuscript.

**Funding:** This work was supported by a grant from the National Natural Science Foundation of China (Grant No. 61801341). This work was also supported by the Research Project of Wuhan University of Technology Chongqing Research Institute (No. YF2021-06).

**Institutional Review Board Statement:** Not applicable.

**Informed Consent Statement:** Not applicable.

**Data Availability Statement:** The simulation data used to support the findings of this study are available from the corresponding author upon request. The video of the experimental work can be found at the following link: https://drive.google.com/drive/folders/1eYFxNSaYYEhY_M0tdM5 5oybCA6sMMgzG?usp=sharing (accessed on 25 August 2022).

**Conflicts of Interest:** The authors declare no conflict of interest.

### References

1. Hird, J.N.; Montaghi, A.; McDermid, G.J.; Kariyeva, J.; Moorman, B.J.; Nielsen, S.E.; McIntosh, A.C.S. Use of Unmanned Aerial Vehicles for Monitoring Recovery of Forest Vegetation on Petroleum Well Sites. *Remote Sens.* **2017**, *9*, 413. [CrossRef]
2. Shao, Z.; Li, C.; Li, D.; Altan, O.; Zhang, L.; Ding, L. An Accurate Matching Method for Projecting Vector Data Into Surveillance Video To Monitor And Protect Cultivated Land. *ISPRS Int. J. Geo-Inf.* **2020**, *9*, 448. [CrossRef]

3. Shen, Q.; Jiang, L.; Xiong, H. Person Tracking and Frontal Face Capture with UAV. In Proceedings of the IEEE 18th International Conference on Communication Technology (ICCT), Chongqing, China, 8–11 October 2018; pp. 1412–1416.
4. Audebert, N.; Le Saux, B.; Lefèvre, S. Beyond Rgb: Very High Resolution Urban Remote Sensing with Multimodal Deep Networks. *ISPRS J. Photogramm. Remote Sens.* **2018**, *140*, 20–32. [CrossRef]
5. Yuan, Z.; Jin, J.; Chen, J.; Sun, L.; Muntean, G.M. ComProSe: Shaping Future Public Safety Communities with ProSe-based UAVs. *IEEE Commun. Mag.* **2017**, *55*, 165–171. [CrossRef]
6. Munawar, H.S.; Ullah, F.; Heravi, A.; Thaheem, M.J.; Maqsoom, A. Inspecting Buildings Using Drones and Computer Vision: A Machine Learning Approach to Detect Cracks and Damages. *Drones* **2021**, *6*, 5. [CrossRef]
7. Kundid Vasić, M.; Papić, V. Improving the Model for Person Detection in Aerial Image Sequences Using the Displacement Vector: A Search and Rescue Scenario. *Drones* **2022**, *6*, 19. [CrossRef]
8. Reckling, W.; Mitasova, H.; Wegmann, K.; Kauffman, G.; Reid, R. Efficient Drone-Based Rare Plant Monitoring Using a Species Distribution Model and AI-Based Object Detection. *Drones* **2021**, *5*, 110. [CrossRef]
9. Redmon, J.; Divvala, S.; Girshick, R.; Farhadi, A. You Only Look Once: Unified, Real-Time Object Detection. In Proceedings of the IEEE Conference on Computer Vision and Pattern Recognition, Las Vegas, NV, USA, 27–30 June 2016; pp. 779–788.
10. Redmon, J.; Farhadi, A. Yolo9000: Better, Faster, Stronger. In Proceedings of the IEEE Conference on Computer Vision and Pattern Recognition, Honolulu, HI, USA, 21–26 July 2017; pp. 7263–7271.
11. Redmon, J.; Farhadi, A. Yolov3: An Incremental Improvement. *arXiv* **2018**, arXiv:1804.02767.
12. Bochkovskiy, A.; Wang, C.Y.; Liao, H.Y.M. Yolov4: Optimal Speed And Accuracy of Object Detection. *arXiv* **2020**, arXiv:2004.10934.
13. Ren, S.; He, K.; Girshick, R.; Sun, J. Faster R-Cnn: Towards Real-Time Object Detection with Region Proposal Networks. *IEEE Trans. Pattern Anal. Mach. Intell.* **2017**, *39*, 1137–1149. [CrossRef] [PubMed]
14. Lin, T.Y.; Maire, M.; Belongie, S.; Hays, J.; Perona, P.; Ramanan, D.; Dollár, P.; Zitnick, C.L. Microsoft Coco: Common Objects in Context. In *European Conference on Computer Vision*; Springer: Cham, Switzerland, 2014; pp. 740–755.
15. Deng, J.; Dong, W.; Socher, R.; Li, L.; Li, K.; Fei-Fei, L. Imagenet: A Large-Scale Hierarchical Image Database. In Proceedings of the IEEE Conference on Computer Vision And Pattern Recognition, Miami, FL, USA, 20–25 June 2009; pp. 248–255.
16. Everingham, M.; Van Gool, L.; Williams, C.K.I.; Winn, J.; Zisserman, A. The Pascal Visual Object Classes (VOC) Challenge. *Int. J. Comput. Vis.* **2010**, *88*, 303–338. [CrossRef]
17. Zhu, P.; Wen, L.; Bian, X.; Ling, H.; Hu, Q. Vision Meets Drones: A Challenge. *arXiv* **2018**, arXiv:1804.07437.
18. Kalra, I.; Singh, M.; Nagpal, S.; Singh, R.; Vatsa, M.; Sujit, P.B. Dronesurf: Benchmark Dataset for Drone-Based Face Recognition. In Proceedings of the 2019 14th IEEE International Conference on Automatic Face & Gesture Recognition (Fg 2019), Lille, France, 14–18 May 2019; pp. 1–7.
19. Glenn, J. Yolov5 Release v6.1. 2022, 2, 7, 10. Available online: https://github.com/ultralytics/yolov5/releases/tag/v6.1 (accessed on 25 August 2022).
20. Zhu, X.; Lyu, S.; Wang, X.; Zhao, Q. Tph-Yolov5: Improved Yolov5 Based on Transformer Prediction Head for Object Detection on Drone-Captured Scenarios. In Proceedings of the IEEE/CVF International Conference on Computer Vision, Montreal, BC, Canada, 11–17 October 2021; pp. 2778–2788.
21. Akyon, F.C.; Altinuc, S.O.; Temizel, A. Slicing Aided Hyper Inference And Fine-Tuning for Small Object Detection. *arXiv* **2022**, arXiv:2202.06934.
22. Zhang, J.; Huang, J.; Chen, X.; Zhang, D. How To Fully Exploit the Abilities of Aerial Image Detectors. In Proceedings of the IEEE/CVF International Conference on Computer Vision Workshops, Seoul, Korea, 27–28 October 2019.
23. Solovyev, R.; Wang, W.; Gabruseva, T. Weighted Boxes Fusion: Ensembling Boxes From Different Object Detection Models. *Image Vis. Comput.* **2021**, *107*, 104117. [CrossRef]
24. Lin, T.Y.; Dollár, P.; Girshick, R.; He, K.; Hariharan, B.; Belongie, S. Feature Pyramid Networks for Object Detection. In Proceedings of the IEEE Conference on Computer Vision And Pattern Recognition, Honolulu, HI, USA, 21–26 July 2017; pp. 2117–2125.
25. Hartigan, J.A.; Wong, M.A. Algorithm as 136: A k-Means Clustering Algorithm. *J. R. Stat. Society. Ser. C (Appl. Stat.)* **1979**, *28*, 100–108. [CrossRef]
26. Comaniciu, D.; Meer, P. Mean Shift: A Robust Approach Toward Feature Space Analysis. *IEEE Trans. Pattern Anal. Mach. Intell.* **2002**, *24*, 603–619. [CrossRef]
27. Wang, Y.; Yang, Y.; Zhao, X. Object Detection Using Clustering Algorithm Adaptive Searching Regions In Aerial Images. In *European Conference on Computer Vision*; Springer: Cham, Switzerland, 2020; pp. 651–664.
28. Padilla, R.; Passos, W.L.; Dias, T.L.B.; Netto, S.L.; Da Silva, E.A.B. A Comparative Analysis of Object Detection Metrics with a Companion Open-Source Toolkit. *Electronics* **2021**, *10*, 279. [CrossRef]
29. Kingma, D.P.; Ba, J. Adam: A Method for Stochastic Optimization. *arXiv* **2014**, arXiv:1412.6980.
30. Deng, S.; Li, S.; Xie, K.; Song, W.; Liao, X.; Hao, A.; Qin, H. A Global-Local Self-Adaptive Network for Drone-View Object Detection. *IEEE Trans. Image Process.* **2020**, *30*, 1556–1569. [CrossRef] [PubMed]
31. Zhang, P.; Zhong, Y.; Li, X. Slimyolov3: Narrower, Faster and Better for Real-Time UAV Applications. In Proceedings of the IEEE/CVF International Conference on Computer Vision Workshops, Seoul, Korea, 27–28 October 2019.

32. Yang, F.; Fan, H.; Chu, P.; Blasch, E.; Ling, H. Clustered Object Detection in Aerial Images. In Proceedings of the IEEE/CVF International Conference on Computer Vision, Seoul, Korea, 27–28 October 2019; pp. 8311–8320.
33. Li, C.; Yang, T.; Zhu, S.; Chen, C.; Guan, S. Density Map Guided Object Detection in Aerial Images. In Proceedings of the IEEE/CVF Conference on Computer Vision and Pattern Recognition Workshops, Seattle, WA, USA, 14–19 June 2020; pp. 190–191.
34. Liao, J.; Piao, Y.; Su, J.; Cai, G.; Huang, X.; Chen, L.; Huang, Z.; Wu, Y. Unsupervised Cluster Guided Object Detection in Aerial Images. *IEEE J. Sel. Top. Appl. Earth Obs. Remote Sens.* **2021**, *14*, 11204–11216. [CrossRef]

*Article*

# Using Classify-While-Scan (CWS) Technology to Enhance Unmanned Air Traffic Management (UTM)

Jiangkun Gong [1] , Deren Li [1], Jun Yan [1,*], Huiping Hu [2] and Deyong Kong [3]

1   State Key Laboratory of Information Engineering in Surveying, Mapping and Remote Sensing, Wuhan University, Wuhan 430072, China
2   Wuhan Geomatics Institute, Wuhan 430022, China
3   School of Information Engineering, Hubei University of Economics, Wuhan 430205, China
*   Correspondence: yanjun_pla@whu.edu.cn; Tel.: +86-027-68778527

**Abstract:** Drone detection radar systems have been verified for supporting unmanned air traffic management (UTM). Here, we propose the concept of classify while scan (CWS) technology to improve the detection performance of drone detection radar systems and then to enhance UTM application. The CWS recognizes the radar data of each radar cell in the radar beam using advanced automatic target recognition (ATR) algorithm and then integrates the recognized results into the tracking unit to obtain the real-time situational awareness results of the whole surveillance area. Real X-band radar data collected in a coastal environment demonstrate significant advancement in a powerful situational awareness scenario in which birds were chasing a ship to feed on fish. CWS technology turns a drone detection radar into a sense-and-alert planform that revolutionizes UTM systems by reducing the Detection Response Time (DRT) in the detection unit.

**Keywords:** automatic target recognition (ATR); classify while scan (CWS); drone detection radar; detection response time (DRT); unmanned air traffic management (UTM)

**Citation:** Gong, J.; Li, D.; Yan, J.; Hu, H.; Kong, D. Using Classify-While-Scan (CWS) Technology to Enhance Unmanned Air Traffic Management (UTM). *Drones* **2022**, 6, 224. https://doi.org/10.3390/drones6090224

Academic Editors: Daobo Wang and Zain Anwar Ali

Received: 17 July 2022
Accepted: 24 August 2022
Published: 27 August 2022

**Publisher's Note:** MDPI stays neutral with regard to jurisdictional claims in published maps and institutional affiliations.

## 1. Introduction

Within the past few years, the number of unmanned aircraft systems (UAS), also often referred to as drones, has increased rapidly. They play essential roles in various tasks, including both civil applications and military applications. Drones are often used for reconnaissance, communication, and even attack missions by military clients. In 2020, drones revealed a new war era using the drone and anti-drone war after they achieved brilliant results in the Nagorno-Karabakh conflict [1,2], such as STM Kargu, Bayraktar TB2, IAI Harop, Orbiter 1K, etc. In addition to the military applications, drones, especially small drones, are popular in many civil projects [3], including entertainment photography, sports recording, infrastructure monitoring, precision agriculture, package delivery, rescue operations, remote mapping, and more. It seems that the era of drones will be the near-term future.

The flush drone applications require urgent counter UAS systems (C-UAS). C-UAS are generally designed to eliminate drone targets, and this includes three steps: (1) detection, (2) tracking and identification, and (3) effector [4,5]. The detection unit generally uses radar systems to detect the possible threat, and then operators identify the threat using EO/IR systems; finally, they utilize the effector unit to defeat the threat. Unlike military applications, some civil applications require that a C-UAS solution replace the defeater unit with a management unit. The typical application is airspace management at airports. An effective unmanned aircraft traffic management (UTM) system has been investigated for managing drones flying at low altitudes around airports [3,6].

The concept of UTM systems originates from the current air traffic management (ATM) system supporting flight operations. Most of the detection workload is taken by ground-based radar systems, such as airport traffic control (ATC) radar [7,8]. The primary identification work is supported by mandatory ACK systems such as the ADS-B system and the TCAS. One of the most challenging problems in designing and applying UTM is that there are neither associated EO/IR systems nor mandatory ADS-B sensors [9]. In this case, the radar system should take the task of identification. If a radar system can automatically detect radar echoes from drones and identify them from other clutters, it can considerably improve the performance of the UTM system. In other words, the drone detection radar needs to be equipped with automatic target recognition (ATR) module.

A radar ATR function is defined as recognizing targets mainly based on radar signals. Primitive target recognition was performed by using the audible representation of the received echoes, and then a trained operator deciphered the information in the sound. Later, radar engineers and scholars conducted design algorithms to do the work automatically. After decades, several schools have recognized radar features that succeed in ATR applications in some cases. Note that the traditional ATR solution contains two procedures, including feature extraction and pattern recognition. Here, we use the feature to classify the interpretations of ATR. The first one is high-range resolution profile (HRRP) technology [10]. Radar transmits ultrawide-band signals and obtains the target profiles in the range direction. The HRRP is a mapping of the shape of the target, and it is processed with a template matching approach in the dataset to obtain the target class. The second one is the micro-Doppler [11]. Micro-Doppler is thought to be the additional Doppler component produced by the micro motions on the target, such as the rotating movements of helicopters' blades, flapping birds' wings, and more. The embedded kinematic/structural information of the target can be measured from the micro-Doppler and then used for registering the target [11]. In addition, the third method could sometimes use tracking information such as speed and trajectory for identifying targets. Although machine learning is popular in recent years [9,12,13], it is a black-box algorithm to process the units of feature extraction and pattern recognition. Since machine learning in ATR applications is an approach over a feature, it is not discussed here. Generally, a machine learning method can process either HRRP data, micro-Doppler data, or tracking data to recognize the target from different background cluttered environments. In addition, all these radar data can be presented via radar images or signals.

Currently, the discussion of applying ATR to drone detection projects or other projects underestimates the value of an ATR algorithm. This paper proposes the higher value of ATR functions, which is the situational awareness (SA) ability using the patented classify while scan (CWS) technology. Section 2 presents the basic introduction of CWS technology and our radar platform. Section 3 describes and analyzes some experimental results, and the application is discussed in Section 4. Finally, our conclusion is presented in Section 5.

## 2. Materials and Methods

### 2.1. Design Principle

Generally, most ATR functions focus on the "point" information. The "point" characteristics are in both time and space dimensions. Spatially, they often recognize radar signals in a single radar resolution cell, and they also focus on current radar echoes over past and future radar signals temporally. They neither use tracking data of the target nor connect the radar signals in the radar resolution cells around the target. To the best of our knowledge, the track while scanning (TWS) could be probably the only technology that uses the time information of the target.

TWS is a mode of radar operation in which the radar scans the surveillance area while the acquired targets are tracked [14]. The significance of TWS is that the radar provides an overall view of the surveillance area and helps maintain better situation awareness (SA). Radar systems equipped with the TWS function sometimes also claim that they are 4D radar, in which time labels of the target are also regarded as the fourth dimension together with traditional 3D information (i.e., azimuth, elevation angle, and slant range). However, it seems that 4D radar is just a sales-promoting buzzword that has nothing to do with the fourth dimension in physics. In our opinion, the real 4D radar should include the target attribute given by an ATR function.

Integrating ATR and TWS is the technology of classify-while-scan (CWS). As shown in Figure 1, CWS processes the raw data in one radar resolution cell and obtains the object's ID. The ID of the target can be either used for recording, displaying on the radar screen, or assisting the tracking unit. With the ID of the target, the tracking unit could easily connect the same ID between the contiguous tracking data. This process can be called track-after-identify (TAI). Then, the CWS function processes radar data in every radar cell and outputs the targets along with the ranged cells in the radar beam. Consequently, the radar beam continues to scan the area and capture the traces of the active targets, and then the whole scenario is presented in a radar display following the scan of the radar beam. Figure 2 demonstrates the basic flowchart of the CWS function.

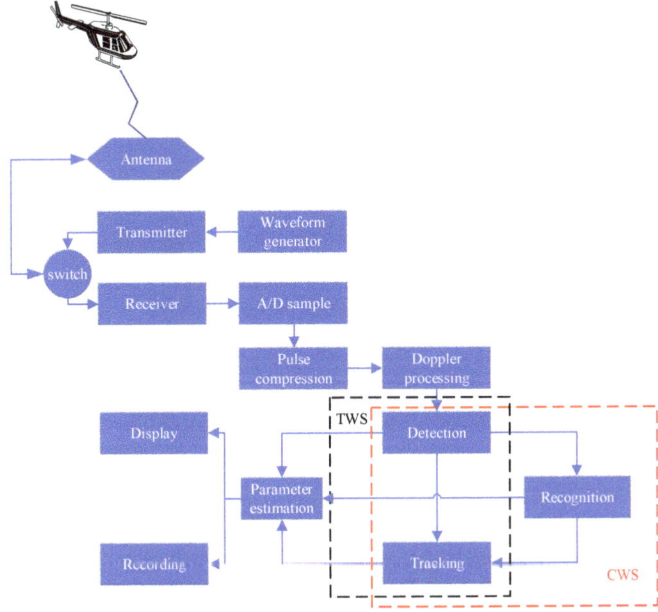


**Figure 1.** The block diagram of radar equipped with CWS and TWS technology.

The specific algorithms in CWS technology can differ with different methods. For example, the detection methods can include an algorithm based on the signal-to-noise ratio (SNR), the moving target indication (MTI), the algorithm extracting the signal-to-clutter ratio (SCR), and more. ATR algorithms can be diverse, such as HRRP, micro-Doppler, and time-frequency analysis. The only difference exists in the tracking algorithm. Traditional tracking algorithms mainly depend on the correlation of Doppler and trajectory. However, CWS provides ID labels of objects in the radar beam, and then tracking algorithms can use ID labels to improve the correlation process and enhance the tracking accuracy. As shown in Figure 2, our CWS processing algorithm contains several steps, including:

(1)    Assume there are N radar range cells in one radar beam;
(2)    Extract the radar signals of objects using our SCR detector [15,16];
(3)    Recognize the radar echoes of objects using radar signal signatures;
(4)    Repeatthe above detection & recognition algorithms and obtain all the IDs in each radar range cell;
(5)    The radar tracking unit uses the recognized results (i.e, IDs of targets) to track targets.

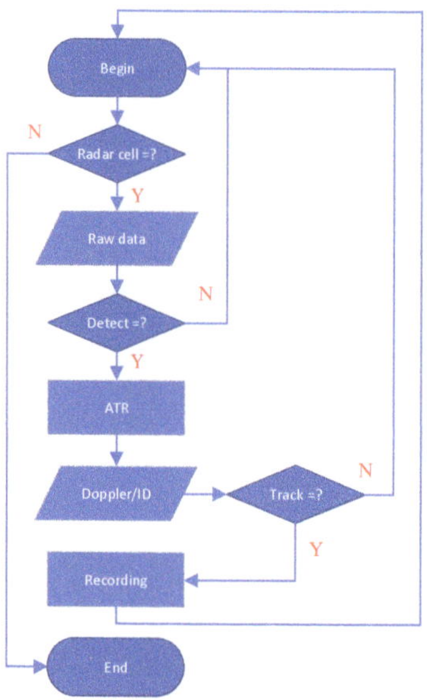


**Figure 2.** The basic flowchart of CWS technology.

### 2.2. Test Conditions

Here, we demonstrate a drone detection radar system and the detection results using CWS technology. We applied the CWS technology to a new coastal surveillance radar. The pulse-Doppler radar works in the X-band. It is a narrow-band radar with a range resolution of approximately 12 m. The radar is equipped with an active, electronically scanned, phased-array antenna and deployed on a rotating table to achieve 360-degree coverage in the azimuth scan. Moreover, it uses digital beam-forming (DBF) technology to obtain multiple radar beams in the pitch direction every time. The flexibly configured rotating speed of the rotating table is between 2 s and 20 s. The detection response time (DRT) is approximately 30 ms, representing that the lag between an echo return, detection, and eventual display is only 30 ms. It is capable of tracking 1000 targets simultaneously. It can recognize different targets, including birds, drones, vehicles, ships, people, helicopters, and others, and then it presents detection results with graphic icons, which label the recognition results, along with the rotating motion of the scanning beam. The numbers around the icons are tracking numbers.

The test was conducted in a coastal area with a cluttered sea environment. The test area is located on the Yellow Sea coast of Qidong, China. The radar is set on the roof of a 12-m tall building, and the sea was scanned horizontally. The initial goal of this project was to develop a prototype drone detection radar. During the project, an infrared sensor and an optical camera were deployed to support the project and used to confirm the recognition results of the ATR function. This project continued for several months in 2020, and some data were extracted in this paper. The sea scale during the test was about Degree 3~5. The height of the wave was 0.5–1.25 m with Degree 3 and 2.50–4.00 m with Degree 5.

We collected radar signals of different types of drones, including a quad-rotor drone, a fixed-wing drone, and a hybrid Vertical Take-off and Landing (VTOL) fixed-wing drone. They were cooperative targets. Some of their parameters are listed in Table 1. Albatross 1 is a homemade fixed-wing drone with only one pusher blade, DJI Phantom 4 is a famous quad-rotor drone with four lifting blades, and TX25A is a large hybrid VTOL fixed-wing drone with one pusher blade and four lifting blades. Photos of these drones are shown in Figure 3. If we refer to NATO's category of drones [4], Albatross 1 is a Microdrone, DJI

Phantom 4 is a Mini-drone, and TX25A is a Small-drone. In addition, some local fishing ships and birds were also our test targets, and we also collected their radar data.

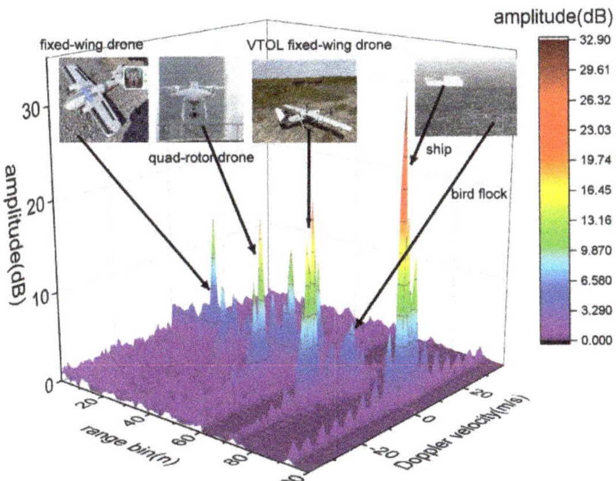


**Figure 3.** The range-Doppler image of radar data in one radar beam containing different objects.

**Table 1.** Parameters of different drones.

| Drone Type | Hybrid VTOL Fixed-Wing | Fixed-Wing | Multi-Rotor |
| --- | --- | --- | --- |
| Model | TX25A | Albatross 1 | Phantom 4 |
| Manufacturer | Harryskydream Inc. | Homemade | DJI Inc. |
| Flight weight | 26 kg | 0.3 kg | 1.38 kg |
| Body size | 197 cm | 80 cm | 40 cm |
| Wing span | 360 cm | 108 cm | 40 cm |
| Cruise speed | 25 m/s | 10 m/s | 15 m/s |
| Rotor number | 5 | 2 | 4 |
| Blade length | 30 cm | 10 cm | 20 cm |
| Aero-frame materials | FRP (Fibre reinforces plastic) | EPP (Expanded polypropylene) | PC (Polycarbonate) |

## 3. Results

It required several steps from an echo return into the eventual display on the radar display. Figures 3–5 show some examples of the whole procedure, and Figure 6 demonstrates a real radar screenshot using CWS technology. Figure 3 shows the range-Doppler images containing several specific objects in one simulated radar beam. Each radar range sector had twenty bins, where the target was in the center of the range sector. The range resolution of a radar bin is 12 m, and the range of a range sector is 240 m. To demonstrate the detection results, we manually jointed five sectors from different radar beams into the detection result in Figure 3. Most clutters were approximately 0 Hz. Since the detection ranges of those targets were different, the amplitudes of the signals were incomparable. Nevertheless, the Doppler signals in each spectrum still differ from each other. We use the SCR detector to extract radar signals of targets in cluttered backgrounds [15]. The SCR detector performs superior to the SNR detector, reducing missed and false alarms when detecting and tracking drones.

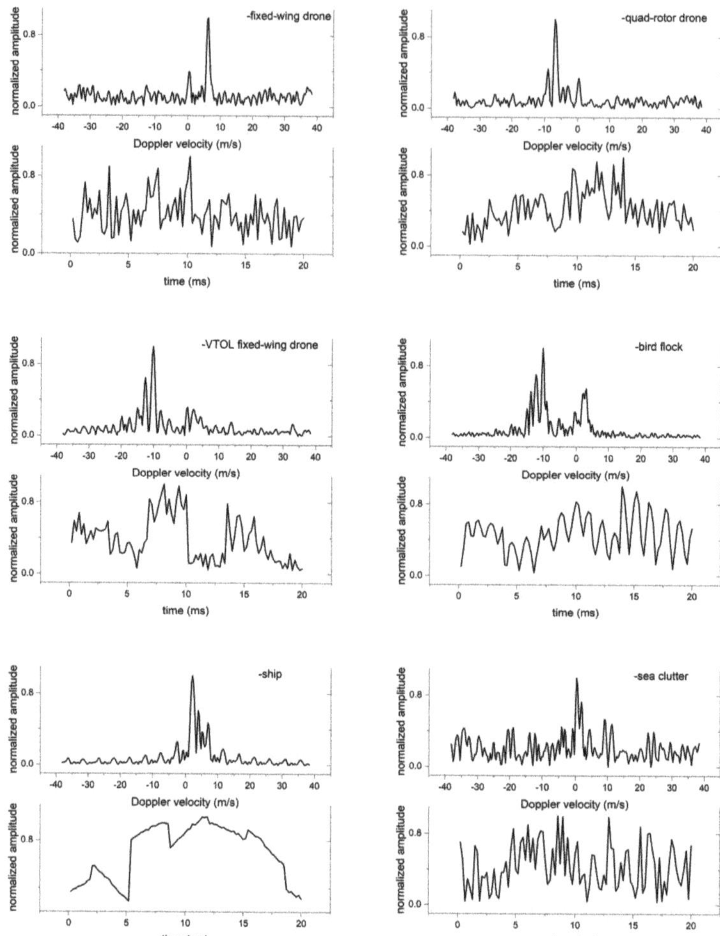

**Figure 4.** Raw time & frequency data of different objects.

The time-frequency characteristics of different objects provide useful signatures for the ATR module. Figure 4 abstracts the raw radar of objects in Figure 5 and plots the time-frequency distributions of each object. The spectra were obtained using conventional Fourier transforms. We normalized the signal amplitudes to remove the interferences posed by the detection ranges. The radar dwell time in the time domain was 20 ms. There were no patterns in the time domains. In addition to the body Doppler, the micro-Doppler also appeared in the spectra, and they seem to have patterns. Then, a patented ATR algorithm is used to analyze the time-frequency characteristics of radar signals from different objects and obtain the recognized results of the targets.

The radar tracking units used the recognized results in each radar beam to enhance the tracking data of every target. Figure 5 shows the tracking data of our targets in Figures 3 and 4. These data were real data collected on different test dates. The height information was calculated using DBF technology; therefore, the accuracy was limited. Nevertheless, the trace of each target still described the moving kinematics of each target. Generally, the mean Doppler velocity values of the fixed-wing drone, quad-rotor drone, VTOL fixed-wing drone, bird flock, and ship were approximately 7.34 m/s, 6.76 m/s, 15.22 m/s, 12.75 m/s, and 3.36 m/s, respectively. The ship has the slowest moving speed, while the VTOL drone has the fastest moving velocity. In addition, the measured height numbers of those targets were approximately 670.07 m, 272.68 m, 100.05 m, 20.81 m, and 0 m, respectively. These measured characteristics were following the natural flight parameters of these objects.

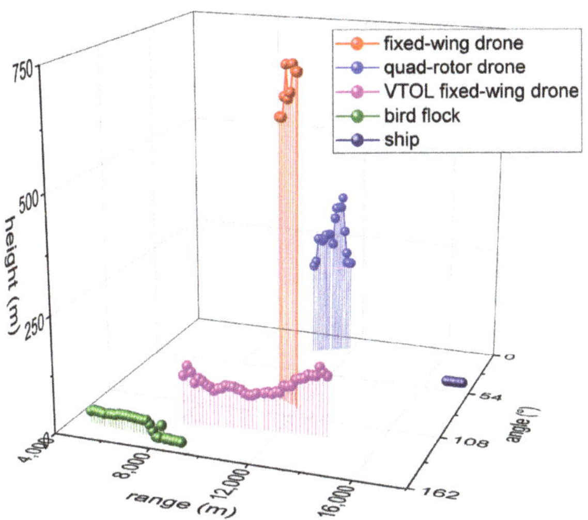


**Figure 5.** The tracking data of different moving targets.

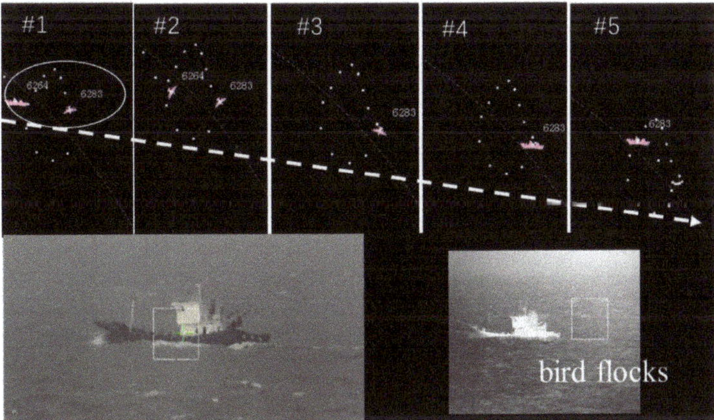


**Figure 6.** A real scenario sensed by situation awareness using the CWS technology, where the birds were chasing the ship to feed on fish.

## 4. Discussion

The most significant benefit provided by CWS technology is situational awareness (SA). Figure 6 provides a typical example of an application of situational awareness where birds chased ships to pick up fish awakened by the ship's propellers. The recognized results were marked using graphic icons, including ships, helicopters, drones, and birds. The thin white dotted lines are the tracking traces of objects. The numbers around the icons were the tracking numbers of objects. CWS enhanced the radar to become a WYSIWYG (What You See Is What You Get) system or a real-time sense-and-alert system. Moreover, the whole scenario in the surveillance area was captured and updated with new data. Note the tracking birds of No. 6264 were flying around the ship of No. 6283; thus, radar echoes from birds were interfering with the detection of the ship. When they appeared in the same radar cell, the technology recognized that these data were sometimes birds while the others were the ship. The birds were flying around the ship because they habitually chase ships in the sea to feed on the fish, which were awakened by the rotating propellers of the ship. In other words, SA provided by CWS described a habit of sea birds.

UTM is a safety system that helps ensure the newest entrant, a variety of drones, into the skies does not collide with buildings, larger aircraft, or one another. UTM is different from the ATM functioned by the airport agencies in some respects. First, UTM cannot use the airport traffic control (ATC) radar systems used by the current ATM system. ATC radar systems are primarily designed to detect and track large, fast-moving aircraft flying in high airspace. Compared to large aircraft, drones generally have small RCS values and slow-moving speeds. In particular, the ATC radar systems were reported to turn to identify and retain targets that move consistently, remain visible from sweep to sweep, and have a ground speed of at least 15.432 m/s, and, as a result, they failed to detect the flock of Canadian geese in the aircraft accident of flight 1549 on 15 January 2009 [17]. The flight characteristics of drones make them potentially undetectable by traditional ATC radar systems. Second, drones may not be equipped with signal transmission sensors, such as RID broadcast receivers and ADS-B. This means that ground-based detection systems could be critical sensors to support the UTM system.

Similar to the ATC in the ATM system, a drone detection radar system is the key role of UTM. Nevertheless, not all drone detection radar systems are suitable. Here, we insist that a 4D radar must detect an object and obtain its 3D position and 1D attribution, provided by an ATR algorithm. Figure 7 demonstrates a typical application, where 4D radar can enhance the real-time situational awareness at a simulated airport. In this way, the drone detection radar can detect and track different types of drones (e.g., quad-rotor drone, VTOL drone, unmanned-helicopter, etc.) over other large aircraft or bird clutters and then support a UTM system. ATR turns the radar data of objects into meaningful IDs of targets. These knowledge-based IDs could also be used for other applications, such as human-machine distributed situational awareness [9]. SA may be a key in UTM applications.

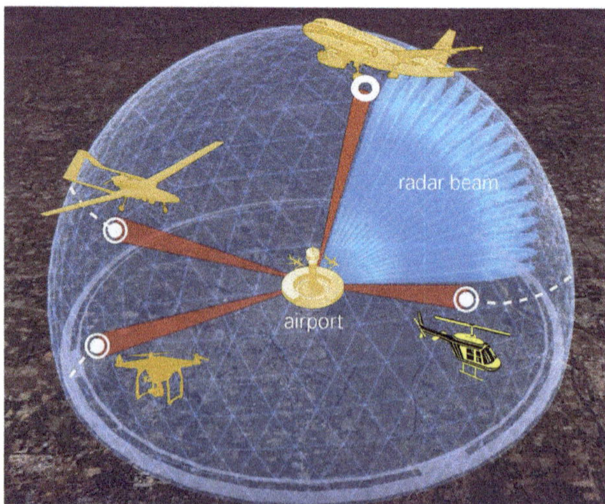


**Figure 7.** Using CWS to enhance UTM at an airport.

Generally, the performance of an ATR algorithm can be described in four tiers:

(1) Tier DETECTION: extracting radar signals of a target in background clutter, such as whether the radar echoes are from a target or a clutter;

(2) Tier CLASSIFICATION: classifying the class of the target, such as whether the radar echoes are from a drone or a bird;

(3) Tier IDENTIFICATION: identifying the type of the target in one class, such as whether the radar echoes are from a fixed-wing drone or a multi-rotor drone;

(4) Tier DESCRIPTION: describing the model of the target, such as whether the radar echoes are from a DJI Phantom 4 drone or a DJI S800 drone.

There are many types of aircraft at airports, so the ATR function should achieve the Tier IDENTIFICATION to support the UTM application (e.g., the example in Figure 7). As a higher tier of ATR function, CWS technology provides valuable situational awareness to support the UTM system. The performance of CWS technology must consider three factors, including the detection range, processing speed, and confidence factor. First, no matter what the algorithm of a CWS function is, it should be independent of the detection range of a target. Otherwise, the drone detection radar could not cover enough airspace at airports. Second, the processing speed determines the DRT. There is always a lag between an echo return and eventual display on the radar screen. The lag is sometimes called the DRT, which contains the time spent on communication and processing. Ideally, DRT should be at a level of milliseconds to fulfill a WYSIWYG system. Practically, the DRT should be shorter than the update interval between the current radar beam and the next radar beam. Last, but not least, the confidence factor determines the recognized result of CWS technology. Unlike the identification probability, the confidence factor describes the trust level of the recognized results calculated by the ATR algorithm of the CWS technology. Besides, the trajectory prediction is a key module in the ATM/UTM system [18–23], which can use the predicted trajectory to reduce the DRT in the tracking unit. In contrast, our CWS technology shortens the DRT in the detection unit. If the CWS technology is deployed in the radar system, it can provide the location of a target for the tracking unit using trajectory prediction algorithms, faster. In total, the shorter DRT and a higher confidence factor of a good CWS technology can result in a better UTM system.

## 5. Conclusions

There is a need for drone detection radar systems to support UTM systems at the airport. Here, we propose CWS technology to improve the detection performance of drone detection radar systems and enhance UTM systems. The CWS processes radar data of each radar cell in the radar beam using the ATR algorithm and then obtains all targets' recognized results. It reduces the DRT in the detection unit of the radar. Moreover, the recognized results can be used to track targets using the TAI algorithm in a surveillance area and then obtain real-time situational awareness (SA) results. The future work will investigate the performance of the TAI algorithm, and the performance of the radar situational awareness (SA) with the CWS and TAI algorithms. With situational awareness, drone detection radar has become a WYSIWYG, bringing revolutionary performance to UTM systems.

**Author Contributions:** Conceptualization, J.Y.; methodology, J.G.; software, D.K.; validation, J.Y.; formal analysis, J.G.; investigation, J.Y.; resources, D.L.; data curation, J.Y.; writing—original draft preparation, J.G.; writing—review and editing, H.H.; visualization, H.H.; supervision, J.Y.; project administration, D.L.; funding acquisition, D.K. All authors have read and agreed to the published version of the manuscript.

**Funding:** This research received some support by the Natural Science Foundation of Hubei Providence (General Program: 2021CFB309).

**Institutional Review Board Statement:** Not applicable.

**Informed Consent Statement:** Not applicable.

**Data Availability Statement:** Some of the data presented in this study may be available on request from the corresponding author. The data are not publicly available due to the internal restriction of the research group.

**Acknowledgments:** We appreciate both the testers during the collection of the data, and we also want to thank the authors whose photographs are reproduced in this study.

**Conflicts of Interest:** The authors declare that they have no conflict of interest.

## References


1. Semercioğlu, H. The New Balance of Power in the Southern Caucasus in the Context of the Nagorno-Karabakh Conflict in 2020. *Res. Stud. Anatolia J.* **2021**, *4*, 49–60.
2. Hecht, E. Drones in the Nagorno-Karabakh War: Analyzing the Data. *Mil. Strateg. Mag.* **2022**, *7*, 31–37 .
3. Jiang, T.; Geller, J.; Ni, D.; Collura, J. Unmanned Aircraft System Traffic Management: Concept of Operation and System Architecture. *Int. J. Transp. Sci. Technol.* **2016**, *5*, 123–135. [CrossRef]
4. Wellig, P.; Speirs, P.; Schuepbach, C.; Oechslin, R.; Renker, M.; Boeniger, U.; Pratisto, H. Radar Systems and Challenges for C-UAV. In Proceedings of the 2018 19th International Radar Symposium (IRS), Bonn, Germany, 20–22 June 2018; pp. 1–8.
5. Watts, A.C.; Ambrosia, V.G.; Hinkley, E.A. Unmanned Aircraft Systems in Remote Sensing and Scientific Research: Classification and Considerations of Use. *Remote Sens.* **2012**, *4*, 1671–1692. [CrossRef]
6. Chin, C.; Gopalakrishnan, K.; Egorov, M.; Evans, A.; Balakrishnan, H. Efficiency and Fairness in Unmanned Air Traffic Flow Management. *IEEE Trans. Intell. Transp. Syst.* **2021**, *22*, 5939–5951. [CrossRef]
7. Nakamura, H.; Matsumoto, Y.; Suzuki, S. Flight Demonstration for Information Sharing to Avoid Collisions between Small Unmanned Aerial Systems (Suass) and Manned Helicopters. *Trans. Jpn. Soc. Aeronaut. Space Sci.* **2019**, *62*, 75–85. [CrossRef]
8. Taylor, J.W.; Brunins, G. Design of a New Airport Surveillance Radar (ASR-9). *Proc. IEEE* **1985**, *73*, 284–289. [CrossRef]
9. Pérez-Castán, J.A.; Pérez-Sanz, L.; Bowen-Varela, J.; Serrano-Mira, L.; Radisic, T.; Feuerle, T. Machine Learning Classification Techniques Applied to Static Air Traffic Conflict Detection. *IOP Conf. Ser. Mater. Sci. Eng.* **2022**, *1226*, 12019. [CrossRef]
10. Tait, P. *Introduction to Radar Target Recognition*; Institution of Electrical Engineers: London, UK, 2006; ISBN 9781849190831.
11. Chen, V.C. *The Micro-Doppler Effect in Radar*; Artech House: Norwood, MA, USA, 2011; ISBN 9781608070572/1608070573.
12. Raval, D.; Hunter, E.; Hudson, S.; Damini, A.; Balaji, B. Convolutional Neural Networks for Classification of Drones Using Radars. *Drones* **2021**, *5*, 149. [CrossRef]
13. Barbaresco, F.; Brooks, D.; Adnet, C. Machine and Deep Learning for Drone Radar Recognition by Micro-Doppler and Kinematic Criteria. In Proceedings of the 2020 IEEE Radar Conference (RadarConf20), Florence, Italy, 21–25 September 2020; pp. 1–6.
14. Bair, G.L.; Zink, E.D. Radar Track-While-Scan Methodologies. In Proceedings of the IEEE Region 5 Conference, 1988: 'Spanning the Peaks of Electrotechnology', Colorado Springs, CO, USA, 21–23 March 1988; pp. 32–37.
15. Gong, J.; Yan, J.; Li, D. The Radar Detection Method Based on Detecting Signal to Clutter Ratio (SCR) in the Spectrum. In Proceedings of the 2019 PhotonIcs & Electromagnetics Research Symposium - Spring (PIERS-Spring), Rome, Italy, 17–20 June 2019; pp. 1876–1882.
16. Gong, J.; Li, D.; Yan, J.; Hu, H.; Kong, D. Comparison of Radar Signatures from a Hybrid VTOL Fixed-Wing Drone and Quad-Rotor Drone. *Drones* **2022**, *6*, 110. [CrossRef]
17. National Transportation Safety Board. *Aircraft Accident Report: Loss of Thrust in Both Engines After Encountering a Flock of Birds and Subsequent Ditching on the Hidson River*; National Transportation Safety Board: Washington, DC, USA, 2010.
18. Wang, X.; Musicki, D.; Ellem, R.; Fletcher, F. Efficient and Enhanced Multi-Target Tracking with Doppler Measurements. *IEEE Trans. Aerosp. Electron. Syst.* **2009**, *45*, 1400–1417. [CrossRef]
19. Pang, Y.; Zhao, X.; Hu, J.; Yan, H.; Liu, Y. Bayesian Spatio-Temporal GrAph TRansformer Network (B-STAR) for Multi-Aircraft Trajectory Prediction. *Knowledge-Based Syst.* **2022**, *249*, 108998. [CrossRef]
20. Pang, Y.; Zhao, X.; Yan, H.; Liu, Y. Data-Driven Trajectory Prediction with Weather Uncertainties: A Bayesian Deep Learning Approach. *Transp. Res. Part C Emerg. Technol.* **2021**, *130*, 103326. [CrossRef]
21. Xu, Z.; Zeng, W.; Chu, X.; Cao, P. Multi-Aircraft Trajectory Collaborative Prediction Based on Social Long Short-Term Memory Network. *Aerospace* **2021**, *8*, 115. [CrossRef]
22. Zeng, W.; Chu, X.; Xu, Z.; Liu, Y.; Quan, Z. Aircraft 4D Trajectory Prediction in Civil Aviation: A Review. *Aerospace* **2022**, *9*, 91. [CrossRef]
23. Pang, B.; Zhao, T.; Xie, X.; Wu, Y.N. Trajectory Prediction with Latent Belief Energy-Based Model. In Proceedings of the IEEE/CVF Conference on Computer Vision and Pattern Recognition (CVPR), Nashville, TN, USA, 20–25 June 2021; pp. 11814–11824.

*drones*

**MDPI**

*Article*

# A Modified YOLOv4 Deep Learning Network for Vision-Based UAV Recognition

Farzaneh Dadrass Javan [1,2,*], Farhad Samadzadegan [2], Mehrnaz Gholamshahi [3] and Farnaz Ashatari Mahini [2]

1    Faculty of Geo-Information Science and Earth Observation (ITC), University of Twente, 7522 NB Enschede, The Netherlands
2    School of Surveying and Geospatial Engineering, College of Engineering, University of Tehran, Tehran 1439957131, Iran; samadz@ut.ac.ir (F.S.); f.ashtari@ut.ac.ir (F.A.M.)
3    Department of Electrical and Computer Engineering, Faculty of Engineering, Kharazmi University, Tehran 1571914911, Iran; mehrnazgholamshahi.khu@gmail.com
*    Correspondence: f.dadrassjavan@utwente.nl

**Abstract:** The use of drones in various applications has now increased, and their popularity among the general public has increased. As a result, the possibility of their misuse and their unauthorized intrusion into important places such as airports and power plants are increasing, threatening public safety. For this reason, accurate and rapid recognition of their types is very important to prevent their misuse and the security problems caused by unauthorized access to them. Performing this operation in visible images is always associated with challenges, such as the small size of the drone, confusion with birds, the presence of hidden areas, and crowded backgrounds. In this paper, a novel and accurate technique with a change in the YOLOv4 network is presented to recognize four types of drones (multirotors, fixed-wing, helicopters, and VTOLs) and to distinguish them from birds using a set of 26,000 visible images. In this network, more precise and detailed semantic features were extracted by changing the number of convolutional layers. The performance of the basic YOLOv4 network was also evaluated on the same dataset, and the proposed model performed better than the basic network in solving the challenges. Compared to the basic YOLOv4 network, the proposed model provides better performance in solving challenges. Additionally, it can perform automated vision-based recognition with a loss of 0.58 in the training phase and 83% F1-score, 83% accuracy, 83% mean Average Precision (mAP), and 84% Intersection over Union (IoU) in the testing phase. These results represent a slight improvement of 4% in these evaluation criteria over the YOLOv4 basic model.

**Keywords:** convolutional neural network CNN; YOLO deep learning; drone; UAV; drone detection; drone recognition

**Citation:** Dadrass Javan, F.; Samadzadegan, F.; Gholamshahi, M.; Ashatari Mahini, F. A Modified YOLOv4 Deep Learning Network for Vision-Based UAV Recognition. *Drones* **2022**, *6*, 160. https://doi.org/10.3390/drones6070160

Academic Editors: Daobo Wang and Zain Anwar Ali

Received: 3 June 2022
Accepted: 24 June 2022
Published: 27 June 2022

**Publisher's Note:** MDPI stays neutral with regard to jurisdictional claims in published maps and institutional affiliations.

## 1. Introduction

Drones are actively used in a variety of fields, including recreational, commercial, security, crisis management, and mapping [1,2]. They are also used in combination with other platforms such as satellites in resource management, agriculture, and environmental protection [3–5]. However, the negligent and the malicious use of these flying vehicles poses a great threat to public safety in sensitive areas such as government buildings, power plants, and refineries [6,7]. For this reason, it is important to recognize drones to prevent them from entering critical infrastructure or ensuring security in large locations such as stadiums [8].

In this study, the recognition of four types of drones was investigated. Conventional drone detection technologies include the use of various sensors such as radar (radio detection and ranging) [9], Lidar (Light Detection and Ranging) [10], acoustic [11], and thermal sensors [12]. In these methods, first, the presence or the absence of the drone in the scene is checked and then the drone type recognition process is performed [13,14]. However, the application of these types of sensors has always been associated with problems such as higher costs and higher energy consumption [15]. In contrast, visible images do not have these problems and are widely used for object recognition and semantic segmentation due to their high resolution [16]. On the other hand, the use of visible images also introduces problems such as light changes within the imagery, the presence of occluded areas, and a crowded background, which necessitates the application of an efficient and comprehensive method for recognition.

Recent advances in deep convolutional neural networks and the appearance of more improved hardware make it possible to use visual information to recognize objects with higher accuracy and speed [17]. Unlike conventional drone detection technologies, the nature of deep learning networks is to perform drone recognition simultaneously. By classifying inputs into several classes, these networks determine the presence, absence, image location, and type of drone class [18]. Among neural networks, the convolutional neural network (CNN) is one of the most important representatives of image recognition and classification. In this network, the input data enters the convolutional layers. The convolution operation is then performed using the network kernel to find similarities. Finally, feature extraction is performed using the resulting feature map [19]. There are different types of convolutional neural networks available such as R-CNN (Region-based CNN) [20], SPPNet (Spatial Pyramid Pooling Network) [21], and Faster-RCNN [22]. In these networks, due to the application of convolutional operations, more features are extracted than in conventional object detection methods and better speed and accuracy are achieved in recognizing objects. The extracted features are essentially descriptors of objects, and as the number of these features increases, object recognition is performed with higher accuracy. In these networks, the proposed regions are first defined using region proposal networks (RPNs) [23]. Then, convolutional filters are applied to these regions, and the extracted features are obtained as the result of the convolutional operation [22]. In other deep learning methods such as SSD (Single Shot MultiBox Detector) [24] and YOLO [25], the image is generally explored, which results in higher accuracy and speed in object recognition as compared to the basic methods [25]. The reason for the higher speed in these methods is the architecture is simpler than in region-based methods. The YOLO network is a method for detecting and for recognizing an object based on CNNs. The YOLO network predicts bounding box coordinates and class probabilities for these boxes, considering the whole image. The fourth edition of the YOLO Network is the YOLOv4 Deep Learning Network, which performs better than previous versions in terms of speed and accuracy [26]. However, the YOLOv4 deep learning network may not be able to overcome some challenges, such as the small size of the drone in different images [16]. In this study, this network could not recognize the drone in some of the challenging images. These challenges include confusing some drones with birds due to their small size, and the presence of drones in crowded backgrounds and hidden areas. Therefore, the YOLOv4 deep learning network was modified to better overcome the challenges of recognizing flying drones. The change in the architecture of this network is the main innovation in this article. Also, 4 types of multirotors, fixed-wings, helicopters, and VTOLs (Vertical Take-Off and Landing) were recognized. Given the need to recognize each type of UAV in different applications, the study of this topic can be considered as another innovation of this paper.

*1.1. Challenges in Drone Recognition*

Drone recognition is always fraught with challenges. Some of the important challenges in this regard are discussed.

1.1.1. Confusion of Drones and Birds

Due to the physical characteristics of drones, they can easily be confused with birds in human eyes. This problem is more challenging when using drones in maritime areas due to the presence of more birds. The similarity between drones and birds and their distinction from each other is shown in Figure 1.

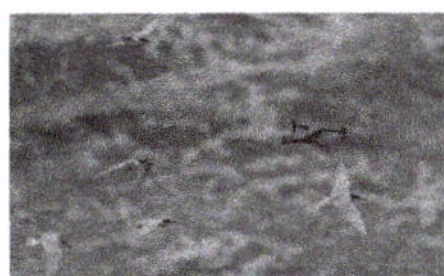

**Figure 1.** Challenges related to confusion with birds in drone recognition.

### 1.1.2. Crowded Background

As it appears from Figure 2, the presence of drones in areas with crowded backgrounds and similar environments has made them more difficult to recognize due to the inability to isolate the background. A crowded background refers to conditions such as the existence of clouds, dust, fog, and fire in the sky.

**Figure 2.** Challenges related to a crowded background in drone recognition.

### 1.1.3. Small Drone Size

The small size of drones makes them difficult to see at longer distances and difficult to quickly and accurately recognize, or they are possibly recognized as birds. Furthermore, the presence of a swarm of UAVs at different scales makes the recognition process more challenging. Figure 3 illustrates some examples of the presence of small drones at different scales.

**Figure 3.** Challenges related to the presence of small drones at different scales.

Drone recognition is always fraught with challenges. For this reason, it is necessary to use a fast, accurate, robust, and efficient method to overcome the challenges and to correctly recognize drones.

## 2. State of the Art Work

Recently, the use of drones has become increasingly popular, and they have been applied to various scientific and commercial purposes in different fields of photogrammetry, surveying, agriculture, natural disaster management, and so on [27]. There are different types of drones, each used for a specific purpose; and, in terms of design technology, application, and physical characteristics, they can be divided into four types: multirotor, helicopter, VTOL, and fixed-wing [28–31]. They are also divided into two scenarios in terms of operation manner in the environment. In the first scenario, drones operate individually, while in the second scenario they fly in combination with others, which are normally known as a swarm of UAVs [32–35].

Because of the enormous potential applications of each type of UAV in meeting the needs of society, the possibility of their misuse has become a major concern for communities. Over the past decade, much of the research has focused on finding efficient and accurate techniques for the recognition of different types of UAVs [12,17,36]. However, sometimes drone recognition is difficult because they are normally flying in challenging environments. Therefore, the recognition of UAVs requires advanced techniques that can recognize them as they fly individually or in swarm mode.

## 3. Related Works

Due to the increasing development of deep neural networks in visual applications, these networks are also used widely for the recognition of objects in visible images [36–38]. In 2019, Nalamati et al. used a collection of visible images to detect small drones and solve their detection challenges. In this work, different CNN-based architectures were used, such as SSD [22], Faster-RCNN with ResNet-101 [22], and Faster-RCNN [22] with Inceptionv2. Based on the results, the R-CNN network with ResNet-101 performs the best in training and testing [39]. In 2019, Unlu et al. used an independent drone detection system, using the YOLOv3 Deep Learning Network. One of the advantages of this system is its cost-effectiveness due to the limited need for GPU memory. This study can detect drones of a small size and at a minimal distance, but it cannot recognize the types of drones [40]. In 2020, Mahdavi et al. detected a drone using a fisheye camera, and three methods of classification were applied: convolutional neural network (CNN), support vector machine (SVM), and nearest-neighbor. The results showed that CNN, SVM, and nearest-neighbor have total accuracy of 95%, 88%, and 80%, respectively. Compared with other classifiers with the same experimental conditions, the accuracy of the convolutional neural network classifier was satisfactory. In this study, only the detection of drones without considering their types and challenges has been investigated [41]. In 2020, Behera et al. detected and classified drones in RGB images using the YOLOv3 network, and they achieved a mAP of 74% after 150 epochs. In this article, only drones were detected at various distances, and the issue of drone recognition and its distinction from birds was not discussed [42]. In 2020 Shi et al. proposed a detection process of the low-altitude drone based on the YOLOv4 deep learning network. They then compared the YOLOv4 detection result with the YOLOv3 and the SSD networks. In this study, the YOLOv4 network performed better than the YOLOv3 and the SSD networks in detecting, recognizing, and identifying three types of drones in terms of mAP and detection speed, achieving 89% mAP [43]. In 2021, Tan Wei Xun et al. detected and tracked a drone using the YOLOv3 deep learning network. In their study, the NVIDIA Jetson TX2 was used to detect drones in real-time. The results of this method show that the proposed YOLOv3 network detects drones of three sizes: small, medium, and large, with an average confidence score of 88% and a confidence score between 60% and 100% [44]. In 2021, Isaac-Medina et al. detected and tracked drones using a set of visible and thermal images and four deep learning network architectures. In this paper, the deep learning networks Faster RCNN, SSD, YOLOv3, and DETR (DEtection TRansformer) are used. Based on the results, all the studied networks were able to detect a small drone at a far distance. But the YOLOv3 deep learning network generally leads to better accuracy (up to 0.986 mAP) and the RCNN network performed better in detecting small drones (up to 0.77 mAP) [45]. In 2021 Singha et al. developed an automatic drone detection system using YOLOv4. They used a dataset of drones and birds to detect drones, and then evaluated the model on two types of drone videos. The results obtained in this study for detecting two types of multirotor drones are: mAP 74.36%, F1-score 0.79, recall 0.68, and precision 0.95 [46]. In 2021, Liu et al. examined three object detection methods, such as YOLOv3, YOLOv4, RetinaNet, and FCOS (Fully Convolutional One-stage Object Detector) networks, on visible image data. To get great accuracy in drone detection, the pruned YOLOv4 model is used to build a sparser, flatter network. The application of the method has improved the detection of small drones and high-speed drones. The pruned YOLOv4, with a pruning rate of 0.8 and a 24-layer pruning, achieved a mAP of 90.5%, an accuracy of 22.8%, a recall of 12.7%, and a processing speed of 60%. However, the challenges of crowded backgrounds, hidden areas, and surveys of multiple drone types have not yet been addressed [16]. In 2022, Samadzadegan et al. detected and recognized drones using YOLOv4 Deep Networks in visible images [47]. This network can recognize multirotor and helicopters directly, and it can differentiate between drones and birds with a mAP of 84%, an IoU of 81%, and an accuracy of 83%. In this paper, the challenges related to recognition have been well addressed, but this method is limited to detecting and to recognizing only two drone types, such as multirotor and helicopter, and it has not detected other types [47].

In this study, to achieve higher accuracy in solving the challenges of drone type recognition in visible images, a modified YOLOv4 network is proposed. Drone recognition challenges include the drone's far distance from the camera, a crowded background, unpredictable movements, and the drone's resemblance to birds. As an independent approach, the proposed modified YOLOv4 deep learning network architecture is capable of recognizing birds and four types of drones: multirotors, fixed-wings, helicopters, and VTOLs. To show the improved results of the new model, its performance is also compared with the base YOLOv4 network.

## 4. Methodology

In this study, a modified network based on the latest version of the YOLO network is proposed. The steps to recognize bird species and four different types of drones are presented in Figure 4. In the first step, the input data was prepared to be ready to enter the proposed network. In the second step, the model was trained to recognize the drones, and the weight file obtained for the testing phase was generated. In the third step, the network was tested to observe how it worked; and, in the last step, the proposed deep learning network was evaluated using evaluation metrics.

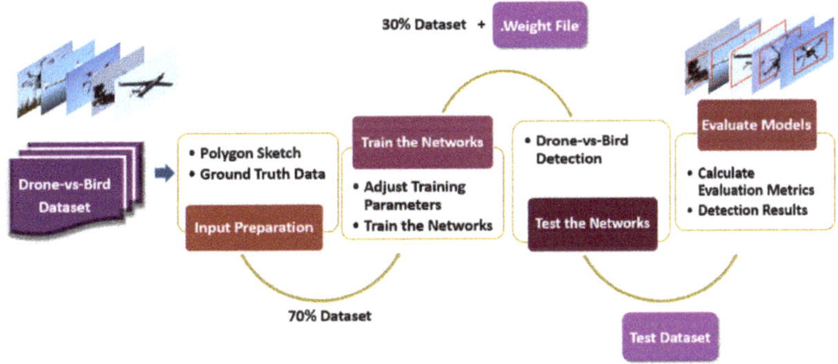


**Figure 4.** Recognition process using implemented modified network.

### 4.1. Input Preparation

Drones are generally divided into four categories: multirotors, fixed-wings, VTOLs, and helicopters. Multirotors are mainly developed in different structures such as octorotor, octo coax wide, hexarotor, and quadrotor. The fixed-wing drone can also be one of the types of fixed wing, plane a-tail, and standard plane, and the VTOL drones are the combination of both the previous versions, including four types of standard VTOL, VTOL duo tailsitter, VTOL quad tailsitter, and plane a-tail. Because of the similarity of the behavior of birds at long distances, this group of datasets belongs to the fifth category of network input data. The schematic drawings of each drone are presented in Figure 5. Thus, in this study, the data was labeled into a total of five classes and prepared for the training phase.

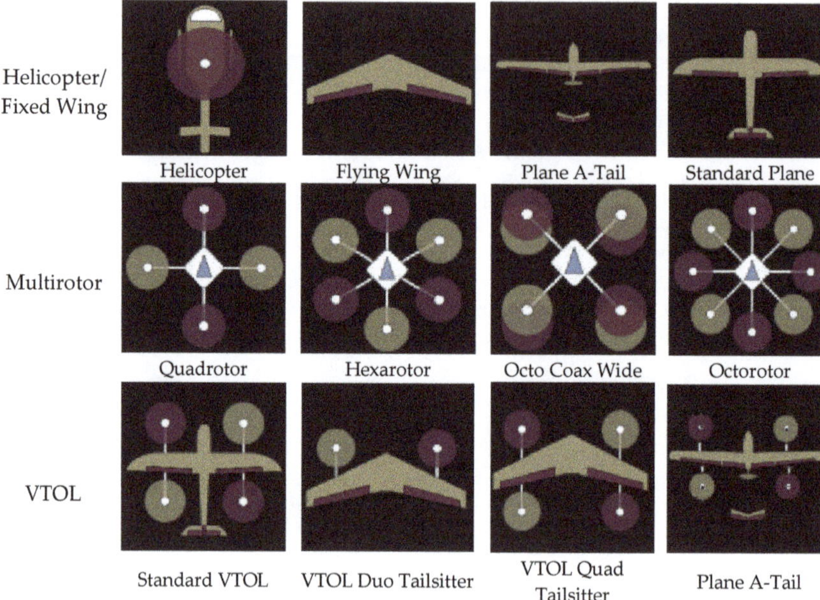



**Figure 5.** Different types of drones as Helicopters, Fixed-Wings, Multirotors, and VTOLs [48].

One strategy for preparing input data is to draw a rectangular bounding box around the object. In this study, the polygon sketch was used to label the drone dataset. Then, the best rectangle containing the object was fitted to the polygon. As shown in Figure 6, drawing the best rectangular bounding box around the drone results in the accurate extraction of the pixels containing the object.

**Figure 6.** The conversion of the polygon to the best-fitted rectangle.

Finally, the bounding box central coordinates, the class number, and its width and its height are normalized in the range of [0, 1], and they are introduced to the next step in the proposed method.

### 4.2. Train the Networks

The YOLOv4 Deep Learning Network is selected as the drone vs. bird recognition network because of its advantages in this area. In addition, this network was modified to improve the performance of the basic network and to better address challenges.

#### 4.2.1. YOLOv4 Deep Learning Network Architecture

According to Figure 7, the YOLOv4 network consists of four main parts: the input of the network, backbone (feature map extractor), neck (feature map collector), and head (results of recognition).

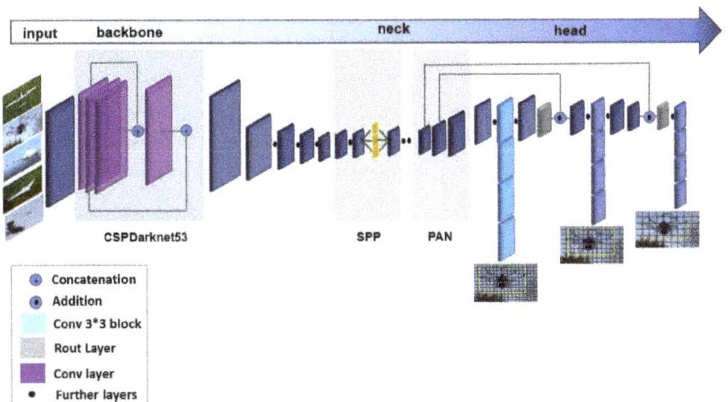

**Figure 7.** The basic YOLOv4 Deep Learning Network Architecture.

The backbone is responsible for extracting the feature map from the previously entered data [26]. In the implemented model, a Darknet53-based network, CSP Darkent-53 [49], is used to extract the feature. CSPDarknet53 enhances the learning process of CNN. The CSPDarknet53 overhead integration pyramid section is connected to improve the receiver field and to differentiate very important context features. Extracting better features leads to increasing the accuracy of detecting drones, recognizing their type, and differentiating them from birds. In the structure of the YOLOv4 network, there are several convolution layers after the backbone. In convolution layers, internal multiplication and feature extraction operations are performed using the obtained feature maps. In this network, a $3 \times 3$ convolutional layer is used after the backbone layer to extract more detailed and accurate features using the Mish activity function. The reason for using the Mish activity function is that this function also considers negative values, solving the problem of overfitting with precise regulatory effects.

The neck receives the feature map created in the backbone stage. This helps to add a layer between the backbone and the head. It consists of a modified Spatial Pyramid Pooling(SPP) and a modified Path Aggregation Network(PAN), both of which are used to gather information to improve accuracy [21,50]. Spatial Pyramid Pooling is an integration layer that removes the constraint of the fixed size of the network input. This layer consists of three pooling, with sizes 256-d, $4 \times 256$-d, $16 \times 256$-d, and (Figure 8). The SPP layer receives the feature map created from the previous convolution layer. It combines features and produces fixed-length outputs, it then connects to fully connected layers, and then enters the improved PAN network. This network is used to aggregate parameters for different detector surfaces instead of feature pyramid networks (FPNs) to detect the object used in YOLOv3.

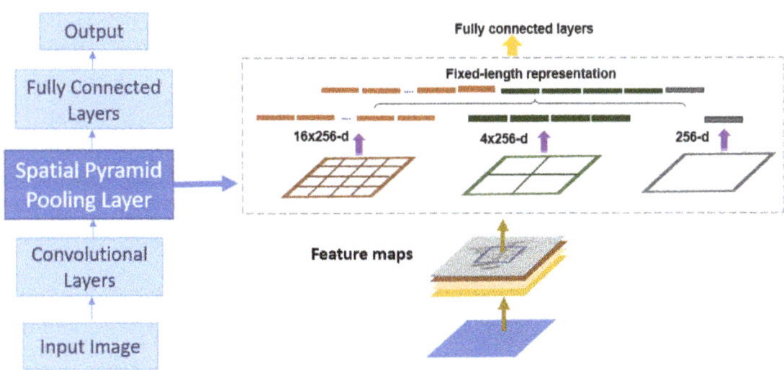

**Figure 8.** Spatial Pyramid Pooling (SPP) block.

The head part is used to classify and to locate the predicted boxes of the proposed network. At this level, the probabilities and the bounding box coordinates (x, y, height, and width) are given. This part uses the YOLOv3 deep learning network architecture, so the output is a tensor containing bounding box coordinates and class probabilities [51].

Bag-of-specials and bag-of-freebies are techniques used in the YOLOv4 algorithm to increase accuracy during and after training [26]. Bag-of-freebies helps to improve recognition during training, without increasing the inference time. This feature uses techniques such as data augmentation and drop blocks. Bag-of-freebies uses CutMin, data augmentation, and DropBlock techniques to increase and to regularize the data. It also uses techniques, such as grid sensitivity elimination, CIoU-loss, CmBN, self-adversarial training, and use of several anchors for one ground truth in detection techniques. Bag-of-specials are techniques that slightly increase the inference time and change the architecture. This feature uses methods such as non-maximum suppression, multi-input residual connections, and cross-stage partial connections (CSP) in the backbone. Additionally, in detection, it uses SPP-block, PAN-path aggregation, mish activation, SAM-block, and DIoU-NMS. Both technologies and their features are useful for training and testing the networks [26].

This network is not suitable for various applications, such as recognition of small objects and objects in cluttered backgrounds and hidden areas. Therefore, in this article, the YOLOv4 network architecture was modified to increase the accuracy of drone recognition and to improve the performance of the model to overcome challenges, which is one of the innovations in this article.

### 4.2.2. The Modified YOLOv4 Deep Learning Network Architecture

In the modified YOLOv4 network, according to Figure 9, three convolutional layers were added to the basic YOLOv4 network architecture after the backbone. Convolutional layers are useful for extracting features from images because these layers deal with spatial redundancy by weight sharing. The addition of these three layers makes the extracted features more exclusive and informative, and it reduces redundancy. This is primarily due to the repeated cascaded convolutions and to information compression by subsampling layers. By reducing redundancy, the network displays a compressed feature about the content of the image. Consequently, by increasing the depth of the network's convolutional layers, more accurate semantic features are extracted, recognition accuracy is increased, and overcoming existing challenges is facilitated [52,53].

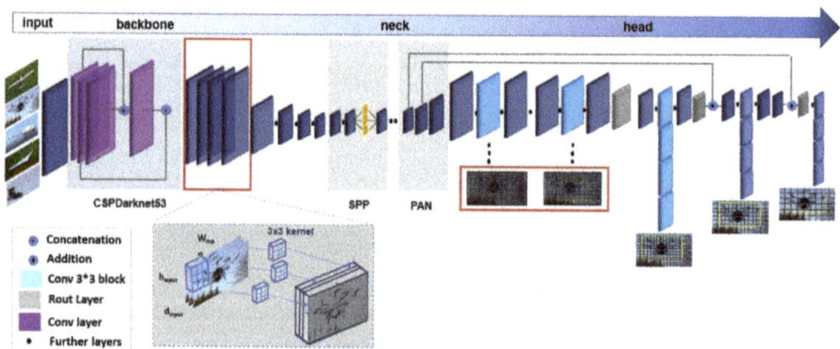


**Figure 9.** Recognition process using modified YOLOv4 network.

In addition, two large-scale convolutional layers were added to the head. In this part, a convolutional layer with a size of 3 × 3 and another with a size of 1 × 1 was used. This 1 × 1 convolutional layer reduced the final depth as well as the volume of network computation. This change balances the modified network to recognize large and small long-range targets.

The implemented network can recognize drones better than the basic YOLOv4 network at long distances with small sizes and against crowded backgrounds and hidden areas. This powerful object recognition model uses a single GPU to provide accurate object recognition. For training these two networks, 70% of the entire dataset was used. The two networks were trained under the same conditions, i.e., with the same data set, the same training parameters, and the same number of iterations. Once the model implementation and training phase are complete, the testing and evaluation phase of the network begins based on the selected evaluation metrics.

### 4.3. Test the Networks

After completing the network training, the network testing process began with 30% of the whole dataset. This stage was to select the best bounding box with the object and to evaluate the network's performance in recognizing drones and birds. The proposed learning network defines a bounding box around the detected drones. Since the drones in the dataset have different sizes and shapes, the proposed model creates multiple bounding boxes to recognize them. However, to select the most suitable bounding box out of the others, the non-maximum suppression (NMS) algorithm must be used [54]. This algorithm was used to remove bounding boxes with lower confidence scores and to select the best drone and bird box. As it appears from Figure 10, the green box is the best box that contains a drone, the other two bounding boxes also cover part of the drone, and they are candidates for drone recognition. In this algorithm, the bounding box with the highest confidence score is selected first and then the boxes with higher overlap with the selected box are removed. This process is continued until no representative boxes contain the drone. The same process is performed for the bird and, finally, the best bounding box is selected.

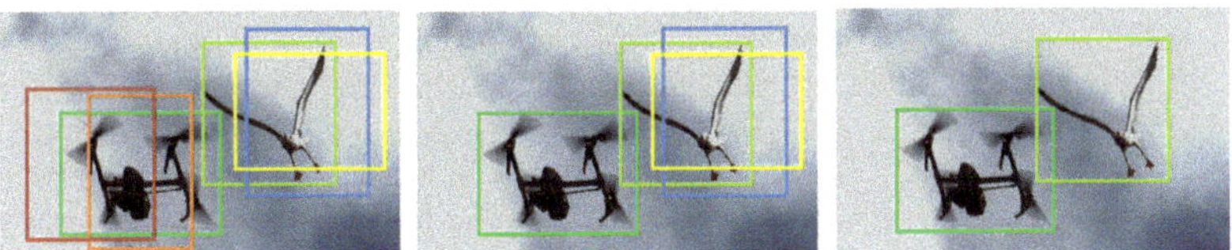

**Figure 10.** Proposed non-maximum suppression (NMS) box selection method.

The steps for performing the non-maximum suppression (NMS) algorithm can be summarized as follows:

- Select the predicted bounding box with the highest confidence level;
- Calculate *IoU* (the intersection and overlap of the selected box and other boxes) (Equation (1));

$$IoU = \frac{Area\ of\ Overlap}{Area\ of\ Union} \tag{1}$$

- Remove boxes with an overlap of more than the default *IoU* threshold of 7% with the selected box;
- Repeat steps 1–3.

In Figure 11, the above algorithm was run twice, and the green boxes were selected as the final bounding boxes containing the drone and the bird.

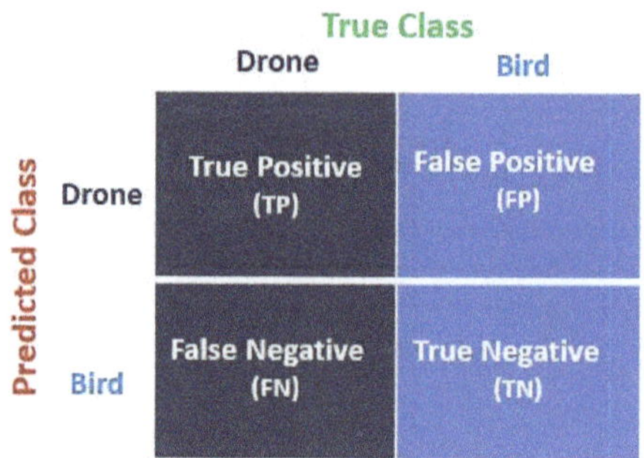



**Figure 11.** Schematic view of the confusion matrix.

*4.4. Evaluation Metrics*

To better understand how the proposed model works, the model was evaluated using mAP, IoU, accuracy, recall, F1-score, and precision.

The mAP is one of the most important evaluation metrics. In this paper, mAP was used as an evaluation metric to recognize drones and birds. It can be claimed that mAP is responsible for comparing the predicted bounding boxes by the network with the ground truth box [19].

The other important parameter that is the key to obtaining recall, F1-score, and precision measures is the confusion matrix. Computing confusion matrix can provide valuable information on the performance of the modified model and the presence of various errors [19]. Therefore, the values of recall, F1-score, and precision can be calculated from the confusion matrix using the values true negative (TN), false negative (FN), true positive (TP), and false positive (FP). The structure of the confusion matrix is illustrated in Figure 11.

The IoU shows the connection between the ground truth bounding box and the predicted bounding box. If the intersection value of these bounding boxes is above the default threshold value of 0.7, the classification is performed correctly (TP, true positive). On the other hand, if the IoU value is below 0.7, it is misdiagnosed (FP, false positive), and if these bounding boxes do not overlap with each other, it is considered a false negative (FN).

Precision means the percentage of positive predictions among predicted classes determined to be positive [19]. Precision, F1-score, and recall metrics are calculated separately for each class of multirotor, helicopter, fixed-wing, VTOL, and bird.

The recall value shows the percentage of positive predictions among all data in the positive class. The F1-score is the mean of values for accuracy and precision, and it can indicate the validity of the classification process. This metric works well for imbalanced data because it takes into account the FN and the FP values [19].

Another metric examined in this study is the overall accuracy of the model [19]. This metric shows the performance of the model in recognizing drones and birds.

## 5. Experiments and Results

To assess the performance of proposed networks in recognizing drones and distinguishing them from birds, the steps of implementing the YOLOv4 deep learning network and the proposed modified network are presented in this section. Moreover, the types of the applied dataset, the network evaluation over different types of images, and the recognition challenges are discussed.

*5.1. Data Preparation*

To begin the training phase, a set of 26,000 visible images were first prepared that included various bird species and four types of drones such as multirotors, helicopters, fixed-wing, and VTOLs (examples are shown in Figure 12). The use of these four drone types in the collected dataset is another innovation of this study. Public images and videos were used to create the dataset. Approximately 70% of these images were used for training and 30% for testing the network. In both phases, the number of images used is the same in each class. To prepare these images, first, a polygon around the target was determined. This task was handled using an efficient and useful Computer Vision Annotation Tool (CVAT) and the input images were categorized into five classes. Based on these boxes, the best rectangles containing the drone were then selected and passed to the network. In this dataset, multirotors are classified in the first class, helicopters in the second class, fixed-wing aircraft in the third class, VTOL in the fourth class, and birds in the fifth class.

**Figure 12.** Included in the drone dataset.

## 5.2. Model Implementation and Training Results

The modified YOLOv4 network and the basic YOLOv4 network were trained using an Nvidia GeForce MX450 Graphics Processing Unit (GPU) hardware with 30,000 iterations. To train these two networks, the settings in the configuration file were changed as follows:

- The input image size was set to $160 \times 160$;
- The subdivision and the batch parameters were changed to 1 and 64, respectively (these settings were made to avoid errors due to lack of memory);
- The learning rate was changed to 0.0005;
- The step parameter was changed to 24,000 and 27,000, (with 80% and 90% of the number of iterations, respectively);
- The size of the filter in three convolutional layers near the YOLO layers was changed to 30 according to the number of classes.

The basic network used for training is the widely used Darknet Framework [55]. Depending on the hardware used, a CUDA (Compute Unified Device Architecture) Toolkit version 10.0, CUDNN (CUDA Deep Neural Network library) version 8.2, Visual Studio 2017, and OpenCV version 4.0.1 are used. During the training, a graph of the number of iterations and loss was plotted as shown in Figure 13. After 30 k iterations (about 3–4 days), the modified implemented method and the basic YOLOv4 model achieved a loss of 0.58 and 0.68, respectively. The obtained values show that the detection loss of the proposed network was lower than that of the basic network, indicating the better performance of the proposed model in the training stage.

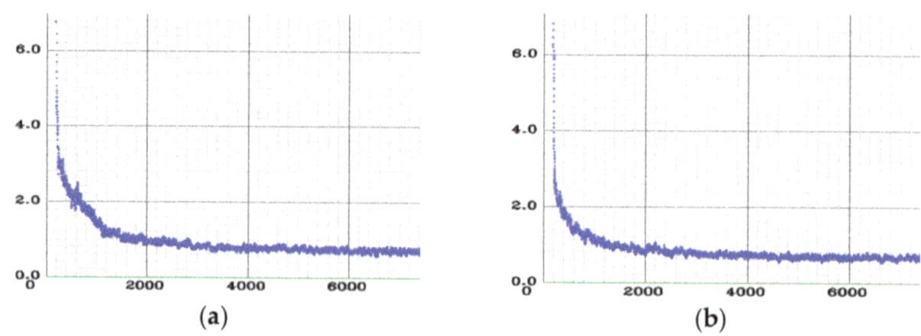

**Figure 13.** The loss graph in the training process; (**a**) The YOLOv4 network. (**b**)The modified YOLOv4 network.

### 5.3. Evaluation of the YOLOv4 and Modified YOLOv4 Models

The modified YOLOv4 and the basic YOLOv4 networks were precisely evaluated using accuracy, mAP, recall, precision, and F1-score in recognizing the drones. Tables 1 and 2 show the evaluation result of the proposed model and the basic YOLOv4 network. Accordingly, the performance of both models was compared with the same dataset, with an IoU default threshold of 0.07 and the same number of iterations. The recognition performance of the modified model was performed in five classes: multi-rotor, helicopter, fixed-wing, VTOL, and birds; and, the evaluation metrics of precision, recall, F1-score, accuracy, mAP, and IoU were increased. These results indicate the better performance of the modified model than the basic YOLOv4 model. In this network, the accuracy was 0.83%, the mAP was 83%, and the IoU was 84%, which was 4% better than the basic model.

**Table 1.** Evaluation results of the basic and Modified YOLOv4 networks.

| Dataset | Model | Num of Images | Precision % | Recall % | F1-Score % |
|---|---|---|---|---|---|
| Bird | YOLOv4 | 1570 | 81 | 87 | 84 |
| | Modified YOLOv4 | | 87 | 90 | 89 |
| Fixed Wing | YOLOv4 | 1570 | 88 | 70 | 78 |
| | Modified YOLOv4 | | 88 | 77 | 82 |
| Helicopter | YOLOv4 | 1570 | 81 | 73 | 77 |
| | Modified YOLOv4 | | 88 | 73 | 80 |
| Multirotor | YOLOv4 | 1570 | 77 | 90 | 83 |
| | Modified YOLOv4 | | 79 | 90 | 84 |
| VTOL | YOLOv4 | 1570 | 72 | 77 | 74 |
| | Modified YOLOv4 | | 74 | 83 | 78 |
| Total | YOLOv4 | 7850 | 80 | 79 | 79 |
| | Modified YOLOv4 | | 83 | 83 | 83 |

**Table 2.** Total evaluation results of the basic and Modified YOLOv4 networks.

| Dataset | Model | Num of Images | Accuracy % | mAP % | IoU % |
|---|---|---|---|---|---|
| Total | YOLOv4 | 7850 | 79 | 79 | 80 |
| | Modified YOLOv4 | | 83 | 83 | 84 |

The comparison of precision, recall, F1-score, and total mAP in all five classes in the two implemented models are graphically presented in Figure 14. The precision was improved in four classes (multirotor, helicopter, VTOL, and bird) and unchanged in the fixed-wing class. In general, these results show the improvement of model performance in the proposed network. By comparing the recall according to Figure 15, its improvement is observed in three classes, and there is no change in the other two classes. This means that the ability and the performance of the second model in recognition were improved. As it appears from Figure 16, the F1-score increased compared to the basic model of YOLOv4, and this also shows the better performance of the modified YOLOv4 network.

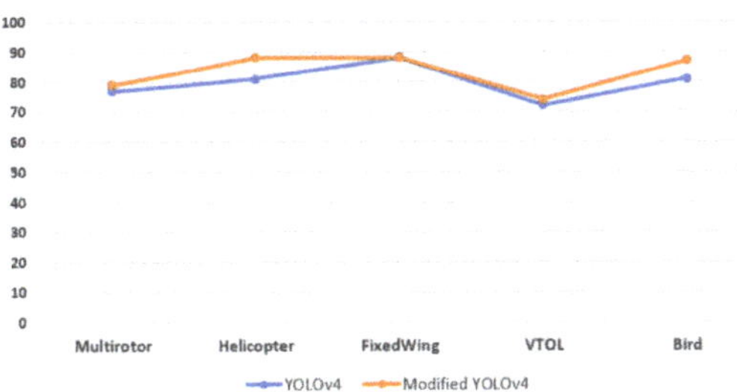

**Figure 14.** The comparison between the precision values in the basic and the modified YOLOv4 models.

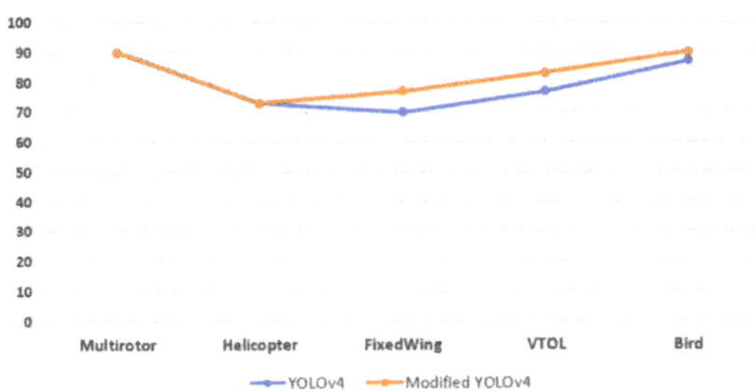

**Figure 15.** The comparison between the recall values in the basic and the modified YOLOv4 models.

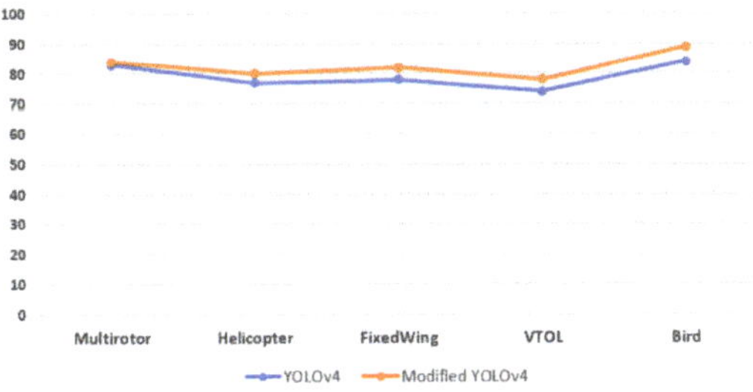

**Figure 16.** The comparison between F1-score in the basic and the modified YOLOv4 models.

Figure 17 shows the results of drone recognition in the basic YOLOv4 network and Figure 18 shows these results with the modified YOLOv4 network. A comparison of the results of the two networks shows an improvement in the accuracy of the bounding boxes and the class probabilities in the modified network. On the other hand, the recognition of four types of multirotor (octorotor, octo coax wide, hexarotor, and quadrotor), three types of fixed-wing (flying wing, plane a-tail, and standard plane), four types of VTOLs (standard VTOL, VTOL duo tailsitter, VTOL quad tailsitter, and plane a-tail), one type of helicopter, and their discrimination from birds were improved in the proposed model.

**Figure 17.** Some samples of drone-vs-bird recognition results with a basic YOLOv4 network.

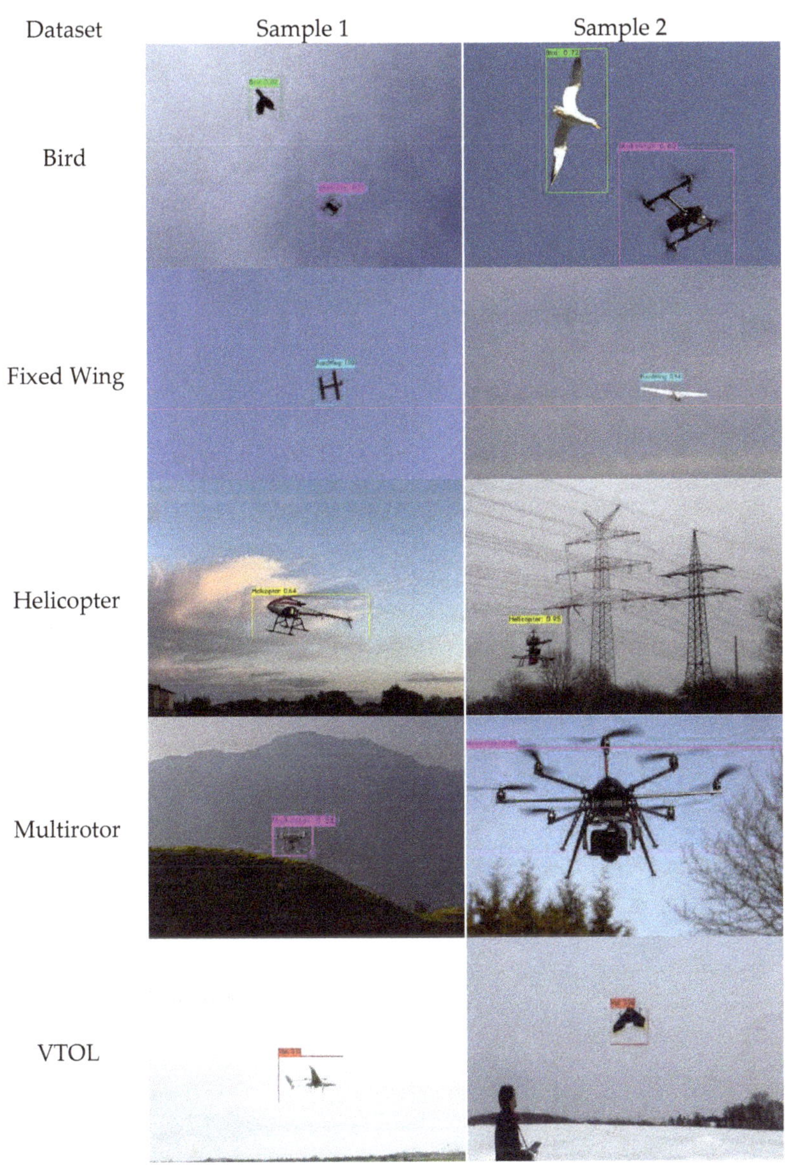

**Figure 18.** Some samples of drone-vs-bird recognition results with a modified YOLOv4 network.

*5.4. Addressing the Challenges in the Modified YOLOv4 Model*

In this study, the YOLOv4 network was modified to improve the recognition results of four types of drones and to differentiate them from birds concerning existing challenges, such as the smaller size of the drones, crowded backgrounds, loss of scalability, and similarity with birds. Figure 19 shows the performance of both the basic and the modified models in overcoming the challenges.

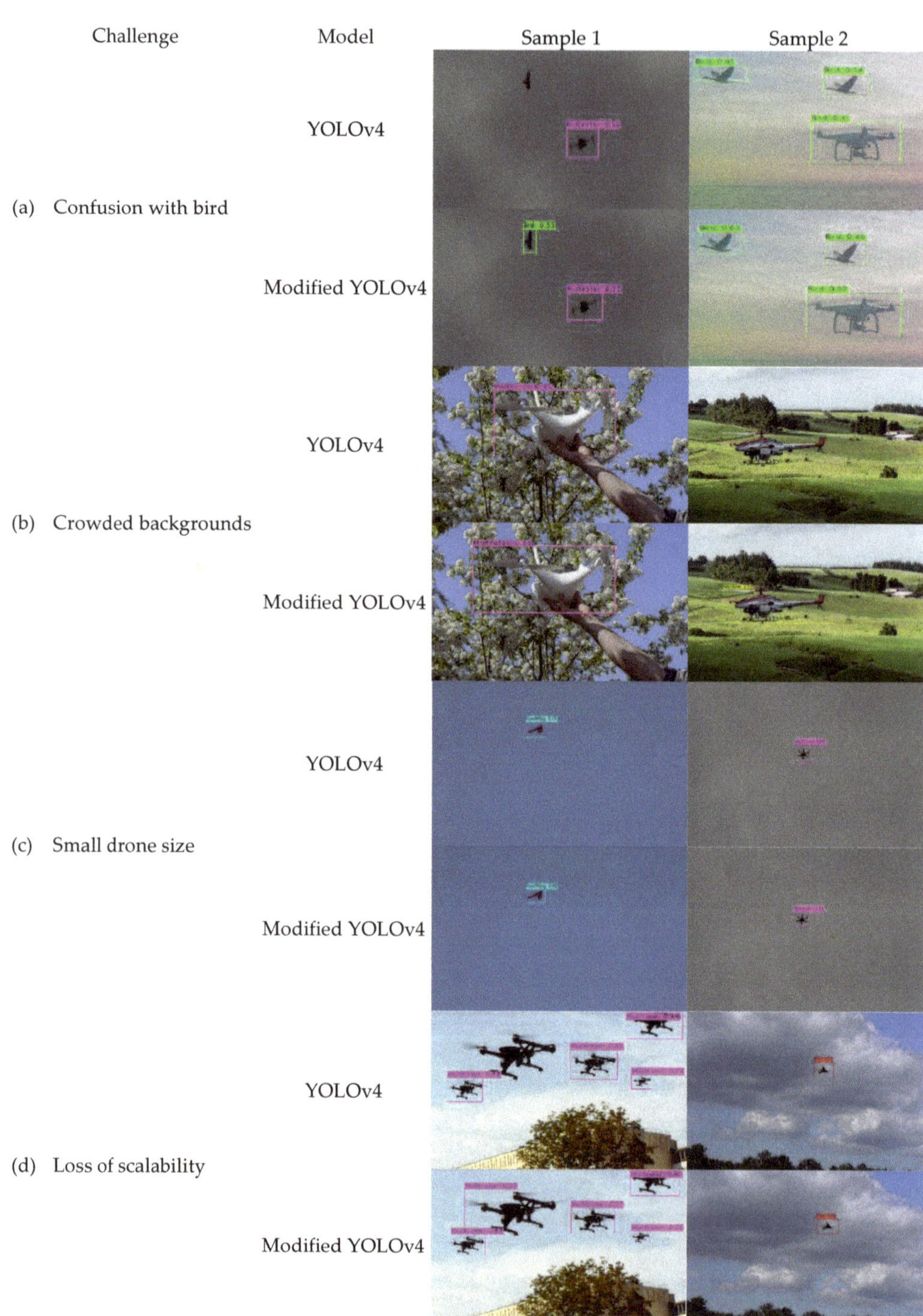

**Figure 19.** Samples of handling challenges in drone recognition.

The first challenge is to recognize the drone and to distinguish it from the bird; the basic model is unable to recognize the bird in some samples and the class probability values are lower than for the modified model. The modified YOLOv4 model can recognize drones and distinguish them from birds by extracting smaller and more accurate features, and it has demonstrated this ability in the evaluated samples (Figure 19). The second challenge studied is the presence of drones in a crowded background, where the modified model can recognize targets with a high prediction probability. In some cases, the base model is unable to recognize the drone or it has a lower probability than the modified model. The third challenge is the small size of the drones and their placement at long ranges, where the modified model performs better than the base model. The modified model can accurately recognize all small drones in the test images. This ability is due to the use of more convolutional layers at the head of the network. The change in network structure does not remarkably affect the execution time of the network training algorithm. The training of the proposed network takes approximately 3 h longer than that of the first network. Finally, the challenge of the detectability swarm of UAVs at different scales is investigated. In this challenge, the base model is unable to recognize drones in the images in some cases, and it generally has lower class probability than the modified model.

Figure 20 illustrates several samples of different drone types in crowded environments with lighting conditions and different weather. As the results show, the modified YOLOv4 network is capable of recognizing different drone types under complex and challenging conditions.

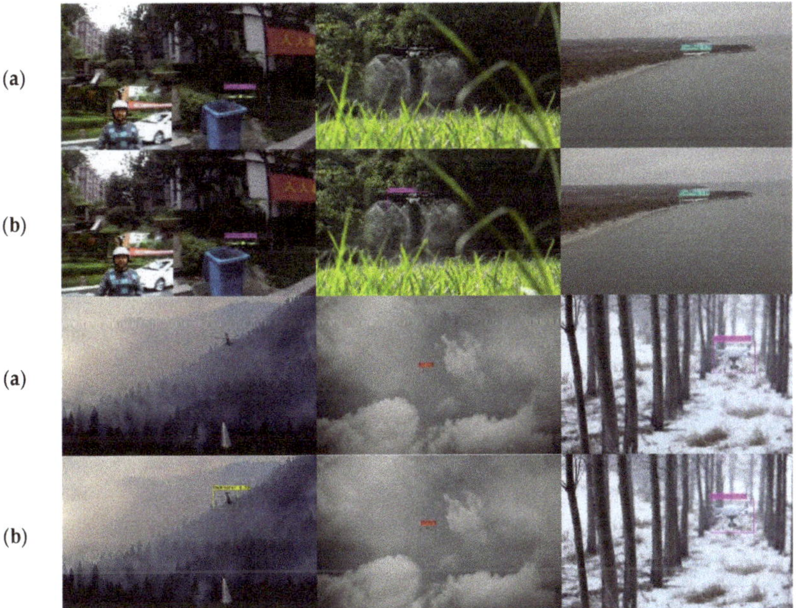

**Figure 20.** Several samples of drone recognition in different light and weather conditions and against a crowded background. (**a**) YOLOv4. (**b**) Modified YOLOv4.

## 6. Discussion

The evaluation metric examined in this study is confusion matrix, IoU, mAP, accuracy, precision, recall, and F1-score. The values obtained from these criteria in the modified network are as follows: 79% precision, 84% F1-score, and 90% recall for the multirotor; 88% precision, 82% F1-score, and 77% recall for the fixed-wing; 88% precision, 80% F1-score, and 73% recall for the helicopter; 74% precision, 78% F1-score, and 83% recall for VTOL; and 87% precision, 89% F1-score, and 90% recall for the bird. Based on the basic YOLOv4 network results, it can be said that the proposed network had an average improvement of 3.4% in precision, 3.2% in the recall, and 3.4% in F1-score values. The overall accuracy, indicating the correct classification of the input data set into five classes, was also investigated. This criterion increased by 4% in the modified network compared to the basic network. The mAP also increased by 4% in the modified network compared to the basic YOLOv4 network. This rate indicates an improvement in the classification of the proposed model in the five classes. Finally, the IoU metric was used to calculate the overlap value of the predicted bounding box with the ground truth bounding box. This metric also increased by 4% in the modified network, indicating the higher accuracy of the predicted bounding box.

In recent years, the use of artificial intelligence and deep learning methods has become one of the most popular and useful methods in object recognition. In 2021, Tan Wei Xun et al. detected and tracked a drone using the YOLOv3. The results of this method show that the proposed YOLOv3 network detects drones of different sizes with a confidence level between 60 and 100% [44]; also this year, drone detection and tracking were performed by Isaac-Medina et al. using a set of visible and thermal images and four deep learning network architectures, including RCNN, SSD, YOLOv3, and DETR. According to the results of this study, it can be said that the YOLOv3 deep learning network performs better than the other models [45]. One of the problems with these studies was the inability to detect small drones over long distances. In addition, an automatic drone detection system using YOLOv4 was developed by Singha et al. In this system, a collection of drone-vs-bird images were used to detect the drone and its evaluation was tested using a video dataset. The results of this work include the high accuracy of the YOLOv4 network [46]. One of the limitations of this study was the lack of recognition of the different types of drones and the lack of challenging images. Moreover, Liu et al. applied three object detection methods, such as YOLOv3, YOLOv4, RetinaNet, and FCOS networks on the drone dataset. To build a scattered, flat network and to get great accuracy in drone detection, the pruned YOLOv4 model was used, which improved the accuracy of drone detection at high speed. However, the study did not address the challenges of crowded background images, hidden areas, and the distance of the drone from the camera [16]. Additionally in 2022, Samadzadegan et al. recognized two types of drones (multirotors and helicopters) and birds using the YOLOv4 deep learning network. However, this paper did not recognize two other drone types such as VTOL and fixed-wing drones, which could be more dangerous than the previous two types. In addition, a rectangular bounding box was used in the data preparation, which resulted in the input of additional information and reduced the accuracy of recognition [47].

Previous studies have focused on drone detection, and a small number of these studies have examined the recognition of different types of drones. Moreover, the challenges of drone recognition, such as the small size of the drone, crowded background, hidden areas, and confusion with birds, have not been comprehensively addressed in these studies. Therefore, it can be said that the unauthorized presence of drones in challenging environments and their inaccurate recognition in sensitive infrastructures is still one of the most important problems in ensuring public safety. The main goal of this research is to recognize four types of drones and to differentiate them from birds at far distances despite challenges, such as the small size of the drone, a crowded background, and the presence of hidden areas.

In this study, the YOLOv4 network was modified to improve drone recognition challenges. A set of visible images with different types of drones and birds in different environments at near and far distances were collected to recognize four types of drones and to differentiate them from birds. Two convolutional layers were added to the head of the YOLOv4 to solve the challenges of small drone recognition. For example, in Figure 20 there are examples of small drones that the basic model was unable to recognize in some cases. However, the modified model recognized them; and, in other cases they operated with less accuracy than the modified network. This result shows the improvement of the current network compared to the basic network for recognizing small drones. Furthermore, to increase the accuracy of feature extraction and precise recognition of drones and birds, three convolutional layers were added after the backbone layer. Adding these layers to the architecture of the basic YOLOv4 network did not change the training time of the network, and it took only a few hours longer than training the basic network. By applying these changes in the network architecture and using extensive datasets, the proposed method was able to recognize all drone types and bird species in challenging environments. In Figure 19, rows (a), (b), and (d) provide examples of difficult images and swarms of UAVs, all of which have higher recognition accuracy in the modified network than in the basic network. Figure 20 also contains other challenging examples where the proposed network performs better.

## 7. Conclusions

As has been noted, UAV recognition in various situations is a complex process; the usual methods and even conventional deep learning network methods do not work well in some cases. In this study, the basic YOLOv4 network was used to recognize drone types and to differentiate them from bird species. To increase recognition accuracy and to better address existing challenges, a novel and modified model of this network was proposed. To train, test, and evaluate these two networks, a collection of 26,000 visible image datasets including four types of UAVs (multirotor, fixed-wing, helicopter, and VTOL) and birds were collected. The comparison of these two models was done using mAP, confusion matrix, IoU, precision, accuracy, F1-score, and recall evaluation metrics. With the modified YOLOv4 model, we achieved 84% IoU, 83% mAPs, and 83% accuracy, which is better than the basic model, and it solved the challenges well. In the future, real-time identification with onboard systems can be studied in addition to drone recognition. Background removal algorithms can also be used to make labeling input data easier and faster. In addition to the multirotor types used, Tri-rotors can also be used to complete the dataset. Additionally, other deep learning networks can be used to compare their results with the results of this modified network.

**Author Contributions:** All authors contributed to the study conception and design. F.D.J. contributed to supervision, reviewing, and validation. F.S. is a drone expert involved in the conceptualization, methodology, and editing of the draft. F.A.M. contributed to programing, visualization, computer vision concepts, writing, and editing. M.G. is involved in software, deep learning concepts, data collection and preparation, and drafting. All authors have read and agreed to the published version of the manuscript.

**Funding:** This research received no external funding.

**Institutional Review Board Statement:** Not applicable.

**Informed Consent Statement:** Not applicable.

**Data Availability Statement:** Not applicable.

**Conflicts of Interest:** The authors have no conflict of interest to disclose.

## References

1. Mueller, M.; Smith, N.; Ghanem, B. *A Benchmark and Simulator for UAV Tracking*; Springer: Berlin/Heidelberg, Germany, 2016; Volume 9905, pp. 445–461.
2. Wu, M.; Xie, W.; Shi, X.; Shao, P.; Shi, Z. Real-time drone detection using deep learning approach. In Proceedings of the International Conference on Machine Learning and Intelligent Communications, Hangzhou, China, 6–8 July 2018; Springer: Berlin/Heidelberg, Germany, 2018.
3. Bansod, B.; Bansod, B.; Singh, R.; Thakur, R.; Singhal, G. A comparision between satellite based and drone based remote sensing technology to achieve sustainable development: A review. *J. Agric. Environ. Int. Dev.* **2017**, *111*, 383–407.
4. Orusa, T.; Orusa, R.; Viani, A.; Carella, E.; Borgogno Mondino, E. Geomatics and EO Data to Support Wildlife Diseases Assessment at Landscape Level: A Pilot Experience to Map Infectious Keratoconjunctivitis in Chamois and Phenological Trends in Aosta Valley (NW Italy). *Remote Sens.* **2020**, *12*, 3542. [CrossRef]

5. Chiu, M.; Xu, X.; Wei, Y.; Huang, Z.; Schwing, A.; Brunner, R.; Khachatrian, H.; Karapetyan, H.; Dozier, I.; Rose, G.; et al. Agriculture-Vision: A Large Aerial Image Database for Agricultural Pattern Analysis. In Proceedings of the IEEE/CVF Conference on Computer Vision and Pattern Recognition, Seattle, WA, USA, 13–19 June 2020.

6. Anwar, M.Z.; Kaleem, Z.; Jamalipour, A. Machine Learning Inspired Sound-Based Amateur Drone Detection for Public Safety Applications. *IEEE Trans. Veh. Technol.* **2019**, *68*, 2526–2534. [CrossRef]

7. Sathyamoorthy, D. A Review of Security Threats of Unmanned Aerial Vehicles and Mitigation Steps. *J. Def. Secur.* **2015**, *6*, 81–97.

8. Yaacoub, J.-P.; Noura, H.; Salman, O.; Chehab, A. Security Analysis of Drones Systems: Attacks, Limitations, and Recommendations. *Internet Things* **2020**, *11*, 100218. [CrossRef]

9. Semkin, V.; Yin, M.; Hu, Y.; Mezzavilla, M.; Rangan, S. Drone Detection and Classification Based on Radar Cross Section Signatures. In Proceedings of the 2020 International Symposium on Antennas and Propagation (ISAP), Osaka, Japan, 25–28 January 2021.

10. Haag, M.U.D.; Bartone, C.G.; Braasch, M.S. Flight-test evaluation of small form-factor LiDAR and radar sensors for sUAS detect-and-avoid applications. In Proceedings of the 2016 IEEE/AIAA 35th Digital Avionics Systems Conference (DASC), Sacramento, CA, USA, 25–29 September 2016.

11. Svanstrom, F.; Englund, C.; Alonso-Fernandez, F. Real-Time Drone Detection and Tracking With Visible, Thermal and Acoustic Sensors. In Proceedings of the 2020 25th International Conference on Pattern Recognition (ICPR), Milan, Italy, 10–15 January 2020.

12. Andraši, P.; Radišić, T.; Muštra, M.; Ivošević, J. Night-time Detection of UAVs using Thermal Infrared Camera. *Transp. Res. Procedia* **2017**, *28*, 183–190. [CrossRef]

13. Nguyen, P.; Ravindranatha, M.; Nguyen, A.; Han, R.; Vu, T. Investigating Cost-effective RF-based Detection of Drones. In Proceedings of the 2nd Workshop on Micro Aerial Vehicle Networks, Systems, and Applications for Civilian Use, Singapore, 26 June 2016; pp. 17–22.

14. Humphreys, T.E. *Statement on the Security Threat Posed by Unmanned Aerial Systems and Possible Countermeasures*; Oversight and Management Efficiency Subcommittee, Homeland Security Committee: Washington, DC, USA, 2015.

15. Drozdowicz, J.; Wielgo, M.; Samczynski, P.; Kulpa, K.; Krzonkalla, J.; Mordzonek, M.; Bryl, M.; Jakielaszek, Z. 35 GHz FMCW drone detection system. In Proceedings of the 2016 17th International Radar Symposium (IRS), Krakow, Poland, 10–12 May 2016.

16. Liu, H.; Fan, K.; Ouyang, Q.; Li, N. Real-time small drones detection based on pruned yolov4. *Sensors* **2021**, *21*, 3374. [CrossRef]

17. Seidaliyeva, U.; Alduraibi, M.; Ilipbayeva, L.; Almagambetov, A. Detection of loaded and unloaded UAV using deep neural network. In Proceedings of the 2020 4th IEEE International Conference on Robotic Computing (IRC), Taichung, Taiwan, 9–11 November 2020.

18. Ashraf, M.; Sultani, W.; Shah, M. Dogfight: Detecting Drones from Drones Videos. In Proceedings of the 2021 IEEE/CVF Conference on Computer Vision and Pattern Recognition (CVPR), Nashville, TN, USA, 20–25 June 2021; pp. 7063–7072.

19. Simonyan, K.; Zisserman, A. Very deep convolutional networks for large-scale image recognition. *arXiv* **2014**, arXiv:1409.1556.

20. Girshick, R.; Donahue, J.; Darrell, T.; Malik, J. Rich Feature Hierarchies for Accurate Object Detection and Semantic Segmentation. In Proceedings of the IEEE Computer Society Conference on Computer Vision and Pattern Recognition, Columbus, OH, USA, 23–28 June 2013.

21. He, K.; Zhang, X.; Ren, S.; Sun, J. Spatial Pyramid Pooling in Deep Convolutional Networks for Visual Recognition. In *Computer Vision–ECCV 2014*; Springer International Publishing: Cham, Switzerland, 2014.

22. Ren, S.; He, K.; Girshick, R.; Sun, J. Faster R-CNN: Towards real-time object detection with region proposal networks. In Proceedings of the 28th International Conference on Neural Information Processing Systems, Montreal, QC, Canada, 7–12 December 2015; MIT Press: Cambridge, MA, USA, 2015; Volume 1, pp. 91–99.

23. Long, J.; Shelhamer, E.; Darrell, T. Fully convolutional networks for semantic segmentation. In Proceedings of the IEEE Conference on Computer Vision and Pattern Recognition, Boston, MA, USA, 7–12 June 2015.

24. Liu, W.; Anguelov, D.; Erhan, D.; Szegedy, C.; Reed, S.; Fu, C.-Y.; Berg, A.C. SSD: Single Shot MultiBox Detector. In Proceedings of the European Conference on Computer Vision, Amsterdam, The Netherlands, 11–14 October 2016; Volume 9905, pp. 21–37.

25. Redmon, J.; Divvala, S.; Girshick, R.; Farhadi, A. You Only Look Once: Unified, Real-Time Object Detection. *arXiv* **2016**, arXiv:1506.02640.

26. Bochkovskiy, A.; Wang, C.-Y.; Liao, H.-Y.M. Yolov4: Optimal speed and accuracy of object detection. *arXiv* **2020**, arXiv:2004.10934.

27. Chaurasia, R.; Mohindru, V. Unmanned aerial vehicle (UAV): A comprehensive survey. In *Unmanned Aerial Vehicles for Internet of Things (IoT) Concepts, Techniques, and Applications*; Wiley: New Delhi, India, 2021; pp. 1–27.

28. Gu, H.; Lyu, X.; Li, Z.; Shen, S.; Zhang, F. Development and experimental verification of a hybrid vertical take-off and landing (VTOL) unmanned aerial vehicle (UAV). In Proceedings of the 2017 International Conference on Unmanned Aircraft Systems (ICUAS), Miami, FL, USA, 13–16 June 2017.

29. Cai, G.; Lum, K.; Chen, B.M.; Lee, T.H. A brief overview on miniature fixed-wing unmanned aerial vehicles. In Proceedings of the IEEE ICCA, Xiamen, China, 9–11 June 2010.

30. Kotarski, D.; Piljek, P.; Pranjić, M.; Grlj, C.G.; Kasać, J. A Modular Multirotor Unmanned Aerial Vehicle Design Approach for Development of an Engineering Education Platform. *Sensors* **2021**, *21*, 2737. [CrossRef] [PubMed]

31. Cai, G.; Chen, B.M.; Lee, T.H.; Lum, K.Y. Comprehensive nonlinear modeling of an unmanned-aerial-vehicle helicopter. In Proceedings of the AIAA Guidance, Navigation and Control Conference and Exhibit, Honolulu, HI, USA, 18–21 August 2008.

32. Qin, B.; Zhang, D.; Tang, S.; Wang, M. Distributed Grouping Cooperative Dynamic Task Assignment Method of UAV Swarm. *Appl. Sci.* **2022**, *12*, 2865. [CrossRef]

33. Shafiq, M.; Ali, Z.A.; Israr, A.; Alkhammash, E.H.; Hadjouni, M. A Multi-Colony Social Learning Approach for the Self-Organization of a Swarm of UAVs. *Drones* **2022**, *6*, 104. [CrossRef]
34. Ali, Z.A.; Han, Z.; Masood, R.J. Collective Motion and Self-Organization of a Swarm of UAVs: A Cluster-Based Architecture. *Sensors* **2021**, *21*, 3820. [CrossRef]
35. Xu, C.; Zhang, K.; Jiang, Y.; Niu, S.; Yang, T.; Song, H. Communication Aware UAV Swarm Surveillance Based on Hierarchical Architecture. *Drones* **2021**, *5*, 33. [CrossRef]
36. Li, Y. Research and application of deep learning in image recognition. In Proceedings of the 2022 IEEE 2nd International Conference on Power, Electronics and Computer Applications (ICPECA), Shenyang, China, 21–23 January 2022; IEEE: Piscataway, NJ, USA, 2022.
37. Pathak, A.R.; Pandey, M.; Rautaray, S. Application of deep learning for object detection. *Procedia Comput. Sci.* **2018**, *132*, 1706–1717. [CrossRef]
38. Deng, L.; Yu, D. Deep Learning: Methods and Applications. In *Foundations and Trends® in Signal Processing*; Now Publishers Inc.: Hanover, NH, USA, 2014; Volume 7, pp. 197–387.
39. Nalamati, M.; Kapoor, A.; Saqib, M.; Sharma, N.; Blumenstein, M. Drone Detection in Long-Range Surveillance Videos. In Proceedings of the 2019 16th IEEE International Conference on Advanced Video and Signal Based Surveillance (AVSS), Taipei, Taiwan, 18–21 September 2019; pp. 1–6.
40. Unlu, E.; Zenou, E.; Riviere, N.; Dupouy, P.-E. Dupouy Deep learning-based strategies for the detection and tracking of drones using several cameras. *IPSJ Trans. Comput. Vis. Appl.* **2019**, *11*, 7. [CrossRef]
41. Mahdavi, F.; Rajabi, R. Drone Detection Using Convolutional Neural Networks. In Proceedings of the 2020 6th Iranian Conference on Signal Processing and Intelligent Systems (ICSPIS), Mashhad, Iran, 23–24 December 2020; IEEE: Piscataway, NJ, USA, 2020.
42. Behera, D.K.; Raj, A.B. Drone detection and classification using deep learning. In Proceedings of the 2020 4th International Conference on Intelligent Computing and Control Systems (ICICCS), Madurai, India, 13–15 May 2020; IEEE: Piscataway, NJ, USA, 2020.
43. Shi, Q.; Li, J. Objects Detection of UAV for Anti-UAV Based on YOLOv4. In Proceedings of the 2020 IEEE 2nd International Conference on Civil Aviation Safety and Information Technology (ICCASIT), Weihai, China, 14–16 October 2020; IEEE: Piscataway, NJ, USA, 2020.
44. Xun, D.T.W.; Lim, Y.L.; Srigrarom, S. Drone detection using YOLOv3 with transfer learning on NVIDIA Jetson TX2. In Proceedings of the 2021 2nd International Symposium on Instrumentation, Control, Artificial Intelligence, and Robotics (ICA-SYMP), Bangkok, Thailand, 20–22 January 2021; IEEE: Piscataway, NJ, USA, 2021.
45. Isaac-Medina, B.K.; Poyser, M.; Organisciak, D. Unmanned aerial vehicle visual detection and tracking using deep neural networks: A performance benchmark. *arXiv* **2021**, arXiv:2103.13933.
46. Singha, S.; Aydin, B. Automated Drone Detection Using YOLOv4. *Drones* **2021**, *5*, 95. [CrossRef]
47. Samadzadegan, F.; Javan, F.D.; Mahini, F.A.; Gholamshahi, M. Detection and Recognition of Drones Based on a Deep Convolutional Neural Network Using Visible Imagery. *Aerospace* **2022**, *9*, 31. [CrossRef]
48. Roche, R. QGroundControl (QC). 2019. Available online: http://qgroundcontrol.com/ (accessed on 20 May 2022).
49. Wang, C.; Liao, H.M.; Wu, Y.; Chen, P.; Hsieh, J.; Yeh, I. CSPNet: A New Backbone that can Enhance Learning Capability of CNN. In Proceedings of the 2020 IEEE/CVF Conference on Computer Vision and Pattern Recognition Workshops (CVPRW), Seattle, WA, USA, 14–19 June 2020.
50. Liu, S.; Qi, L.; Qin, H.; Shi, J.; Jia, J. Path aggregation network for instance segmentation. In Proceedings of the IEEE Conference on Computer Vision and Pattern Recognition, Salt Lake City, UT, USA, 18–23 June 2018.
51. Redmon, J.; Farhadi, A. Yolov3: An incremental improvement. *arXiv* **2018**, arXiv:1804.02767.
52. Ñanculef, R.; Radeva, P.; Balocco, S. Training Convolutional Nets to Detect Calcified Plaque in IVUS Sequences. In *Intravascular Ultrasound*; Elsevier: Amsterdam, The Netherlands, 2020; pp. 141–158.
53. Wang, L.; Lee, C.-Y.; Tu, Z.; Lazebnik, S. Training deeper convolutional networks with deep supervision. *arXiv* **2015**, arXiv:1505.02496.
54. Hosang, J.; Benenson, R.; Schiele, B. Learning non-maximum suppression. In Proceedings of the IEEE Conference on Computer Vision and Pattern Recognition, Honolulu, HI, USA, 21–26 July 2017.
55. Redmon, J. Darknet: Open Source Neural Networks in C. 2013–2016. Available online: http://pjreddie.com/darknet/ (accessed on 20 May 2022).

 *drones*

*Article*

# Anti-Occlusion UAV Tracking Algorithm with a Low-Altitude Complex Background by Integrating Attention Mechanism


Chuanyun Wang [1,*], Zhongrui Shi [2], Linlin Meng [1], Jingjing Wang [3], Tian Wang [4], Qian Gao [1] and Ershen Wang [5]

1. College of Artificial Intelligence, Shenyang Aerospace University, Shenyang 110136, China; menglinlin@stu.sau.edu.cn (L.M.); gaoqian@buaa.edu.cn (Q.G.)
2. School of Computer Science, Shenyang Aerospace University, Shenyang 110136, China; shizhongrui@stu.sau.edu.cn
3. China Academic of Electronics and Information Technology, Beijing 100041, China; wangjingjing@cetccloud.com
4. Institute of Artificial Intelligence, Beihang University, Beijing 100191, China; wangtian@buaa.edu.cn
5. School of Electronic and Information Engineering, Shenyang Aerospace University, Shenyang 110136, China; wes2016@sau.edu.cn
* Correspondence: wangcy0301@sau.edu.cn



**Citation:** Wang, C.; Shi, Z.; Meng, L.; Wang, J.; Wang, T.; Gao, Q.; Wang, E. Anti-Occlusion UAV Tracking Algorithm with a Low-Altitude Complex Background by Integrating Attention Mechanism. *Drones* **2022**, 6, 149. https://doi.org/10.3390/drones6060149

Academic Editors: Daobo Wang and Zain Anwar Ali

Received: 11 May 2022
Accepted: 14 June 2022
Published: 16 June 2022

**Publisher's Note:** MDPI stays neutral with regard to jurisdictional claims in published maps and institutional affiliations.





**Abstract:** In recent years, the increasing number of unmanned aerial vehicles (UAVs) in the low-altitude airspace have not only brought convenience to people's work and life, but also great threats and challenges. In the process of UAV detection and tracking, there are common problems such as target deformation, target occlusion, and targets being submerged by complex background clutter. This paper proposes an anti-occlusion UAV tracking algorithm for low-altitude complex backgrounds by integrating an attention mechanism that mainly solves the problems of complex backgrounds and occlusion when tracking UAVs. First, extracted features are enhanced by using the SeNet attention mechanism. Second, the occlusion-sensing module is used to judge whether the target is occluded. If the target is not occluded, tracking continues. Otherwise, the LSTM trajectory prediction network is used to predict the UAV position of subsequent frames by using the UAV flight trajectory before occlusion. This study was verified on the OTB-100, GOT-10k and integrated UAV datasets. The accuracy and success rate of integrated UAV datasets were 79% and 50.5% respectively, which were 10.6% and 4.9% higher than those of the SiamCAM algorithm. Experimental results show that the algorithm could robustly track a small UAV in a low-altitude complex background.

**Keywords:** unmanned aerial vehicle; target tracking; attention mechanism; anti-occlusion; location prediction


## 1. Introduction

In recent years, with the rapid development of the UAV industry and the continuous improvement of artificial intelligence, UAVs have been widely used in public security, disaster relief, photogrammetry, news broadcasts, travel, and other fields, bringing great convenience to production and social life. However, the increasing number of UAVs in low-altitude airspace and the frequent occurrence of various illegal flight incidents have brought great threats and challenges to aviation flight, security, confidentiality protection and privacy protection [1].

In order to detect UAV in the low-altitude airspace as early and as far as possible using computer vision, it is often necessary to implement the long-distance detection and tracking of UAVs [2], which cause small imaging sizes and weak signals [3]. At the same time, the flight altitude of UAV in low-altitude airspace is very low, often only dozens to hundreds of meters, and the surrounding environment of this altitude is relatively complex, such as trees, buildings, and walls, which may lead to the visual tracking of UAV being interfered by strong clutter, occlusion, and other factors, resulting in tracking drift and loss. Therefore, it is important and urgent to find a robust tracking algorithm against background clutter interference and occlusion for UAV tracking in low-altitude airspace.

On the basis of combining existing UAV visual tracking technology and referring to the network structure of ATOM [4], this paper proposes an anti-occlusion target tracking algorithm by integrating the SeNet [5] attention mechanism to solve the complex background and occlusion problems during tracking, which achieved good performance. First, the SeNet attention mechanism was introduced into the original feature extraction network to enhance the extracted features, which effectively improved the performance of subsequent tracking process. Second, an occlusion-sensing model was designed to judge the state of the target. Lastly, the LSTM [6] trajectory prediction network was used to predict the UAV position according to the target state. This study was verified on the OTB-100 [7], GOT-10k [8] and integrated UAV datasets. Experimental results show that the proposed algorithm could effectively reduce the influence of low-altitude complex environments on the target and robustly track a UAV.

The main contributions of this paper are:

1. In order to solve the problem of UAVs in the low-altitude airspace being easy to be submerged in complex background clutter, the SeNet attention mechanism was used in the backbone to improve the correlation between feature channels, enhance the feature of the target, and reduce the influence of background clutter.

2. In order to solve the occlusion problem in low-altitude airspace during flight, an occlusion judgment mechanism is proposed to judge whether the target is occluded.

3. When the target is occluded, the LSTM trajectory prediction network is used to predict the flight trajectory of the aircraft, so as to achieve robust tracking and stop the template update to improve tracking accuracy.

## 2. Related Works

Existing target tracking algorithms can be roughly divided into two categories: One is the traditional target tracking method based on correlation filtering [9–13], which uses the response diagram between the template frame and the detection frame after Fourier transform to determine the target of the detection frame. The other is the deep-learning target tracking method based on a convolutional neural network [14–17], which obtains the features of the target by convolutional operation on the images of the template and detection frames, and then obtains the tracking target by similarity matching. This section reviews the related work of researchers in recent years.

### 2.1. Algorithm Based on Correlation Filter

The tracker based on a discriminant correlation filter (DCF) can effectively use limited data and enhance the training set by using all shifts of local training samples in the learning process. The method based on DCF trains the least-squares regression to predict the target confidence score by using the characteristics of cyclic correlation and fast Fourier transform (FFT) in the learning and detection steps [18]. Mosse [9] was the first pioneering work to propose correlation filter for tracking that uses a random affine set of samples from a single initial frame transformation to construct a minimal output sum of the squares' filter. KCF reduces storage and calculation [6] by several orders of magnitude by diagonalizing the cyclic data matrix with discrete Fourier transform. The periodic assumption of KCF also introduces an unnecessary boundary effect, which seriously reduces the quality of tracking model. SRDCF introduces a spatial regularization component in the learning process in order to reduce the boundary effect, which punishes them according to the spatial position of the correlation filter coefficients [19]. In addition, there are several excellent trackers based on correlation filters, such as STRCF [20] and ECO [8]. They usually divide tracking into two stages: feature extraction and target classification, so end-to-end training is not possible. The objective function in the target classification module in ATOM [13] is based on the mean square error, like the discriminant correlation filtering method, but it is established on a two-layer fully convolutional neural network. ATOM returns the size of the target with IoU-Net. Although ATOM has achieved effective performance, it is sometimes wrong in size estimation because the predicted joint cross (IoU) may be inaccurate, which leads to tracking failure, especially in a cluttered background.

## *2.2. Algorithm Based on Siamese Network*

In recent years, the visual tracker based on a Siamese network has attracted much attention due to its good balance between tracking performance and efficiency [21]. A Siamese network learns similarity measure functions offline from image pairs, and transforms a tracking task into a template matching task. SiamFC [21] uses a large number of templates and search areas for sample matching in offline training. During online tracking, the template and search regions are correlated in the feature space through forward propagation, and the target position is determined according to the peak position of the correlated response. SiamRPN [12] adds a region proposal network (RPN) to obtain various aspect ratio candidate target frames. It interprets the template branch in a Siamese network as a training parameter, predicts the kernel of a local detection task, and regards the tracking task as a one-time local detection task. SiamMask [22] added a segmented branch based on SiamFC and SiamRPN. The size and shape of the target are obtained, and the tracking results are refined according to the mask of the position of the maximal classification score. One disadvantage of Siamese method is that it ignores the context information around the template, and only extracts the template information from the initial target area.

Due to unrestricted video conditions such as illumination changes and viewpoint changes, the appearance of subsequent targets may be greatly different from that of the initial target. Therefore, the previously proposed Siamese-based tracker degenerates when similar disturbances and object appearance changes occur, which leads to tracking drift and failure. In order to overcome the shortcomings of the Siamese method, DiMP [23] trains a discriminant classifier online and separates the target from the background. This model is derived from a discriminant learning loss by designing a special optimization process, which predicts a strong model in several iterations. The tracker continuously collects positive and negative samples in the tracking process when the target has sufficient confidence prediction, and the classifier template is updated online when 20 frames of target are tracked or a disturbance peak is detected to deal with the appearance change.

So far, many researchers have studied the occlusion problem. However, most of the research is based on the correlation filtering algorithm using handcrafted features, and the effect is not very good. As shown in Figure 1, occlusion may lead to target loss, so it is difficult to achieve accurate tracking through the method based on a Siamese network in actual industrial production. Target redetection algorithms are more used to solve the occlusion problem in tracking. However, the premise is that other cameras must capture unoccluded targets at a time of occlusion, which means that a target requires at least two or more cameras; that is, the number of cameras should at least double. Therefore, it is necessary to propose a cheap method to solve the problem of target occlusion.

**Figure 1.** Diagram of tracking failure caused by occlusion.

## 3. Anti-Occlusion UAV Tracking Algorithm by Integrating Attention Mechanism

In order to solve the problem of complex background and occlusion in low-altitude UAV tracking, this paper proposes a single-target tracking algorithm with attention mechanism and anti-occlusion ability. Extracted convolutional features are enhanced to solve the problem of complex backgrounds by adding a squeeze-and-excitation (SE) module to feature extraction network for feature optimization. Combining target tracking and UAV flight trajectory prediction, the trajectory prediction module is started to predict the position of a UAV when it is occluded. In this paper, the ATOM algorithm is improved. The sequence and exception (SE) module was added to the feature extraction part, and the occlusion-sensing and trajectory-prediction modules were added to the tracking process to realize the robust tracking of low-altitude UAVs.

### 3.1. Squeeze-and-Excitation (SE) Module

A squeeze-and-excitation (SE) module is an attention method to improve the correlation between feature channels, and enhance target features. By introducing the SeNet attention mechanism, the representation of targets in the channel dimension is enhanced. At the same time, by emphasizing the target and suppressing background information, adjusting the parameters in the network, it shows obvious advantages in image classification. Therefore, the SeNet attention mechanism was added to the ResNet-18 feature extraction network and the final output to enhance the extracted target features in order to solve the complex background problem of low-altitude UAVs.

As shown in Figure 2, squeeze-and-excitation (SE) modules generate different weight coefficients for each channel according to relationship between feature channels, multiplying the previous features and adding them to the original features to achieve the purpose of enhancing features.

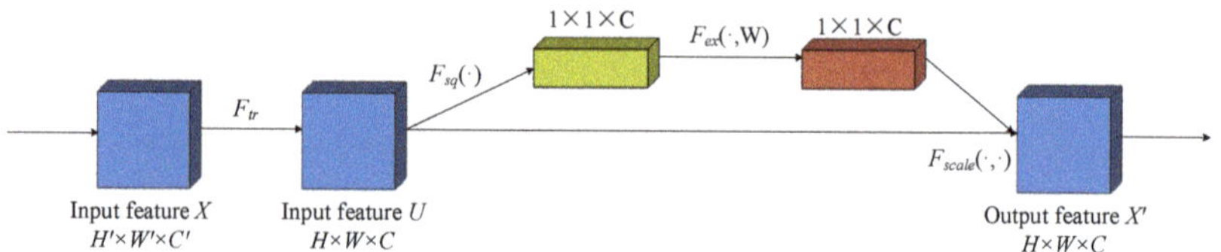


**Figure 2.** Squeeze-and-excitation (SE) module.

As shown in Figure 2, the process of the SeNet attention mechanism is as follows: First, the extracted feature $X \in \mathbb{R}^{H' \times W' \times C'}$ is mapped to $U \in \mathbb{R}^{H \times W \times C}$ by transforming function $Ftr$. Then, the global information of each channel is represented by a channel feature description value through global average pooling $Fsq(\cdot)$, and the channel feature description value is adaptively calibrated by $Fex(\cdot, W)$ to render the weight more accurate. Lastly, the enhanced feature $Y \in \mathbb{R}^{H \times W \times C}$ is obtained by multiplying the weight value and the original feature by $Fscale(\cdot, \cdot)$.

Specifically, $Ftr$ is treated as a convolutional operator, $V = [v_1, v_2, \cdots, and v_C]$ represents the set of learned filter kernels, where $v_i$ represents the parameters of the $i - th$ filter. So, the output of $X$ through $Ftr$ is $U = [u_1, u_2, \cdots, u_C]$,

$$u_i = v_i * X = \sum_{t=1}^{C'} v_i^t * x^t \tag{1}$$

where $*$ represents a convolutional operation, and $v_i^t$ is a two-dimensional spatial kernel that represents the channel corresponding to a single channel in $X$.

Global average pooling $Fsq(\cdot)$: In order to better represent the features of all channels without losing any features, global average pooling is used for the feature information of each channel, and the feature information of the channel is expressed as a value. $z_i$ represents the feature description value of each channel, which is expressed as

$$z_i = F_{sq}(u_i) = \frac{1}{H \times W} \sum_{r=1}^{H} \sum_{c=1}^{W} u_i(r, c) \tag{2}$$

where $u_i$ is a feature in the $i$-th channel.

Adaptive calibration $Fex(\cdot, W)$: two fully connected layers are used to fully exploit the correlation between channels. First, the number of channels is reduced to $C/r$ through a fully connected layer to reduce the amount of calculation, and the ReLU function is used to activate the output. Then, the number of channels is again restored to $C$ through a fully connected layer, and a sigmoid activation function is adopted to output. This process is expressed as

$$s = F_{ex}(z, W) = \sigma(g(z, W)) = \sigma(W_2\delta(W_1 z)) \tag{3}$$

where $\delta$ is the ReLU activation function, $\sigma$ is the sigmoid activation function, and $W_1 \in \mathbb{R}^{\frac{C}{r} \times C}, W_2 \in \mathbb{R}^{C \times \frac{C}{r}}$.

Lastly, enhanced features on the channel are obtained by multiplying the weight coefficient through the fully connected layers with the previous features.

$$y_i = F_{scale}(u_i, s_i) = s_i u_i \tag{4}$$

where $y_i$ is the feature of the $i-th$ channel after weight multiplication, $Y \in [y_1, y_2, \cdots, y'_C]$ is the enhanced feature through the channel.

Inspired by SeNet, combined with the characteristics of ResNet-18 network, a ResNet-18 network combined with SeNet is proposed. On the basis of the original ResNet-18 network, the SeNet layer was added behind each dense block to realize the utilization of the attention mechanism of ResNet-18 network channel. The specific network framework is shown in Figure 3.

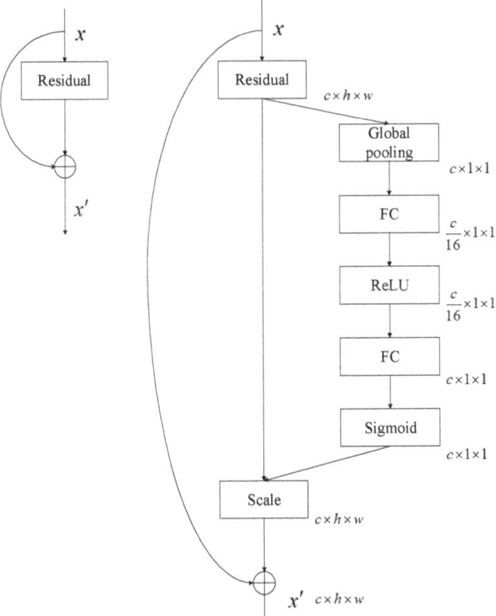


**Figure 3.** Se-ResNet network by integrating SeNet.

At the same time, after features are extracted from the Se-ResNet network, SeNet is used again to enhance the extracted features to solve the complex background problem in the tracking of low-altitude UAV. The feature extraction network is shown in Figure 4.

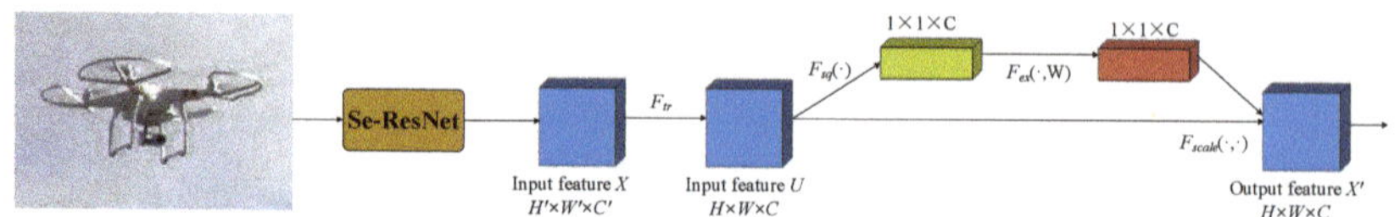


**Figure 4.** Feature extraction network by integrating attention mechanism.

### 3.2. Occlusion-Sensing Module

An occlusion-sensing module is proposed to determine whether a target is occluded. The Gaussian response map is obtained by cross-correlation between the feature map from the feature extraction network in the search area and the target frame. Response values within a certain range are found, and the position is recorded as set $A$. The Euclidean distance between the target and the elements in set $A$ is calculated, and the average value is obtained. If the distance is greater than the threshold set by the algorithm, it is determined as an occlusion.

In this study, the feature response diagram of a feature extraction network was analyzed through the visualization of the training process. When occlusion occurs, the response graph fluctuates and the response peak is not prominent. On the basis of this phenomenon, an occlusion-sensing module is proposed to accurately determine whether the target is occluded, as shown in Figure 5.

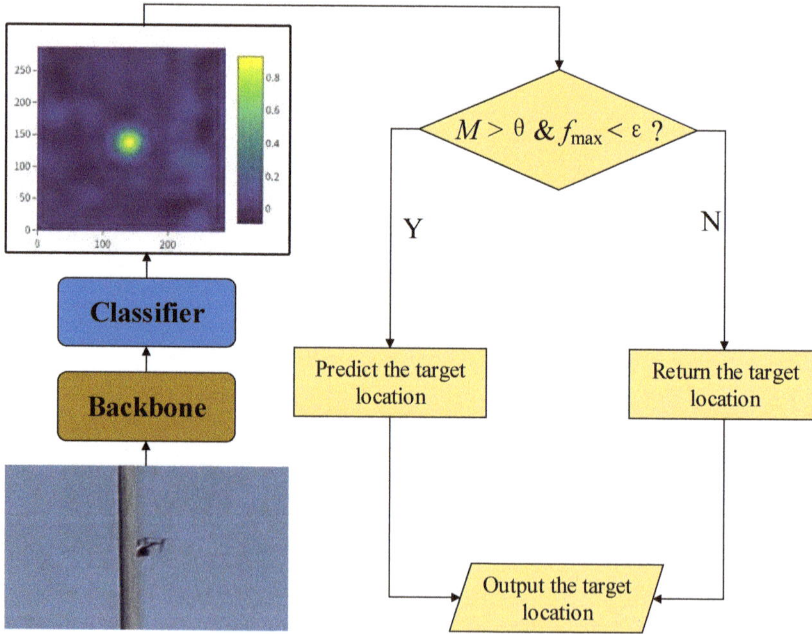


**Figure 5.** Occlusion-sensing module.

First, points with a certain range of response values in the response diagram are obtained, and the position set is denoted as $A$.

$$A = \{(i,j)|(\hat{r}(i,j) > \eta_1 mean(\hat{r}))and(\hat{r}(r,j) < \eta_2 max(\hat{r}))\} \tag{5}$$

where $\hat{r}$ is the current frame response graph, and $mean(\hat{r})$ is the average response graph.

Average occlusion distance metric $M_O$ is defined as

$$M_O = \frac{1}{n} \sum_{(i,j) \in A} \sqrt{(i-m)^2 + (j-n)^2} \tag{6}$$

where $n$ represents the number of points contained in set $A$, and $(m,n)$ represents the location of the peak response.

Figure 6 shows the target response diagram after using the proposed occlusion strategy. Three common response diagrams of a tracking target state are given, namely, no occlusion, partial occlusion, and complete occlusion. The dark points in the figure represent the points in set $A$. With the increase in the occlusion degree of the target, the response diagram dramatically changed, the number of points in set $A$ increased, the average occlusion distance metric $M_O$ also increased, and multiple peaks appeared in the response diagram. Therefore, average occlusion distance measure $M_O$ could reflect the occlusion state of the target to a certain extent.

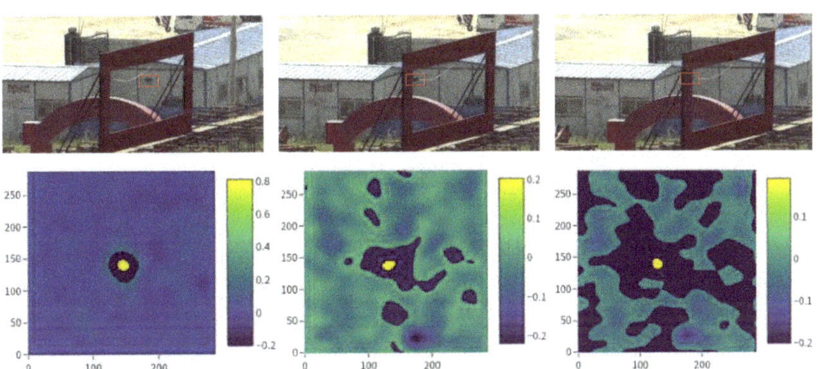

**Figure 6.** Response diagram under occlusion-sensing module.

In view of the phenomenon described in Figure 6, an occlusion-sensing module is proposed to discriminate the occlusion of UAVs during tracking. The specific process is shown in Figure 5. Feature extraction is performed on the target frame, and a mask operation is performed on the extracted features and the trained template to obtain the Gaussian response diagram. When global mean occlusion distance $D_O$ is greater than the set threshold $\theta$, and the peak $f_{max}$ of the response graph is less than set threshold $\varepsilon$, the UAV is judged to be occluded. The trajectory prediction module is called to predict the next position of the UAV, and the template update is stopped to prevent the template from being occluded.

### 3.3. UAV Trajectory Prediction Based on LSTM

The traditional Kalman filter algorithm has achieved good results in terms of trajectory prediction and has been applied in engineering. However, the Kalman filter is only applicable to tracking linear moving targets. The single-target tracking problem with different trajectory types is difficult. The measurement value is uncertain, especially when the target is occluded or has disappeared, so it is difficult to effectively predict in this case.

Most target trajectories do not follow the linear principle in common UAV flight videos, which hinders the Kalman filter from predicting trajectories well, while long short-term memory (LSTM) performs better. LSTM is more suitable for solving the prediction problem of a nonlinear motion trajectory because it benefits from its internal mechanism. For example, the Social-LSTM algorithm achieved good trajectory prediction performance. In view of the diversification of target trajectories, a trajectory prediction model is proposed by improving the LSTM algorithm. The central coordinates of the historical frame before occlusion are used as the input of the trajectory prediction model, trajectory samples are generated by LSTM, and the next prediction position of the target is obtained, which solves the problem of tracking failure when the UAV is occluded.

The space coordinate of UAV at the time $t$ is $(x_t, y_t)$, in which the time $t = 1$ to $t = t_{obs}$ is observable, and the corresponding observable trajectory is represented as $(x_1, y_1), (x_2, y_2), \cdots,$ $(x_{obs}, y_{obs})$. Time $t = t_{pred}$ is the prediction time, and the corresponding prediction coordinates is represented as $(\hat{x}_{pred}, \hat{y}_{pred})$.

The trajectory prediction network based on LSTM proposed in this paper is shown in Figure 7. With the historical flight trajectory of UAV as the input, the predicted flight trajectory is output after a LSTM encoder and decoder.

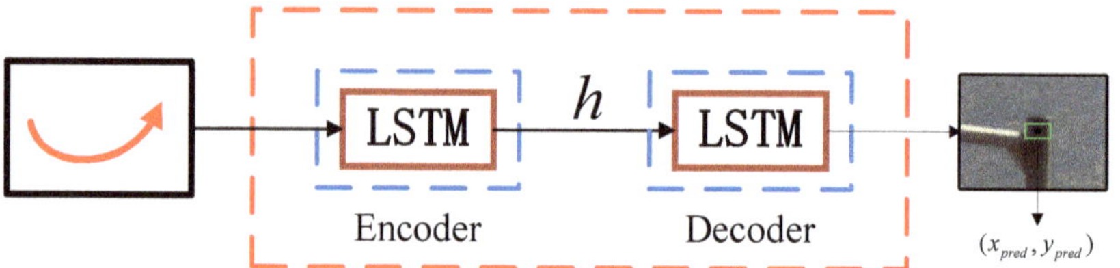


**Figure 7.** Trajectory prediction network based on LSTM.

Let $h^t$ represent the hidden state of LSTM at time $t$, which is used to predict the distribution of target position $(\hat{x}_{t+1}, \hat{y}_{t+1})$ at time $t + 1$. Assuming that it obeys binary Gaussian distribution, mean $\mu_{t+1}$, standard deviation $\sigma_{t+1}$ and correlation coefficient $\rho_{t+1}$ are predicted by weight matrix $W_p$. Then, prediction coordinate $(\hat{x}_t, \hat{y}_t)$ at time $t$ is:

$$(\hat{x}_t, \hat{y}_t) \sim (\mu_t, \sigma_t, \rho_t) \tag{7}$$

The parameters of the model are learnt by minimizing the negative logarithmic likelihood function:

$$[\mu_t, \sigma_t, \rho_t] = W_p h^{t-1} \tag{8}$$

$$L(W_e, W_l, W_p) = - \sum_{t=T_{obs}+1}^{T_{pred}} \log(P(x_t, y_t | \sigma_t, \mu_t, \rho_t)) \tag{9}$$

The model is trained by minimizing the loss for all trajectories in the training dataset, where $W_l$ is the network weight of the LSTM, and $W_e$ is the weight of the position coordinates. Because this article only predicts the trajectory of the UAV, there is no relationship with other trajectories, and there is no need to calculate the weight associated with other trajectories, so the weight $W_e$ of the position coordinate iwass set to 1.

### 3.4. Comprehensive Scheme and Algorithm Implementation

Anti-occlusion target tracking for UAVs integrating the SeNet attention mechanism is proposed considering the SeNet attention module, occlusion-sensing module, and flight trajectory prediction above, as shown in Algorithm 1. Global and local information is fused for feature enhancement by using the SeNet attention mechanism. The occlusion-sensing module is used to determine whether the target is occluded. When the target is occluded, the LSTM algorithm is used to predict the target position. The whole process of the algorithm is shown in Figure 8.

---

**Algorithm 1** Proposed UAV tracking algorithm.

---

**Input:** Target position *pos* and the size of bounding box *rect* in the first frame.
**Output:** Target position $pos_i$ and the size of bounding box $rect_i$ in the $i - th$ frame.
1: Initialize $N_{image}$, *pooling*, $t$, $\epsilon$.
2: **for** $i = 2$ to $N_{image}$ **do**
3:  Extract the area of *pooling* $*$ *rect* size as search area with $pos_{i-1}$ coordinates in the $i - th$ frame.
4:  Extract features in search area by the backbone.
5:  Generate response graph using classified regression filter.
6:  Calculate $A$ and $M_O$ using Equations (5) and (6).
7:  **if** $M_O > \theta$   *and*   $f_{max} < \epsilon$ **then**
8:   Call LSTM trajectory prediction algorithm, enter $[pos_{i-t}, ..., pos_{i-1}]$, and output $pos_i$.
9:  **else**
10:   Output classification regression filter response graph corresponding position $pos_i$.
11:  **end if**
12:  Extract multiple bounding boxes of different scales with $pos_i$ as the coordinate origin, and calculate the $IoU$ scores. The bounding box with the highest score corresponds to the $pos_i$ and $rect_i$ of the target in the $i - th$ frame.
13: **end for**

---

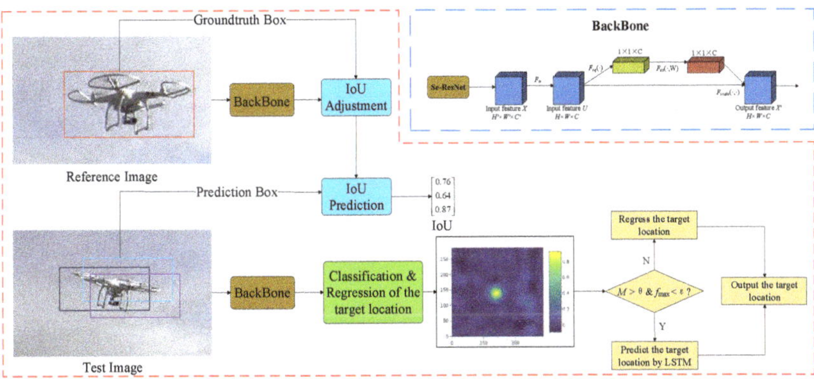


**Figure 8.** Comprehensive scheme of the proposed UAV tracking algorithm.

## 4. Experimental Results and Analysis

In this paper, some UAV datasets in Drone-vs.-Birds [24] and LaSOT [25] were integrated to form UAV datasets. Experimental verification was carried out on the OTB-100, GOT-10k, and integrated UAV datasets to verify the effect of the improved algorithm proposed in this paper, and the tracking-success and precision plots were used for evaluation.

### 4.1. Experimental Environment and Parameters Setting

The algorithm was implemented in Python 3.7 with the PyTorch framework. The experimental computer operating system was Ubuntu 180.4 64-bit, CPU InterCore i7-9700k, the main frequency was 3.60 GHz, with 16 GB memory, NVIDIA GeForce RTX2080Ti, and 11 GB memory. In the training process, some LaSOT and GOT-10k-train dataset are used as the training set, and the part of GOT-10k-train dataset that does not participate in the training is used as the verification set. The pretraining parameters on ImageNet are used in the backbone. By training the network, the common features in the visual tracking process are learned for the following tracking. In the tracking process, the occlusion threshold is set to $\theta = 12$, $\epsilon = 0.1$.

### 4.2. Comparison and Analysis of Experimental Results

#### 4.2.1. Experiment on OTB-100 Dataset

The OTB-100 dataset contains 100 different video sequences. The coordinates of the target and the size of the bounding box in the sequence are manually labeled, and are relatively accurate. The dataset contains 25% gray images, which pose a challenge to the algorithm on the basis of color feature tracking.

The proposed algorithm was tested on OTB-100 and compared with four advanced trackers, namely, the Siamfc, Dimp, Prdimp, and ATOM algorithms. Figure 9 shows the precision and success plots of the five algorithms on the OTB-100 dataset. The precision and success rate of the proposed algorithm were improved compared with the second algorithm after adding the SeNet attention mechanism and anti-occlusion module.

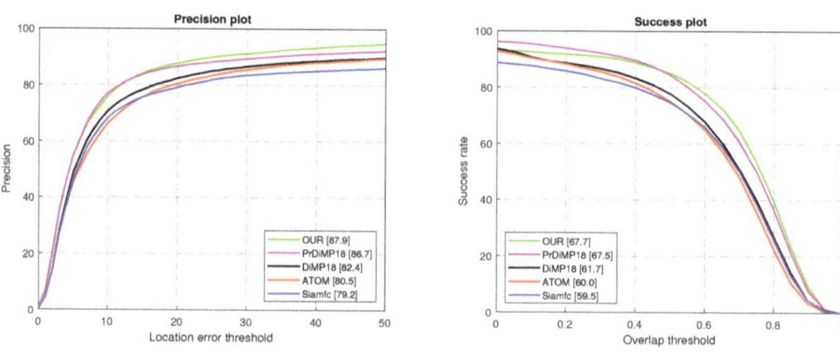


**Figure 9.** Precision and success plots on OTB-100 dataset.

The tracking results of some video sequences of the OTB-100 dataset are shown in Figure 10. The ATOM and Siamfc algorithms lost the target if the occlusion time was too long. However, the algorithm proposed in this paper could effectively resist occlusion with the use of the anti-occlusion module. Furthermore, for short-term occluded targets, the algorithm proposed in this paper tracked the target position more accurately than other baseline algorithms did because of the LSTM prediction module.

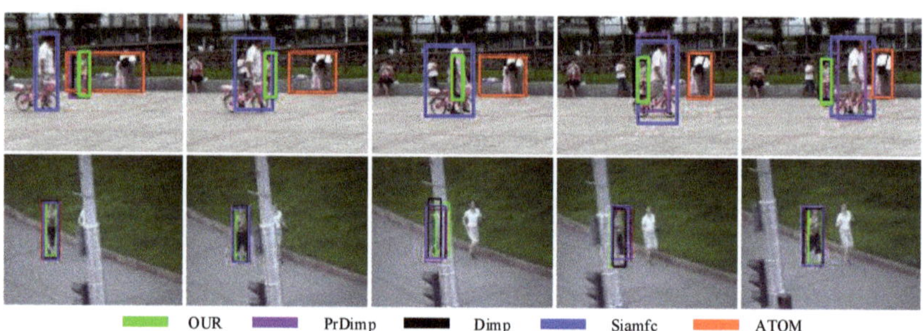


**Figure 10.** Tracking results of video sequences of the OTB-100 dataset.

#### 4.2.2. Experiment on GOT-10k Dataset

The GOT-10k dataset contains video sequences of more than 10,000 moving targets in the real world, in which more than 1.5 million targets are manually marked in location and bounding box. The GOT-10k test set contains 84 target categories and 32 moving target categories, without overlap between the training set and test set. Therefore, GOT-10k-val for testing is not affected by GOT-10k-train for training.

The proposed algorithm was also compared with the Siamfc, Dimp, Prdimp and ATOM algorithms on the GOT-10k dataset. Figure 11 shows the precision and success plots of the five algorithms on the GOT-10k dataset. After adding the SeNet attention mechanism and anti-occlusion module, the accuracy and success rate of the proposed algorithm were 59.9% and 73.1%, respectively, which were 8.3% and 3.7% higher than those of the second algorithm.

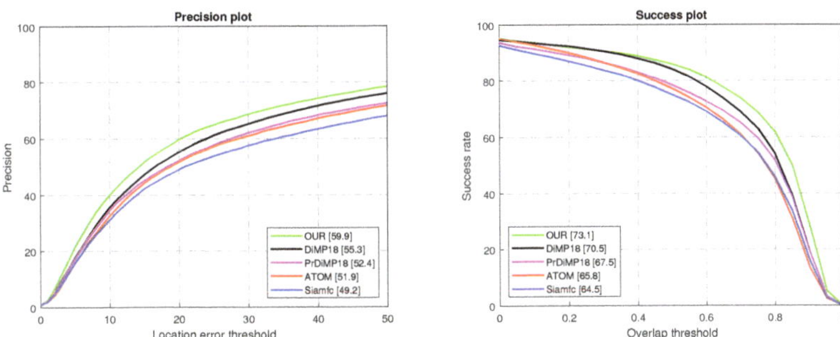

**Figure 11.** Precision and success plots on GOT-10k dataset.

The visualization of part of the video sequence tracking results of the GOT-10k-val dataset is shown in Figure 12. The ATOM and Siamfc algorithms were not accurate in predicting the target scale in a complex background. With the use of the SeNet attention mechanism, the algorithm proposed in this paper was more accurate for the scale regression of the target than other baseline algorithms were.

**Figure 12.** Tracking results of video sequences of the GOT-10k dataset.

### 4.2.3. Experiment on Integrated UAV Dataset

Drone-vs.-Birds is a target detection dataset used to distinguish between UAVs and birds with video sequences of UAVs and birds. This study uses its UAV video sequence and the UAV video sequence of the LaSOT dataset to form a dataset for UAV tracking to verify the proposed algorithm. UAV video sequences in the Drone-vs.-Birds and LaSOT datasets were combined into a dataset for UAV tracking to verify the algorithm proposed in this study.

Figure 13 shows the precision and success plots of the Siamfc, Dimp, Prdimp, ATOM, and proposed algorithms on the integrated UAV dataset. The accuracy and success rate of the proposed algorithm were 79% and 50.5%, which are 10.6% and 4.9% higher than those of the second algorithm.

The visualization of the partial tracking process is shown in Figure 14. When occlusion occurred, the Siamfc and ATOM algorithms may have lost the target and failed in tracking. With the use of SeNet attention mechanism and anti-occlusion module, the algorithm proposed in this paper could achieve better tracking results.

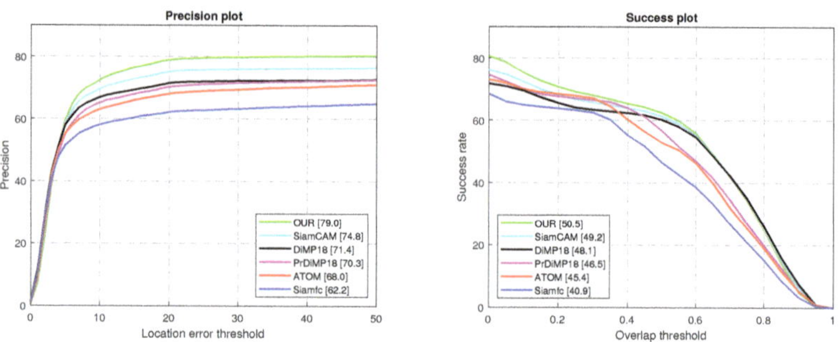

**Figure 13.** Precision and success plots on integrated UAV dataset.

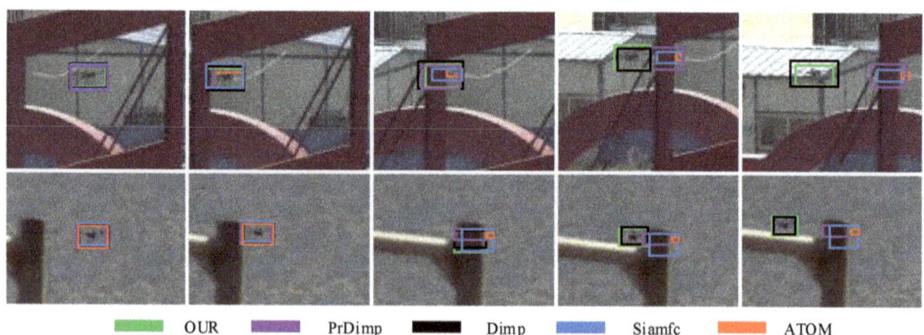


**Figure 14.** Tracking results under occlusion of the integrated UAV dataset.

## 5. Conclusions

Aiming at the problems of complex background and occlusion of UAVs in low-altitude airspace during flight, an anti-occlusion UAV tracking algorithm with an integrated attention mechanism was proposed. In this algorithm, the SeNet attention mechanism is introduced to fuse global and local information for feature enhancement to solve the problem of complex backgrounds. The occlusion-sensing module was designed to determine whether the target is occluded, and if the target is occluded, the LSTM algorithm is used to predict the target position to solve the occlusion problem. By validating on three different datasets, the method proposed in this paper achieved good results and tracked UAVs well. However, with the addition of SeNet attention mechanism and anti-occlusion module, the algorithm parameters increased and the amount of calculation increased, resulting in a decrease in the running speed of the algorithm. The running speed on the GPU 2080ti server was 49 fps/s, which basically achieves real-time tracking. Further improving the tracking speed and performance of the algorithm without reducing its accuracy is future research work.

**Author Contributions:** Conceptualization, C.W. and J.W.; methodology, Z.S.; investigation, Q.G.; resources, C.W. and T.W.; writing—original draft preparation, Z.S.; writing—review and editing, C.W. and L.M.; visualization, E.W.; supervision, C.W. and Q.G. All authors have read and agreed to the published version of the manuscript.

**Funding:** This research was funded by National Natural Science Foundation of China with grant Nos. 61703287 and 62173237, Scientific Research Program of Liaoning Provincial Education Department of China with grant Nos. LJKZ0218 and JYT2020045, Young and middle-aged Science and Technology Innovation Talents Project of Shenyang of China with grant No. RC210401 and Liaoning Provincial Key R&D Program of China with grant No. 2020JH2/10100045.

**Institutional Review Board Statement:** Not applicable.

**Informed Consent Statement:** Not applicable.

**Acknowledgments:** The authors would like to thank the reviewers and editors for their valuable comments and suggestions.

**Conflicts of Interest:** The authors declare no conflict of interest.

# References


1. Tapsall, B.T. Using crowd sourcing to combat potentially illegal or dangerous UAV operations. In Proceedings of the Unmanned/Unattended Sensors and Sensor Networks XII. SPIE, Edinburgh, UK, 27 September 2016; Volume 9986, pp. 23–28.
2. Jin, H.; Wu, Y.; Xu, G.; Wu, Z. Research on an Urban Low-Altitude Target Detection Method Based on Image Classification. *Electronics* **2022**, *11*, 657. [CrossRef]
3. Liu, C.; Xu, S.; Zhang, B. Aerial Small Object Tracking with Transformers. In Proceedings of the 2021 IEEE International Conference on Unmanned Systems (ICUS), Beijing, China, 15–17 October 2021; pp. 954–959.
4. Danelljan, M.; Bhat, G.; Khan, F.S.; Felsberg, M. ATOM: Accurate Tracking by Overlap Maximization. In Proceedings of the 2019 IEEE/CVF Conference on Computer Vision and Pattern Recognition (CVPR), Long Beach, CA, USA, 15–20 June 2019; pp. 4655–4664.
5. Hu, J.; Shen, L.; Sun, G. Squeeze-and-excitation networks. In Proceedings of the IEEE Conference on Computer Vision and Pattern Recognition (CVPR), Salt Lake City, UT, USA, 18–23 June 2018; pp. 7132–7141.
6. Hochreiter, S.; Schmidhuber, J. Long Short-Term Memory. *Neural Comput.* **1997**, *9*, 1735–1780. [CrossRef]
7. Wu, Y.; Yang, M. Object Tracking Benchmark. *IEEE Trans. Pattern Anal. Mach. Intell.* **2015**, *37*, 1834–1848. [CrossRef]
8. Huang, L.; Zhao, X.; Huang, K. GOT-10k: A Large High-Diversity Benchmark for Generic Object Tracking in the Wild. *IEEE Trans. Pattern Anal. Mach. Intell.* **2021**, *43*, 1562–1577. [CrossRef]
9. Yin, H.P.; Chen, B.; Chai, Y.; Liu, Z.D. Vision-based object detection and tracking: A review. *Acta Autom. Sin.* **2016**, *42*, 1466–1489.
10. Li, Y.; Zhu, J. A Scale Adaptive Kernel Correlation Filter Tracker with Feature Integration. In Proceedings of the 2014 European Conference on Computer Vision (ECCV), Zurich, Switzerland, 6–12 September 2014; pp. 254–265.
11. Henriques, J.F.; Caseiro, R.; Martins, P.; Batista, J. High-Speed Tracking with Kernelized Correlation Filters. *IEEE Trans. Pattern Anal. Mach. Intell.* **2015**, *37*, 583–596. [CrossRef] [PubMed]
12. Danelljan, M.; Robinson, A.; Shahbaz Khan, F.; Felsberg, M. Beyond correlation filters: Learning continuous convolution operators for visual tracking. In *European Conference on Computer Vision*; Springer: Cham, Switzerland, 2016; pp. 472–488.
13. Danelljan, M.; Bhat, G.; Khan, F.S.; Felsberg, M. ECO: Efficient Convolution Operators for Tracking. In Proceedings of the 2017 IEEE Conference on Computer Vision and Pattern Recognition (CVPR), Honolulu, HI, USA, 21–26 July 2017; pp. 6931–6939.
14. Bolme, D.S.; Beveridge, J.R.; Draper, B.A.; Lui, Y.M. Visual object tracking using adaptive correlation filters. In Proceedings of the 2010 IEEE Conference on Computer Vision and Pattern Recognition (CVPR), San Francisco, CA, USA, 13–18 June 2010; pp. 2544–2550.
15. Held, D.; Thrun, S.; Savarese, S. Learning to Track at 100 FPS with Deep Regression Networks. In Proceedings of the 2016 European Conference on Computer Vision (ECCV), Amsterdam, The Netherlands, 11–14 October 2016; pp. 749–765.
16. Valmadre, J.; Bertinetto, L.; Henriques, J.; Vedaldi, A.; Torr, P.H. End-to-end representation learning for correlation filter based tracking. In Proceedings of the IEEE Conference on Computer Vision and Pattern Recognition, Honolulu, HI, USA, 21–26 July 2017; pp. 2805–2813.
17. Li, B.; Yan, J.; Wu, W.; Zhu, Z.; Hu, X. High performance visual tracking with siamese region proposal network. In Proceedings of the IEEE Conference on Computer Vision and Pattern Recognition, Salt Lake City, UT, USA, 18–23 June 2018; pp. 8971–8980.
18. Gladh, S.; Danelljan, M.; Khan, F.S.; Felsberg, M. Deep motion features for visual tracking. In Proceedings of the 2016 International Conference on Pattern Recognition (ICPR), Cancun, Mexico, 4–8 December 2016; pp. 1243–1248.
19. Danelljan, M.; Hager, G.; Shahbaz Khan, F.; Felsberg, M. Learning Spatially Regularized Correlation Filters for Visual Tracking. In Proceedings of the 2015 IEEE International Conference on Computer Vision (ICCV), Santiago, Chile, 7–13 December 2015; pp. 4310–4318.
20. Li, F.; Tian, C.; Zuo, W.; Zhang, L.; Yang, M.H. Learning spatial-temporal regularized correlation filters for visual tracking. In Proceedings of the IEEE Conference on Computer Vision and Pattern Recognition, Salt Lake City, UT, USA, 18–23 June 2018; pp. 4904–4913.
21. Bertinetto, L.; Valmadre, J.; Henriques, J.F.; Vedaldi, A.; Torr, P.H. Fully-convolutional siamese networks for object tracking. In *Proceedings of the European Conference on Computer Vision*; Springer: Cham, Switzerland, 2016; pp. 850–865.
22. Wang, Q.; Zhang, L.; Bertinetto, L.; Hu, W.; Torr, P.H. Fast online object tracking and segmentation: A unifying approach. In Proceedings of the IEEE/CVF Conference on Computer Vision and Pattern Recognition, Long Beach, CA, USA, 15–20 June 2019; pp. 1328–1338.
23. Bhat, G.; Danelljan, M.; Gool, L.V.; Timofte, R. Learning Discriminative Model Prediction for Trackingh. In Proceedings of the 2019 IEEE/CVF International Conference on Computer Vision (ICCV), Seoul, Korea, 27–28 October 2019; pp. 6181–6190.

24. Coluccia, A.; Fascista, A.; Schumann, A.; Sommer, L.; Dimou, A.; Zarpalas, D.; Akyon, F.C.; Eryuksel, O.; Ozfuttu, K.A.; Altinuc, S.O.; et al. Drone-vs-Bird Detection Challenge at IEEE AVSS2021. In Proceedings of the 2021 17th IEEE International Conference on Advanced Video and Signal Based Surveillance (AVSS), Washington, DC, USA, 16–19 November 2021; pp. 1–8.
25. Fan, H.; Lin, L.; Yang, F.; Chu, P.; Deng, G.; Yu, S.; Bai, H.; Xu, Y.; Liao, C.; Ling, H. Lasot: A high-quality benchmark for large-scale single object tracking. In Proceedings of the IEEE/CVF Conference on Computer Vision and Pattern Recognition, Long Beach, CA, USA, 15–20 June 2019; pp. 5374–5383.

*Review*

# Optimization Methods Applied to Motion Planning of Unmanned Aerial Vehicles: A Review

Amber Israr [1], Zain Anwar Ali [1,*], Eman H. Alkhammash [2] and Jari Juhani Jussila [3]

1   Electronic Engineering Department, Sir Syed University of Engineering & Technology, Karachi 75300, Pakistan; aisrar@ssuet.edu.pk
2   Department of Computer Science, College of Computers and Information Technology, Taif University, Taif 21944, Saudi Arabia; eman.kms@tu.edu.sa
3   HAMK Design, Factory, Häme University of Applied Sciences, 13100 Hämeenlinna, Finland; jari.jussila@hamk.fi
*   Correspondence: zaali@ssuet.edu.pk

**Abstract:** A system that can fly off and touches down to execute particular tasks is a flying robot. Nowadays, these flying robots are capable of flying without human control and make decisions according to the situation with the help of onboard sensors and controllers. Among flying robots, Unmanned Aerial Vehicles (UAVs) are highly attractive and applicable for military and civilian purposes. These applications require motion planning of UAVs along with collision avoidance protocols to get better robustness and a faster convergence rate to meet the target. Further, the optimization algorithm improves the performance of the system and minimizes the convergence error. In this survey, diverse scholarly articles were gathered to highlight the motion planning for UAVs that use bio-inspired algorithms. This study will assist researchers in understanding the latest work done in the motion planning of UAVs through various optimization techniques. Moreover, this review presents the contributions and limitations of every article to show the effectiveness of the proposed work.

**Keywords:** unmanned aerial vehicle; motion planning; optimization techniques

**Citation:** Israr, A.; Ali, Z.A.; Alkhammash, E.H.; Jussila, J.J. Optimization Methods Applied to Motion Planning of Unmanned Aerial Vehicles: A Review. *Drones* **2022**, *6*, 126. https://doi.org/10.3390/drones6050126

Academic Editors: Kamesh Namuduri and Oleg Yakimenko

Received: 17 March 2022
Accepted: 10 May 2022
Published: 13 May 2022

**Publisher's Note:** MDPI stays neutral with regard to jurisdictional claims in published maps and institutional affiliations.

## 1. Introduction

Flourishing high-tech innovations are making aerial robots an integral part of our daily lives. There are extensive research and analyses on flying robots that possess the mobility given by flight [1,2]. Among these, Unmanned Aerial vehicles (UAVs) are vastly used flying robots due to these distinguishing advantages over others, i.e., budget-friendly, small-sized, lighter in weight, and portable. Moreover, the state-of-the-art characteristics of UAVs are position controlling, sensor employment, auto-level application, structure monitoring, etc. [3–5]. It also has a diverse array of applications, whether in the military or civilian sectors [6]. There are two primary models of UAVs; one is fixed-wing, and the other one is multi-rotor UAVs. The essentials of UAV performance are higher in complex tasks or uncertain environments. Usually, a single UAV has a small size, which limits its volume of sensing, communication, and computation [7]. Thus, cooperative UAVs working together have more benefits and potential results in comparison to a single UAV [8]. A few of them are cost and operation time reduction, low failure of missions, and achievement of higher flexibility, survivability, configurability, and multi-tasks capability [9].

**Background:** It is one of the utmost evolving technologies from the 18th century and is advancing till now. At first, in 1849, Montgolfier's French brothers and Austrians employed unmanned balloons filled with bombs [10]. The development of UAVs with cameras occurred in 1860, which helped with vigilance [11]. In 1917, Charles F. Kettering invented an Aerial Torpedo and named unmanned balloons bugs. The Royal Navy tested a radio-controlled pilotless aircraft during the 1930s [12]. The 1940s were marked by operation

Aphrodite, in which a formation of UAVs with handheld control took place for the first time and radio control-based Queen Bee was developed. A few of them were Pioneer, Predator, Ryan fire bee, etc. In 2003, Amazon started using UAVs commercially [13].

**Related Work:** Extensive analysis of various core issues on UAVs related to motion planning under different circumstances and environments [14]. To design motion control protocols and select path planning techniques, many problems and factors require serious considerations [15]. Numerous researchers have proposed distributed consensus-based motion controls for results with efficacy and accuracy. Some developed leader-follower strategies for efficient outcomes [16]. Some analyses have used bio-inspired algorithms for better path planning with minimal run time. Many employed hybrid algorithms for optimal path planning and achieved a reduction in cost and convergence time [17].

**Motivation and Contribution:** The motivation for this paper is to assemble various strategies used in different research together in a single place. This will help researchers select the best strategy for their required missions while comparing the explorations and exploitations of all the strategies. To overcome the hurdles of different limitations, uncertain disturbances, and complexities, appropriate strategies are essential. This makes the system more stable and efficient and reduces the convergence rate and cost. The prime contributions of this review paper are:

A. The evaluation of the challenges faced by UAVs under different scenarios.
B. Summarizing various promising motion planning techniques and algorithms for determining the optimum path for UAVs.
C. To gather the contributions and limitations presented in each article.

This review is based on the research studies and publications from reputed authors in the field of motion planning techniques used for UAVs over the last three years.

**Organization:** The layout of this paper has many sections, of which Section 2 discusses the challenges that a UAV faces. Section 3 reviews recent developments in motion control and path planning mechanisms. Section 4 evaluates the motion planning and optimization algorithms. Section 5 presents the discussion. Section 6 provides the conclusion, and Section 7 gives directions for future work.

## 2. Challenges in Unmanned Aerial Vehicles

There are extensive investigations regarding UAVs, but still, they face various challenges. The prime challenges that all the researchers face include the selection of UAVs with appropriate path planning that is suitable for the mission [18]. Then, forming efficient motion control and achieves optimal path planning. Moreover, employing proper techniques for navigation and communication so that obstacle avoidance and collision avoidance are possible. Along with this certification, regulation and human-machine interface issues are of much importance. Below are some of the challenges that require serious consideration:

### 2.1. Navigation and Guidance

UAVs have to track their mobility by measuring their distances, making maps, and sensing physical surroundings. To determine the positions of aerial robots, it is essential to develop a navigation system, which is automatic and does not require human interventions [19]. These robots are for flying at higher altitudes and under different environments and hazards. Therefore, the safety and reliability of the system to operate properly are major challenges.

### 2.2. Obstacle Detection and Avoidance

The navigation of UAVs is much influenced by obstacles and collisions. Providing UAVs with an ideal environment is not a viable option. Obstacles that come in the path can be avoided. Moreover, the performances of multiple aerial robots are more beneficial and efficient than a single flying robot. Working in groups can result in collisions. UAVs must be furnished with algorithms or techniques that can handle these issues [20].

### 2.3. Shape and Size

Nowadays, UAVs are widely used for different purposes. They are required to fly at different levels with different ranges. Some have to stay for a longer period to accomplish their missions. Some use runways for flying and landing. Some have to pass through narrow areas. To solve all these issues, it is necessary to consider the appropriate shapes and sizes of UAVs according to the missions [21]. Figure 1 shows some of these challenges faced by UAV [22].

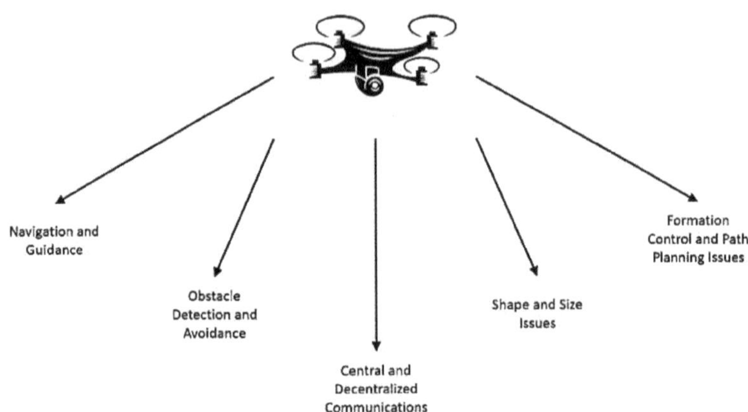


**Figure 1.** Challenges in UAVs [22].

### 2.4. Formation Control and Path Planning Issues

### 2.4.1. Formation Control Issues

There are numerous studies on motion control, but it still lacks and requires consideration and further handling. For example, there stands a need to tackle distributed levels with their effects properly. Similarly, machine learning and reinforcement learning require a longer time for the online learning period and huge data sets for offline training. Therefore, the integration of artificial intelligence (AI) techniques into control protocols is essential. One more challenge in motion control protocol is its robustness, which is highly influenced by environmental disturbances [23].

### 2.4.2. Path Planning Issues

Path planning is to obtain a path for UAVs from the starting to the goal point in such a way that they will carry out their tasks efficiently. UAVs require optimal paths that satisfy their performance constraints and ensure collision avoidance. Such optimal and dynamic paths consume less time and energy. Path planning is a global optimization problem that requires various technologies and algorithms to be integrated [24].

Among all the challenges, the most crucial is path planning and motion control for UAVs. These require considerations so that the UAV can perform well during tasks under any environmental conditions. Several research centers, academies, and industries are analyzing the aforementioned challenges and trying to overcome these issues by developing more improved strategies. Section 3 reviews the development of various protocols and techniques used for the above challenges.

## 3. Recent Developments in UAVs

UAV technology is expanding due to technological innovations. UAVs are becoming more affordable and easy to use, which enhances their application in diverse areas [6]. This paper reviews the strength and development of navigation, communication, shape and size,

collision avoidance, motion control methods, and path planning techniques. It deliberates how they provide solutions to challenging problems while making a considerable impact.

*3.1. Developments in Navigation and Guidance of UAVs*

Navigation technology is quite significant for UAV flight control. Various developed navigation technologies possess different features. Such as satellite, geometric, integrated, Doppler, and inertial navigations. Different purposes require different navigation technologies. The main navigation systems for UAVs are a tactical or medium range navigation system and a high-altitude long-endurance navigation system [25,26]. Development in navigation can be evaluated as:

D.   **High-performance Navigation with Data Fusion:** Navigation uses a Kalman filter; China introduced a data fusion mechanism using this filtering technology. This data fusion is improved by using AI technology. It helps to determine the flight status and guarantees the normal flight of UAVs.

E.   **New Inertial Navigation System:** Many researchers rendered services to develop optical fiber inertial navigation and laser inertial navigation. Improvement was required by the industry. The widely used silicon micro resonant accelerometer helps in UAV navigation. It simplifies the weight and volume, consumes less energy, and refines flight pliability.

F.   **Intelligent Navigation Ability:** An emergency navigation system utilizes various adaptive technologies along with mission characteristics and modes. Moreover, information technology is applied to boost the UAV technology and upgrade the navigation system.

*3.2. Developments in Shape and Size of UAVs*

Earlier, UAVs were applicable for military purposes only, but now they are used for various tasks. This is all due to the rapid progress in developing UAVs with a wide range of shapes and sizes [27]. Different UAVs are utilized for different purposes. According to physical types, we have fixed-wing and multi-rotor UAVs.

**Fixed-Wing UAVs:** These UAVs possess only one long wing on any body's side and require a runway or a broad and flat area. These can consume less battery; therefore, they can stay in the air for maximum hours. They are widely used for long-distance purposes, especially for military surveillance.

**Multi-Rotor UAVs:** These UAVs are built up with multiple propellers and rotors and do not require a runway for vertical flying and landing. With more rotors, the position of UAVs can be controlled in a better way. Mostly quad-rotors are used for small and regular-sized UAVs. Similarly, UAVs are classified based on their sizes into micro or mini-UAVs, tactical UAVs, strategic UAVs, and special-task UAVs.

**Micro and Mini-UAVs:** Many missions require small UAVs. Such as surveillance inside buildings, Nuclear, Biological, and Chemical (NBC) sampling, the agricultural sector, and broadcast industries. Micro and mini-UAVs were developed for these purposes. The take-off weight of a micro-UAV is 0.1 kg, and a mini-UAV is less than 30 kg. Both fly below 300 m with less than 2 h of endurance. The communication range is up to 10 km.

**Tactical UAVs:** Missions such as search and rescue operations, mine detection, communication relays, and NBC sampling use tactical UAVs. They can have a take-off weight of up to 1500 kg. Tactical UAVs can fly up to 8000 m with an endurance of up to 48 h. The communication range is around 10–500 km.

**Strategic UAVs:** For airport security, communication relays, intercept vehicles, and RSTA, strategic UAVs are highly suitable. They can have a maximum take-off weight of around 12,500 kg. They can fly up to 20,000 m with 48 h of endurance. The communication range is more than 2000 km.

*3.3. Developments in Collision Avoidance of UAVs*

A collision usually occurs between a UAV and its neighboring UAV or an obstacle whenever there is less distance between them. A collision avoidance system (CAS) makes sure that no collision takes place with any stationary or moving obstacle [28]. The CAS first requires the perception phase and is then followed by the action phase.

**Perception Phase:** CAS detects an obstacle in this phase while utilizing various active or passive sensors according to their functionality principle. Active sensors possess their sources for wave emission or light transmission along with the receiver or detector. The most-used active sensors include radars, sonar, and LiDARs. All of these use minimum processing power, give a quick response, are less affected by weather, scan bigger portions in minimum time, and can return various parameters of the obstacles effectively. Whereas passive sensors are only capable of reading the emitted energy from another source such as the sun. Widely used passive sensors are visual or optical cameras and infrared (IR) or thermal cameras. The image formed by a visual camera requires visual light, whereas a thermal camera requires IR light.

**Action Phase:** This phase utilizes four prime strategies for collision avoidance. These are geometric, force-field, optimized, and sense and avoid methods. The geometric approach utilizes the information about the location and velocity of the UAV along with its obstacle or neighbors. This is performed by trajectory simulation in which nodes are reformed for collision avoidance. In force-field, the approach manipulates the attractive or repulsive forces to avoid collisions. In the optimized method, the parameters of obstacles, which are already known, are utilized for route optimization. In the sense and avoid technique, runtime decisions are made based on obstacle avoidance. The development in CAS helps in simple tasks by warning the vehicle operator and in complex tasks partially or completely controlling the system for collision avoidance.

*3.4. Developments in Formation Control Protocols of UAVs*

Formation control aims to generate control signals, which pilot UAVs to form a specific shape. Along with the architecture of motion control, the developed strategies for obtaining it are of much importance [29].

**Formation Control Design:** Motion controls of UAVs require a flow of information within its team; therefore, it uses communication architectures.

There may be a lack of availability of global information in a single UAV for a whole operation. Due to its restricted capabilities to compute and communicate, centralized architecture is considered or used rarely. Decentralized architecture is preferred more for multi-UAV systems and uses the consensus algorithm technique for designing it. It is based on local interactions with the neighbors while maintaining a certain distance.

**Formation Control Strategies:** Various developed control approaches are discussed here that aid the researchers and possess certain benefits and limitations. They are:

i. **Leader-Follower Strategy:** As obvious from its title, this approach assigns one UAV as a leader, while the remaining UAVs as followers in a group. The mission information remains with the leader only while the followers chase their leader with pre-designed spaces. The major benefit of this strategy is that it can be implemented simply and easily. Due to leader dependency, this strategy faces single-point failures. This limitation can be compensated by assigning multi-leaders and virtual leaders.

ii. **Behavior-based Strategy:** This approach produces control signals, which consider several mission essentials, by adding various vector functions. Its greatest merit is that it is highly adaptable to any unknown environment. Its demerit is the requirement to model it mathematically, which leads to difficulty in analyzing system stabilities.

iii. **Virtual Structure Strategy:** This approach considers rigid structure for the desired shape of the group of UAVs. To achieve the desired shape, there is a need to fly each UAV towards its corresponding virtual node. Abilities to maintain the formation and fault-tolerance are its greatest advantages. This approach faces failure when the detection of a UAV is faulty in the formation. The compensation for this faulty UAV requires reconfiguration of the formation shape. This approach calls for a strong ability to compute, which is a disadvantage of this approach.

*3.5. Developments in Path Planning Techniques of UAVs*

Path planning aims to design a flight path towards a target with fewer chances of being demolished while facing limitations. Extensive research proposed different methods that overcome the path planning complexity of UAVs. To design algorithms for path planning, certain parameters, such as obstacles, the environment, and constraints, require selection with considerations [30]. The approaches employed for path planning have classifications based on their features and methodology.

## 4. Motion Planning and Optimization

*4.1. Motion Planning*

In robotics, motion planning refers to the act of dissolving a specified mobility goal into distinct motions. However, it is used to fulfill movement limitations while also potentially optimizing some components of the motion. However, motion planning is the challenge of planning for a vehicle that operates in areas with a high number of objects, performing actions to move through the environment as well as modify the configuration of the objects [31]. Even though the motion planning situation has arisen in continuous C-space, the calculation is discrete. As a result, we need a means to "discretize" the problem if we want an algorithmic solution. As a result, there are mainly two types of planning, combinatorial planning and sampling-based planning.

### 4.1.1. Combinatorial Motion Planning

Combinatorial Motion Planning is a type of motion planning that involves more than one approach to achieve the task, as shown in Figure 2. Although combinatorial motion planning discovers the pathways through the continuous configuration space, by using these strategies, researchers obtain a better result. The effective combination of algorithms is commonly based on bio-inspired algorithms with different approaches.

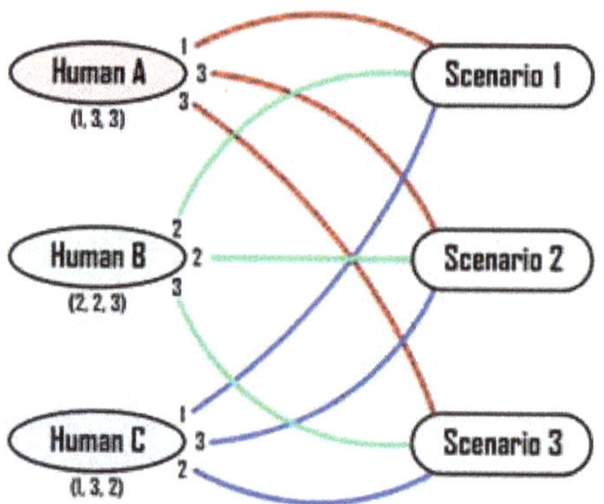


**Figure 2.** Three tasks Combinatorial Optimization example [32].

### 4.1.2. Sampling-Based Motion Planning

Random selection is used in sampling-based motion planning to build a graph or tree (path) in C-space on which queries (start/goal configurations) can be solved, as shown in Figure 3. To increase planner performance, we look at a variety of general-purpose strategies. At times over the past years, sampling-based path planning algorithms, such as Probabilistic Road Maps (PRM) and Rapidly Exploring Random Trees (RRT), have been demonstrated to perform effectively in reality and to provide theoretical assurances such as probabilistic completeness.

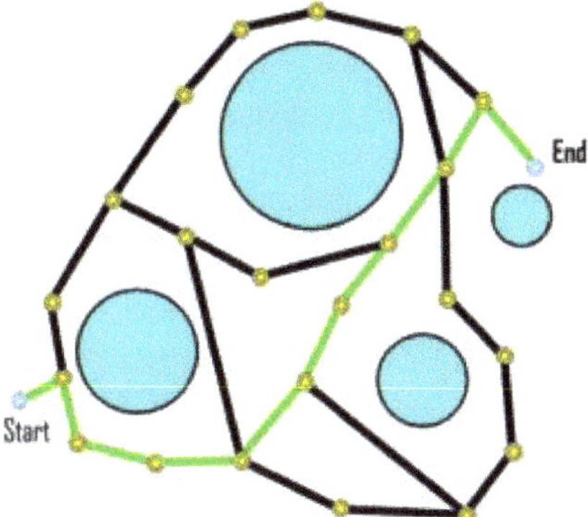


**Figure 3.** Sampling-based motion planning in the complex environment [33].

*4.2. Optimization Approach in Motion Planning*

The world has a desire for optimization concerning every natural phenomenon and its aspects. Therefore, many researchers developed optimization methods for multi-dimensional problems in various areas. These algorithms provide optimum solutions to the motion planning problems of UAVs, such as reducing production costs, convergence rate, energy consumption, and enhancing strength, efficiency, and reliability. The optimization algorithms are classified into biological algorithms, physical algorithms, and geographical algorithms, as presented in Figure 4 [34,35]. Biological algorithms have further classifications, namely swarm-based and evolution-based algorithms.

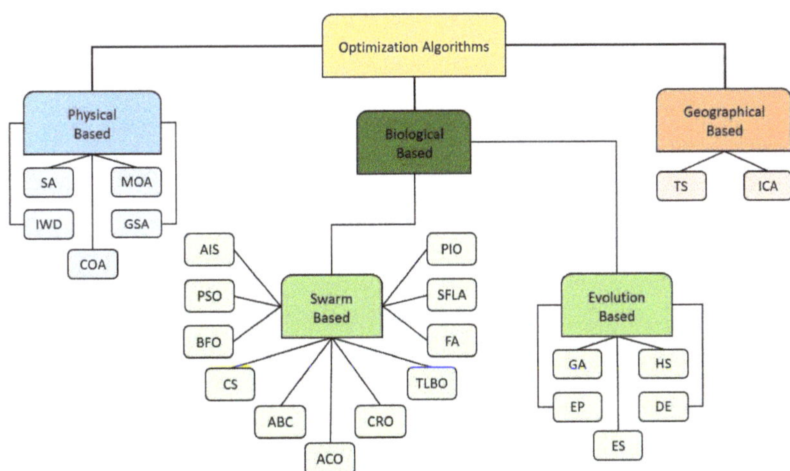


**Figure 4.** Classification of Optimization algorithms [36].

Biological Algorithms

Bionic researchers on a natural pattern developed nature-based algorithms and termed them biological algorithms. These are stemmed according to the correspondence between biological evolution and activities. The prime benefit of biological algorithms is their strength to tackle static as well as dynamic threats and ensure offline working. Without classifying these algorithms into further groups, we can label them as memetic algorithms. On the contrary, we can classify these algorithms into two categories, evolution-based algorithms and swarm-based algorithms [37].

A. Evolution-Based Algorithms

An evolution-based algorithm provides an optimal path for UAVs with consideration of three aspects. These aspects include travel distance, cost incurred, and path reliability cost to track that path. These evolutionary algorithms choose practical and achievable solutions randomly as the first generation and consider the parameters later to explain which randomly selected feasible solutions are appropriate or not. For determining curved paths with essential aspects in 3D terrain; an offline path planner with an evolutionary algorithm is required [38]. By taking aspects into account, for example, beeline to destination, min-max distance related to targets, and topographical obstacles free tracks, one can display the B-spline curve as a flying path. Some examples of these algorithms include Genetic algorithm (GA), Evolutionary Programming (EP), Evolutionary Strategy (ES), Differential Evolution algorithm (DE), and Harmony Search algorithm (HS).

GA gives the best optimal results in search space using three steps selection, crossover, and mutation. Besides its benefits, sometimes it gives long and premature convergence and loses optimal results. Moreover, it is not applied to real-time data. In 1990, Fogel introduced a technique called EP. It reaches optimal results after many iterations. Similarly, another evolutionary algorithm is ES, which uses specified principles in optimization problems. DE employs real coding instead of binary coding. It refines the final path while reducing the computational cost. The evolutionary algorithm that mimics a musician's improvisation process is the HS algorithm. It shows promising results in optimization problems. It is further improved with various versions.

B. Swarm-Based Algorithms

Nature-based along with population-based algorithms evolved into swarm-based algorithms [39]. The swarm represents the combined behavior of all the agents. Agents in a swarm have limited capabilities, but working together, they achieve the given tasks while being at distances. As a result of which, fast, low cost, and optimal solutions are obtained even in the uncertainties and complexities. Some examples of these algorithms include Artificial Immune System (AIS), Particle Swarm Optimization (PSO), Bacteria Foraging Optimization algorithm (BFO), Cuckoo Search algorithm (CS), Artificial Bee Colony algorithm (ABC), Ant Colony Optimization algorithm (ACO), Coral Reef Optimization algorithm (CRO), Teaching–Learning Based Optimization algorithm (TLBO), Firefly algorithm (FA), Shuffled Frog Leaping algorithm (SFLA), and Pigeon-Inspired Optimization (PIO).

AIS is an intelligent swarm-based algorithm that is modeled on the natural principles of the immune system of humans. It has the characteristics of the immune system of memory and learning to utilize for solving problems. It gives adequate trajectories in path planning with less computation. The development of PSO is based on the mobility theory of an insect crowd. In the layout of this fact-finding approach, every solo particle in the crowd recognizes the points given by the last swarm and produces a velocity vector towards the target point. The key benefit of this algorithm is that it is capable of obtaining optimal path planning in 3D, whereas its disadvantages are premature convergence and high time complexity. Passino introduced an algorithm based on the foraging behavior of Escherichia coli bacteria that lies in human intestines. He labeled this intelligent algorithm as BFO.

It provides rapid convergence and a global search. The CS algorithm replaces the average solutions and applies the solution that is potentially better. The ABC algorithm provides solutions to various optimization problems having constraints. The ACO algorithm is based on depositing characteristics of ants during food search and proved to be a meta-heuristic technique to derive the shortest path while dealing with continuous and multi-objective path planning issues. The CRO algorithm works efficiently with many advantages for difficult optimization problems. The TLBO algorithm requires minimum computational memory and can be employed easily. FA works efficiently for multimodal optimization problems. It finds the best location for UAVs with less energy consumption. SFLA depends on frogs' clusters that are looking for food. It gathers the best frog, which can give local optimum and evolves the frog with inaccurate positions. It continues making iterations until the accomplishment of an optimal path with better convergence. PIO works via sharing information and striving among all to quickly achieve the optimal global solution.

C.    Physical Algorithms

Heuristic algorithms that imitate physical laws and processes of nature are known as physical algorithms. These algorithms copy the physical conduct and characteristics of matter [40]. These are applicable for non-linear, high-dimensional, multimodal as well as complex optimization problems. There is very little research available on physical algorithms. These are categorized as Simulated Annealing (SA), Gravitational Search algorithm (GSA), Chaotic Optimization algorithm (COA), Intelligent Water Drops algorithm (IWD), and Magnetic Optimization algorithm (MOA). SA is suggested after a technique, annealing in metallurgy. It is employed for more complex computational optimization problems and gives approximate global optimum within a fixed time. GSA is a newly introduced algorithm that mimics laws of motion and gravitational law. It is applied to optimization problems with various functions. COA is an easily implemented and powerful mechanism that can escape convergence to a local optimum within a short time. The IWD algorithm is based on how natural rivers can find the best paths among many probable paths to their ultimate destination. MOA, a newly emerging algorithm, is derived from the basic principles of magnetism. The dual function of this algorithm can balance the disadvantages against the advantages in optimization problems.

D.    Geographical Algorithms

The meta-heuristic algorithms that give random outcomes in geographical search space are labeled as geographical algorithms [41]. Some of the geographical algorithms are the Tabu Search algorithm (TS) and Imperialistic Competition algorithm (ICA). The TS algorithm determines an optimal solution among various feasible solutions. Its memory can recall the recent optimal solution and guide the search to trace the previous solutions. It is employed for optimization problems in various areas. Another geographical algorithm for the global best solution in optimization is ICA. It imitates sociopolitical imperialist competition. It involves imperialistic competition among empires along with assimilation and revolution of colonies and so on. Due to its robust searching ability, it provides many benefits in optimization problems.

Among all the aforementioned algorithms, most are based on the swarm. These population-based algorithms are robust at obtaining better global solutions via their cooperative and self-adaptive abilities. These algorithms are employed for solving challenging issues of UAVs. This review paper gives details on a comparison of the aforesaid algorithms used for motion control and path planning of UAVs.

## 5. Related Review

To succeed, most motion planning approaches necessitate the use of appropriate optimization algorithms. These strategies can be used on a single UAV as well as a group of UAVs or a swarm of UAVs. When several UAV missions are viable for civilian objectives, a nature-inspired algorithm is required for control and optimization. Table 1 presents a detailed overview of the manuscripts related to motion planning problems of UAVs. The review also helps scholars with the optimization techniques applied to single or multiple UAVs.

**Table 1.** A detailed study on the Motion Planning of UAVs using the optimization approach.

| Ref. | Topic | Optimization Approach | UAV Type | Contributions | Limitations |
|---|---|---|---|---|---|
| [42] | "Collision free 4D path planning for multiple UAVs based on spatial refined voting mechanism and PSO approach" | PSO | Multiple | • Enhances searching ability and improves velocity.<br>• Gives collision-free paths. | • Returns to initial points in extreme conditions. |
| [43] | "Dynamic Discrete Pigeon-inspired Optimization for Multi-UAV Cooperative Search-attack Mission Planning" | D$^2$PIO | Multiple | • Ability to switch task.<br>• Superior performance in discrete environment. | • Frequent switching led to incomplete tasks.<br>• Computational cost is higher due to population size. |
| [44] | "MVO-Based Path Planning Scheme with Coordination of UAVs in 3-D Environment" | MA | Multiple | • Gives optimized path costs.<br>• Maintains coordination. | • Do not give dynamic obstacles.<br>• Does not consider hardware-oriented constraints. |
| [45] | "UAV trajectory optimization for Minimum Time Search with communication constraints and collision avoidance" | ACO | Single | • Detects the target quickly.<br>• Maintains connection with GCS and avoids collision. | • Greater computational time.<br>• A mandatory ground connection is needed to obtain desired results. |
| [46] | "Efficient path planning for UAV formation via comprehensively improved particle swarm optimization" | IPSO | Multiple | • Boosts the convergence rate.<br>• Improves the solution optimality. | • Does not allow path re-planning with moving and unexpected obstacles. |
| [47] | "Secrecy improvement via a joint optimization of UAV relay flight path and transmit power" | PSO | Single | • Enhances the secrecy capacity.<br>• Allows optimum position flying. | • Needs further improvement for full-duplex relaying. |
| [48] | "Trajectory Planning for UAV Based on Improved ACO Algorithm" | MACO | Multiple | • Optimized initial trajectory<br>• Proposed trajectory correction schemes for collision avoidance. | • No real-time trajectory planning used. |
| [49] | "Optimized Path-Planning in Continuous Spaces for Unmanned Aerial Vehicles Using Meta-Heuristics" | DE PSO GA | Multiple | • Less computation for first feasible path.<br>• DE overtakes PSO and GA in convergence. | • Work exists for static environment only.<br>• No real-time implementation. |
| [50] | "Multi-UAVs trajectory and mission cooperative planning based on the Markov model" | SA | Multiple | • Improves drone survivability.<br>• Solves multi-aircraft mission planning problems. | • Needs NP problem exploration. |

**Table 1.** *Cont.*

| Ref. | Topic | Optimization Approach | UAV Type | Contributions | Limitations |
|---|---|---|---|---|---|
| [51] | "PSO-based Minimum-time Motion Planning for Multiple Vehicles Under Acceleration and Velocity Limitations" | PSO | Multiple | • Minimizes the travelling time for slowest UAV. • Reduces the parameters for mathematical modeling. | • No control law for motion tracking. • Only applied to selected vehicles. |
| [52] | "Information fusion estimation-based path following control of quad-rotor UAVs subjected to Gaussian random disturbance" | GIFC | Single | • Reduces the design complexity. • Allows trajectory tracking with high accuracy. | • Contains a huge amount of matrix inversion operations. |
| [53] | "3D multi-UAV cooperative velocity-aware motion planning" | A* | Multiple | • Shows a higher possibility of reaching destinations. • Reduces time costs and paths. | • Does not serve complex missions and more UAVs. |
| [54] | "Unmanned aerial vehicle swarm distributed cooperation method based on situation awareness consensus and its information processing mechanism" | SDCM | Multiple | • Works efficiently in a complex and antagonistic mission environment. • Obtains the mission essentials at a bearable cost. | • On a larger scale, communication topology and management mode changes. |
| [55] | "A co-optimal coverage path planning method for aerial scanning of complex structures" | CCPP PSO | Multiple | • Optimizes path efficiency and inspection quality. • Provides improved flexible options. | • The exponential growth of complexity occurs as the problem size increases. • Needs uniform configuration spaces. |
| [56] | "A novel hybrid grey wolf optimizer algorithm for unmanned aerial vehicle (UAV) path planning" | Hybrid GWO | Single | • Generates smooth flight routes. • Accelerates the rate of convergence and retains the ability to explore. | • The optimal value is lower than GWO, SA, and SOS. • Execution time is higher than GWO in all cases. |
| [57] | "Continuous-Time Trajectory Optimization for Decentralized Multi-Robot Navigation" | DA | Multiple | • Generates collision-free trajectories. • Reduces jerk and time. | • Robustness and scalability can fail sometimes. • It has dynamic speed limits. |

**Table 1.** *Cont.*

| Ref. | Topic | Optimization Approach | UAV Type | Contributions | Limitations |
|---|---|---|---|---|---|
| [58] | "A Self-Heuristic Ant-Based Method for Path Planning of Unmanned Aerial Vehicle in Complex 3-D Space with Dense U-Type Obstacles" | SHA | Single | • The number of retreats reduced significantly. <br> • Time analysis enhanced compared to basic ACO. | • Applied to static obstacles only. <br> • Actual taboo nodes are not used. |
| [59] | "A novel mission planning method for UAVs' course of action" | TDRS | Single | • Generates multiple schemes automatically. <br> • Completes tasks in a shorter time. | • Time optimization is essential for war scenarios. <br> • Variations in threat and utilization factors. |
| [60] | "A multi-objective pigeon-inspired optimization approach to UAV distributed flocking among obstacles" | Improved MPIO | Single | • Guarantees stable and collision-free flocking. <br> • Prior environmental details and the number of UAVs are essential. | • Lacks convergence analysis. <br> • Deadlocks can occur. <br> • Emergency conditions and dynamic obstacles are not tested. |
| [61] | "Application of the ACO algorithm for UAV path planning" | ACO | Single | • Intermediate waypoint concept introduced for ACO. <br> • Improved fitness value. | • Search space is bigger due to ACO hunting procedure. <br> • Higher computational complexity. |
| [62] | "A method of feasible trajectory planning for UAV formation based on bi-directional fast search tree" | Bi-RRT | Single | • Solves the minimum efficiency of compound models in complicated environments. <br> • Yields safe and efficient formation and obstacle avoidance. | • GA algorithm has a smoother path than Bi-RRT. <br> • Can move very close to an obstacle. |
| [63] | "Towards a PDE-based large-scale decentralized solution for path planning of UAVs in shared airspace" | PDE | Single | • Ensures collision-free and optimal path flight safety. <br> • Proves to be computationally efficient. | • Does not allow UAVs to share their trajectories during the mission. |
| [64] | "Optimized multi-UAV cooperative path planning under the complex confrontation environment" | Improved GWO | Multiple | • Minimizes fuel costs and threats. <br> • Proves to be effective in cooperative path planning. | • The average distance of most UAVs is greater. |

**Table 1.** *Cont.*

| Ref. | Topic | Optimization Approach | UAV Type | Contributions | Limitations |
|---|---|---|---|---|---|
| [65] | "A constrained differential evolution algorithm to solve UAV path planning in disaster scenarios" | CDE | Single | • Refines the limitations. • Continues the investigations. | • Used only unconstrained optimization problems. |
| [66] | "A novel reinforcement learning-based grey wolf optimizer algorithm for unmanned aerial vehicles (UAVs) path planning" | GWO | Single | • Achieves effective and feasible routes smoothly. • Enables each UAV to perform operations independently. | • Not efficient in solving other sorts of an issue at the same time while introducing another algorithm. |
| [67] | "Synergistic path planning of multi-UAVs for air pollution detection of ships in ports" | PSO | Multiple | • Detects air pollution efficiently. • Guarantees reduction of ship emissions. | • Does not cover air control and wind speed influences. • Lacks large-scale data testing. |
| [68] | "An intelligent cooperative mission planning scheme of UAV swarm in uncertain dynamic environment" | HAPF ACO | Multiple | • Enhances searching abilities. • Executes tasks and avoids collisions and obstacles efficiently. | • Aims cooperative search-attacks at homogeneous UAVs only. |
| [69] | "Path planning of multiple UAVs with online changing tasks by an ORPFOA algorithm" | ORPFOA | Multiple | • Solves tasks efficiently with task preference and swapping tasks. • Determines optimal paths smoothly. | • Needs more reduction in running time. • It has some complex computations. |
| [70] | "Path Planning for Multi-UAV Formation Rendezvous Based on Distributed Cooperative Particle Swarm Optimization" | DCPSO | Multiple | • All UAVs arrived simultaneously without collision. • It avoids all types of obstacles. | • It cannot be used in real-time scenarios. • It takes more time to avoid collisions. |
| [71] | "A Performance Study of Bio-Inspired Algorithms in Autonomous Landing of Unmanned Aerial Vehicle" | BOA MFO ABC | Single | • MFO obtains the best points with minimal run time and error. • Gives bearable accuracy. | • Error is not optimized. |
| [72] | "UAVs path planning architecture for effective medical emergency response in future networks" | CVRP PSO ACO GA | Single | • CVRP outperforms with the least runtime and minimal cost and enhanced capacities. • Achieves the proper navigation. | • Lacks benchmark solutions. Does not consider real-time or complex scenarios. |

**Table 1.** *Cont.*

| Ref. | Topic | Optimization Approach | UAV Type | Contributions | Limitations |
|---|---|---|---|---|---|
| [73] | "Path planning of multiple UAVs using MMACO and DE algorithm in dynamic environment" | MMACO DE | Multiple | • Increases the robustness. • Preserves the global convergence speed. | • In multi-colonies, one colony follows same path as basic ACO. |
| [74] | "Multi-UAV coordination control by chaotic grey wolf optimization-based distributed MPC with event-triggered strategy" | Chaotic GWO | Multiple | • Gives efficiency in computations. • Enhances the global search mobility convergence speed. | • Stability conditions are not analyzed. • Has limited communication. |
| [75] | "Collective Motion and Self-Organization of a Swarm of UAVs: A Cluster-Based Architecture" | PSO | Multiple | • Gives fast connectivity and convergence. • Assures stability with fewer turns. | • Not implemented on hardware. • Focused on a specific scenario. |
| [76] | "A Cluster-Based Hierarchical-Approach for the Path Planning of Swarm" | MMACO | Multiple | • Gives superior performance. • Gives an optimal path with better convergence. | • Variation in the optimization costs in colonies 2 and 3 is neglected. |
| [77] | "Cooperative Path Planning of Multiple UAVs by using Max-Min Ant Colony Optimization along with Cauchy Mutant Operator" | MMACO CM | Multiple | • Finds the optimal routes with the shortest distance. • Avoids collision. | • Enhances the system complexity. |
| [78] | "A multi-strategy pigeon-inspired optimization approach to active disturbance rejection control parameters tuning for vertical take-off and landing fixed-wing UAV" | MPIO | Single | • Proves to be superior among all algorithms to solve multi-dimensional searching issues. • It converges faster and exploits in a better way. | • Altitude fluctuation is still present. • Immature result after 2nd iteration. |
| [79] | "Landing route planning method for micro drones based on hybrid optimization algorithm" | DO | Multiple | • Shows stronger convergence both locally and globally. • Yields better outcomes than both single algorithms. | • Speeds up convergence after orthogonal learning. |
| [80] | "Energy Efficient Neuro-Fuzzy Cluster-based Topology Construction with Metaheuristic Route Planning Algorithm for Unmanned Aerial Vehicles" | QALO | Single | • Gives more energy-efficient results, more rounds, higher throughput, and lower average delay results. • Selects optimal routes. | • Does not manage resources optimally. |

**Table 1.** *Cont.*

| Ref. | Topic | Optimization Approach | UAV Type | Contributions | Limitations |
|---|---|---|---|---|---|
| [81] | "Coordinated path following control of fixed-wing unmanned aerial vehicles in wind" | CPFC | Single | • Attains leaderless synchronization.<br>• Satisfies UAVs' constraints and upper bound path following errors. | • Requires better simulation of the external environment and the wireless communications. |
| [82] | "A diversified group teaching optimization algorithm with segment-based fitness strategy for unmanned aerial vehicle route planning" | GTO | Single | • Gives faster convergence.<br>• Handles all the complex constrained problems. | • Parameters need automatic adjustments. |
| [83] | "Coverage path planning for multiple unmanned aerial vehicles in maritime search and rescue operations" | RSH | Multiple | • Gives optimal results in a shorter time.<br>• Robust to strong wind. | • Does not provide exact solutions for larger instances. |
| [84] | "Hybrid FWPS cooperation algorithm based unmanned aerial vehicle constrained path planning" | FWPSALC | Single | • Produces high and superior quality solutions.<br>• Handles constraints in a better way. | • Gives poor performance for fewer number of particles or a large number of fireworks. |
| [85] | "Safety-enhanced UAV path planning with spherical vector-based particle swarm optimization" | PSO | Single | • Reduces the cost function.<br>• Gives the shortest and smoothest paths with fast convergence. | • Faces premature convergence. |

In 2019, Yang et al. [42] proposed a spatial refined voting mechanism and PSO algorithm that gave a 4D-space path planning that was collision-free and obstacle-free for multi-UAVs. Duan et al. [43] used a dynamic discrete pigeon-inspired optimization technique for search attack missions by using distributed path generation and central tasks mission. Jain et al. [44] suggested MVO and Munkres algorithms for the path planning and coordination of multiples, it compared the results with the results of BBO and GSO and concluded that the proposed algorithm is highly efficient in reducing execution time and finding optimized path costs. Pérez-Carabaza et al. [45] worked on optimizing trajectories for UAVs that used less time in searching for targets, avoided collisions, and maintained communication. Then, there is a comparison of this MMAS-based algorithm with GA and CEO, and it yielded better results than they yield. Shao et al. [46] used comprehensively modified PSO for the path planning of UAVs. This method gave a faster and improved convergence rate and solution optimality when compared with SPSO and MGA.

Mah et al. [47] suggested a joint optimization method that gave the best secrecy performance to combat eavesdropping on the flight path and transmits power and gave superior results to the max SNR method. Bo Li et al. [48] designed an improved ACO algorithm based on the metropolis criterion and predicted three trajectory corrections schemes for collision avoidance protocols and the inscribed circle method for smoothness. Geovanni et al. [49] proposed an optimized path planning method using a meta-heuristic in the continuous 3D environment. The study also minimizes the path length in the presence of static obstacles by manipulating control inputs. Ning et al. [50] solved the task-planning issue of multi-target and multi-aircraft by proposing a two-layer mission-planning model depending on the annealing and TS algorithms. Lihua et al. [51] gave an online priority configuration algorithm for the UAV swarm flight in an environment having compounded obstacles and showed superiority in cost of energy and time in simulation results.

In 2020, Xu et al. [52] solved the LQG problem of quad-rotor UAVs by presenting a Gaussian information fusion control (GIFC) method that allowed accurate trajectory tracking and reduced the design complexity. Hu et al. [53] proposed a 3D multi-UAV cooperative velocity-aware motion planning using VeACA2D and VeACA3D. While comparing with LyCL and PALyCL, this algorithm gave higher possibilities of reaching the destination while following shorter paths and reduced time costs. Gao and li [54] considered the distributed cooperation approach formed on situation awareness consensus and its details processing method for UAV swarms. Shang et al. [55] linked a co-optimal coverage path planning method with a PSO algorithm for aerial scanning of compounded models. Qu et al. [56] evaluated a novel hybrid grey wolf optimizer algorithm with MSOS and gave better and improved results for UAV path planning in a complex environment.

Krishnan et al. [57] optimized the continuous-time trajectory by combining a decentralized algorithm with third-order dynamics that helped robots to re-plan trajectories. Zhang et al. [58] introduced an ant-based self-heuristic method for path planning of multi-UAVs. In this study, the authors used U-shaped dense complex 3D space to reduce the confusion of obstacle detection. It reduces the deadlock state with a two-stage strategy. Zhou et al. [59] utilized the multi-string chromosome genetic and cuckoo search algorithms to improve the MDLS algorithm. This improved algorithm proved that it had a better global optimization capability and diversified scheme options, and completed tasks in a shorter time as compared to the simplified MDLS. Qiu and Duan [60] developed an improved MPIO formulated on hierarchical learning behavior that gave improved distributed flocking among obstacles. Comparison with MPIO and NSGA-II showed that the improved MPIO proved to be more suitablefort handling the various-objective optimization and obstacle avoidance for UAV flocking.

Konatowski and Pawłowski [61] presented a path planning for UAVs with the help of ACO. It uses waypoints along its path with unknown parameters. The proposed work reduces the computational time and obtains the optimal route. Huang and Sun [62] detailed an approach to feasible trajectory planning formation that depends on a bi-directional fast search tree for UAVs. Radmanesh et al. [63] applied a PDE-based large-scale decentralized approach and compared it with centralized and sequential approaches to obtain collision-free and optimal path planning of multiple UAVs. Xu et al. [64] linked the grey wolf optimizer algorithm with the PSO algorithm to achieve cooperative path planning of multi-UAVs under the threats of ground radar, missiles, and terrain. Yu et al. [65] introduced an improved constrained differential evolution algorithm that reduced the fitness functions and satisfied the three constraints, namely, height, angle, and slope of UAVs.

Later, this improved algorithm was compared with FIDE, DE variants, RankDE, CMODE, and $(\mu + \gamma) - $ CDE and proved that the proposed CDE generated more optimal paths smoothly. Qu et al. [66] used a reinforcement learning-based grey wolf optimizer algorithm. Then, compared the outcomes with the results of GWO, MGWO, EEGWO, and IGWO algorithms and concluded that the proposed RLGWO gives better, feasible, and effective path planning for UAVs. Shen et al. [67] solved the air pollution detection problem for ships in ports and evaluated a synergistic path planning of multiple UAVs. He suggested an improved PSO algorithm with a Tabu Search (TS) table, proved the efficient detection of air pollution, and ensured less emission by ships.

Zhen et al. [68] gave an improved method that is a hybrid artificial potential field with ant colony optimization (HAPF-ACO) method that executes tasks and avoids collisions and obstacles efficiently for the cooperative mission planning of fixed-wing UAVs. The results were compared with ACOAPF and PSO algorithms that proved the suggested algorithm to be highly efficient in task execution. Li et al. [69] detailed an ORPFOA algorithm that allows online changing tasks for optimal path planning of multi-UAVs for solving faster and giving higher optimization. Then, the outcomes of this suggested algorithm were compared with GWO, PSO, PIO, PSOGSA, PPPIO, and FOA. The proposed algorithm gave faster convergence and optimization than the others.

Shao et al. [70] obtained multi-UAV path planning by using the distributed cooperative PSO approach. This study presents a complex dynamic environment with a higher success rate of 0.9 compared to CCGA. Ilango and R. [71] studied Bio-inspired algorithms and analyzed their performance in the autonomous landing of UAVs. Wu et al. [72] applied a new method to UAVs that is based on consensus theory for their formation control as well as obstacle avoidance.

In 2021, recent research by Ali et al. [73] developed a multi-colonies optimization and combined MMACO and DE techniques for the cooperative path planning of many UAVs in a dynamic environment. WANG et al. [74] proposed an MPC framework along with Chaotic Grey Wolf Optimization (CGWO) and an event-triggered approach to give UAV coordination control and trajectory tracking. Ali et al. [75] used combined movement along with the reflexivity of a UAV swarm via the cluster-based technique by combining the PSO algorithm with the MAS. It showed better convergence and durability. Shafiq et al. [76] suggested a cluster-based hierarchical approach for control and path planning. It quickly finds the optimal path along with the minimal costs. Ali et al. [77] applied a hybrid algorithm of the max-min ant colony optimization algorithm with CM operators on multiple UAVs for collective path planning. It gives the optimal global solution in minimum time. He and Duan [78] considered flying, as well as touching down, issues and suggested an improved PIO for tuning the parameters of ADRC. Liang et al. [79] developed an optimal route planning for the landing of micro-UAVs using hybrid optimization algorithms with orthogonal learning.

Pustokhina et al. [80] designed clustering that is energy efficient and plans optimal routes by developing Energy Efficient Neuro-Fuzzy Cluster-based Topology Construction with the MRP technique for UAVs. Chen et al. [81] suggested a coordination strategy for fixed-wing UAVs with wind disturbances and developed a hardware-in-the-loop (HIL) simulation. Jiang et al. [82] worked on path planning for UAVs under various obstacles and proposed a diversified group teaching optimization algorithm with a segment-based fitness approach that has better global exploration ability. Cho et al. [83] gave a coverage path planning strategy with two phases for multi-UAVs that helped in searching and rescuing in maritime environment. Zhang et al. [84] presented a hybrid FWPSALC mechanism for the path planning method for UAVs that proved to be robust in searching and handling constraints and had a better speed convergence. Phung and Ha. [85–88] developed a novel technique with spherical vector-based particle swarm optimization (SPSO) that ensures safety, feasibility, and optimal paths and gives results better than classic PSO, QPSO, θ-PSO, and various other algorithms.

## 6. Discussion

The most crucial challenge in the field of UAVs is efficient motion planning. It requires a state-of-the-art optimization method to counter issues. This research evaluates various challenges faced by UAVs and all the current designs of motion planning techniques. The recent developments discussed the results in high adaptable ability, cost and time reductions in task executions, energy efficiency, obstacles, and collision avoidance.

While reviewing various motion planning approaches, it became evident that most of the researchers preferred to use an optimization approach with nature-inspired algorithms. While discussing numerous categories of path planning strategies, it appears that hybrid algorithms give better performance. These improved and optimized algorithms overcome the limitations of numerical and analytical techniques. By analyzing the manuscript, it can be concluded that the best optimization approaches are swarm-based due to their exceptional ability to solve complex issues with their simplified approach.

## 7. Conclusions

UAVs are flying machines that possess safe and task-oriented mobility in the presence of uncertainties with the help of modified techniques and the latest technological developments. The autonomous capability of these machines is also advancing and upgrading to provide efficient flying and stable formation in dynamic environments. However, motion planning issues in UAVs are most challenging among scholars. In this article, a detailed comparative study on the motion planning issues and achievements of UAVs has been presented, along with the limitations of each article. The study also presents recent challenges in all possible categories of UAVs to highlight the importance of UAVs in our society along with their developments and state-of-the-art work performed in the last 3 years.

## 8. Future Work

There is a very bound analysis in the comparison field of motion planning and optimization algorithms that exists already and the determination of the best among them. To deploy the multiple UAV systems in a finer way, various challenges and possibilities need more exploration, as well as a reduction in exploitations. Leads for future work are to model different swarm-based intelligent optimization approaches with high accuracy and efficiency and further feasible algorithms for 3D-path planning strategies.

**Author Contributions:** Conceptualization, Z.A.A.; methodology, A.I.; software, E.H.A.; validation, E.H.A.; investigation, A.I.; data curation, E.H.A.; writing—original draft preparation, A.I. and J.J.J.; writing—review and editing, E.H.A. and A.I.; supervision, Z.A.A.; project administration, Z.A.A.; funding acquisition, J.J.J. All authors have read and agreed to the published version of the manuscript.

**Funding:** This research was supported by the European Regional Development project Green Smart Services in Developing Circular Economy SMEs (A77472).

**Data Availability Statement:** All the data are in the article.

**Conflicts of Interest:** The authors declare no conflict of interest.

## Abbreviations

| Acronyms | Definitions |
| --- | --- |
| UAV | Unmanned Aerial Vehicles |
| AI | Artificial Intelligence |
| P2P | Point-to-Point |
| MAC | Medium Access Control |
| IETF | Internet Engineering Task Force |
| MAVLink | Micro Air Vehicle Link |
| NBC | Nuclear, Biological, and Chemical |
| CAS | Collision Avoidance System |
| IR | InfraRed |
| GA | Genetic algorithm |
| EP | Evolutionary Programming |
| ES | Evolutionary Strategy |
| DE | Differential Evolution |
| HS | Harmony Search |
| AIS | Artificial Immune System |
| PSO | Particle Swarm Optimization |
| BFO | Bacteria Foraging Optimization |
| CS | Cuckoo Search |
| ABC | Artificial Bee Colony |
| ACO | Ant Colony Optimization |

| | |
|---|---|
| CRO | Coral Reef Optimization |
| TLBO | Teaching-Learning Based Optimization |
| FA | Firefly algorithm |
| SFLA | Shuffled Frog Leaping algorithm |
| PIO | Pigeon Inspired Optimization |
| SA | Simulated Annealing |
| GSA | Gravitational Search algorithm |
| COA | Chaotic Optimization algorithm |
| IWD | Intelligent Water Drops |
| MOA | Magnetic Optimization |
| TS | Tabu Search algorithm |
| ICA | Imperialistic Competition algorithm |
| MACO | Metropolis Criterion ACO |
| MA | Munkres algorithm |
| GIFC | Gaussian information fusion control |
| DA | Decentralized algorithm |
| SHA | Self-Heuristic Ant |
| TDRS | Task Decomposition Recourse Scheduling |
| CDE | Constraint Differential Evolution |
| PDE | Partial Differential Equation |
| DCPSO | Distributed Cooperative Particle Swarm Optimization |
| DO | Dragonfly Optimization |
| QALO | Quantum Ant Lion Optimization |
| CPFC | Coordinated Path Following Control strategy |
| RSH | Randomized Search Heuristic |
| GTO | Group Teaching Optimization |
| SDCM | Swarm Distributed Cooperation Method |
| MFO | Moth Flame Optimization |
| BOA | Bat Optimization algorithm |

## References


1. Wang, Z.; Liu, R.; Liu, Q.; Thompson, J.S.; Kadoch, M. Energy-efficient data collection and device positioning in UAV-assisted IoT. *IEEE Internet Things J.* **2019**, *7*, 1122–1139. [CrossRef]
2. Ouns, B.; Abrassart, A.; Garcia, F.; Larrieu, N. A mobility model for UAV ad hoc network. In Proceedings of the 2014 International Conference on Unmanned Aircraft Systems (ICUAS), Orlando, FL, USA, 27–30 May 2014; pp. 383–388.
3. Kuntz, R.R.; Kienitz, K.H.; Brandão, M.P. Development of a multi-purpose portable electrical UAV system, fixed & rotative wing. In Proceedings of the 2011 Aerospace Conference, Big Sky, MT, USA, 5–12 March 2011; pp. 1–9.
4. Carrivick, J.L.; Smith, M.W.; Quincey, D.J.; Carver, S.J. Developments in budget remote sensing for the geosciences. *Geol. Today* **2013**, *29*, 138–143. [CrossRef]
5. Taeyoung, L.; Leok, M.; McClamroch, N.H. Nonlinear robust tracking control of a quadrotor UAV on SE (3). *Asian J. Control* **2013**, *15*, 391–408.
6. Ali, K.N.; Brohi, S.N.; Jhanjhi, N.Z. UAV's applications, architecture, security issues and attack scenarios: A survey. In *Intelligent Computing and Innovation on Data Science*; Springer: Singapore, 2020; pp. 753–760.
7. Thammawichai, M.; Baliyarasimhuni, S.P.; Kerrigan, E.C.; Sousa, J.B. Optimizing communication and computation for multi-UAV information gathering applications. *IEEE Trans. Aerosp. Electron. Syst.* **2017**, *54*, 601–615. [CrossRef]
8. Sun, J.; Tang, J.; Lao, S. Collision avoidance for cooperative UAVs with optimized artificial potential field algorithm. *IEEE Access* **2017**, *5*, 18382–18390. [CrossRef]
9. Hu, N.; Tian, Z.; Sun, Y.; Yin, L.; Zhao, B.; Du, X.; Guizani, N. Building agile and resilient uav networks based on sdn and blockchain. *IEEE Netw.* **2021**, *35*, 57–63. [CrossRef]
10. Ziegler, C.A. Weapons development in context: The case of the World War I balloon bomber. *Technol. Cult.* **1994**, *35*, 750–767. [CrossRef]
11. Bertacchi, A.; Giannini, V.; di Franco, C.; Silvestri, N. Using unmanned aerial vehicles for vegetation mapping and identification of botanical species in wetlands. *Landsc. Ecol. Eng.* **2019**, *15*, 231–240. [CrossRef]
12. Keane, J.F.; Carr, S.S. A brief history of early unmanned aircraft. *Johns Hopkins APL Tech. Dig.* **2013**, *32*, 558–571.
13. Jung, S.; Kim, H. Analysis of amazon prime air uav delivery service. *J. Knowl. Inf. Technol. Syst.* **2017**, *12*, 253–266.
14. Yan, F.; Liu, Y.; Xiao, J. Path planning in complex 3D environments using a probabilistic roadmap method. *Int. J. Autom. Comput.* **2013**, *10*, 525–533. [CrossRef]

15. Ahmed, S.; Mohamed, A.; Harras, K.; Kholief, M.; Mesbah, S. Energy efficient path planning techniques for UAV-based systems with space discretization. In Proceedings of the 2016 IEEE Wireless Communications and Networking Conference, Doha, Qatar, 3–6 April 2016; pp. 1–6.
16. Pachter, M.; D'Azzo, J.J.; Dargan, J.L. Automatic formation flight control. *J. Guid. Control Dyn.* **1994**, *17*, 1380–1383. [CrossRef]
17. Nguyen, H.T.; Quyen, T.V.; Nguyen, C.V.; Le, A.M.; Tran, H.T.; Nguyen, M.T. Control algorithms for UAVs: A comprehensive survey. *EAI Endorsed Trans. Ind. Netw. Intell. Syst.* **2020**, *7*, 164586. [CrossRef]
18. Stentz, A. Optimal and efficient path planning for partially known environments. In *Intelligent Unmanned Ground Vehicles*; Springer: Boston, MA, USA, 1997; pp. 203–220.
19. Lu, Y.; Xue, Z.; Xia, G.; Zhang, L. A survey on vision-based UAV navigation. *Geo-Spat. Inf. Sci.* **2018**, *21*, 21–32. [CrossRef]
20. Fraga-Lamas, P.; Ramos, L.; Mondéjar-Guerra, V.; Fernández-Caramés, T.M. A review on IoT deep learning UAV systems for autonomous obstacle detection and collision avoidance. *Remote Sens.* **2019**, *11*, 2144. [CrossRef]
21. Cai, G.; Chen, B.M.; Lee, T.H. An overview on development of miniature unmanned rotorcraft systems. *Front. Electr. Electron. Eng. China* **2010**, *5*, 1–14. [CrossRef]
22. Azoulay, R.; Haddad, Y.; Reches, S. Machine Learning Methods for Management UAV Flocks-a Survey. *IEEE Access* **2021**, *9*, 139146–139175. [CrossRef]
23. Do, H.T.; Hua, H.T.; Nguyen, M.T.; Nguyen, C.V.; Nguyen, H.T.T.; Nguyen, H.T.; Nguyen, N.T.T. Formation control algorithms for multiple-UAVs: A comprehensive survey. *EAI Endorsed Trans. Ind. Netw. Intell. Syst.* **2021**, *8*, e3. [CrossRef]
24. Aggarwal, S.; Kumar, N. Path planning techniques for unmanned aerial vehicles: A review, solutions, and challenges. *Comput. Commun.* **2020**, *149*, 270–299. [CrossRef]
25. Zhang, J.; Liu, W.; Wu, Y. Novel technique for vision-based UAV navigation. *IEEE Trans. Aerosp. Electron. Syst.* **2011**, *47*, 2731–2741. [CrossRef]
26. Ali, Z.A.; Wang, D.B.; Loya, M.S. SURF and LA with RGB Vector Space Based Detection and Monitoring of Manholes with an Application to Tri-Rotor UAS Images. *Int. J. Eng. Technol.* **2017**, *9*, 32.
27. Craighead, J.; Murphy, R.; Burke, J.; Goldiez, B. A survey of commercial & open source unmanned vehicle simulators. In Proceedings of the Proceedings 2007 IEEE International Conference on Robotics and Automation, Rome, Italy, 10–14 April 2007; pp. 852–857.
28. Park, J.-W.; Oh, H.-Y.; Tahk, M.-I. UAV collision avoidance based on geometric approach. In Proceedings of the 2008 SICE Annual Conference, Tokyo, Japan, 20–22 August 2008; pp. 2122–2126.
29. Anderson, B.; Fidan, B.; Yu, C.; Walle, D. UAV formation control: Theory and application. In *Recent Advances in Learning and Control*; Springer: London, UK, 2008; pp. 15–33.
30. Bortoff, S.A. Path planning for UAVs. In Proceedings of the 2000 American Control Conference, ACC (IEEE Cat. No. 00CH36334), Chicago, IL, USA, 28–30 June 2000; Volume 1, pp. 364–368.
31. Du, T.; Noel, E.; Burdick, J.W. Robotic motion planning in dynamic, cluttered, uncertain environments. In Proceedings of the 2010 IEEE International Conference on Robotics and Automation, Anchorage, AK, USA, 3–7 May 2010; pp. 966–973.
32. Malik, W.; Rathinam, S.; Darbha, S.; Jeffcoat, D. Combinatorial motion planning of multiple vehicle systems. In Proceedings of the 45th IEEE Conference on Decision and Control, San Diego, CA, USA, 13–15 December 2006; pp. 5299–5304.
33. Lindemann, S.R.; LaValle, S.M. Current issues in sampling-based motion planning. In *Robotics Research. The Eleventh International Symposium*; Springer: Berlin/Heidelberg, Germany, 2005; pp. 36–54.
34. Ferguson, D.; Howard, T.M.; Likhachev, M. Motion planning in urban environments. *J. Field Robot.* **2008**, *25*, 939–960. [CrossRef]
35. Overmars, M.H. *A Random Approach to Motion Planning*; Department of Computer Science, Utrecht University: Utrecht, The Netherlands, 1992.
36. Behera, S.; Sahoo, S.; Pati, B.B. A review on optimization algorithms and application to wind energy integration to grid. *Renew. Sustain. Energy Rev.* **2015**, *48*, 214–227. [CrossRef]
37. Iztok, F., Jr.; Yang, X.; Fister, I.; Brest, J.; Fister, D. A brief review of nature-inspired algorithms for optimization. *arXiv* **2013**, arXiv:1307.4186.
38. Rathbun, D.; Kragelund, S.; Pongpunwattana, A.; Capozzi, B. An evolution based path planning algorithm for autonomous motion of a UAV through uncertain environments. In Proceedings of the 21st Digital Avionics Systems Conference, Irvine, CA, USA, 27–31 October 2002; Volume 2, p. 8D2.
39. Yang, X.-S. Swarm-based metaheuristic algorithms and no-freelunch theorems. In *Theory and New Applications of Swarm Intelligence*; Intech: Rijeka, Croatia, 2012.
40. Beheshti, Z.; Shamsuddin, S.M.H. A review of population-based meta-heuristic algorithms. *Int. J. Adv. Soft Comput. Appl.* **2013**, *5*, 1–35.
41. Kanza, Y.; Safra, E.; Sagiv, Y.; Doytsher, Y. Heuristic algorithms for route-search queries over geographical data. In Proceedings of the 16th ACM SIGSPATIAL International Conference on Advances in Geographic Information Systems, Irvine, CA, USA, 5–7 November 2008; pp. 1–10.
42. Liu, Y.; Zhang, X.; Zhang, Y.; Guan, X. Collision free 4D path planning for multiple UAVs based on spatial refined voting mechanism and PSO approach. *Chin. J. Aeronaut.* **2019**, *32*, 1504–1519. [CrossRef]
43. Duan, H.; Zhao, J.; Deng, Y.; Shi, Y.; Ding, X. Dynamic discrete pigeon-inspired optimization for multi-UAV cooperative search-attack mission planning. *IEEE Trans. Aerosp. Electron. Syst.* **2020**, *57*, 706–720. [CrossRef]

44. Jain, G.; Yadav, G.; Prakash, D.; Shukla, A.; Tiwari, R. MVO-based path planning scheme with coordination of UAVs in 3-D environment. *J. Comput. Sci.* **2019**, *37*, 101016. [CrossRef]
45. Pérez-Carabaza, S.; Scherer, J.; Rinner, B.; López-Orozco, J.A.; Besada-Portas, E. UAV trajectory optimization for Minimum Time Search with communication constraints and collision avoidance. *Eng. Appl. Artif. Intell.* **2019**, *85*, 357–371. [CrossRef]
46. Shao, S.; Peng, Y.; He, C.; Du, Y. Efficient path planning for UAV formation via comprehensively improved particle swarm optimization. *ISA Trans.* **2020**, *97*, 415–430. [CrossRef]
47. Mah, M.-C.; Lim, H.-E.; Tan, A.W. Secrecy improvement via joint optimization of UAV relay flight path and transmit power. *Veh. Commun.* **2020**, *23*, 100217. [CrossRef]
48. Li, B.; Qi, X.; Yu, B.; Liu, L. Trajectory planning for UAV based on improved ACO algorithm. *IEEE Access* **2019**, *8*, 2995–3006. [CrossRef]
49. Flores-Caballero, G.; Rodríguez-Molina, A.; Aldape-Pérez, M.; Villarreal-Cervantes, M.G. Optimized path-planning in continuous spaces for unmanned aerial vehicles using meta-heuristics. *IEEE Access* **2020**, *8*, 176774–176788. [CrossRef]
50. Ning, Q.; Tao, G.; Chen, B.; Lei, Y.; Yan, H.; Zhao, C. Multi-UAVs trajectory and mission cooperative planning based on the Markov model. *Phys. Commun.* **2019**, *35*, 100717. [CrossRef]
51. Pamosoaji, A.K.; Piao, M.; Hong, K. PSO-based minimum-time motion planning for multiple vehicles under acceleration and velocity limitations. *Int. J. Control Autom. Syst.* **2019**, *17*, 2610–2623. [CrossRef]
52. Xu, Q.; Wang, Z.; Zhen, Z. Information fusion estimation-based path following control of quadrotor UAVs subjected to Gaussian random disturbance. *ISA Trans.* **2020**, *99*, 84–94. [CrossRef] [PubMed]
53. Hu, Y.; Yao, Y.; Ren, Q.; Zhou, X. 3D multi-UAV cooperative velocity-aware motion planning. *Future Gener. Comput. Syst.* **2020**, *102*, 762–774. [CrossRef]
54. Gao, Y.; Li, D. Unmanned aerial vehicle swarm distributed cooperation method based on situation awareness consensus and its information processing mechanism. *Knowl. Based Syst.* **2020**, *188*, 105034. [CrossRef]
55. Shang, Z.; Bradley, J.; Shen, Z. A co-optimal coverage path planning method for aerial scanning of complex structures. *Expert Syst. Appl.* **2020**, *158*, 113535. [CrossRef]
56. Qu, C.; Gai, W.; Zhang, J.; Zhong, M. A novel hybrid grey wolf optimizer algorithm for unmanned aerial vehicle (UAV) path planning. *Knowl. Based Syst.* **2020**, *194*, 105530. [CrossRef]
57. Krishnan, S.; Rajagopalan, G.A.; Kandhasamy, S.; Shanmugavel, M. Continuous-Time Trajectory Optimization for Decentralized Multi-Robot Navigation. *IFAC-Pap. OnLine* **2020**, *53*, 494–499. [CrossRef]
58. Zhang, C.; Hu, C.; Feng, J.; Liu, Z.; Zhou, Y.; Zhang, Z. A self-heuristic ant-based method for path planning of unmanned aerial vehicle in complex 3-D space with dense U-type obstacles. *IEEE Access* **2019**, *7*, 150775–150791. [CrossRef]
59. Zhou, Y.; Zhao, H.; Chen, J.; Jia, Y. A novel mission planning method for UAVs' course of action. *Comput. Commun.* **2020**, *152*, 345–356. [CrossRef]
60. Qiu, H.; Duan, H. A multi-objective pigeon-inspired optimization approach to UAV distributed flocking among obstacles. *Inf. Sci.* **2020**, *509*, 515–529. [CrossRef]
61. Konatowski, S.; Pawłowski, P. Application of the ACO algorithm for UAV path planning. *Prz. Elektrotechniczny* **2019**, *95*, 115–118. [CrossRef]
62. Huang, J.; Sun, W. A method of feasible trajectory planning for UAV formation based on bi-directional fast search tree. *Optik* **2020**, *221*, 165213. [CrossRef]
63. Radmanesh, R.; Kumar, M.; French, D.; Casbeer, D. Towards a PDE-based large-scale decentralized solution for path planning of UAVs in shared airspace. *Aerosp. Sci. Technol.* **2020**, *105*, 105965. [CrossRef]
64. Xu, C.; Xu, M.; Yin, C. Optimized multi-UAV cooperative path planning under the complex confrontation environment. *Comput. Commun.* **2020**, *162*, 196–203. [CrossRef]
65. Yu, X.; Li, C.; Zhou, J. A constrained differential evolution algorithm to solve UAV path planning in disaster scenarios. *Knowl. Based Syst.* **2020**, *204*, 106209. [CrossRef]
66. Qu, C.; Gai, W.; Zhong, M.; Zhang, J. A novel reinforcement learning based grey wolf optimizer algorithm for unmanned aerial vehicles (UAVs) path planning. *Appl. Soft Comput.* **2020**, *89*, 106099. [CrossRef]
67. Shen, L.; Wang, Y.; Liu, K.; Yang, Z.; Shi, X.; Yang, X.; Jing, K. Synergistic path planning of multi-UAVs for air pollution detection of ships in ports. *Transp. Res. Part E Logist. Transp. Rev.* **2020**, *144*, 102128. [CrossRef]
68. Zhen, Z.; Chen, Y.; Wen, L.; Han, B. An intelligent cooperative mission planning scheme of UAV swarm in uncertain dynamic environment. *Aerosp. Sci. Technol.* **2020**, *100*, 105826. [CrossRef]
69. Li, K.; Ge, F.; Han, Y.; Xu, W. Path planning of multiple UAVs with online changing tasks by an ORPFOA algorithm. *Eng. Appl. Artif. Intell.* **2020**, *94*, 103807. [CrossRef]
70. Shao, Z.; Yan, F.; Zhou, Z.; Zhu, X. Path planning for multi-UAV formation rendezvous based on distributed cooperative particle swarm optimization. *Appl. Sci.* **2019**, *9*, 2621. [CrossRef]
71. Ilango, H.S.; Ramanathan, R. A Performance Study of Bio-Inspired Algorithms in Autonomous Landing of Unmanned Aerial Vehicle. *Procedia Comput. Sci.* **2020**, *171*, 1449–1458. [CrossRef]
72. Khan, S.I.; Qadir, Z.; Munawar, H.S.; Nayak, S.R.; Budati, A.K.; Verma, K.D.; Prakash, D. UAVs path planning architecture for effective medical emergency response in future networks. *Phys. Commun.* **2021**, *47*, 101337. [CrossRef]

73. Ali, Z.A.; Zhangang, H.; Zhengru, D. Path planning of multiple UAVs using MMACO and DE algorithm in dynamic environment. *Meas. Control* **2020**, 0020294020915727. [CrossRef]
74. Wang, Y.; Zhang, T.; Cai, Z.; Zhao, J.; Wu, K. Multi-UAV coordination control by chaotic grey wolf optimization based distributed MPC with event-triggered strategy. *Chin. J. Aeronaut.* **2020**, *33*, 2877–2897. [CrossRef]
75. Ali, Z.A.; Han, Z.; Masood, R.J. Collective Motion and Self-Organization of a Swarm of UAVs: A Cluster-Based Architecture. *Sensors* **2021**, *21*, 3820. [CrossRef]
76. Shafiq, M.; Ali, Z.A.; Alkhammash, E.H. A cluster-based hierarchical-approach for the path planning of swarm. *Appl. Sci.* **2021**, *11*, 6864. [CrossRef]
77. Ali, Z.A.; Zhangang, H.; Hang, W.B. Cooperative path planning of multiple UAVs by using max–min ant colony optimization along with cauchy mutant operator. *Fluct. Noise Lett.* **2021**, *20*, 2150002. [CrossRef]
78. He, H.; Duan, H. A multi-strategy pigeon-inspired optimization approach to active disturbance rejection control parameters tuning for vertical take-off and landing fixed-wing UAV. *Chin. J. Aeronaut.* **2021**, *35*, 19–30. [CrossRef]
79. Liang, S.; Song, B.; Xue, D. Landing route planning method for micro drones based on hybrid optimization algorithm. *Biomim. Intell. Robot.* **2021**, *1*, 100003. [CrossRef]
80. Pustokhina, I.V.; Pustokhin, D.A.; Lydia, E.L.; Elhoseny, M.; Shankar, K. Energy Efficient Neuro-Fuzzy Cluster based Topology Construction with Metaheuristic Route Planning Algorithm for Unmanned Aerial Vehicles. *Comput. Netw.* **2021**, *107*, 108214. [CrossRef]
81. Chen, H.; Wang, X.; Shen, L.; Yu, Y. Coordinated path following control of fixed-wing unmanned aerial vehicles in wind. *ISA Trans.* **2021**, *122*, 260–270. [CrossRef]
82. Jiang, Y.; Wu, Q.; Zhang, G.; Zhu, S.; Xing, W. A diversified group teaching optimization algorithm with segment-based fitness strategy for unmanned aerial vehicle route planning. *Expert Syst. Appl.* **2021**, *185*, 115690. [CrossRef]
83. Cho, S.W.; Park, H.J.; Lee, H.; Shim, D.H.; Kim, S. Coverage path planning for multiple unmanned aerial vehicles in maritime search and rescue operations. *Comput. Ind. Eng.* **2021**, *161*, 107612. [CrossRef]
84. Zhang, X.; Xia, S.; Zhang, T.; Li, X. Hybrid FWPS cooperation algorithm based unmanned aerial vehicle constrained path planning. *Aerosp. Sci. Technol.* **2021**, *118*, 107004. [CrossRef]
85. Phung, M.D.; Ha, Q.P. Safety-enhanced UAV path planning with spherical vector-based particle swarm optimization. *Appl. Soft Comput.* **2021**, *107*, 107376. [CrossRef]
86. Suo, W.; Wang, M.; Zhang, D.; Qu, Z.; Yu, L. Formation Control Technology of Fixed-Wing UAV Swarm Based on Distributed Ad Hoc Network. *Appl. Sci.* **2022**, *12*, 535. [CrossRef]
87. Zong, Q.; Wang, D.; Shao, S.; Zhang, B.; Han, Y. Research status and development of multi UAV coordinated formation flight control. *J. Harbin Inst. Technol.* **2017**, *49*, 1–14.
88. Ambroziak, L.; Ciężkowski, M. Virtual Electric Dipole Field Applied to Autonomous Formation Flight Control of Unmanned Aerial Vehicles. *Sensors* **2021**, *21*, 4540. [CrossRef]

*Article*

# Drones Classification by the Use of a Multifunctional Radar and Micro-Doppler Analysis


**Mauro Leonardi** [1,*], **Gianluca Ligresti** [1] **and Emilio Piracci** [2]

1. Department of Electronic Engineering, Tor Vergata University, Via del Politecnico 1, 00133 Rome, Italy; ligrestigianluca@gmail.com
2. Rheinmetall Italia S.p.A, Via Affile 102, 00131 Rome, Italy; e.piracci@rheinmetall.it
* Correspondence: mauro.leonardi@uniroma2.it



**Abstract:** The classification of targets by the use of radars has received great interest in recent years, in particular in defence and military applications, in which the development of sensor systems that are able to identify and classify threatening targets is a mandatory requirement. In the specific case of drones, several classification techniques have already been proposed and, up to now, the most effective technique was considered to be micro-Doppler analysis used in conjunction with machine learning tools. The micro-Doppler signatures of targets are usually represented in the form of the spectrogram, that is a time–frequency diagram that is obtained by performing a short-time Fourier transform (STFT) on the radar return signal. Moreover, frequently it is possible to extract useful information that can also be used in the classification task from the spectrogram of a target. The main aim of the paper is comparing different ways to exploit the drone's micro-Doppler analysis on different stages of a multifunctional radar. Three different classification approaches are compared: classic spectrogram-based classification; spectrum-based classification in which the received signal from the target is picked up after the moving target detector (MTD); and features-based classification, in which the received signal from the target undergoes the detection step after the MTD, after which discriminating features are extracted and used as input to the classifier. To compare the three approaches, a theoretical model for the radar return signal of different types of drone and aerial target is developed, validated by comparison with real recorded data, and used to simulate the targets. Results show that the third approach (features-based) not only has better performance than the others but also is the one that requires less modification and less processing power in a modern multifunctional radar because it reuses most of the processing facility already present.

**Keywords:** drone; micro-Doppler; radar; target; classification





**Citation:** Leonardi, M.; Ligresti, G.; Piracci, E. Drones Classification by the Use of a Multifunctional Radar and Micro-Doppler Analysis. *Drones* **2022**, *6*, 124. https://doi.org/10.3390/drones6050124

Academic Editors: Daobo Wang and Zain Anwar Ali

Received: 29 April 2022
Accepted: 7 May 2022
Published: 11 May 2022

**Publisher's Note:** MDPI stays neutral with regard to jurisdictional claims in published maps and institutional affiliations.




## 1. Introduction

The identification of targets by radar has become a subject of great interest in recent years. The main motivations for the growth of this interest are related to the increased number of applications in which the target identification and classification could be useful: the ability to classify and identify targets is an important aspect in air traffic surveillance and in modern military applications. These applications require sensor systems able to identify threatening targets with high reliability and precision [1]. Thus, the target classification activity, that consists of giving to the system the ability to associate an object to a given class of targets, is a main area of development in both civil and defence systems.

Concerning the specific case of drone classification, radars are capable of detecting at longer ranges than other sensors and perform reliably in all weather conditions at any time of the day [2]. Moreover, modern multifunctional radars (MFR) have been developed recently and are able to perform several operations by dedicating specifically adapted waveforms to different tasks, including the target classification task [3,4]. Thus, with modern radar technologies, it is possible to perform the target classification task along with all the other classic radar operations, such as surveillance and tracking.

Concerning military and defence applications, several types of threat exist and today is mandatory to distinguish between aircraft and drones. In particular, fixed-wings (FW) aircraft and rotary-wings (RW) aircraft, such as helicopters, must be distinguished from drones. In fact, the increased military and civil use of drones and the possibility to use them as threatening weapons have caused drone's detection and identification to be an important matter of public safety.

Radar sensors for drone tracking and classification have been extensively studied in the past [5–7] and several target identification and classification techniques are discussed in the literature [8]. Up to now, the most effective technique was considered to be the micro-Doppler analysis [9–12] that, used in conjunction with machine learning classification tools, allows researchers to solve target classification problems [13–17].

A micro-Doppler signature of a target is created when specific components of an object move with respect to the main body of the object itself, such as the rotating blades of a flying helicopter [9] or of a drone.

The micro-Doppler signatures of targets are usually represented in the form of the spectrogram, that is a time–frequency diagram that is obtained by performing a short-time Fourier transform (STFT) on the radar return signal [10]. Using a spectrogram, it is usually possible to extract useful information that can be used in the classification task, such as the spectrum width, that allows researchers to distinguish the rotary wings from the fixed wings, or the time distance between the vertical lines in the spectrograms, referred to as blade flashes, that is related to the rotation rate of the propeller blades.

Even in the case of the drones, that have very short and thin blades, the rotation of the propeller blades is sufficient to generate clear micro-Doppler signatures. In particular, the works in [5,6,18] showed that an analysis of the radar return can be used to distinguish drones from birds.

The final scope of this work is finding the best way to integrate a target classification task (with particular attention to the distinction of the drone from the other objects) into a multifunctional (not dedicated) radar. In multifunctional radars, the classification task is only one of the tasks of the radar (e.g., together with surveillance tasks, tracking tasks, etc.) and it is subjected to many constraints, such as, for example, a maximum time on-target, the computational power, the signal processing capabilities, etc. On the other hand, in MFR, some signal filters or signal processing facilities are already present for other purposes (see the following section) and can be reused for the target classification task without any additional cost.

To better understand which is the best possible approach and where to perform the target classification in a multifunctional radar, a comparison of three possible methods based on neural network classification applied at different stages of the radar processing chain was performed, also taking into account the processing block already present in the radar.

Summing up, the main aims and contributions of this paper are:

- A detailed model for the received radar signal from a drone is derived from the model presented in [19], where only a single blade was considered: in the new model, a given number of blades, rotors, and the body are considered, and the elevation angle under which the drone is seen by the radar is taken into account. The proposed model fits with real recorded data found in the literature and with signals recorded by a real multifunctional by Rheinmetall. Finally, this model is useful to train the machine learning algorithms and to simulate the radar signals;

- Three different approaches of classification suitable for a multifunctional radar are compared: exploiting the spectrogram of the target, exploiting the spectrum of the target, and exploiting a small number of features (extracted from the spectrum). Each approach is analytically derived from the signal model and a neural network classifier is trained and tested. The comparison of the different radar drone classification methods is the goal of the paper and, to the best knowledge of the authors, this has not been performed before for this type of radar. To simplify the comparison and to better understand the results, a well-known and simple neural network classifier is used; a more complex and high-performance classification algorithm will be selected in a future work.

The paper is organized as follows. In Section 2, the radar signal model is derived and the radar processing approach is introduced. In Section 3, the proposed model is compared with some past recorded data and with real data coming from the real multifunctional radar, where the classification task will be implemented. In Section 4, the three different approaches are evaluated using a neural network classifier and, last, some discussions and conclusions are reported in Section 5.

## 2. Radar Signal Model for Drones and Aircraft

In order to derive the characteristics of the drone to be used for the micro-Doppler-based target classification, the first step is the development of a mathematical model of the radar return.

Simply representing a generic radar signal with its analytic vector, as follows:

$$S_{TX}(t) = Ae^{j\omega} \tag{1}$$

the echo of a simple scatterer is usually represented as follows (neglecting the delay due to the propagation):

$$S_{RX}(t) = A_{RX}e^{j(\omega+\omega_d)} \tag{2}$$

where $\omega = 2\pi f_o$ and $\omega_d = 2\pi f_d$. $A$ and $f_0$ are the amplitude and the frequency of the transmitted waveform and $A_{RX}$ and $f_d$ are the received amplitude (related to the distance of the target and to its capability to reflect the radio frequency signals) and the Doppler frequency due to the target radial velocity.

A mathematical model of the signal back-scattered by a single rotating rotor blade can be derived, as discussed in [19], modelling a blade as a stiff rod of a length, $L$, that changes its orientation in time due to its rotation velocity (see Figure 1).

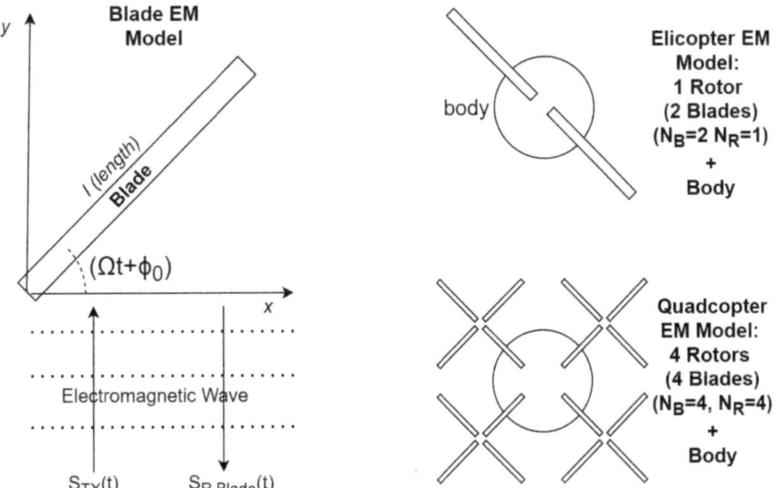


**Figure 1.** Geometrical models to compute the radar returned signal: single blade (on the left), single rotor helicopter, and quadcopter on the right.

During the time, $t$, it is seen under different angles by the radar:

$$s_{R,Blade}(t) = \int_0^{L\cos(\Omega t + \phi_0)} \exp\left\{ j\omega \left( t - \frac{2x\tan(\Omega t + \phi_0)}{c} \right) \right\} dx =$$
$$= \exp\{j\omega t\} \int_0^{L\cos(\Omega t + \phi_0)} \exp\left\{ -j\frac{4\pi}{\lambda} x \tan(\Omega t + \phi_0) \right\} dx \tag{3}$$

with $\omega$ as the angular frequency of the radar transmitted wave, $t$ as the time, $c$ as the velocity of light, $\lambda$ as the wavelength, $\phi_0$ as the initial phase of the rotor blade, and $\Omega$ as the rotation rate of the rotor blade.

The model directly computes the total received signal of each part of the rod, by coherent integration on its length. The position of the rod change (and also the integral limits) due to its rotation.

In this paper, the model in (3) is improved by considering also the dependence on the elevation angle $\beta$ under which the rod is seen and, moving to base-band:

$$s_{R,Blade}(t) = \int_0^{L\cos(\Omega t + \phi_0)} \exp\left\{ -j\frac{4\pi}{\lambda} x \cos\beta \tan(\Omega t + \phi_0) \right\} dx \tag{4}$$

Now, considering a rotor with $N_B$ blades, in which each blade has its own initial rotation angle $\phi_{0,k}$:

$$s_{R,Rotor}(t) = \sum_{k=0}^{N_B-1} \int_0^{L\cos(\Omega t + \phi_{0,k})} \exp\left\{ -j\frac{4\pi}{\lambda} x \cos\beta \tan\left( \Omega t + \phi_{0,k} + k\frac{2\pi}{N_B} \right) \right\} dx =$$
$$\sum_{k=0}^{N_B-1} \frac{\lambda}{j4\pi \cos\beta \tan\left( \Omega_r t + \phi_{0,k} + k\frac{2\pi}{N_B} \right)} \cdot \left( 1 - \exp\left\{ -j\frac{4\pi}{\lambda} L \cos\beta \sin\left( \Omega_r t + \phi_{0,k} + k\frac{2\pi}{N_B} \right) \right\} \right) \tag{5}$$

Before generalizing the model of a rotor to the model of a drone it must be noted that this formulation is more general and accurate then many other rotor models that are usually based on the derivation given in [9]. In fact, this model not only includes the dependency on the vertical aspect angle of the blade, but also the includes dependency of the reflected signal on the blade horizontal aspect angle. In particular, the reduction in the blade reflecting area is taken into account solving the integral in the interval $[0 \cdots L\cos(\Omega t + \phi_0)]$.

Observing both the formulations, it is possible to understand that the instantaneous Doppler frequency shift induced by the $k$-th blade is a sinusoidal contribution, representing the fact that the Doppler frequency is modulated by the rotation rate $\Omega$ through sinusoidal functions. Moreover, concerning the amplitude of the received signal in (5), the maximum level occurs when the blade is orthogonal to the radar beam, then the amplitude tends to drop sharply while the blade rotates. When the blade is orthogonal to the radar, it is possible to appreciate the presence of *blade flashes*, that are vertical lines in the time–frequency diagram of the rotating blades.

Finally, concerning the spectrum of a rotor with $N_b$ blade, it can be represented, as described in [20], with the following expression:

$$S_R(f) = \sum_k c_{N_Bk} \delta(f - f_D - N_B k f_r) \tag{6}$$

with $f$ as the frequency of the transmitted signal, $f_D$ as the Doppler frequency, and $f_r = \Omega/(2\pi)$ as the frequency of rotation of the blade. The coefficient $c_{Nk}$ has a complex expression that depends on the number of blades, on the blade length, on the angle between the plane of rotation of the blade and the line of sight of the radar and on the wavelength. The important consideration is that, according to the expression in (6), the spectrum of rotating rotor blades is a train of Diracs, being the received signal periodic in the time-domain. Thus, the distance between the frequency Diracs in the spectrum is related to the distance between the blade flashes in the spectrogram and, consequently, to the rotation rate $\Omega$. It follows that, in principle, the information derivable from the phase term of the returned signal can be used to derive the target characteristics and, taking some measurements on the spectrogram or on the spectrum of the target, it is possible to estimate some of the physical parameters representative for the target. A summary of the relationship between the signal, spectrogram, and spectrum parameters with the target physical characteristics is reported in Table 1.

**Table 1.** Relationship between target physical characteristics and the signal, spectrogram, and spectrum parameters.

| Target Physical Characteristic | Signal, Spectrogram, Spectrum Parameters |
|---|---|
| Rotation Rate, $\Omega$ | Time Period of Blade Flashes in spectrogram, $T_c = \frac{2\pi}{N_B \cdot \Omega}$ |
| Blade Length, $L$ | Frequency distance between spectrum Diracs, $\Delta F = \frac{N_B \cdot \Omega}{2\pi}$ |
| Number of Blades, $N_B$ | Maximum Doppler Shift, $f_{D,max} = \frac{2 \cdot L \cdot \Omega}{\lambda}$ |
| Blade Tip Velocity, $V_{tip} = L \cdot \Omega$ | |

Coming back to the model in (5), it can be used to represent a drone with more than one rotor (see Figure 1): considering a given number of rotors $N_R$, with $N_B$ blades each, (5) becomes:

$$
s_{R,Rotors}(t) = \sum_{r=1}^{N_R} \sum_{k=0}^{N_B-1} \exp\left\{-j\frac{4\pi}{\lambda}\delta_r\right\} \frac{\lambda}{j4\pi \cos\beta \tan\left(\Omega_r t + \phi_{0,r} + k\frac{2\pi}{N_B}\right)} \cdot
$$
$$
\cdot \left(1 - \exp\left\{j\frac{4\pi}{\lambda}L \cos\beta \sin\left(\Omega_r t + \phi_{0,r} + k\frac{2\pi}{N_B}\right)\right\}\right)
\tag{7}
$$

This model is valid for a generic continuous-wave signals reflected by a target.

In order to match the model in (7) to the case of a multifunctional pulsed radar, the typical radar processing chain and radar waveform parameters must be considered.

The generic radar processing chain is reported in Figure 2. The radar transmits a sequence of pulses and their echoes (produced by the target) are received from the radar receiver. In a modern radar, the echoes of the same target are processed together to improve the radar performance. Each echo of the target is related to the target properties.

Having $N$ echoes coming from the same target, the most used approach is to process them together with the following steps (after the base-band conversions):

- Cancel the echoes from the ground by the use of a moving target indicator (MTI) that is a sort of notch filter centred at zero frequency;
- Distinguish between targets that have different radial velocities (moving target detector MTD) that is typically implemented by the use of a fast Fourier transform of the incoming signals;
- Last, the detection phase, compare the output of the MTD with a threshold to declare whether the target is present or not at a given Doppler frequency (constant false alarm rate—CFAR—detector).

In this context, the classification of the target can be performed at different levels exploiting part of the processing block already present in a multifunctional radar.

The first proposed approach (the most used in the literature [10]) is the spectrogram-based classification, in which the received signal from the target goes directly into a separated chain that implements the STFT processor (block in green in Figure 2). Once the spectrogram of the received signal is available, it can be used to classify the target with any classification algorithm. This approach needs a dedicated processing chain that computes the STFT from the raw data.

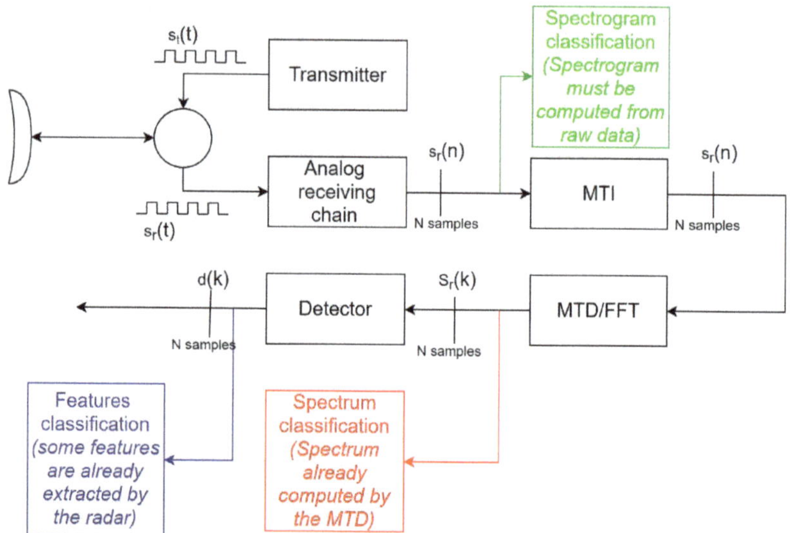


**Figure 2.** Typical radar chain, composed of MTI, MTD, and CFAR, and related points, in which the spectrogram, spectrum, and features classification can be implemented.

The second proposed approach is a spectrum-based classification (block in red in Figure 2), in which the signal representing the target is picked up after the moving target indicator (MTI) and moving target detector (MTD) blocks. This can be performed because, in its most simple implementation, the MTD gives the discrete Fourier transform samples of the received signal, that is, it gives the target spectrum, which is useful to perform the classification task.

The third proposed approach is a features-based classification, in which discriminating features are extracted after the radar detection mechanism. In that case, the received signal from the target undergoes into all the processing steps. The CFAR is used to appropriately set a threshold on the received signal spectrum and to declare whether a target is present and, if there, in which velocity windows (i.e., which spectrum line) it falls. Observing in which spectrum lines the detections fall, it is still possible to extract some features (for example, the spectrum width, etc.) that can be used to classify the target (block in blue in Figure 2).

As described before, pulsed radars transmit pulses with nominal pulse width, $\tau$, of the order of microseconds, at regular intervals in time called pulse repetition time—$PRT$—of the order of milliseconds. In that case, confusing the pulse shape with a Dirac, the echo signal from the target can be considered as sampled with sampling rate equal to PRT (see Figure 3):

$$s_{R,Rotors}(nPRT) = \sum_{r=1}^{N_R} \sum_{k=0}^{N_B-1} \exp\left\{-j\frac{4\pi}{\lambda}\delta_r\right\} \frac{\lambda}{j4\pi\cos\beta\tan\left(\Omega_r(nPRT) + \phi_{0,r} + k\frac{2\pi}{N_B}\right)} \cdot$$
$$\cdot \left(1 - \exp\left\{j\frac{4\pi}{\lambda}L\cos\beta\sin\left(\Omega_r(nPRT) + \phi_{0,r} + k\frac{2\pi}{N_B}\right)\right\}\right) \tag{8}$$

Finally, the non-rotating part of the drone must also be considered and, adding the body with its own Doppler frequency, $f_D$, due to the target radial velocity, the model of the received signal becomes:

$$s_{R,Target}(nPRT) = \sqrt{\sigma_{body}}\exp\{j\omega_D t\} + \sqrt{\sigma_{blade}}\exp\{j2\omega_D t\} \cdot \sum_{r=1}^{N_R}\sum_{k=0}^{N_B-1}\exp\left\{-j\frac{4\pi}{\lambda}\delta_r\right\} \cdot$$
$$\cdot \frac{\lambda}{j4\pi\cos\beta\tan\phi_{k,r}(nPRT)}\left[1 - \exp\left(j\frac{4\pi}{\lambda}L\cos\beta\sin\phi_{k,r}(nPRT)\right)\right] \tag{9}$$

with

$$\phi_{k,r}(t) = 2\pi\Omega_r t + \phi_{0,r} + k\frac{2\pi}{N_B} \tag{10}$$

In (9), the terms $\sigma_{blade}$ and $\sigma_{body}$ represent the fractional part of the total RCS of the blades and of the body of the drone, respectively, and $\omega_D = 2\pi f_D$. The radar cross section (RCS) of the target is the capability of the target to back-scatter the incoming signal [3], it depends on several parameters, such as the direction of the incident wave, the polarization of the incident wave, the material of which the target is made, the wavelength, and many others.

As shown in the following, this model can be used to represent both drones and aircraft by appropriately defining their parameters, such as the number of rotors, the length of the blades, the rotation rate of the rotors, the relative dimension (in terms of RCS) of the body and of the blades, etc.

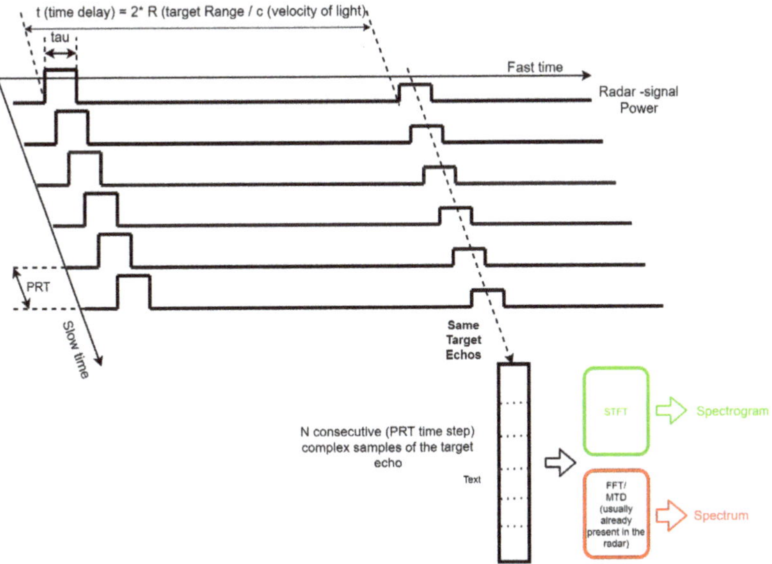


**Figure 3.** Pictorial view of the radar echoes of a given target and. The radar signal can be folded every PRT and represented on the fast-slow time plane. Fixing the time from the transmitted pulse (fast time) is possible to fix the target distance that we are processing. Collection of the echoes along the slow time is possible to extract a time series representing the same target useful to estimate its spectrum and/or spectrogram.

## 3. Model Comparison with Real Recorded Data

The proposed model was evaluated by comparison with both data coming from the analysis of the literature and real recorded data. In particular, once defined the type of target, the model can be used to simulate the target echo and the related spectrogram and/or spectrum.

A comparison of a real *T-REX 450* single-rotor helicopter drone with $N_B = 2$ blades, from the work in [21], with the simulated helicopter drone spectrogram, by using the proposed model, is shown in Figure 4. The real *T-REX 450* helicopter drone is characterized by a blade length about $L = 45$ cm and a rotation rate of the rotor blades about $\Omega = 40$ rev/s. In the spectrograms, it is possible to note the sinusoidal modulation due to the blades rotation; moreover, when the blade is orthogonal to the radar view, a blade flash appears, while for other angles, the amplitude drops sharply. As mentioned before, the amplitude of the sinusoidal modulation is related to the blade tip and the distance between the blade flashes is related to the rotation rate of the rotor blades.

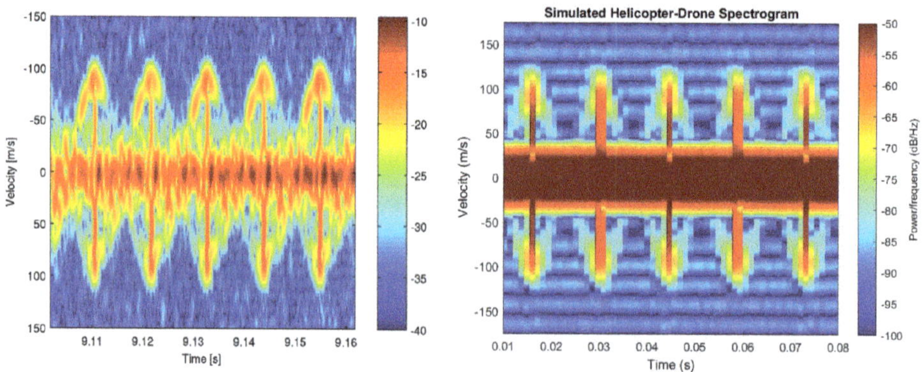

**Figure 4.** Real and simulated spectrogram of the helicopter drone. Adapted from [21].

Another comparison is reported in Figure 5 where the spectrogram of a real quadcopter drone with $N_B = 2$ blades for each rotor, from the work in [22], is compared with the simulated quadcopter drone spectrogram, by using the proposed model. The real quadcopter drone is characterized by a blade length about $L = 10$ cm and a rotation rate of the rotor blades about $\Omega = 100$ rev/s. In the spectrograms it is possible to appreciate the contribution of the main body, centred at the Doppler frequency $f_D = 0$ Hz, and the micro-Doppler contributions of the blades represented by the blade flashes.

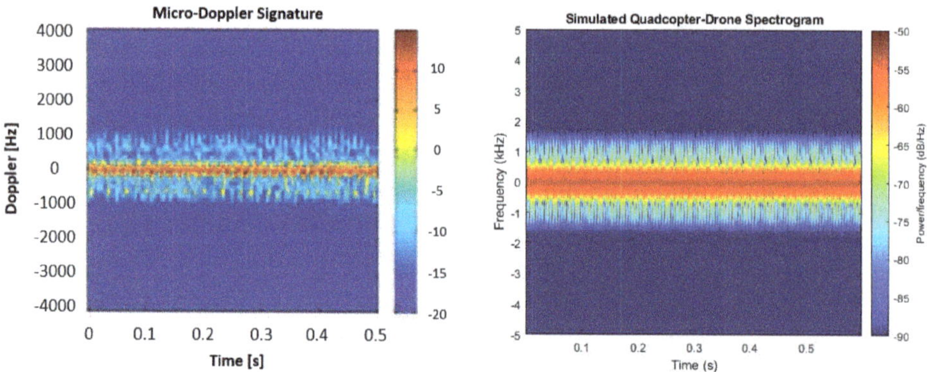

**Figure 5.** Real and simulated spectrogram of the quadcopter drone. Adapted from [22].

Note that, in general, the time resolution $\Delta t$ and the frequency resolution $\Delta f$ that characterize the spectrogram of a target are related to the radar waveform parameters introduced in the previous section. In particular, by computing the STFT with a window of $N$ pulses that is shifted of $N/2$ pulses at each FFT computation, the spectrogram resolutions have the following expressions:

$$\Delta t = \frac{N}{2 \cdot PRF} = \frac{N \cdot PRT}{2} \tag{11}$$

$$\Delta f = \frac{PRF}{N} = \frac{1}{N \cdot PRT} \tag{12}$$

With fixed $N$, if the PRT is increased, then the time resolution gets worse, up to the loss of the blade flashes information in the spectrogram; instead, if the PRT is reduced, then the frequency resolution gets worse, up to the loss of the ability to discriminate the micro-Doppler contributions from the Doppler contribution of the body.

Thus, in general, it is important to find the best trade-off on the radar waveform parameters in order to preserve the characteristics of the spectrogram. In the particular case of a multifunction radar the time available to perform the task and the PRF have both an upper limit and this will limit the resolution of the spectrum and spectrogram that can be obtained with the radar (for example see the spectrograms reported in Figure 6, where the PRF and total time of observation has some restriction due to the radar setting).

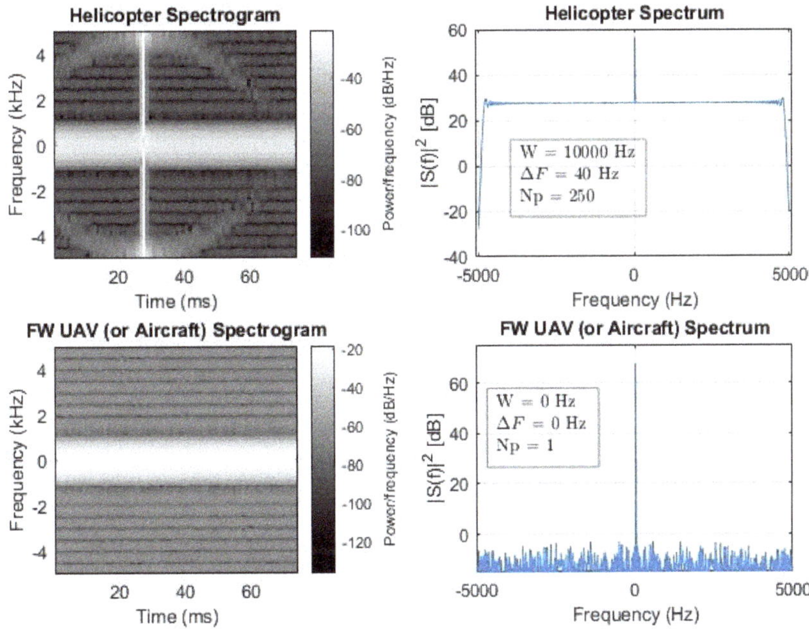


**Figure 6.** *Cont.*

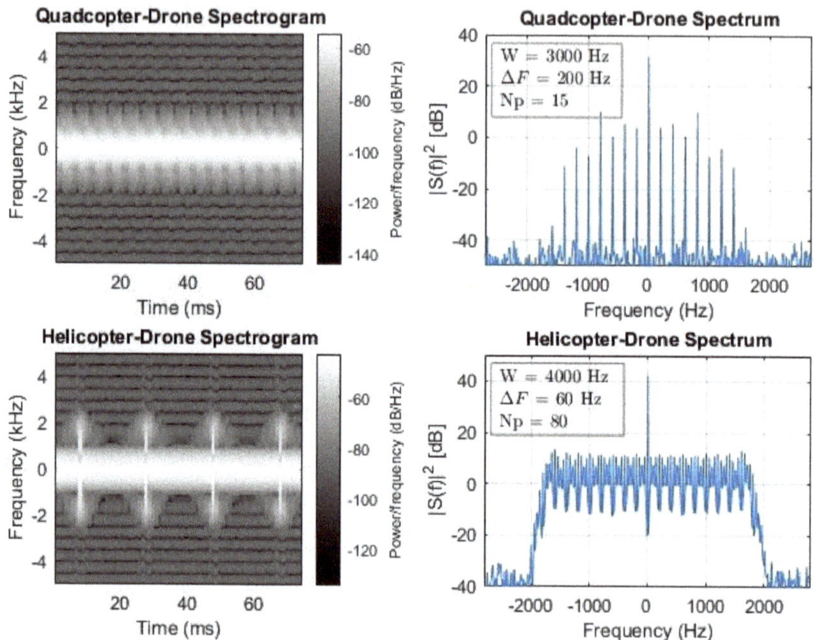



**Figure 6.** Simulated helicopter spectrogram, spectrum, and features (first row); simulated fixed-wings UAV (or aircraft) spectrogram, spectrum, and features (second row); simulated quadcopter drone spectrogram, spectrum, and features (third row); and simulated helicopter drone spectrogram, spectrum, and features (fourth row).

An additional analysis, based on real recorded data, was carried out by exploiting recorded data of a real DJI Matrice 600 drone (shown in Figure 7), coming from an S-band multifunctional radar. The measurement campaign was performed on December 2021 during clear and sunny days. The DJI Matrice 600 drone model type has the characteristics reported in Table 2.

**Figure 7.** DJI Matrix 600 drone model.

**Table 2.** DJI Matrix 600 drone's main characteristics.

| Characteristics | DJI Matrix 600 |
| --- | --- |
| Weight [kg] | 10 |
| Max Speed [m/s] | 30 |
| Max Ascent Speed [m/s] | 5 |
| Max Descent Speed [m/s] | 3 |
| Hovering Time [min] | 30 |
| Max Angular Velocity [deg/s] | 250 |
| Diagonal Wheelbase [mm] | 1100 |

A comparison between the real drone spectrum and the drone spectrum that was obtained using the proposed model is shown in Figure 8. The spectra in Figure 8 were obtained after the radar MTI elaboration and integrating more detections.

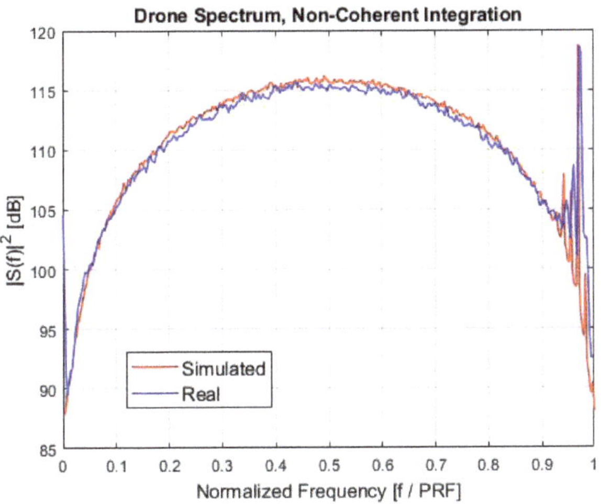


**Figure 8.** Simulated spectrum (red) vs. real spectrum (blue) after non-coherent integration.

In Figure 8, the drone spectrum shows both the body contribution, that is the main peak centred on the Doppler frequency related to the drone's flight, and the blades micro-Doppler contributions, that are the secondary peaks around the main body peak. Moreover, by measuring the frequency distance between the secondary peaks, it is possible to derive the rotation rate of the rotor blades of the drone, by using the equations reported in Table 1. In this case, $\Omega$ is about 50 rev/s, that is compatible with the DJI Matrice 600 model type.

## 4. Radar Processing and Target Classification

The developed mathematical model can be used to compute the reference signal for each target type to be classified. Then, computing the spectrogram and the spectrum of each type of target, it is possible to have a reference dataset for any possible machine learning classification algorithm.

The target classes that are considered in this paper are:

- Class 1: helicopter;
- Class 2: fixed-wings (FW) UAV or aircraft;
- Class 3: quadcopter drone (drone 1);
- Class 4: helicopter drone (drone 2).

The machine learning tool that is considered for the classification is a classic feed-forward neural network (FFNNET) with 1 hidden layer composed of 10 neurons. The number of generated signals to train the neural network is $N_{train} = 5000$, while the number of generated signals to test the neural network is $N_{test} = 1000$. The classification results that are reported in the following are the average results that are obtained after 5 trainings of the neural network. As said, the FFNNET is trained and tested with three types of input representing the three possible classification algorithms at different stages of the radar processing chain:

- Spectrogram samples generated from the raw signal coming from the radar chain;
- Spectrum samples generated from the MTD processor;
- Features extracted after the detector.

Concerning the third approach, the following features are considered:

- Spectrum width ($W$): it allows researchers to discriminate between rotary-wing targets and fixed-wing targets. In fact, a rotary-wing target has rotating blades that introduce micro-Doppler contributions in the received signal, that make the spectrum wide. Instead, a fixed-wing target is always characterized by a narrow spectrum because no micro-Doppler contributions are present. The spectrum width feature is computed by measuring the maximum distances (in frequency) between the detections of the same target;
- Distance between two detections ($\Delta F$): it allows researchers to discriminate between different categories of rotary-wing targets, such as helicopters and drones. In fact, the spectrum of a rotary-wing targets is a train of Diracs and the frequency distance between these Diracs is directly related to the rotation rate of the rotor blades; so, by measuring the frequency distance between the peaks in the spectrum, it is possible to classify the specific type of rotary-wing target;
- Number of peaks over threshold ($N_p$): in addition to the frequency distance between Diracs, this feature may allow researchers to discriminate between different categories of rotary-wing targets.

In Table 3, the physical parameters used to generate the four classes are reported, and more details on the neural network used in the simulation are reported in Figure 9.

Note that in training the Doppler frequency of the target body is set to $f_D = 0$ Hz, in order to avoid that the neural network identifies the Doppler of the target body as a discriminating feature for classification. Moreover, each target parameter is varied in a range of about $\pm10\%$ around the reference value, in order to take care of the variability that can characterize different drones in the same class or different flight configurations. Last, but not least, the training dataset is generated by considering an additive white Gaussian noise (with $SNR = 50$ dB) as regularization noise, in order to avoid that the neural network identifies as features the systematic errors that can be produced during the signal generation. Examples of the simulated spectrogram, spectrum, and features after detections of each target, used in the training phase, are shown in Figure 6.

**Table 3.** Target parameters considered in the neural network training phase. * In the case of UAV, the contribution of the propeller is considered negligible due to its relative small dimension and due to the fact that its plane of rotation is perpendicular to the line of sigh of the radar, producing almost zero Doppler frequency.

| Target Class | $N_R$ | $N_B$ | $L$ [m] | $\Omega$ [rev/s] | $RCS_{tot}$ [m²] |
|---|---|---|---|---|---|
| Helicopter | 1 | 2 | $8 \pm 10\%$ | $5 \pm 10\%$ | $3 \pm 10\%$ |
| Fixed-Wings UAV * (or aircraft) | - | - | - | - | $10 \pm 10\%$ |
| Quadcopter drone | 4 | 2 | $0.1 \pm 10\%$ | $100 \pm 10\%$ | $0.01 \pm 10\%$ |
| Helicopter drone | 1 | 2 | $0.6 \pm 10\%$ | $25 \pm 10\%$ | $0.1 \pm 10\%$ |

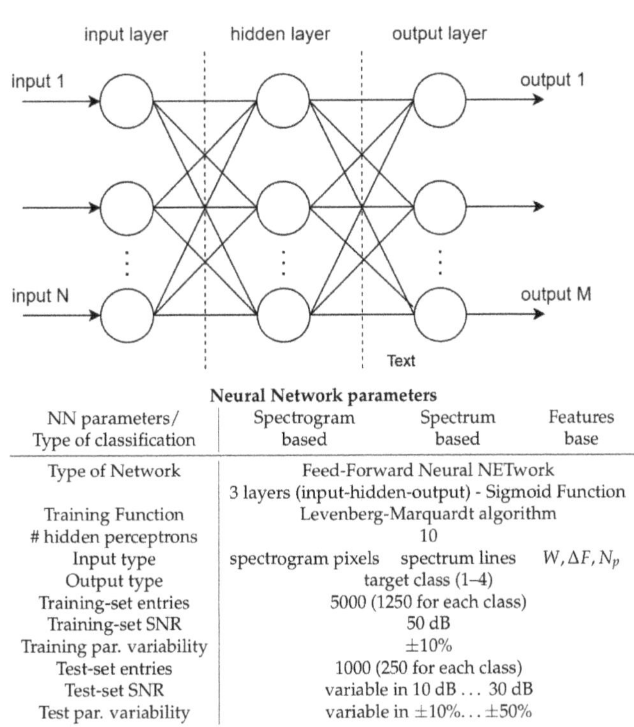


**Neural Network parameters**

| NN parameters/ Type of classification | Spectrogram based | Spectrum based | Features base |
|---|---|---|---|
| Type of Network | | Feed-Forward Neural NETwork | |
| | | 3 layers (input-hidden-output) - Sigmoid Function | |
| Training Function | | Levenberg-Marquardt algorithm | |
| # hidden perceptrons | | 10 | |
| Input type | spectrogram pixels | spectrum lines | $W, \Delta F, N_p$ |
| Output type | | target class (1–4) | |
| Training-set entries | | 5000 (1250 for each class) | |
| Training-set SNR | | 50 dB | |
| Training par. variability | | $\pm 10\%$ | |
| Test-set entries | | 1000 (250 for each class) | |
| Test-set SNR | | variable in 10 dB ... 30 dB | |
| Test par. variability | | variable in $\pm 10\%$... $\pm 50\%$ | |

**Figure 9.** Details of the neural network used in the simulation (block diagram and parameters).

Concerning the test dataset, the spectrogram and the spectrum of each target are generated by adding higher noise to the measurements (SNR in the range of 10 to 30 dB) and more variability to the physical parameters representing the targets (from 10% to 50%) and, finally, changing the observation time from 25 to 100 ms.

Moreover, in the test phase, the Doppler frequency of the target body is assumed to be different from zero, in order to take care of the fact that the real target can have its own velocity that must be estimated at the radar side. The target central frequency estimation is an important preprocessing step to be carried out before the classification, in order to remove it from the received signal and to come back to features that are similar to the ones used in the training phase. In any case, the residual error due to the not perfect estimation of the central frequency of the incoming signal becomes negligible after this preprocessing step.

The values of $f_D$ are imposed starting from the typical velocities of the targets of interest and using the following classic Doppler formula: $f_D = 2v_r/\lambda$.

Examples of classification results are reported in Tables 4–6, that report the confusion matrices for the three classification blocks, for a time on-target $t_{min} = 75$ ms, for a $SNR$ in training equal to 50 dB, and for a $SNR$ in the test equal to 30 dB. Table 4 represents the confusion matrix in the spectrogram case, Table 5 represents the confusion matrix in the spectrum case, and Table 6 represents the confusion matrix in the features case.

**Table 4.** Confusion matrix: classification results with spectrogram samples, $SNR = 30$ dB in test, $t_{min} = 75$ ms.

| REAL/PRED | HELICOPTER | FW-UAV | QUAD-DRONE | HEL-DRONE | |
|---|---|---|---|---|---|
| HELICOPTER | 1215 | 35 | 0 | 0 | 97.2% |
| FW-UAV | 0 | 1250 | 0 | 0 | 100% |
| QUAD-DRONE | 0 | 0 | 750 | 500 | 60.0% |
| HEL-DRONE | 0 | 0 | 250 | 1000 | 80.0% |
| | 100% | 97.3% | 75.0% | 66.7% | 84.3% |

**Table 5.** Confusion matrix: classification results with spectrum samples, $SNR = 30$ dB in test, $t_{min} = 75$ ms.

| REAL/PRED | HELICOPTER | FW-UAV | QUAD-DRONE | HEL-DRONE | |
|---|---|---|---|---|---|
| HELICOPTER | 989 | 261 | 0 | 0 | 79.1% |
| FW-UAV | 93 | 1157 | 0 | 0 | 92.6% |
| QUAD-DRONE | 0 | 0 | 1153 | 97 | 92.2% |
| HEL-DRONE | 0 | 1 | 277 | 972 | 77.7% |
| | 91.4% | 76.1% | 80.6% | 90.9% | 85.4% |

**Table 6.** Confusion matrix: classification results with features, $SNR = 30$ dB in test, $t_{min} = 75$ ms.

| REAL/PRED | HELICOPTER | FW-UAV | QUAD-DRONE | HEL-DRONE | |
|---|---|---|---|---|---|
| HELICOPTER | 1034 | 206 | 8 | 2 | 82.7% |
| FW-UAV | 0 | 1241 | 9 | 0 | 99.3% |
| QUAD-DRONE | 0 | 158 | 1092 | 0 | 87.4% |
| HEL-DRONE | 0 | 1 | 58 | 1191 | 95.3% |
| | 100% | 77.3% | 93.6% | 99.8% | 91.1% |

The correct classification rate results to be 84.3% in the spectrogram case, 85.4% in the spectrum case, and 91.1% in the features case. Thus, the features-based classification allows researchers to reach slightly better performances than the spectrogram-based and spectrum-based classification, that instead have similar performances. However, by looking at the content of the confusion matrices, it is possible to note that some confusion is present, in particular in the distinction of the quadcopter drone with the helicopter drone and in the distinction of the helicopter with the FW-UAV (or aircraft). A deeper evaluation is carried out by evaluating the the $F1$-score for each configuration, by varying the SNR, the parameter variation span, and the time on-target in the test. The $F1$-score is defined as [23]:

$$F1 = \frac{2}{C} \sum_{c=1}^{C} \frac{P_c \cdot R_c}{P_c + R_c} \tag{13}$$

where $C$ represents the total number of classes (four in this paper). In Equation (13), $P_c$ is the precision of the $c$-th class and $R_c$ is the recall of the $c$-th class, defined as:

$$P_c = \frac{TP_c}{TP_c + FP_c}; \qquad R_c = \frac{TP_c}{TP_c + FN_c} \tag{14}$$

where $TP_c$ is the number of true positives, $FP_c$ is the number of false positives and $FN_c$ is the number of false negatives, all of the $c$-th class.

With reference to the confusion matrix in Table 5, concerning the helicopter class, the recall is $R_{hel} = 79.1\%$, the precision is $P_{hel} = 91.4\%$, and the relative F1-score is $F1_{hel} = 84.8\%$. The same can be carried out for the other classes and, by applying the formula in Equation (13), it is possible to derive the overall $F1$-score.

Performances when varying the $SNR$ in test are reported in Figure 10. It is possible to notice that, as expected, increasing the $SNR$ improves the performances of all the classifiers. In particular, for $SNR < 20$ dB, the most performing approach results to be the one in which the spectrogram samples are given as input to the neural network. Instead, when the $SNR$ is sufficiently high to correctly reconstruct the spectrograms and spectra of the targets, the performances of the three approaches become comparable and stable around 85%, with the approach of the features that is slightly better than the spectrogram and spectrum approaches.

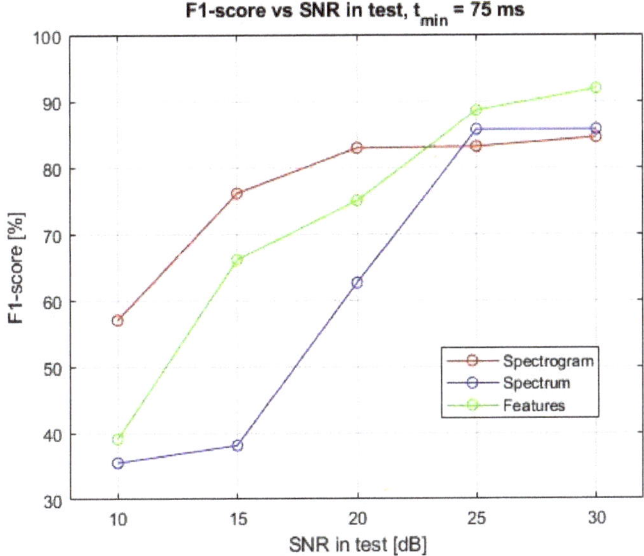


**Figure 10.** $F1$-score vs. SNR in test, time on-target $t_{min} = 75$ ms.

As mentioned before, the three approaches are also compared by varying the range of values of the target parameters in test, in order to emulate different drone models or flight behaviours. The $F1$-score versus the parameters variation in test is reported in Figure 11. As expected, the performances of the classifier get worse if the range of variation of the target parameters increases. In fact, the greater is the range of values in which a parameter varies, the more two different categories of targets tend to be confused. As an example, consider the case of the two drones: when the blade length of a quadcopter drone is varied of about $\pm 50\%$ with respect to its reference value, it is highly probable that the received signal from the quadcopter drone is similar to the received signal from the helicopter drone, leading to subsequent classification errors. It is also interesting to notice that, with the parameters variation rate fixed, the spectrogram and spectrum approaches have an almost equivalent behaviour, while the approach of the features gives much better classification results. In any case, the classification performance is still good also with a parameter variation of about 40%.

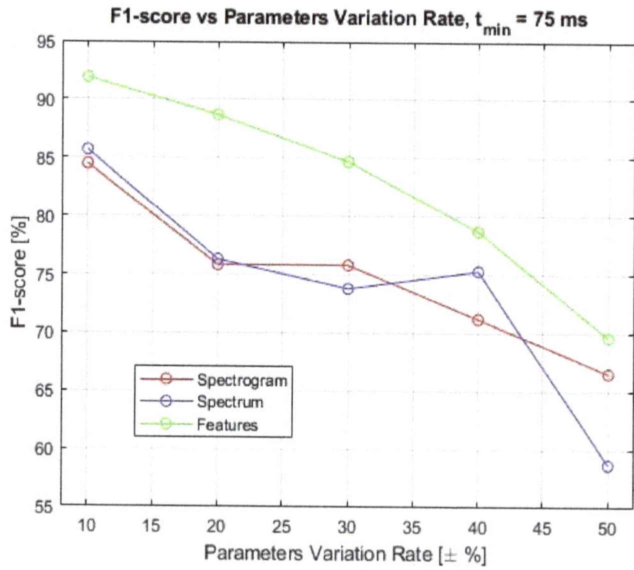

**Figure 11.** $F1$-score vs. parameters variation rate, time on-target $t_{min} = 75$ ms.

Last, the $F1$-score of the three approaches has been evaluated by varying the observation time (time on-target), as reported in Figure 12.

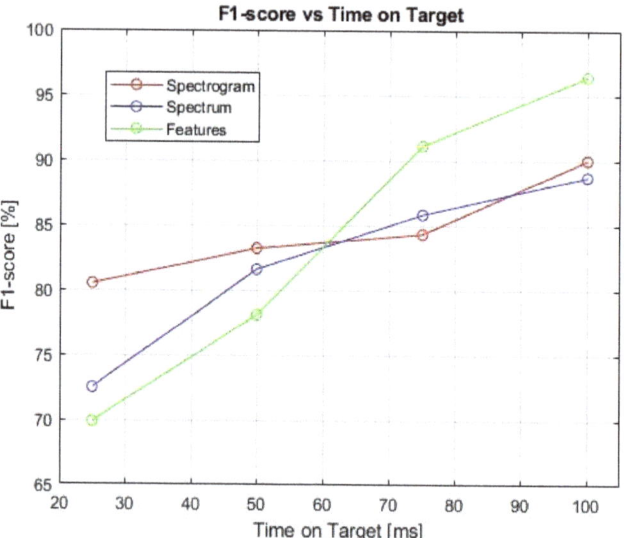

**Figure 12.** $F1$-score vs. time on-target.

It is possible to notice that, as expected, the classification performances of all the three approaches improve when the time on-target increases. In fact, when the observation time is longer, the spectrogram and the spectrum of each target are reconstructed in a better way and, as a consequence, the classification performances improve. In particular, for a $t_{min} = 100$ ms, the spectrogram-based approach and the spectrum-based approach allow researchers to reach an $F1$-score of about 90%, while the features-based approach has better results, with $F1$-score above 95%. However, it is important to remember that also time on-target is subject to a trade-off between the minimization of the task time and the classification performance.

Finally, it is clear that the approach of extracting features from the target spectrum and using them as input to a neural network is the most promising.

## 5. Conclusions

The aim of this paper was the exploration of possible ways to introduce drone and other flying object classification tasks into a multifunctional radar.

In order to reach this goal, a mathematical model of the received signal from a generic drone, having rotary parts or not, was developed. The proposed model can be specialized to any desired target category by simply adjusting its parameters. In fact, even though this paper considers just four target categories (helicopter, fixed-wings UAV (or aircraft), quadcopter drone, and helicopter drone), the developed model can be easily extended to other target categories, such as ballistic targets or other types of drone. Having the models of the targets, it is possible to use them for the training phase of any classifier tool.

The main goal of the paper was the comparison of the classification performance of three different approaches, from low-level to high-level signatures, with respect to the typical radar chain.

Simulation results shows that the three method lead to a similar results but the best classification performance is obtained, for the different tested scenarios, with the features-based classification after the radar detection mechanism and can be chosen as the preferred approach. Moreover, apart from the merely classification results, the features-based approach has also other advantages with regards to the other ones:

- It allows a very fast training of the neural network and a better reliability in the results, because, in general, neural networks perform better when the features to be processed are a small representative number;
- The features-based approach is the one that require the lowest additional processing power in the radar chain. In fact, it exploits all the huge signal processing (filtering, fast Fourier transformation, and thresholding) already present in any modern radar. On the contrary, additional signal processing blocks at lower level of the radar chain, as the spectrogram and spectrum computations are necessary for the other two approaches.

Finally, it is important to note that the proposed model can be used to include other classes in the classification task and that the classification method (FFNNET) can be substituted with any other more performing classification tool. In the authors' opinion, this last step was not necessary at this level due to the need to maintain the results of the proposed comparison generally, which are simple to understand and not dependent on a specific optimized classification algorithm.

Future steps of the work will be the collection of a data-set of real recorded features of different drones to confirm the obtained results and to select a less generic classification method, tailored on the specific radar and feature characteristics.

**Author Contributions:** M.L. proposed the idea and gave the theoretical support, conceptualizations, and methodology; G.L. implemented the algorithm and contributed to the algorithm evaluation; E.P. supported the implementation and the comparison with real recorded data. All authors have read and agreed to the published version of the manuscript.

**Funding:** This research received no external funding.

**Institutional Review Board Statement:** Not applicable.

**Informed Consent Statement:** Not applicable.

**Data Availability Statement:** Not applicable.

**Conflicts of Interest:** The authors declare no conflict of interest.

## References


1. Neri, F. *Introduction to Electronic Defense Systems*; Artech House: Norwood, MA, USA, 2018.
2. Fell, B. Basic Radar Concepts: An Introduction to Radar for Optical Engineers. In *Effective Utilization of Optics in Radar Systems*; International Society for Optics and Photonics: Huntsville, AL, USA, 1977.
3. Skolnik, M. *Radar Handbook*, 2nd ed.; McGraw-Hill: New York, NY, USA, 1990.
4. Barton, D. *Radar System Analysis and Modeling*; Artech House Radar Library: Norwood, MA, USA, 2005.
5. Fuhrmann, L.; Biallawons, O.; Klare, J.; Panhuber, R.; Klenke, R.; Ender, J. Micro-Doppler analysis and classification of UAVs at Ka band. In Proceedings of the 2017 18th International Radar Symposium (IRS), Prague, Czech Republic, 28–30 June 2017; pp. 1–9. [CrossRef]
6. Harmanny, R.I.A.; de Wit, J.J.M.; Cabic, G.P. Radar micro-Doppler feature extraction using the spectrogram and the cepstrogram. In Proceedings of the 2014 11th European Radar Conference, Rome, Italy, 8–10 October 2014; pp. 165–168. [CrossRef]
7. Molchanov, P.; Harmanny, R.I.; de Wit, J.J.; Egiazarian, K.; Astola, J. Classification of small UAVs and birds by micro-Doppler signatures. *Int. J. Microw. Wirel. Technol.* **2014**, *6*, 435–444. [CrossRef]
8. Galati, G. *Advanced Radar Techniques and Systems*; Peter Peregrinus Ltd., on behalf of the Institution of Electrical Engineers: London, UK, 1993.
9. Chen, V. *The Micro-Doppler Effect in Radar*; Artech House: Norwood, MA, USA, 2011.
10. Chen, V.; Tahmoush, D. *Radar Micro-Doppler Signatures: Processing and Applications*; IET Radar, Sonar and Navigation Series 34; IET Digital Library: London, UK, 2014.
11. Zhang, Q. *Micro-Doppler Characteristics of Radar Targets*; Butterworth-Heinemann: Woburn, MA, USA; Elsevier: Amsterdam, The Netherlands, 2016.
12. Gu, J.C. *Short-Range Micro-Motion Sensing with Radar Technology*; IET Control, Robotics and Sensors Series 125; IET Digital Library: London, UK, 2019.
13. Kim, K.; Kim, J.H. Polynomial Regression Predistortion for Phase Error Calibration in X-Band SAR. *IEEE Geosci. Remote Sens. Lett.* **2022**, *19*, 4002705. [CrossRef]
14. Brooks, D.; Schwander, O.; Barbaresco, F.; Schneider, J.Y.; Cord, M. Deep Learning and Information Geometry for Drone Micro-Doppler Radar Classification. In Proceedings of the 2020 IEEE Radar Conference (RadarConf20), Florence, Italy, 21–25 September 2020; pp. 1–6. [CrossRef]
15. Brooks, D.A.; Schwander, O.; Barbaresco, F.; Schneider, J.Y.; Cord, M. Complex-valued neural networks for fully-temporal micro-Doppler classification. In Proceedings of the 2019 20th International Radar Symposium (IRS), Ulm, Germany, 26–28 June 2019; pp. 1–10. [CrossRef]
16. Brooks, D.A.; Schwander, O.; Barbaresco, F.; Schneider, J.Y.; Cord, M. Temporal Deep Learning for Drone Micro-Doppler Classification. In Proceedings of the 2018 19th International Radar Symposium (IRS), Bonn, Germany, 20–22 June 2018; pp. 1–10. [CrossRef]
17. Gérard, J.; Tomasik, J.; Morisseau, C.; Rimmel, A.; Vieillard, G. Micro-Doppler Signal Representation for Drone Classification by Deep Learning. In Proceedings of the 2020 28th European Signal Processing Conference (EUSIPCO), Amsterdam, The Netherlands, 18–21 January 2021; pp. 1561–1565. [CrossRef]
18. Fioranelli, F. Monostatic and Bistatic Radar Measurements of Birds and Micro-Drone. In Proceedings of the IEEE Radar Conference (RadarConf), Philadelphia, PA, USA, 2–6 May 2016.
19. Misiurewicz, J.; Kulpa, K.; Czekala, Z. Analysis of recorded helicopter echo. In Proceedings of the Radar 97 (Conf. Publ. No. 449), Edinburgh, UK, 14–16 October 1997; pp. 449–453. [CrossRef]
20. Martin, J.; Mulgrew, B. Analysis of the theoretical radar return signal form aircraft propeller blades. In Proceedings of the IEEE International Conference on Radar, Arlington, VA, USA, 7–10 May 1990; pp. 569–572. [CrossRef]
21. Bjorklund, S. Target Detection and Classification of Small Drones by Boosting on Radar Micro-Doppler. In Proceedings of the 15th European Radar Conference (EuRAD), Madrid, Spain, 26–28 September 2018.
22. Zhang, P.; Yang, L.; Chen, G.; Li, G. Classification of drones based on micro-Doppler signatures with dual-band radar sensors. In Proceedings of the 2017 Progress in Electromagnetics Research Symposium-Fall (PIERS-FALL), Singapore, 19–22 November 2017; pp. 638–643. [CrossRef]
23. Raval, D.; Hunter, E.; Hudson, S.; Damini, A.; Balaji, B. Convolutional Neural Networks for Classification of Drones Using Radars. *Drones* **2021**, *5*, 149. [CrossRef]

 *drones*

MDPI

*Article*

# Multi-Target Association for UAVs Based on Triangular Topological Sequence

Xudong Li, Lizhen Wu, Yifeng Niu * and Aitong Ma

College of Intelligence Science and Technology, National University of Defense Technology, Changsha 410078, China; lixudong@nudt.edu.cn (X.L.); lzwu@nudt.edu.cn (L.W.); maaitong@nudt.edu.cn (A.M.)
* Correspondence: niuyifeng@nudt.edu.cn

**Abstract:** Multi-UAV cooperative systems are highly regarded in the field of cooperative multi-target localization and tracking due to their advantages of wide coverage and multi-dimensional perception. However, due to the similarity of target visual characteristics and the limitation of UAV sensor resolution, it is difficult for UAVs to correctly distinguish targets that are visually similar to their associations. Incorrect correlation matching between targets will result in incorrect localization and tracking of multiple targets by multiple UAVs. In order to solve the association problem of targets with similar visual characteristics and reduce the localization and tracking errors caused by target association errors, based on the relative positions of the targets, the paper proposes a globally consistent target association algorithm for multiple UAV vision sensors based on triangular topological sequences. In contrast to Siamese neural networks and trajectory correlation, the relative position relationship between targets is used to distinguish and correlate targets with similar visual features and trajectories. The sequence of neighboring triangles of targets is constructed using the relative position relationship, and the feature is a specific triangular network. Moreover, a method for calculating topological sequence similarity with similar transformation invariance is proposed, as well as a two-step optimal association method that considers global objective association consistency. The results of flight experiments indicate that the algorithm achieves an association accuracy of 84.63%, and that two-step association is 12.83% more accurate than single-step association. Through this work, the multi-target association problem with similar or even identical visual characteristics can be solved in the task of cooperative surveillance and tracking of suspicious vehicles on the ground by multiple UAVs.

**Keywords:** multi-target association; topological sequences; triangular networks; global consistency; similar transformation invariance

**Citation:** Li, X.; Wu, L.; Niu, Y.; Ma, A. Multi-Target Association for UAVs Based on Triangular Topological Sequence. *Drones* **2022**, *6*, 119. https://doi.org/10.3390/drones6050119

Academic Editors: Daobo Wang and Zain Anwar Ali

Received: 13 April 2022
Accepted: 3 May 2022
Published: 7 May 2022

**Publisher's Note:** MDPI stays neutral with regard to jurisdictional claims in published maps and institutional affiliations.

## 1. Introduction

Multi-UAV cooperative technology has advanced rapidly in recent years. UAVs have been widely used in the fields of surveillance [1], reconnaissance [2], tracking and positioning [3–7] due to their superior remote perception capabilities. In the multi-target localization problem, the triangulation method [8] is a frequently used method, which first matches the line of sight of the UAV platform pointing at the target [9]. When the view axis is incorrectly matched, it inevitably leads to incorrect positioning results. In real application scenarios, the limitations of high UAV flight altitude and low sensor resolution add the difficulties to target detection and classification. In particular, targets with relatively similar visual features cannot be distinguished and classified by conventional vision-based detection algorithms. These reasons can lead to a large number of incorrect associations, which can produce incorrect targeting results [10,11]. Due to the importance of multi-target correlation problem in the field of localization and tracking, the multi-target association problem has become a key problem to be solved in the field of cooperative multi-UAV detection [12,13]. The main objective of this work is to solve the association problem of targets with similar visual features, and to propose an association method that does not use the visual features of the targets but only the relative position relationship between the targets to improve the accuracy of association of such targets.

Three primary methods can be used to solve the target association problem in UAV sensors: image, trajectory, and relative position [14]. The image-based methods require clear images of the target and a variety of image features [15]. Since UAVs are so maneuverable, the difficulty of extracting target trajectories is also significantly increased [16]. To distinguish and associate targets, relative position-based methods rely heavily on the topological and geometric relationships between them. This method does not require a high-resolution image and does not make use of other data from the UAV [17]. In this paper, we propose a method for constructing a specific topological network using the relative position relationship between targets, which is inspired by topological association methods used in the field of radar detection technology [18]. As illustrated in Figure 1, the method distinguishes and associates the targets based on their adjacent triangular topological sequences. This method does not require the extraction of target image and trajectory features and does not make use of the background of the images or the UAV's attitude information.

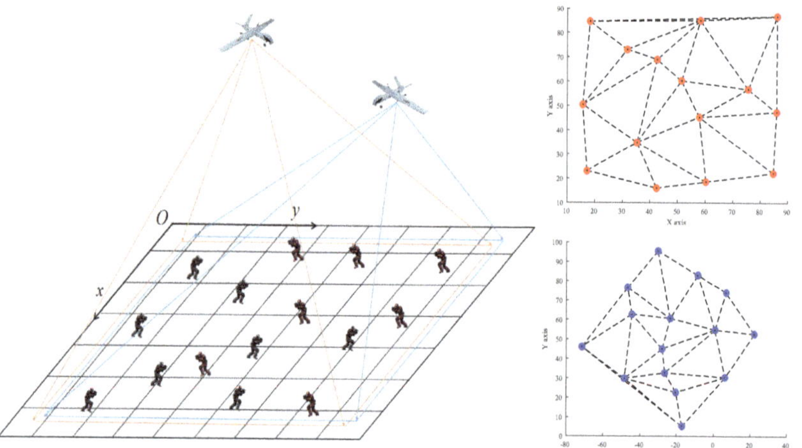

**Figure 1.** A schematic diagram of multi-UAV target association based on a triangular topological sequence.

Considering the issues raised in this paper and the shortcomings of existing multi-target association algorithms, this paper proposes a globally consistent two-step multi-target association algorithm that solves the multi-sensor multi-target association problem solely using the relative position features of multiple targets. First, this paper triangulates all targets and determines their triangular topological characteristics based on their connectivity relationships in the triangular network. Second, to avoid the dependence of features on the coordinate system, this paper calculates the similarity of two triangles using the difference in triangle angles. Meanwhile, the feature similarity with similar transformation invariance is determined by calculating the weighted mean of all neighboring triangles' similarity. Finally, considering the global goal association's consistency, a two-step completion association is proposed. The partial correlation results from the first step are used to solve the transformation relationship between the two sensor coordinate systems. After eliminating the incorrect association in the first step, the transformation matrix is used to investigate possible associations in the unassociated results. The two-step association method not only improves the algorithm's correct association rate, but also enables it to perform better when some targets are not detected. These benefits are confirmed in simulation experiments.

The method's primary contribution to innovation is that it augments the geometric topology-based association method with the following three enhancements. The first is consistency of global association, which is achieved by combining knowledge about multi-view geometry, which has the same mapping relationship for two photos taken from the same viewpoint, and which is considered in this paper. The second is to rely on specific rules rather than empirical thresholds for the construction of topological features. Ref. [18] used thresholds to determine which neighbors should be included in the constructed features, which is not an easy task in practice. By comparison, the structure formed through triangular dissection is more scientific. Thirdly, because the method for calculating feature similarity is similar in terms of transformation invariance, i.e., the same structure remains unchanged after rotation, translation, and scaling, the similarity values obtained using this method are identical. The algorithm's primary contributions are as follows:

(1) Rather than relying on image and trajectory features, it distinguishes and associates identical targets within a single context using their relative positions.
(2) The method for calculating similarity exhibits similar transformation invariance. The method used to generate feature sequences is a non-empirical one.
(3) The association process considers the target coordinates' global consistency.
(4) The algorithms are applicable to vision and vision sensors, as well as vision and infrared sensors, as well as infrared and infrared sensors.

The remainder of this paper is organized as follows: Section 2 discusses some recent research on multi-objective association algorithms. Section 3 describes the multi-objective association algorithm proposed in this paper, which entails the construction of target detection, triangular topological networks, the calculation of the target similarity matrix using neighboring triangular sequences, and a two-step optimal association algorithm based on global consistency. Section 4 conducts simulation experiments to determine the algorithm's association accuracy under various conditions and comparison tests to determine the algorithm's association accuracy in comparison to various other algorithms. Section 5 summarizes the entire paper and provides an outlook on future work.

## 2. Related Works

Three primary algorithms for multi-objective association are described in the published work [14]. They include methods for associating trajectory similarity [16,19], image matching algorithms based on deep neural networks [20,21], and methods for associating reference topology [18]. Additionally, some researchers have attempted to process target trajectories using neural networks [22]. Trajectory-based association, in general, necessitates fixed sensor positions and high-quality extracted trajectories. Correlating images based on their clarity and feature density is difficult for drones flying at high altitudes to accomplish.

The methods for trajectory correlation primarily analyze data using probabilistic and statistically relevant methods. The Nearest Neighbor method was proposed by Singer and Kanyuch [23]. Then, Singer et al. [24] advanced the idea of trajectory association by introducing the concept of hypothesis testing and establishing a weighted trajectory association algorithm under the assumption that the estimation errors are mutually independent. On this basis, Bar-Shalom [25] modified this algorithm's distance metric to remove the requirement that the estimation errors of trajectory sequences be independent of one another and proposed a modified weighted method. Chang et al. [26] introduced the concept of allocation to operations research to extend the weighted method and proposed the classical allocation method for solving the trajectory association problem. The outcomes of trajectory association-based methods are significant and widely applied in the field of visual perception [14]. However, extracting high-quality target trajectories from maneuverable UAV image sensors is challenging.

In terms of Siamese neural network research, Ref. [27] pioneered the concept of a Siamese network based on dual target–environment data streams. Ref. [15] proposed a Siamese network architecture with dual data streams capable of extracting spatial and temporal information from RGB frames and optical flow vectors, respectively, to match pedestrians to distinct data frames. Ref. [28] achieved vehicle association between cameras at various locations by utilizing a license plate-body dual data stream Siamese neural network. Ref. [29] proposed a novel relative geometry-aware Siamese neural network that improves the performance of deep learning-based methods by exploiting the relative geometric constraints between images explicitly. To address the issue of target association robustness, Ref. [30] added a SE-block and a temporal attention mechanism to the framework of the Siamese neural network to enhance the network's discriminative ability and the tracker's recognition ability. The preceding conclusions are predicated on the assumption that the sensor images are of high quality and that the image features of the targets are distinguishable.

The reference topology-based approach solves the multi-target association problem in terms of different trajectories and images, without relying on any image features and without requiring a consistent sensor coordinate system [18,31]. Yue et al. [32] pioneered the reference topology association method. On this basis, Ref. [33] switched to a topological sequence rather than a single reference topology. Zhang et al. [34] correlated topological sequences using the gray correlation method. Instead of a sequence, Wu et al. [35] attempted to construct a more complex target topological matrix. Hao et al. [36] used triangles as the basic topology instead of all neighboring targets in cases such as dense multi-target and formation targets. Hasan et al. [37] used the reference topology to solve the asynchronous trajectory association problem. In other words, Li et al. [38] proposed a topological sequence association method based on a one-dimensional position distribution of multiple targets for determining the characteristics of visual sensors in air-ground systems and applied the method to air-ground systems for the first time. You et al. [39] recovered lost objects using temporal topological constraints based on the continuity and consistency of the multi-target topology in continuous frames. Ref. [17] groups targets according to their positions and velocities before applying a topology-based group target tracking algorithm to the tracking problem of dense target groups. Ref. [35] proposes a method for target association based on the target mutual support topology model, in which the mutual support matrix is defined, and the correct target relationship is determined using the optimal method with constraints. While research on reference topology is expanding, the following issues persist with the current results: (1) since topological features cannot adapt to coordinate system transformations, the reference target must be determined empirically; (2) existing work is primarily concerned with local topology and neglects global consistency; (3) in practice, the correlation error rate is significantly less than the simulation effect.

### 3. Materials and Methods

The algorithm described in this paper is divided into three sections: (1) detecting the target and constructing a triangular network; (2) computing topological similarity; and (3) a two-step optimal association algorithm. The algorithm's main flow is depicted in Figure 2, and this paper will use TTS as an abbreviation for proposed association algorithm based on *Triangular Topological Sequence*. The targets are first detected using vision images from *Sensor_A* and *Sensor_B*, and then their pixel coordinates $\{z_i^a\}_{i=1}^{n_a}$ and $\{z_i^b\}_{i=1}^{n_b}$ are extracted. Their pixel coordinates $\{z_i^a\}_{i=1}^{n_a}$ and $\{z_i^b\}_{i=1}^{n_b}$ are used to construct a specific triangular network, and their target's topological triangular sequence $T_i^a = \{\Delta_{ijk}^b\}$ and $T_i^b = \{\Delta_{ijk}^b\}$ are extracted. Then, using a similarity calculation method based on the triangle angle, the similarity of the target triangular topological sequence is calculated, and the similarity matrix $W_{m \times n}$, which contains the degree of similarity between all combinations of $m$ targets in *Sensor_A* and $n$ targets in *Sensor_B*, is obtained. Finally, in two steps, the target association is completed. The first step processes the similarity matrix $W_{m \times n}$ and extracts the global consistency matrix $\widehat{H}$ to obtain partial target association results $R$. The second step employs the global consistency matrix $\widehat{H}$ to eliminate incorrect associations discovered in the first step's association results $R$ and to identify possible new associations $R_1$ with the remaining unassociated targets under the new law. $R'$ is the result obtained by eliminating the wrong association in $R$.

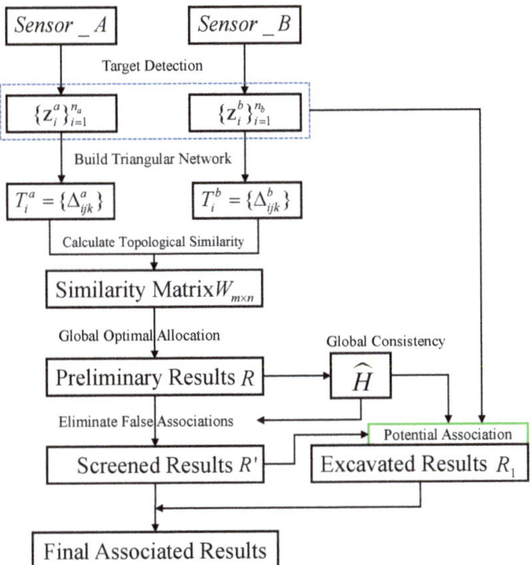

Figure 2 flow chart content:



**Figure 2.** The overall flow chart of multi-objective two-step association algorithm based on triangular topological sequence.

### 3.1. Constructing Triangular Topological Sequences

The input raw data are the visual image of the UAV, and an efficient target detection algorithm is needed to extract the relative position relationship of multiple targets. Many object detection algorithms are introduced in survey [40–43]. Considering the detection speed and accuracy of the algorithms mentioned in the above survey, this paper chooses yolov5 [44] as the target detection algorithm. Figure 3 illustrates the results of two UAV vision sensors' multi-target detection. The results are parsed to extract the set of target points, and a triangular topology network is constructed.

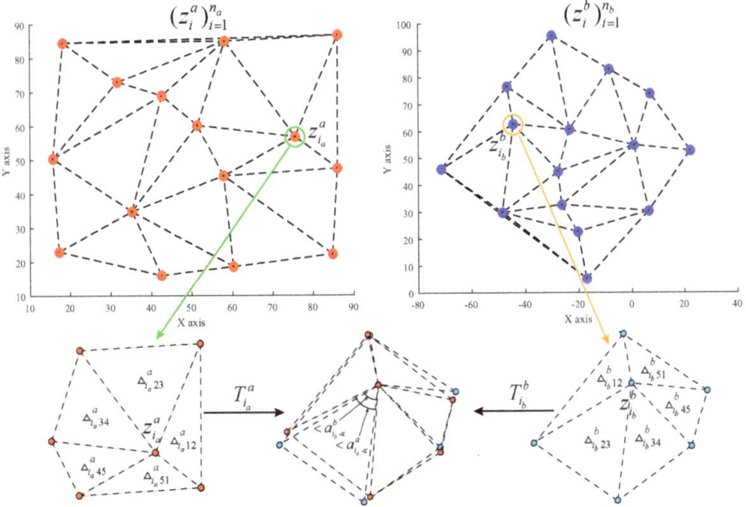



**Figure 3.** Construction of triangular topological features for a multi-target point set and calculation of topological similarity.

In this paper, the Cartesian coordinate systems $U_a$ and $U_b$ are established with reference to the pixel coordinate system of the vision sensor. Meanwhile, the real coordinates of target $i$ in the coordinate system $U_s$ at time $t$ can be expressed as $x_i^s$. Due to the image resolution and the detection algorithm, the target pixel coordinates typically carry an observation error $\tilde{x}_i^s$. Thus, the target coordinates of the detected target in sensor $s$ in the coordinate system $U_s$ with observation error can be expressed as:

$$z_i^s = x_i^s + \tilde{x}_i^s. \tag{1}$$

where $\tilde{x}_i^s \sim U(-P_i^s, P_i^s)$ is the observation error of the target $i$ pixel coordinates in sensor $s$. Let $(z_i^s)_{i=1}^{n_s} = \{z_1^s, z_2^s, \cdots, z_{(n_s)}^s\}$ be the set of pixel coordinates of all targets in sensor $s$ at moment $t$, where $n_s$ denotes the total number of targets detected in sensor $s$. According to [45], pictures taken at various angles for the same scene have a unique perspective transformation relationship, i.e.,

$$\exists \hat{H}, s.t.\ z_i^a = \hat{H} \cdot z_i^b. \tag{2}$$

where $\hat{H}$ is the optimal estimate transformation matrix from $U_a$ to $U_b$.

The following topological features are constructed for $(z_i^s)_{i=1}^{n_s} = \{z_1^s, z_2^s, \cdots, z_{(n_s)}^s\}$. Ref. [18] used empirical thresholds to determine neighboring nodes based on their distance, which can be used only within the same coordinate system. This paper employs a method of feature construction that is not dependent on empirical thresholds or coordinate systems. First, *Delaunay* triangulation [46] is used to create a unique triangular network from $(z_i^s)_{i=1}^{n_s}$. Due to the nature of *Delaunay* triangulation, which is closest, unique, and regional, each point in the constructed triangular network forms a triangle with only the nearest finite point. When a point is removed from the mesh, it affects only the triangles associated with that point. Due to the uniqueness of the result of Delaunay triangulation and the fact that the similarity transformation does not alter the ratio of angles and side lengths of any triangle in the network, the similarity transformation has no effect on the connectivity of points in the triangular network.

In the constructed network, let $L$ be the set of all connected lines in the network and $l_{jk} \in L$ denote the line connecting $z_j^s$ and $z_k^s$. For target $i$, all triangles adjacent to $z_i^s$ constitute the triangular topological features of target $i$:

$$\begin{aligned} T_i^s &= \{\triangle_{ijk}^s | l_{ij}, l_{ik}, l_{jk} \in L; j = 1, \cdots, n_i^s; k = 2, \cdots, n_i^s, 1\} \\ &= \{\triangle_{i12}^s, \triangle_{i23}^s, \cdots, \triangle_{in_i^s 1}^s\}. \end{aligned} \tag{3}$$

where $n_i^s$ denotes the total number of reference points connected to target $i$ in sensor $s$, and $\triangle_{ijk}^s$ denotes the triangle composed of $z_i^s, z_j^s, z_k^s$.

### 3.2. Calculate the Similarity of the Triangular Topological Sequence

For the triangular topological sequence depicted in Figure 3, we introduce a method for calculating the similarity between a single triangle and the triangular topological sequence $T_i^s$. To calculate triangle similarity, Ref. [36] devised a method based on triangle side lengths, but this method is inapplicable when the scales of the two coordinate systems are different. To address this issue, this paper develops a more discriminating similarity calculation function based on the difference in triangle angles. A function for calculating triangle similarity that has a high degree of differentiation should exhibit the following characteristics: (1) The similarity value decreases monotonically as the difference between the two triangles increases. (2) The smaller the triangle difference, the larger the change in similarity caused by the unit change in the triangle. That is, a concave function that decreases monotonically. Let the three angles of $\triangle_{i_sjk}^s$ be $\angle\alpha_i^s$, $\angle\alpha_j^s$, and $\angle\alpha_k$. Then, the difference between $\triangle_{i_ajk}^a$ and $\triangle_{i_bjk}^b$ can be expressed as $\triangle_{i_a i_b} = \sum_{p=i,j,k} |\angle\alpha_p^a - \angle\alpha_p^b|$. Thus, the similarity of the $x$th triangle of $T_{i_a}^a$ and $T_{i_b}^b$ is calculated as:

$$\omega_{i_a i_b - x} = 1 - ln\left(1 + \lambda \cdot \frac{\triangle \alpha_{i_a i_b}}{180}\right). \tag{4}$$

where $\lambda$ adjusts the range of values of $\omega$. In this paper, we take $\lambda = 1.72$. $\mathcal{W}_{m \times n}$ is the similarity association matrix of $m$ targets in sensor $a$ and $n$ targets in sensor $b$, where $\omega_{i_a i_b}$ is the element of the $i_a$th row and $i_b$th column of $\mathcal{W}_{m \times n}$, the similarity between $z_{i_a}^a$ and $z_{i_b}^b$:

$$\omega_{i_a i_b} = \sum_{x=1}^{min\left(n_{i_a}^a, n_{i_b}^b\right)} \frac{\left(\angle \alpha_{i_a - x}^a + \angle \alpha_{i_b - x}^b\right)}{\alpha_{i_a i_b}} \cdot \omega_{i_a i_b - x}. \tag{5}$$

$$\alpha_{i_a i_b} = \sum_{x=1}^{n_{i_a}^a} \angle \alpha_{i_a - x}^a + \sum_{y=1}^{n_{i_b}^b} \angle \alpha_{i_b - y}^b. \tag{6}$$

where $\angle \alpha_{i_s - x}^s$ is the $\angle \alpha_i^s$ of the xth triangle in $T_{i_s}^s$. This method can be described as taking a weighted average of the similarity of the neighboring triangles. The angle occupied by the neighboring triangle $\triangle_{ijk}^s$ around the point $z_i^s$ is used as the weight of $\triangle_{ijk}^s$.

### 3.3. Two-Step Global Constraint-Based Association Algorithm

This section discusses the two-step association algorithm based on the similarity association matrix $\mathcal{W}_{m \times n}$ and the consistency of global association, where the consistency of global association has not been considered in previous works. The global association consistency entails that all association targets in $(z_i^a)_{i=1}^{n_a}$ and $(z_i^b)_{i=1}^{n_b}$ satisfy Equation (2).

In the first step of association, the similarity matrix $\mathcal{W}_{m \times n}$ that contains local feature information is processed, and the association result of the first step is derived. The principle of global optimal allocation is as follows: (1) only one element from each row of the matrix is chosen; (2) only one element is chosen from each column of the matrix; (3) the greatest possible sum of all selected elements' values. In the first step, all selected pairs in $\mathcal{W}_{m \times n}$ form the global optimal association result $R$ in the first step.

As shown in Figure 4, in practice, the triangular topological sequence of some targets produces local variations due to partial target occlusion and the influence of observation errors $\widetilde{x}_i^s$. For the case of multi-target formation, the targets have a similar relative position relationship with each other, so the triangular topological features of the targets are similar, and the similarity difference is small. These factors are not conducive to distinguishing the triangular topological sequence, so that the result of the global optimal assignment principle contains some wrong matches. Among the total association results, the wrong match has the different $\hat{H}$ from the other associations. For the case where most of the correct association results contain a few incorrect associations, the *Ransac* algorithm can be used to compute the transformation matrix $\hat{H}$ based on $R$ and eliminate the incorrect associations. The result that contains only the correct association after algorithm processing is denoted as $R'$.

The second association step is to find potential associations among unassociated target points using $\hat{H}$ and $R'$. $(z_{i_a}^a, z_{i_b}^b) \in R'$ indicates that $z_{i_a}^a$ and $z_{i_b}^b$ are accurately associated. Considering the effect of observation error $\widetilde{x}_i^s$, for $\forall z_{i_b}^b \notin R'$, $z_{i_a}^{ba} = \hat{H} \cdot z_{i_b}^b$, if $\exists z_{i_a}^a \notin R'$ is the closest point in $(z_i^a)_{i=1}^{n_a}$ at distance $z_{i_a}^{ba}$ and satisfies the following formula:

$$\mathcal{W}(\triangle_{i_a j_a k_a}^a, \triangle_{i_b j_b k_b}^b) = \max_x (\omega_{i_a i_b - x}).$$
$$\left(z_{j_a}^a, z_{j_b}^b\right) \in R', \left(z_{k_a}^a, z_{k_b}^b\right) \in R'. \tag{7}$$

$z_{i_a}^a$ and $z_{i_b}^b$ are determined as the new correct association, when $\mathcal{W}(\triangle_{i_a j_a k_a}^a, \triangle_{i_b j_b k_b}^b)$ denotes the similarity of $\triangle_{i_a j_a k_a}^a$ and $\triangle_{i_b j_b k_b}^b$.

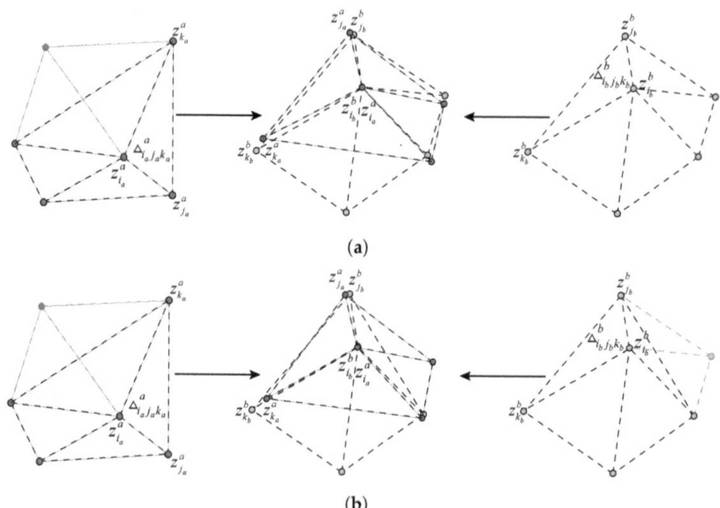

**(a)**

**(b)**

**Figure 4.** (**a**) Topological sequences when a few targets are missing; (**b**) topological sequence when most targets are missing.

In the next experiment part, this paper uses association precision and accuracy as metrics for evaluating the experimental results:

$$Precision = \frac{TP}{TP+FP}, \quad Accuracy = \frac{TP+TN}{TP+TN+FP+FN}. \tag{8}$$

where $TP$ is the number of correct associations predicted as correct associations, $TN$ is the number of incorrect associations predicted as incorrect associations, $FP$ is the number of incorrect associations predicted as associations, and $FN$ is the number of correct associations predicted as incorrect associations.

## 4. Results

This section simulates and tests the proposed algorithm, with the simulation test consisting of three major sections. The first section contains a series of tests to determine the correlation's precision under various experimental conditions. Several experimental conditions were established in the following manner: (1) the sensor coordinate systems have a similar transformation relationship; (2) specific targets are obscured or missed; (3) the experimental area contains a variable number of targets; and (4) errors in observation. The second section compares the correlation accuracy of the proposed TTS, the neighboring reference topological feature association algorithm (RTF) [18], and the triangle topological feature association algorithm (TTF) [36,47] under the four experimental conditions mentioned previously. The third part compares the proposed TTS to the RTF [18], TTF [36,47], and the Siamese neural network method [48,49] for the association accuracy and precision in a physical flight test.

Figure 5 shows the experiments of target detection using yolov5 [44] for UAV vision images in this paper, and the algorithm is fast and accurate for UAV platforms. However, the algorithm also cannot achieve the detection of all targets at all times. Figure 5a shows vehicles moving at high speed on the road, and Figure 5b shows a dense and neatly distributed vehicle in a square, where there are two undetected targets.

(a)           (b)

**Figure 5.** Two scenarios for testing the target detection algorithm, (**a**) is high-speed moving vehicles, and (**b**) is dense and neatly distributed targets.

### 4.1. Testing on the Proposed Algorithm

The first section examines the effect of various experimental conditions on association accuracy and compares the accuracy of single-step and two-step association in this algorithm. As illustrated in Figure 1, multiple targets are randomly distributed throughout the scene and are monitored by two UAV sensors. We set a square area with a shape of 100 m × 100 m, and then fly drones equipped with visual sensors over it. Considering the effect of the number of targets on the association algorithm, we varied the number of targets in the experimental region at intervals of ten in the range of 10 to 100. We created a set of basic tests with a total of 12 combinations for each target quantity setup, which included three different rotation angles of 30°, 75°, and 135°, two scale transformations of 0.5× and 50×, and two translational transformations of 10 m and 50 m.

Figure 6a shows the results of the experiments on the algorithm's similar invariance. The experiments were repeated several times with the same experimental setup, and 10% of the targets were not detected simultaneously by both sensors in each experiment due to occlusion or other factors. The two-step association algorithm's experimental results were analyzed statistically (shown in Table 1), and the mean values of association precision were 98.40%, 98.39%, and 98.34% for three different rotation angles, 98.44% and 98.28% for two scale transformations, and 98.37% and 98.39% for two translational transformation experiments. The mean value of association precision for all results was 98.38%, with 58.90% of experiments achieving 100% correct association. The data indicate that the association precision under different similar transformations is within the overall mean ±0.1% (shown in Figure 6c), indicating that the similar transformation relationship between coordinates has no significant effect on the association precision and that the algorithm has similar transformation invariance. The effect of single-step and double-step association is depicted in Figure 6b, with single-step precision significantly lower than double-step precision. Single-step associations have an average precision of 63.3%, which is significantly less than that of two-step associations. Single-step association precision decreases as the number of targets increases and is greatest when the number of targets is small. The experimental results demonstrate that the association method proposed in this paper has comparable transformation invariance and that the two-step association significantly outperforms the single-step association.

Figure 7 presents the experimental results of this paper for some target loss conditions. Target loss occurs when a target is not detected due to obstructions or the detection algorithm, most commonly certain randomly existing undetected targets and targets in a portion of the area obscured by obstacles. We repeated the 12 base experiments several times at 5% intervals, varying each of the 11 loss rates from 0% to 50%. The statistical data of the results of the two-step correlation experiment are depicted in Figure 7a. When the target is obscured within the local area, the experimental precision exceeds 98% and remains stable when the missing rate is less than 20%. When the missing rate exceeds 20%, the precision of the correlation decreases linearly. When the target is missed by 40%, the correlation precision falls below 30%. The experimental precision is greater than 98% for randomly distributed undetected targets and remains stable at a missing rate of 10% or less. When the rate of missing data exceeds 10%, the correlation precision decreases linearly. Correlation precision decreases to less than 30% at a target missing rate of 30%. In this case, a missing rate of less than 10% has little effect on the association algorithm's precision, whereas the random presence of undetected targets has a greater effect on the algorithm's precision when the missing rate exceeds 10%. In Figure 7a, the precision of the single-step association is significantly lower than the precision of the two-step association, and the two-step association has a significant improvement in association precision. The results of one experiment are depicted in Figure 7b, where the precision of two-step correlation is either extremely high or extremely low. To ensure that the algorithm maintains a higher correlation precision more frequently, the following analysis of this phenomenon is made: The critical step in the two-step association is to identify the global association consistency and estimate the value of $\hat{H}$. When the value of $\hat{H}$ is estimated accurately, incorrect associations are eliminated, correct associations are increased, and association precision are improved. When the value of $\hat{H}$ is estimated incorrectly, the correct association is reduced, and the incorrect association is increased. Analysis of the experimental data reveals that the two-step association algorithm has 88.10% probability of finding the correct $\hat{H}$ when the single-step association precision is more than 30%.

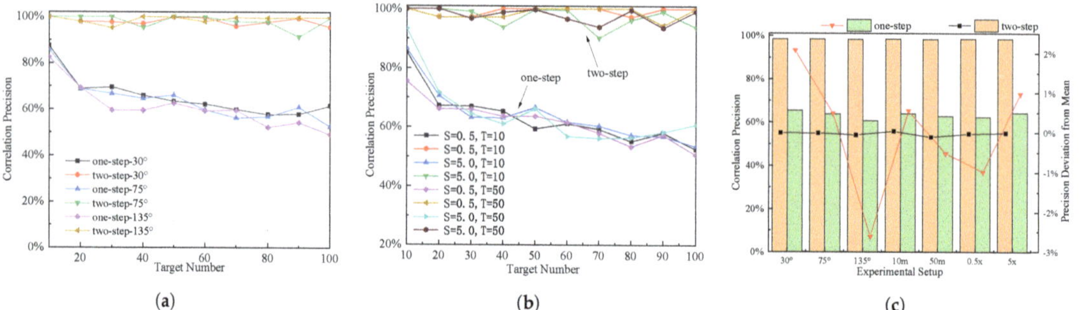

Figure 6. (a) The effect of the number of targets in the region on the correlation precision at three rotation angles; (b) the effect of the number of targets in the experimental area on the correlation precision under different scale transformation and translational transformation conditions; (c) association precision histogram and standard deviation curve of the proposed TTS under various conditions.

Table 1. Statistics of association precision under different similarity transformation experiments.

| Setup | Rotation | | | Translation | | Scale | |
|---|---|---|---|---|---|---|---|
| | 30° | 75° | 135° | 10 m | 50 m | 0.5× | 5× |
| Two-Step (%) | 98.40 | 98.39 | 98.34 | 98.44 | 98.28 | 98.37 | 98.39 |
| One-Step (%) | 65.37 | 63.78 | 60.68 | 63.84 | 62.77 | 62.30 | 64.26 |

The effect of uniformly distributed observation error $\tilde{x}_i^s \sim \mathcal{U}(-P_i^s, P_i^s)$ on the correlation precision is illustrated in Figure 8. In this paper, we set 31 values of $P_i^s$ in the range of 0 3 m at 0.1 m intervals and conduct 12 basic experiments used in Figure 6 under each value to study the trend of algorithm association precision with observation errors. Assuming that the size of the target is 1.5 m, the multiplier of the error relative to the target is used for the horizontal axis. Figure 8a depicts the effect of observation error on average correlation precision. When the observation error is less than 1.5 times, the correlation precision fluctuates steadily above 90% with no decreasing trend. When the observation error exceeds 1.5 times, the correlation precision rapidly decreases to approximately 70%, and the fluctuation increases. In comparison, the precision of the single-step correlation was always around 60%, indicating a slowing trend with insignificant fluctuations. Obviously, the two-step correlation algorithm has advantages. Figure 8b illustrates the effect of observation error on the correlation's precision in one experiment. According to the data in the figure, the two-step association algorithm maintains an association precision close to 100% for observation errors less than 1.5 times, whereas the single-step association algorithm maintains an association precision of around 60%, which is always less than the two-step association. When observation errors exceed 1.5 times, the two-step correlation method begins to exhibit increased fluctuation and the correlation precision decreases further. The experimental results indicate that, while smaller observation errors have a smaller effect on the correlation method proposed in this paper, observation errors greater than 1.5 times have a greater effect on correlation precision, which may be due to observation errors causing $\hat{H}$ estimation errors.

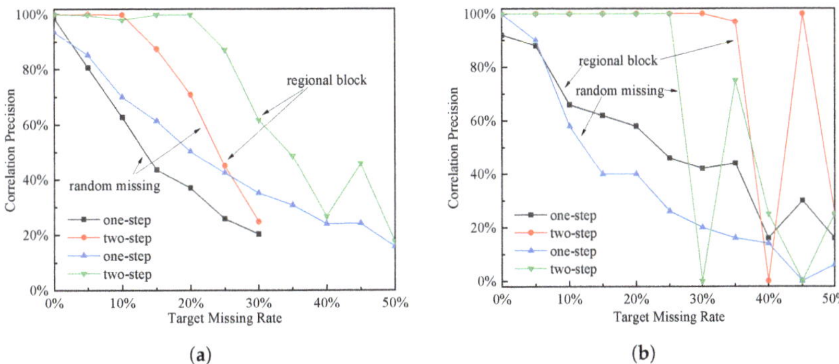


**Figure 7.** (**a**) The effect of target missing rate of statistical results on correlation precision; (**b**) the effect of target missing rate on association precision in a specific experiment.

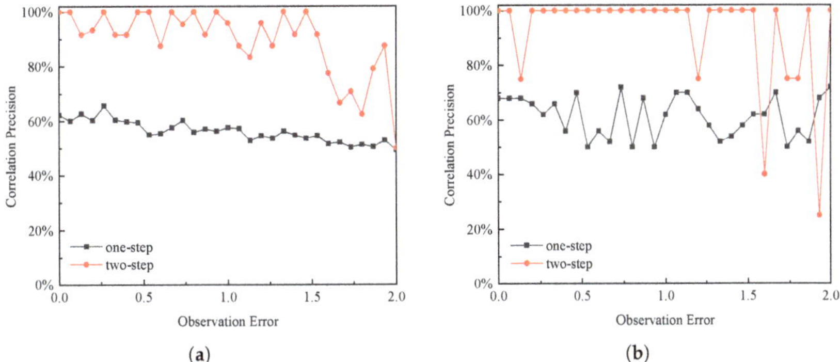


**Figure 8.** (**a**) The effect of observation error on correlation precision; (**b**) the effect of observation error on correlation precision in a specific experiment.

### 4.2. Simulation Comparison Experiments

The second section compares TTS's association accuracy to proposed TTS, RTF, and TTF under the same experimental conditions. We conducted another series of experiments on three different dimensions: (1) certain components are not detected; (2) the error associated with the observation is encoded in the coordinates; (3) the coordinate system of the sensor is scaled differently. The association accuracy variation curves for the three algorithms in different scale sensor coordinate systems are depicted in Figure 9a. Because all three algorithms compared in this paper are rotation and translation invariant, this experiment considers only cases with different scales. According to the experimental results, the correlation accuracy of the current algorithm remains stable at around 98% across a range of scales, whereas the accuracy of the other two algorithms decreases significantly as the coordinate system scales change. This demonstrates that the present algorithm possesses similar transformation invariance to that of other algorithms. Figure 9b illustrates the performance of the three algorithms in the presence of missing targets due to occlusion by a local obstacle. The experiments demonstrate that, as the occluded area increases, the association accuracy decreases. However, the algorithm described in this paper always maintains a high association accuracy, which is on average 19.88% and 27.46% higher than the association accuracy of the other two algorithms. The trend of association accuracy for the three algorithms with varying observation errors is depicted in Figure 9c. The results indicate that, when the observation error is less than 0.73 times of target size, the algorithm in this paper maintains the highest association accuracy. When the observation error is greater than 0.73 times, the results of TTS are lower than the results of TTF. Within the range of observation error distributions of 0 to 2 times, this algorithm's association accuracy is always greater than that of RTF, by an average of 5.2%. When compared to the data in Figure 9, the TTS algorithm outperforms the other algorithms in all three experimental conditions, and its overall performance is optimal.

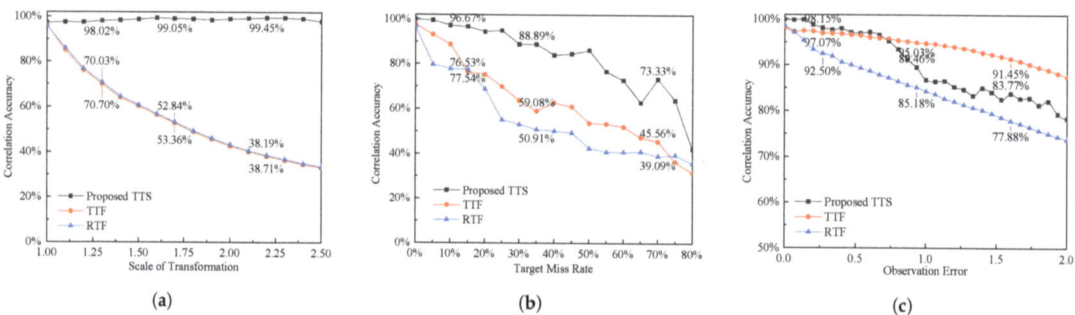

**Figure 9.** The proposed TTS comparison experiments with other algorithms, (**a**) is the variation curve of the effect on association accuracy at different scale transformations (**b**) the variation curve of the effect of target occlusion rate on association accuracy (**c**) is the variation curve of the effect of different observation errors on association accuracy.

### 4.3. Physical Comparison Experiments

The third section compares the proposed TTS to the RTF [18], TTF [36,47], and the Siamese neural network method [48,49] for the association accuracy and precision in the scenario shown in Figure 10.

Figure 10a depicts the algorithm described in this article being tested on a road with heavy traffic and vehicles traveling at high speeds. Two DJI drones are used in this article to capture video above the road. The two drones flew continuously between 80 m and 100 m in altitude, following a predetermined trajectory to observe the vehicles on the road from various angles. The targets are associated using the proposed TTS, RTF [18], TTF [36,47] and Siamese neural networks [49], respectively, and the association precision and accuracy are analyzed (shown in Figure 11). To develop the Siamese neural network's recognition model, we collected training data on the opposite road.

The proposed TTF, RTF, and TTF are compared in Figure 10a scenario in Figure 11a,b, with experimental data for 50 s shown in Table 2. The RTF algorithm must implement the specified threshold value, and Table 3 contains information about the threshold value and association results obtained during the experiments. The proposed two-step TTF achieves an association accuracy of 84.63%, while RTF and TTF achieve only 63.51% and 68.63%, respectively. The proposed two-step TTF achieves a correlation accuracy of 67.87%, while the RTF and TTF achieve only 20.31% and 61.81%, respectively. The proposed TTS's single-step association accuracy and accuracy are 71.80% and 79.99%, respectively, which are higher than the RTF and TTF algorithms' corresponding data. The experimental data demonstrate that the proposed algorithm outperforms comparable algorithms implemented in this paper in terms of association accuracy and association accuracy, with a significant improvement in association accuracy for the second-step association. The curves in Figure 11a,b demonstrate how significantly the association results of all three algorithms vary in actual flight experiments. While the algorithm performs better most of the time, there are times when it performs poorly.

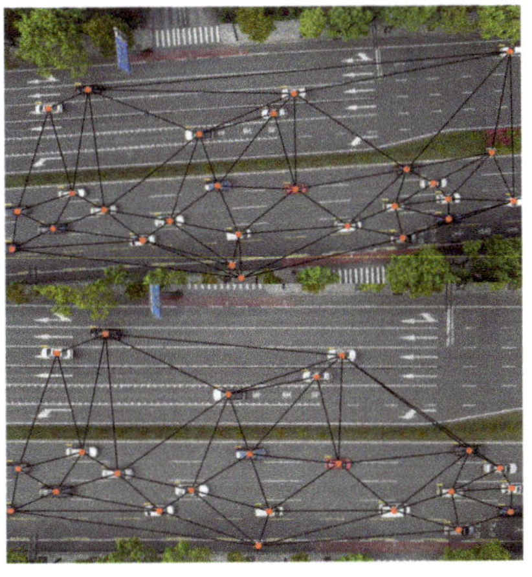
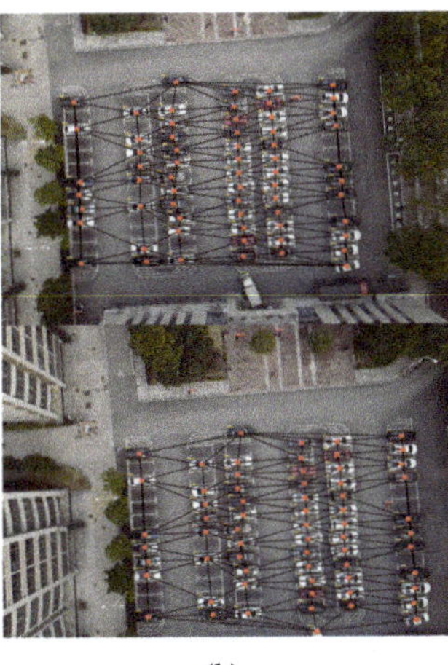

(a)        (b)        (c)

**Figure 10.** The results of target detection and triangular network construction in three flight test scenarios. (**a**) is a scenario with a large number of randomly distributed moving targets. (**b**) is a scenario with a large number of neatly arranged stationary targets. (**c**) is a scene with a small number of neatly arranged stationary targets.

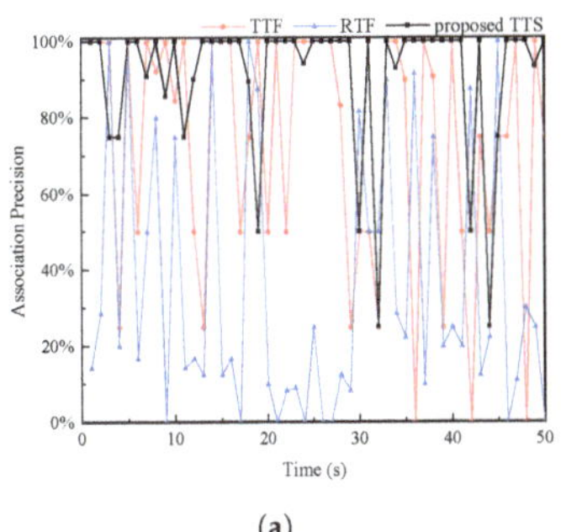
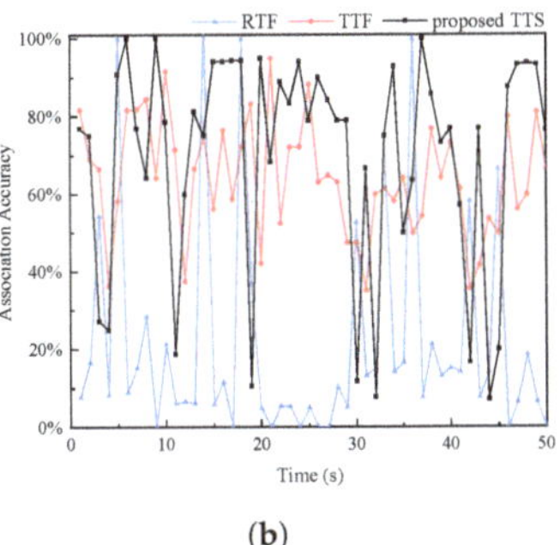
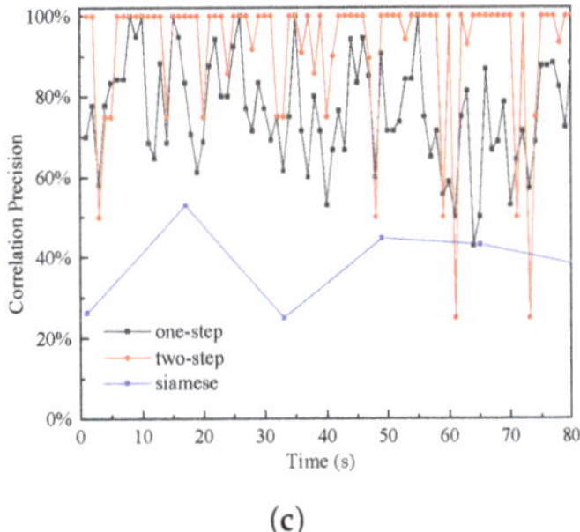

(a)        (b)        (c)

**Figure 11.** (**a**,**b**) is the comparison of the proposed TTS with TTF and RTF in the scene shown in Figure 10a. (**c**) is precision curves of the association between the proposed TTS and Siamese neural networks in real flight experiments.

**Table 2.** Statistical data of experimental results of proposed TTF, RTF, and TTF comparative experiments.

| Index \ Algorithm | Proposed TTS | RTF | TTF |
|---|---|---|---|
| Two-Step-Precision | 84.63% | 63.51% | 68.63% |
| One-Step-Precision | 71.80% | 38.61% | 55.66% |
| Two-Step-Accuracy | 67.87% | 20.31% | 61.81% |
| One-Step-Accuracy | 79.99% | 41.45% | 42.87% |

**Table 3.** Correlation result for different thresholds in the RTF method.

| Index \ Threshold | 150 | 250 | 300 | 400 | 500 | 700 | 800 | 1000 |
|---|---|---|---|---|---|---|---|---|
| Precision | 0% | 36.98% | 42.74% | 41.47% | 53.27% | 61.08% | 63.51% | 43.72% |
| Accuracy | 0% | 15.53% | 24.08% | 22.33% | 21.69% | 19.76% | 20.31% | 12.50% |

The algorithm's association results are depicted in Figure 10 and Table 4. The precision of two-step association is generally greater than that of single-step association, and the Siamese neural network does not perform as well in practical tests as the algorithm described in this paper. The mean precision of two-step associations is 84.63%, single-step associations are 71.80%, and Siamese neural network associations are 38.25%. The algorithm described in this paper outperforms the Siamese neural network-based approach in terms of association precision. In terms of association accuracy, the algorithm in this paper achieves an average of 67.87%, while the Siamese neural network achieves an average of 38.25%, indicating that the algorithm in this paper outperforms the Siamese neural network. They demonstrate that the indicators produced by this algorithm are superior to those produced by the Siamese neural network, and that the two-step association results produced by this algorithm are superior to the single-step association results.

**Table 4.** Statistical table of flight test results of proposed TTS and Siamese neural network in two typical scenarios.

| Scenes \ Index | One-Step-Pre | Two-Step-Pre | Two-Step-Acc | Siamese-Acc |
|---|---|---|---|---|
| Fast Moving (Figure 10a) | 71.80% | 84.63% | 67.87% | 38.25% |
| Dense Stationary (Figure 10b) | 67.19% | 100.00% | 93.75% | 5.08% |

Table 5 summarizes the major advantages and disadvantages of the four algorithms used in this paper based on the analysis of the aforementioned experimental results. Figures 6 and 9a demonstrate that TTS transforms similarly to RTF and TTF without distortion. As illustrated in Table 3, RTF must rely on empirical thresholds, whereas TTS does not. To demonstrate the validity of global consistency, the improved correctness of two-step correlation is obtained. In comparison to Siamese neural networks, TTS, RTF, and TTF do not require consideration of the target's image features to complete the association. Among them, this paper considers non-empiricality, similar transformation invariance, and global consistency for the first time, which is a novel contribution to the field.

**Table 5.** Comparison of advantages and disadvantages of proposed TTS, RTF, and TTF and Siamese neural networks.

| Characteristic \ Algorithm | Proposed | TTS | RTF | TTF | Siamese |
|---|---|---|---|---|---|
| Similarity Transformation Invariance | ✓ | | X | X | ✓ |
| Non-empirical | ✓ | | X | ✓ | ✓ |
| Global Consistency | ✓ | | X | X | X |
| Not Using Image Features | ✓ | | ✓ | ✓ | X |

Additionally, the algorithm described in this paper has some limitations. In the scenario depicted in Figure 10b, where targets are densely aligned within the scene, the topological features constructed by the peripheral targets are more extreme and unsuitable for association, and small changes in the features can result in large changes. Indeed, this paper's algorithm is applicable to a more decentralized target. The Siamese neural network outperforms the algorithm in this paper in terms of association for the two columns of neatly aligned targets shown in Figure 10c. In addition, this algorithm is influenced by global consistency matrix $\hat{H}$, in which, when solved incorrectly, the association effect of this method significantly decreases.

The preceding experiments are divided into three sections: the algorithm's association precision is evaluated in simulation experiments under a variety of experimental conditions, and the association accuracy of multiple topological association methods is compared. Additionally, the results of this algorithm's association with Siamese neural networks are compared under flight test conditions. Experiments demonstrate that the algorithm described in this article has a high correlation accuracy and correctness, similar transformation invariance, and performs well when the observation contains an error, and the target is true. The present algorithm associates significantly better than the Siamese neural network in the flight test and performs admirably in practical applications. Simultaneously, the experiment revealed several flaws in the algorithm described in this paper. First, when the global consistency matrix is solved incorrectly, the algorithm's association effect is weakened. Second, due to the law of triangular dissection's limitation, the topological characteristics of the target construction with neatly aligned boundaries are more extreme and unsuitable for target association.

## 5. Conclusions

The paper proposes a new multi-target association algorithm based on triangular topology sequences that utilizes global consistency and a two-step process to improve association accuracy. The algorithm constructs the feature sequence using a specific triangular network and exhibits similar transformation invariance. The experimental results demonstrate that the proposed multi-target association method achieves a high level of accuracy and precision, and that the two-step association method improves the association effect significantly. Under three experimental conditions, this algorithm had the best association effect when compared to other algorithms. However, the algorithm described in this paper has some limitations. When targets are neatly aligned, association results for targets on the periphery are poor. The results of the association are strongly influenced by the global consistency matrix, and any error in solving the consistency matrix results in a significant reduction in the association's accuracy.

The work presented in this paper can be applied to problems involving correlation in UAV cluster reconnaissance of multiple ground targets, particularly when the visual characteristics of the targets are very similar. For instance, drones are used in conjunction to monitor many suspicious vehicles that share similar characteristics. The following will be improved in future work: We will consider the category information associated with the targets, and we will construct topological networks containing various types of targets and associate them using topological sequences containing various types of targets.

**Author Contributions:** X.L. was the main author of the work in this paper. X.L. conceived the methodology of this paper, designed the experiments, collected and analyzed the data, and wrote the paper; L.W. was responsible for directing the writing of the paper and the design of the experiments; Y.N. was responsible for managing the project and providing financial support; A.M. was responsible for assisting in the completion of the experiments. All authors have read and agreed to the published version of the manuscript.

**Funding:** This work was supported in part by the National Natural Science Foundation of China under Grant No. 61876187.

**Acknowledgments:** Thanks to the UAV Teaching and Research Department, Institute of Unmanned Systems, College of Intelligent Science, National University of Defense Technology for providing the support of the experimental platform.

**Conflicts of Interest:** The authors declare no conflict of interest.

## References


1. Yeom, S.; Nam, D.H. Moving Vehicle Tracking with a Moving Drone Based on Track Association. *Appl. Sci.* **2021**, *11*, 4046. [CrossRef]
2. Zheng, Y.J.; Du, Y.C.; Ling, H.F.; Sheng, W.G.; Chen, S.Y. Evolutionary Collaborative Human-UAV Search for Escaped Criminals. *IEEE Trans. Evol. Comput.* **2019**, *24*, 217–231. [CrossRef]
3. Wang, J.; Han, L.; Dong, X. Distributed sliding mode control for time-varying formation tracking of multi-UAV system with a dynamic leader. *Aerosp. Sci. Technol.* **2021**, *111*, 106549. [CrossRef]
4. Bayerlein, H.; Theile, M.; Caccamo, M. Multi-UAV path planning for wireless data harvesting with deep reinforcement learning. *IEEE Open J. Commun. Soc.* **2021**, *2*, 1171–1187. [CrossRef]
5. Ali, Z.A.; Zhangang, H. Multi-unmanned aerial vehicle swarm formation control using hybrid strategy. *Trans. Inst. Meas. Control* **2021**, *43*, 2689–2701. [CrossRef]
6. Zhang, Z.; Xu, X.; Cui, J.; Meng, W. Multi-UAV Area Coverage Based on Relative Localization: Algorithms and Optimal UAV Placement. *Sensors* **2021**, *21*, 2400. [CrossRef]
7. Wang, X.; Yang, L.T.; Meng, D.; Dong, M.; Ota, K.; Wang, H. Multi-UAV Cooperative Localization for Marine Targets Based on Weighted Subspace Fitting in SAGIN Environment. *IEEE Internet Things J.* **2021**, *9*, 5708–5718. [CrossRef]
8. Bai, G.B.; Liu, J.H.; Song, Y.M.; Zuo, Y.J. Two-UAV intersection localization system based on the airborne optoelectronic platform. *Sensors* **2017**, *17*, 98. [CrossRef]
9. Hinas, A.; Roberts, J.M.; Gonzalez, F. Vision-Based Target Finding and Inspection of a Ground Target Using a Multirotor UAV System. *Sensors* **2017**, *17*, 2929. [CrossRef] [PubMed]
10. Lin, B.; Wu, L.; Niu, Y. End-to-end vision-based cooperative target geo-localization for multiple micro UAVs. *J. Intell. Robot. Syst.* **2022**, *accepted*.
11. Sheng, H.; Zhang, Y.; Chen, J.; Xiong, Z.; Zhang, J. Heterogeneous Association Graph Fusion for Target Association in Multiple Object Tracking. *IEEE Trans. Circuits Syst. Video Technol.* **2018**, *29*, 3269–3280. [CrossRef]
12. Lee, M.H.; Yeom, S. Multiple target detection and tracking on urban roads with a drone. *J. Intell. Fuzzy Syst.* **2018**, *35*, 6071–6078. [CrossRef]
13. Shekh, S.; Auton, J.C.; Wiggins, M.W. The Effects of Cue Utilization and Target-Related Information on Target Detection during a Simulated Drone Search and Rescue Task. *Proc. Hum. Factors Ergon. Soc. Annu. Meet.* **2018**, *62*, 227–231. [CrossRef]
14. Rakai, L.; Song, H.; Sun, S.; Zhang, W.; Yang, Y. Data association in multiple object tracking: A survey of recent techniques. *Expert Syst. Appl.* **2022**, *192*, 116300. [CrossRef]
15. Chung, D.; Tahboub, K.; Delp, E.J. A Two Stream Siamese Convolutional Neural Network for Person Re-identification. In Proceedings of the IEEE International Conference on Computer Vision (ICCV), Venice, Italy, 22–29 October 2017; pp. 1992–2000.
16. Angle, R.B.; Streit, R.L.; Efe, M. Multiple Target Tracking With Unresolved Measurements. *IEEE Signal Process. Lett.* **2021**, *28*, 319–323. [CrossRef]
17. Zhang, Y.; Liu, M.; Liu, X.; Wu, T. A group target tracking algorithm based on topology. *J. Phys. Conf. Ser.* **2020**, *1544*, 012025. [CrossRef]
18. Tian, W.; Wang, Y.; Shan, X.; Yang, J. Track-to-Track Association for Biased Data Based on the Reference Topology Feature. *IEEE Signal Process. Lett.* **2014**, *21*, 449–453. [CrossRef]
19. Tokta, A.; Hocaoglu, A.K. Sensor Bias Estimation for Track-to-Track Association. *IEEE Signal Process. Lett.* **2019**, *26*, 1426–1430. [CrossRef]
20. An, N.; Qi Yan, W. Multitarget Tracking Using Siamese Neural Networks. *ACM Trans. Multimid. Comput. Commun. Appl.* **2021**, *17*, 75. [CrossRef]
21. Valmadre, J.; Bertinetto, L.; Henriques, J. End-to-end representation learning for correlation filter based tracking. In Proceedings of the IEEE Conference on Computer Vision and Pattern Recognition, Honolulu, HI, USA, 21–26 July 2017; pp. 2805–2813.
22. Yoon, K.; Kim, D.; Yoon, Y.C. Data Association for Multi-Object Tracking via Deep Neural Networks. *Sensors* **2019**, *19*, 559. [CrossRef]
23. Kanyuck, A.J.; Singer, R.A. Correlation of Multiple-Site Track Data. *IEEE Trans. Aerosp. Electron. Syst.* **1970**, 180–187. [CrossRef]
24. Singer, R.A.; Kanyuck, A.J. Computer control of multiple site track correlation. *Automatica* **1971**, *7*, 455–463. [CrossRef]
25. Bar-Shalom, Y. On the track-to-track correlation problem. *IEEE Trans. Autom. Control* **1981**, *26*, 571–572. [CrossRef]
26. Chang, C.; Youens, L. Measurement correlation for multiple sensor tracking in a dense target environment. *IEEE Trans. Autom. Control* **1982**, *27*, 1250–1252. [CrossRef]
27. Zagoruyko, S.; Komodakis, N. Learning to compare image patches via convolutional neural networks. In Proceedings of the IEEE Conference on Computer Vision and Pattern Recognition (CVPR), Boston, MA, USA, 7–12 June 2015; pp. 4353–4361.
28. de Oliveira, I.O.; Fonseca, K.V.O.; Minetto, R.A. A Two-Stream Siamese Neural Network for Vehicle Re-Identification by Using Non-Overlapping Cameras. In Proceedings of the IEEE International Conference on Image Processing (ICIP), Taipei, Taiwan, 22–25 September 2019; pp. 669–673.
29. Li, Q.; Zhu, J.; Cao, R. Relative geometry-aware Siamese neural network for 6DOF camera relocalization. *Neurocomputing* **2021**, *426*, 134–146. [CrossRef]
30. Pang, H.; Xuan, Q.; Xie, M.; Liu, C.; Li, Z. Research on Target Tracking Algorithm Based on Siamese Neural Network. *Mob. Inf. Syst.* **2021**, *2021*, 6645629. [CrossRef]

31. Qi, L.; He, Y.; Dong, K.; Liu, J. Multi-radar anti-bias track association based on the reference topology feature. *Iet Radar Sonar Navig.* **2018**, *12*, 366–372. [CrossRef]

32. Yue, S.;Yue, W.; Shu, W.; Xiu, S. Fuzzy Data Association based on Target Topology of Reference. *J. Natl. Univ. Def. Technol.* **2006**, *28*, 105–109.

33. Ze, W.; Shu, R.; Xi, L. Topology Sequence Based Track Correlation Algorithm. *Acta Aeronaut. Astronaut. Sin.* **2009**, *30*, 1937–1942.

34. Yu, Z.; Guo, W.; Cheng, G.; Lei, C. Gray Track Correlation Algorithm Based on Topology Sequence Method. *Electron. Opt. Control* **2013**, *20*, 1–5.

35. Wu, H.; Li, L.; Zhang, K. Track Association Method Based on Target Mutual-Support of Topology. In Proceedings of the 2020 IEEE 9th Joint International Information Technology and Artificial Intelligence Conference (ITAIC), Chongqing, China, 11–13 December 2020; Volume 9, pp. 2078–2082.

36. Hao, Z.; Chula, S. Algorithm of Multi-feature Track Association Based on Topology. *Command. Inf. Syst. Technol.* **2020**, *11*, 83–88.

37. Sönmez, H.H.; Hocaoğlu, A.K. Asynchronous track-to-track association algorithm based on reference topology feature. *Signal Image Video Process.* **2021**, *16*, 789–796. [CrossRef]

38. Li, X.; Wu, L.; Niu, Y.; Jia, S.; Lin, B. Topological Similarity-Based Multi-Target Correlation Localization for Aerial-Ground Systems. *Guid. Navig. Control* **2021**, *1*, 2150016. [CrossRef]

39. You, S.; Yao, H.; Xu, C. Multi-Object Tracking with Spatial-Temporal Topology-based Detector. *IEEE Trans. Circuits Syst. Video Technol.* **2015**, *14*, 12. [CrossRef]

40. Oliveira, B.D.A.; Pereira, L.G.R.; Bresolin, T. A review of deep learning algorithms for computer vision systems in livestock. *Livest. Sci.* **2021**, *253*, 104700. [CrossRef]

41. Jiang, P.; Ergu, D.; Liu, F.; Cai, Y.; Ma, B. A Review of Yolo Algorithm Developments. *Procedia Comput. Sci.* **2022**, *199*, 1066–1073. [CrossRef]

42. Liu, L.; Ouyang, W.; Wang, X.; Fieguth, P.; Chen, J.; Liu, X.; Pietikäinen, M. Deep Learning for Generic Object Detection: A Survey. *Int. J. Comput. Vis.* **2020**, *128*, 261–318. [CrossRef]

43. Zou, Z.; Shi, Z.; Guo, Y.; Ye, J. Object Detection in 20 Years: A Survey. *arXiv* **2019**, arXiv:1905.05055.

44. Ultralytics/yolov5. Available online: https://github.com/ultralytics/yolov5 (accessed on 31 March 2022).

45. Hartley, R.; Zisserman, R. 2D Projective Geometry and Transformations. In *Multi View Geometry in Computer Vision*; Machinery Industry Press: Beijing, China, 2020; pp. 28–33.

46. De Berg, M.T.; Van Kreveld, M.; Overmars, M.; Schwarzkopf, O. Delaunay Triangulation: Height Interpolation. In *Computational Geometry: Algorithms and Applications*, 3rd ed.; Springer: Dordrecht, The Netherlands, 2008; pp. 241–264.

47. Zhe, Y.; Chong, H.; Chen, L.; Min, C. Data Association Based on Target Topology. *J. Syst. Simul.* **2008**, *20*, 2357–2360.

48. Chopra, S.; Hadsell, R.; LeCun, Y. Learning a Similarity Metric Discriminatively, with Application to Face Verification. In Proceedings of the IEEE Computer Society Conference on Computer Vision and Pattern Recognition(CVPR), San Diego, CA, USA, 20–25 June 2005; pp. 539–546.

49. Bubbliiiing, Bubbliiiing/Siamese-Keras. Available online: https://github.com/bubbliiiing/Siamese-keras (accessed on 10 March 2022).

*drones*

*Technical Note*

# The Development of a Visual Tracking System for a Drone to Follow an Omnidirectional Mobile Robot

Jie-Tong Zou * and Xiang-Yin Dai

Department of Aeronautical Engineering, National Formosa University, Yunlin County 632301, Taiwan; 40530149@gm.nfu.edu.tw
* Correspondence: scott@nfu.edu.tw; Tel.: +886-5-6315556

**Abstract:** This research aims to develop a visual tracking system for a UAV which guides a drone to track a mobile robot and accurately land on it when it stops moving. Two LEDs with different colors were installed on the bottom of the drone. The visual tracking system on the mobile robot can detect the heading angle and the distance between the drone and mobile robot. The heading angle and flight velocity in the pitch and roll direction of the drone were modified by PID control, so that the flying speed and angle are more accurate, and the drone can land quickly. The PID tuning parameters were also adjusted according to the height of the drone. The embedded system on the mobile robot, which is equipped with Linux Ubuntu and processes images with OpenCV, can send the control command (SDK 2.0) to the Tello EDU drone through WIFI with UDP Protocol. The drone can auto-track the mobile robot. After the mobile robot stops, the drone can land on the top of the mobile robot. From the experimental results, the drone can take off from the top of the mobile robot, visually track the mobile robot, and finally land on the top of the mobile robot accurately.

**Keywords:** visual tracking system; embedded system; drone; omnidirectional mobile robot

**Citation:** Zou, J.-T.; Dai, X.-Y. The Development of a Visual Tracking System for a Drone to Follow an Omnidirectional Mobile Robot. *Drones* **2022**, *6*, 113. https://doi.org/10.3390/drones6050113

Academic Editors: Daobo Wang and Zain Anwar Ali

Received: 2 April 2022
Accepted: 25 April 2022
Published: 29 April 2022

**Publisher's Note:** MDPI stays neutral with regard to jurisdictional claims in published maps and institutional affiliations.

## 1. Introduction

In recent years the drone industry has seen a boom, and drones are widely applied because they are cheap, light, and safe. A drone positions itself with Lidar, GPS, or an optical flow sensor so that it can fly with autonomy and stability. During the 1980s, research on robotic image recognition began with the rapid development of computer hardware. Then, in the 1990s, with faster computers and advanced cameras, drones were equipped with image recognition. For example, helicopters equipped with vision-based tracking technology are already in use, as studied in [1]. Since GPS signals cannot be received in indoor environments, the positioning method is mostly based on image-based optical flow positioning. In [2], the Lucas–Kanade (LK) algorithm for optical flow localization combines it with the drone tracking of a particular color to realize the drone localization and automatic flight indoors. In [3–5], the image recognition algorithm is designed to recognize ground markers using the camera mounted on the bottom of the drone to achieve automatic and precise landing of the drone. In [6], the developed algorithm was able to detect and track an object with a certain shape on AR. Drone quadcopter, which could follow the line, was able to predict the turn and also to make a turn on the corners. In [7], to solve the problem with complex structures and low resource utilization of a traditional target tracking UAV system based on visual guidance, this paper discusses an implementation method of a quadrotor UAV target tracking system based on OpenMV. In [8], a small drone is taken as an applicable platform, and relevant theoretical research and comprehensive experiments are carried out around monocular vision feature point extractions and matching, UAV target tracking strategies, etc. Finally, a target tracking system will be built in the future to realize real-time face tracking.

This research aims to enable a drone to track a mobile robot only with optical flow sensor and image processes, without the help of GPS, and then to land on the robot when it stops moving.

## 2. Architecture of Visual Tracking System

### 2.1. Drone

The drone used in the experiment is Tello EDU, as shown in Figure 1a, which positions itself with an optical flow sensor and can be controlled by SDK commands [9]. Red and blue LED lights are installed on the bottom of the drone, as in Figure 1b, so that the camera on the mobile robot can detect where the drone is.

(a)

(b)

**Figure 1.** Tello EDU and two LEDs with different colors were installed on the bottom of the drone. (**a**) Top view; (**b**) Bottom view.

### 2.2. Omnidirectional Mobile Robot

Many wheeled mobile robots are equipped with two differential driving wheels. Since these robots possess two degrees-of-freedom (DOFs), they can rotate about any point, but cannot perform holonomic motion including sideways motion [10]. To increase the mobility of this mobile robot, three omnidirectional wheels driven by three DC servo motors are assembled on the robot platform (see Figure 2). The omnidirectional mobile robot can move in an arbitrary direction without changing the direction of the wheels.

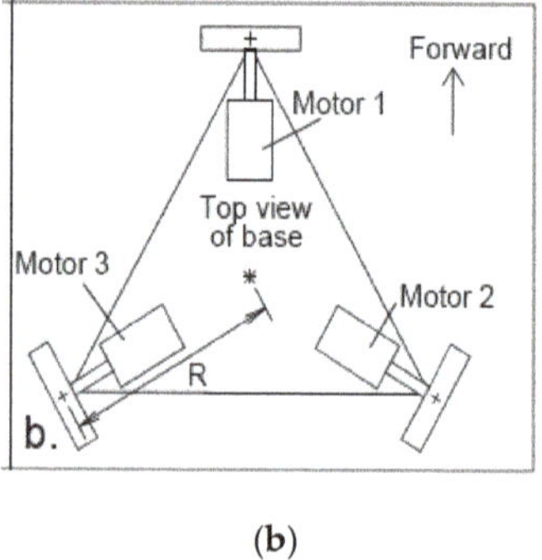


(a)

(b)

**Figure 2.** (**a**) Structure of omnidirectional wheel; (**b**) motor layout of robot platform.

The three-wheeled omnidirectional mobile robots are capable of achieving three DOF motions by driving three independent actuators [11,12], but they may have stability problems due to the triangular contact area with the ground, especially when traveling on a ramp with a high center of gravity, owing to the payload they carry.

Figure 2a is the structure of an omnidirectional wheel, and Figure 2b is the motor layout of the robot platform. The relationship of motor speed and robot moving speed is shown as:

$$V_1 = \omega_1 r = V_x + \omega_p R$$
$$V_2 = \omega_2 r = -0.5V_x + 0.867V_y + \omega_p R \tag{1}$$
$$V_3 = \omega_3 r = -0.5V_x - 0.867V_y + \omega_p R$$

where:

$V_i$ = velocity of wheel $i$;

$\omega_i$ = rotation speed of motor $i$;

$\omega_p$ = rotation speed of robot;

$r$ = radius of wheel;

$R$ = distance from wheel to the center of the platform.

Hardware of the proposed system is shown in Figure 3. The mobile robot is adapted from an omnidirectional robot and reads remote control signals with Arduino Due. A web camera is installed on top of the mobile robot to detect the red and blue LEDs on the bottom of the drone, as in Figure 4a,b. A single board computer, Up Board, is installed inside the robot, as shown in Figure 4c. Up Board is equipped with Linux Ubuntu and processes images with OpenCV. It can calculate the heading angle and distance between the mobile robot and the drone, and then send control commands to the drone through UDP Protocol with a Wi-Fi module.

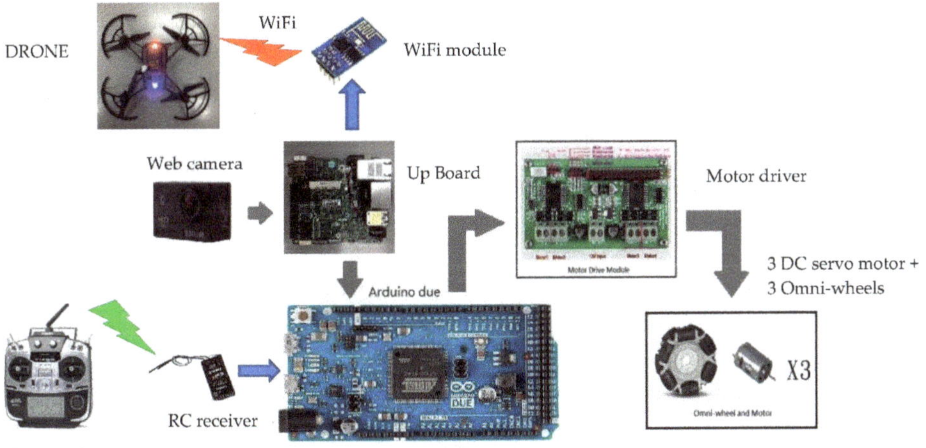


**Figure 3.** Hardware of the proposed system.

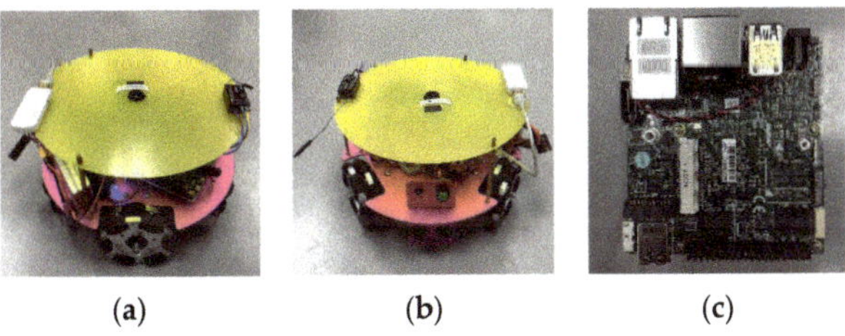

(a)     (b)     (c)

**Figure 4.** (a,b) Omnidirectional mobile robot installed with camera; (c) the embedded system (Up Board).

### 3. Positioning Drone through Image Processing

Usually, objects in images have distinct colors (hues) and luminosities, so that these features can be used to separate different areas of the image. In the RGB representation the hue and the luminosity are expressed as a linear combination of the R, G, B channels, whereas they correspond to single channels of the HSV image (the Hue and the Value channels).

The code for image processing is programmed with Python and cited from the open-source library OpenCV. First, images with RGB color space are converted to those with HSV color space so that a color can be easily presented with numbers, as the conversion algorithm in Equations (2)–(4). Suppose R, G, and B are the value of red, green, and blue in a color, and the value is a real number between 0 and 1. Suppose max is the biggest number among R, G, and B, and suppose min is the smallest number among R, G, and B.

$$
h = \begin{cases}
0^{\circ} & , \text{ if } \max = \min \\
60^{\circ} \times \frac{G-B}{\max-\min} + 0^{\circ}, & \text{ if } \max = R \text{ and } G \geq B \\
60^{\circ} \times \frac{G-B}{\max-\min} + 360^{\circ}, & \text{ if } \max = R \text{ and } G < B. \\
60^{\circ} \times \frac{B-R}{\max-\min} + 120^{\circ}, & \text{ if } \max = G \\
60^{\circ} \times \frac{R-G}{\max-\min} + 240^{\circ}, & \text{ if } \max = B
\end{cases}
\tag{2}
$$

$$
s = \begin{cases}
0, & \text{ if } \max = 0 \\
\frac{\max-\min}{\max} = 1 - \frac{\min}{\max}, & \text{ otherwise}
\end{cases}
\tag{3}
$$

$$
v = \max \tag{4}
$$

At different altitudes, the LEDs on the drone would manifest different levels of brightness and chroma, so the threshold value cannot be fixed, as in study [13]. A self-tuning threshold is applied to make linear adjustments according to the height of the drone. Images are converted to white and black binary images through image thresholding to filter out red and blue LEDs, as in Figure 5a,b. Images are also put through medium filter, erosion, and dilation to filter noise, as in Figure 5c, even if there is still noise, as in Figure 6a–c. For this situation, we designed a robust enough image recognition algorithm. By utilizing the distance proximity of red and blue LEDs, the red and blue LEDs on the bottom of the drone are accurately identified under noise interference. The flow chart of the algorithm is shown in Figure 7. First, the binary images of the red and blue LEDs are dilated so that the areas of the red and blue LEDs overlap, and then the two images are added (see Figure 8a). After dilation, the binary images of the red and blue LEDs are then AND operated to produce the overlapping part of the red and blue LEDs after dilation (see Figure 8b). Calculating the contours and image moments of Figure 8a,b, the center point of the contours is calculated from the image moments (see Figure 8c,d). Let all contours in the added image after dilation be $M_i$. Let the contour of the overlapping part after AND operation be S, and the center of this contour be $S_{center}$. If there is noise in the environment, there will be multiple $M_i$. So we check whether $S_{center}$ is inside $M_i$ to confirm which $M_i$ is the contour formed by adding up the binary images of red and blue LEDs after dilatation. This contour is called $M_{LED}$. Based on the previously derived $M_{LED}$, find which two center points are inside the $M_{LED}$, then the correct center point of the red and blue LEDs can be determined (see Figure 8e).

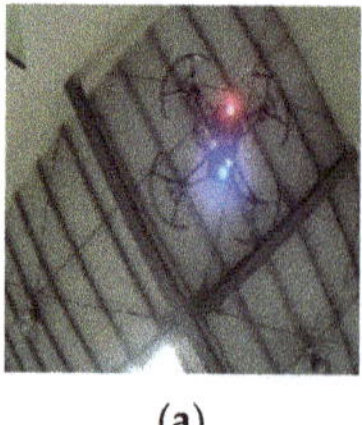
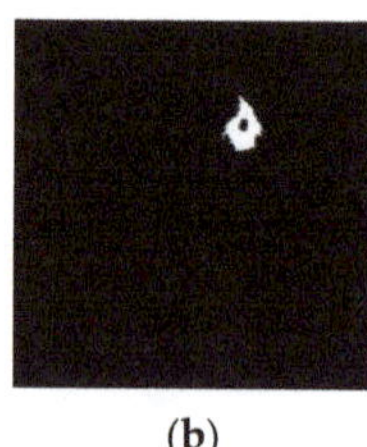
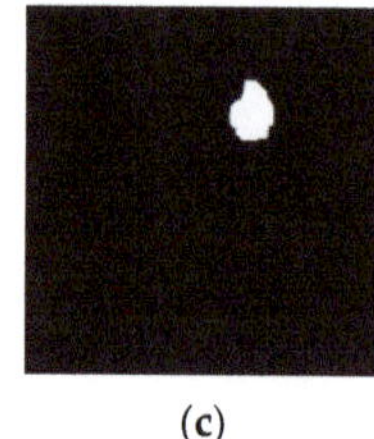

(a)   (b)   (c)

**Figure 5.** (**a**) Original image; (**b**) red LED after thresholding; (**c**) red LED after median filter, erosion, and dilation.

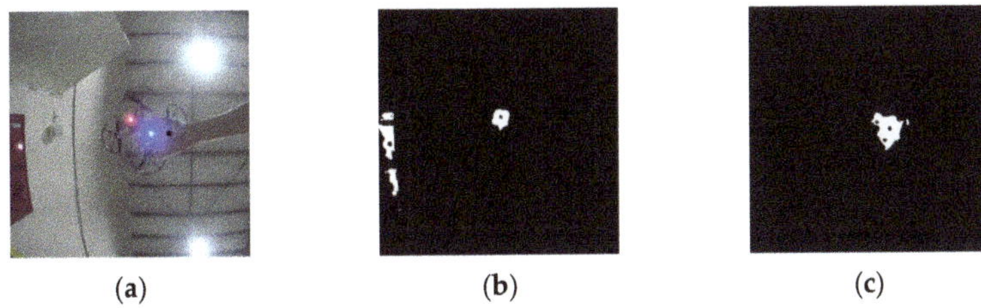



**Figure 6.** (**a**) Original image; (**b**,**c**) images of red and blue LEDs after thresholding, median filter, erosion, and dilation but with noise.

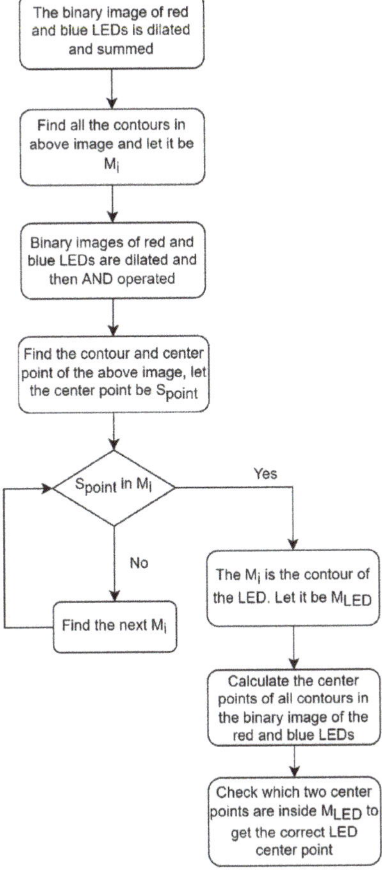



**Figure 7.** Flow chart of the algorithm to recognize red and blue LEDs.

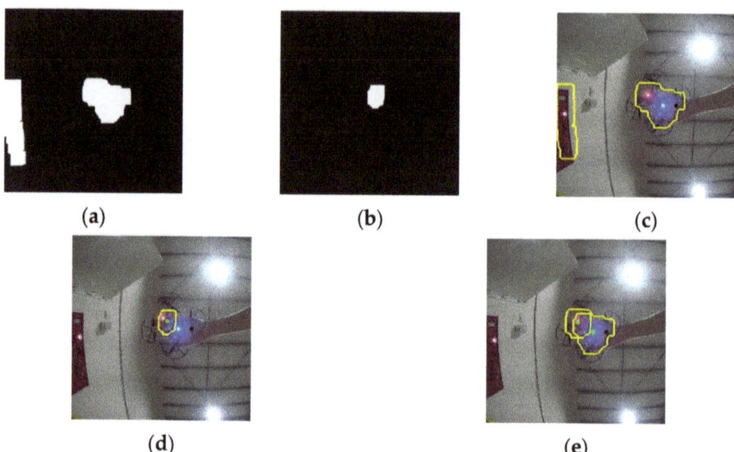

**Figure 8.** (**a**) The binary images of the red and blue LEDs are dilated and added together; (**b**) the binary images of the red and blue LEDs are dilated and operated to obtain the overlap; (**c**) all contours in Figure 8a; (**d**) contours and center points in Figure 8b (marked in green); (**e**) the correct contours and center points of the red and blue LEDs.

There are two coordinate systems in Figure 9: $(X_r, Y_r)$ is the coordinate system of the camera on the mobile robot, and $(X_d, Y_d)$ is the moving coordinate system of the drone. After confirming the contour of red LEDs and blue LEDs, we can figure out the "Image Moments" of red and blue LEDs, which can be used to calculate the coordinate of the central point, denoted as $(x_1, y_1)$ and $(x_2, y_2)$. With it, the distance between the two LEDs, denoted as $d_1$, as in Figure 9, can be calculated (refer to Equation (5)), and that distance $d_1$ can be used to judge the height of the drone. In addition, the central point of the line between the two LEDs represents the central point of the drone, denoted as $(x_c, y_c)$ (refer to Equation (6)). By calculating the distance between the central point $(x_c, y_c)$ of the drone and the central point $(O)$ of the camera, denoted as d (refer to Equation (7)), the process variable of PID control can be worked out. With the line through two LEDs as the axis, the rotating angle of two LEDs, which is also the heading angle of the drone, denoted as $\varphi$, can be calculated. The angle between $d$ and $X_r$-axis, denoted as $\theta$, can serve as the components for the pitch and roll of the drone.

$$d_1 = \sqrt{(x_1 - x_2)^2 + (y_1 - y_2)^2} \tag{5}$$

$$x_c = \frac{x_1 + x_2}{2}, \ y_c = \frac{y_1 + y_2}{2} \tag{6}$$

$$d = \sqrt{(x_c - 320)^2 + (y_c - 240)^2} \tag{7}$$

where:
$(X_r, Y_r)$: the coordinate system of the camera on mobile robot;
$O$: the origin of the coordinate system $(X_r, Y_r)$, and $O$ is also the center point (320,240) of the camera with $640 \times 480$ resolution;
$(X_d, Y_d)$: the moving coordinate system of the drone;
$(x_1, y_1)$: the central point of the red LED;
$(x_2, y_2)$: the central point of the blue LED;
$d_1$ : the distance between the two LEDs;
$(x_c, y_c)$: the central point of the drone;
$d$: the distance between the central point $(x_c, y_c)$ of the drone and that of camera $(O)$;
$\varphi$ : the heading angle of the drone;

$\theta$ : the angle between $d$ and $X_r$-axis.

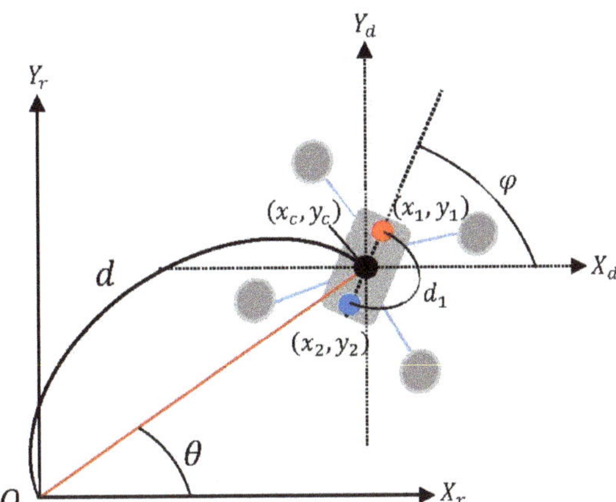

**Figure 9.** Diagram for calculating distance and angle of drone.

## 4. Guidance Law

Through image processing, we can obtain $d$, $d_1$, $\varphi$, and $\theta$, and a guidance law can be developed. The guidance law can direct a drone to track a mobile robot. Tello EDU can follow SDK 2.0 commands and perform various kinds of simple and quick actions. For example, the "rc" command in SDK 2.0 command is used, as in Figure 10, to control the four moving directions of the drone: pitch, roll, yaw, and throttle (height).

| | Set remote controller control via four channels. |
|---|---|
| rc a b c d | "a" = left/right (-100-100)<br>"b" = forward/backward (-100-100)<br>"c" = up/down (-100-100)<br>"d" = yaw (-100-100) |

**Figure 10.** The description of rc command with SDK 2.0 of Tello EDU [9].

### 4.1. PID Control

PID (proportional-integral-differential) control is the most frequently used industrial control algorithm because it is effective, widely applicable, and simple to operate. It can also be applied on drones [14]. A popular method for tuning PID controllers is the Ziegler–Nichols method [15]. This method starts by zeroing the integral gain ($K_I$) and differential gain ($K_D$) and then raising the proportional gain ($K_P$) until the system is unstable. The value of $K_P$ at the point of instability is called $K'_P$; the oscillation period is $T_C$. The $P$, $I$, and $D$ gains are set as Equations (8)–(10).

$$K_P = 0.6K'_P \tag{8}$$

$$K_I = \frac{2}{T_C} \tag{9}$$

$$K_D = \frac{T_C}{8} \tag{10}$$

This research adjusts the flight velocity in pitch and roll direction with a PID control algorithm. With $u_p(t)$ as the output, the distance between the drone and mobile robot (d) is the process variable, the setpoint (target value) is zero, and process variable minus setpoint is the error value, represented as $e_p(t)$. PID control includes three kinds: proportional, integral, and derivative. Proportional control considers current error, and the error value will be multiplied by positive constant $K_{P_p}$. When the value of $K_{P_p}$ increases, the response speed of the system will become faster. However, when the value of $K_{P_p}$ becomes too large, fluctuation of the process variable will happen. Integral control considers that the past error and the sum of the past error value multiplied by positive constant $K_{i_p}$ can be used to eliminate the steady state error. Derivative control considers future error, calculating the first order derivative of error, which will be multiplied by positive constant $K_{d_p}$, thereby predicting the possibilities of error changes and overcoming the delay of the controlled subject, as in Equation (11). When a drone is tracking a mobile robot, the drone is required to react quickly, so higher $K_{P_p}$ and $K_{d_p}$ values are needed. Firstly, the Ziegler–Nichols method was used to tune the PID controller. After some tuning form experimental results, I set $K_{P_p} = 0.14$, $K_{i_p} = 0.15$, $K_{d_p} = 0.13$. However, this will make a drone extremely sensitive to changes in the moving speed of the mobile robot and cause it to overreact to a very low error value, which will lead to a not so smooth landing. To solve this problem, the $K_{P_p}$ and $K_{d_p}$ values are adjusted according to the height of the drone. If the height of the drone is lower than 60 cm, set $K_{P_p} = 0.075$ and $K_{d_p} = 0.05$, so that the drone does not overreact. Apart from this problem, it takes a while for the drone to fly above the mobile robot when the mobile robot stops moving, because the guidance law cannot predict the flying direction of the drone very accurately. The heading angle of the drone is modified by PID control, as in Equation (12), so that the flying angle is more accurate, and the drone can land quickly.

$$u_p(t) = K_{P_p}e_p(t) + K_{i_p}\int_0^t e_p(\tau)d\tau + K_{d_p}\frac{de_p(t)}{dt} \tag{11}$$

$$u_h(t) = K_{P_h}e_h(t) + K_{i_h}\int_0^t e_h(\tau)d\tau + K_{d_h}\frac{de_h(t)}{dt} \tag{12}$$

*4.2. Logic of Guidance Law*

As shown in Figure 11, the angle between d and $X_r$-axis ($\theta$) is used to determine which zone (partition 1) in the ($X_r$, $Y_r$) coordinate system the drone is located in. The heading angle of the drone ($\varphi$) is used to determine which zone (partition 2) in the ($X_d$, $Y_d$) coordinate system the nose of the drone is facing. For example, in Figure 11, $\theta$ is in partition 1a ($0 < \theta < 90°$), and $\varphi$ is in partition 2b ($45° < \varphi < 135°$).

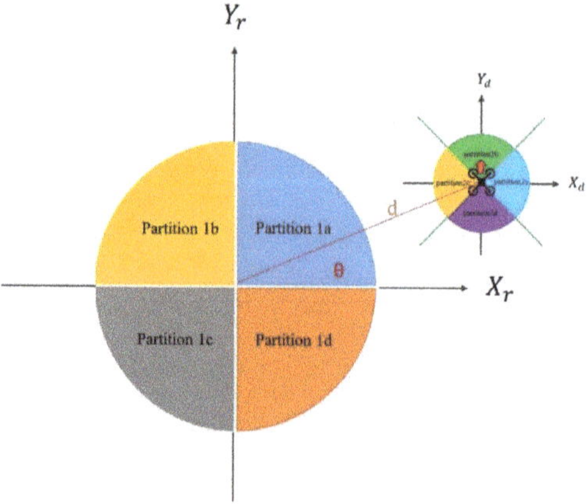


**Figure 11.** $\theta$ is used to determine which zone (partition 1) in the ($X_r$, $Y_r$) coordinate system the drone is located in. $\varphi$ is used to determine which zone (partition 2) in the ($X_d$, $Y_d$) coordinate system the nose of the drone is facing.

The flow chart of the guidance law is shown in Figure 12. According to the angle ($\theta$) and the heading angle $\varphi$, gained through image processing, the component for the pitch and roll movement of the drone can be figured out from the guidance law in Figure 12. The component for the pitch and roll multiplied by the output of PID control is the flight velocity in the pitch and roll direction.

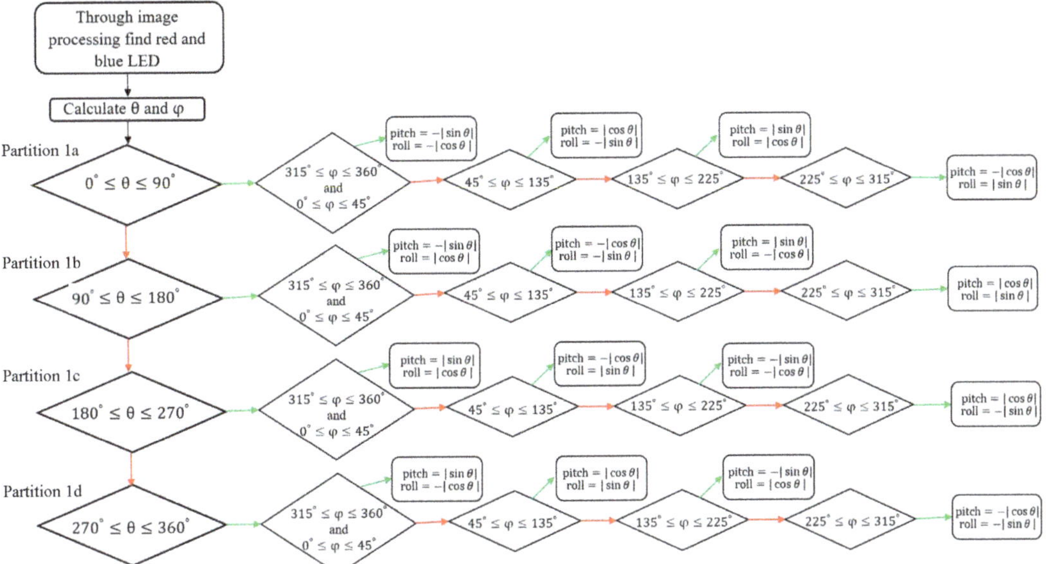

**Figure 12.** Flow chart of the guidance law.

## 5. Experimental Results

### 5.1. Monitoring of Image Processing

This research processes images using a UP Board, which processes 30 images per second. The image processing can be monitored through Windows remote desktop connection, as in Figure 13.

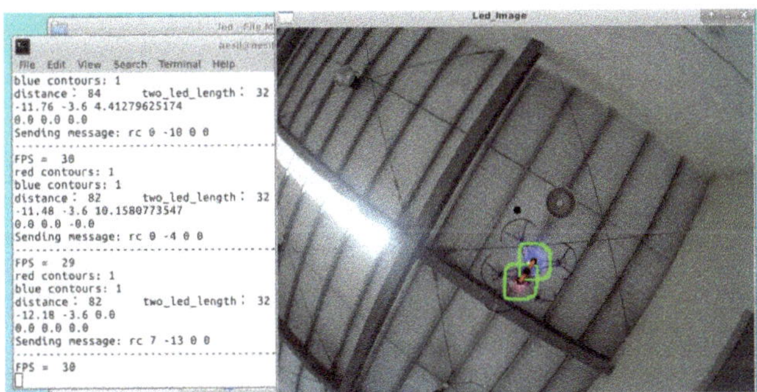

**Figure 13.** Monitor image processing, red and blue LEDs indicated by the green circle.

### 5.2. Experimental Results of Visual Tracking

With appropriate PID parameters, the mobile robot can move in any direction with radio control. The drone can accurately track the mobile robot and land on top of it when it stops moving. The experimental results of visual tracking are shown in Figure 14, and the experimental video can be seen on YouTube (https://www.youtube.com/watch?v=kRorTz26XSg) (accessed on 3 January 2020).

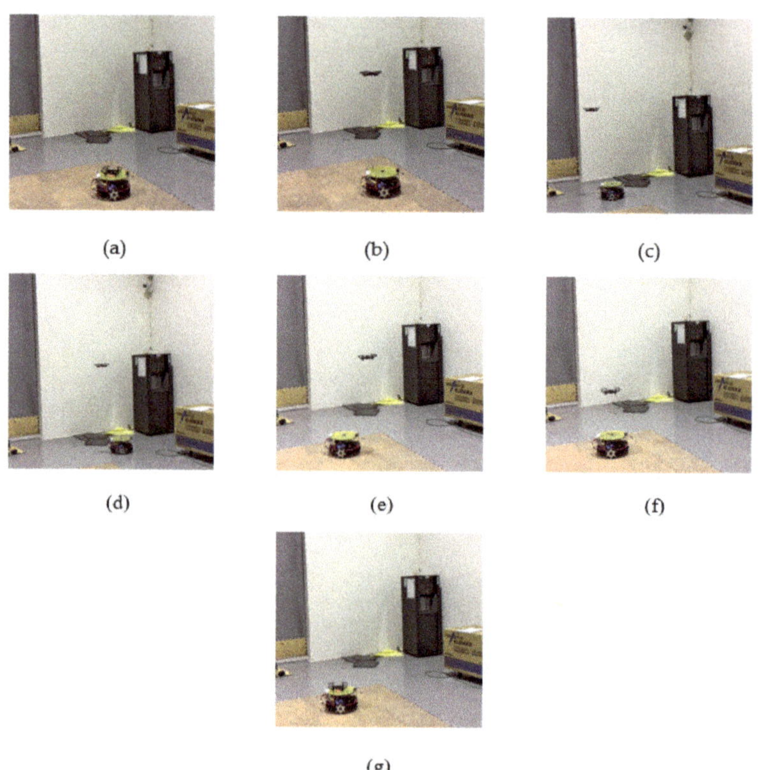

**Figure 14.** (**a,b**) Takeoff from the top of the mobile robot; (**c,d**) tracking the mobile robot; (**e–g**) when the mobile robot stops moving, the drone can land on the top of the mobile robot accurately (https://www.youtube.com/watch?v=kRorTz26XSg) (accessed on 3 January 2020).

1.  Figure 14a,b: The drone takes off from the top of the mobile robot.
2.  Figure 14c,d: The drone visually tracks the mobile robot.
3.  Figure 14e–g: When the mobile robot stops moving, the drone can land on the top of the mobile robot accurately.

## 6. Conclusions

This research developed a system that enables a drone to track a mobile robot via image processing and to land on top of the mobile robot when it stops moving. The web camera on the mobile robot can capture blue and red LEDs which can determine the heading angle and the distance between the drone and mobile robot. The heading angle and flight velocity in the pitch and roll direction of the drone are modified by PID control, so that the flying speed and angle are more accurate, and the drone can land quickly. Firstly, the Ziegler–Nichols method was used to manually tune the PID controller initially, and make fine PID tuning with experimental results. The PID tuning parameters were also adjusted according to the height of the drone.

The embedded system (Up Board) on the mobile robot, which is equipped with Linux Ubuntu and processes images with OpenCV, can send the control command (SDK 2.0) to the Tello EDU drone through WIFI with UDP Protocol. The guidance law can direct the drone to track the mobile robot. Finally, the drone can land on the top of the mobile robot when it stops moving.

The proposed system can also guide drones to land on a wireless charger via image processing and can be applied to the auto-tracking of certain mobile objects. In the future, the accuracy with which a drone recognizes a certain object in a complicated background should be heightened, so that the image recognition technology can be more reliable.

**Author Contributions:** Conceptualization, J.-T.Z.; methodology, J.-T.Z. and X.-Y.D.; software, X.-Y.D.; validation, X.-Y.D.; formal analysis, J.-T.Z. and X.-Y.D.; investigation, X.-Y.D.; resources, J.-T.Z.; data curation, X.-Y.D.; writing—original draft preparation, X.-Y.D.; writing—review and editing, J.-T.Z.; visualization, X.-Y.D.; supervision, J.-T.Z.; project administration, J.-T.Z.; funding acquisition, J.-T.Z. All authors have read and agreed to the published version of the manuscript.

**Funding:** This research was funded by National Formosa University.

**Institutional Review Board Statement:** Not applicable.

**Informed Consent Statement:** Not applicable.

**Data Availability Statement:** Not applicable.

**Conflicts of Interest:** The authors declare no conflict of interest.

## References

1. Choi, J.H.; Lee, W.-S.; Bang, H. Helicopter Guidance for Vision-based Tracking and Landing on a Moving Ground Target. In Proceedings of the 11th International Conference on Control, Automation and Systems, Gyeonggi-do, Korea, 26–29 October 2011; pp. 867–872.
2. Granillo, O.D.M.; Beltrán, Z.Z. Real-Time Drone (UAV) Trajectory Generation and Tracking by Optical Flow. In Proceedings of the 2018 International Conference on Mechatronics, Electronics and Automotive Engineering, Cuernavaca, Mexico, 26–29 November 2018; pp. 38–43.
3. Shim, T.; Bang, H. Autonomous Landing of UAV Using Vision Based Approach and PID Controller Based Outer Loop. In Proceedings of the 18th International Conference on Control, Automation and Systems, PyeongChang, Korea, 17–20 October 2018; pp. 876–879.
4. Tanaka, H.; Matsumoto, Y. Autonomous Drone Guidance and Landing System Using AR/high-accuracy Hybrid Markers. In Proceedings of the 2019 IEEE 8th Global Conference on Consumer Electronics, Osaka, Japan, 15–18 October 2019; pp. 598–599.
5. Liu, R.; Yi, J.; Zhang, Y.; Zhou, B.; Zheng, W.; Wu, H.; Cao, S.; Mu, J. Vision-guided autonomous landing of multirotor UAV on fixed landing marker. In Proceedings of the 2020 IEEE International Conference on Artificial Intelligence and Computer Applications, Dalian, China, 27–29 June 2020; pp. 455–458.
6. Boudjit, K.; Larbes, C. Detection and implementation autonomous target tracking with a Quadrotor AR.Drone. In Proceedings of the 12th International Conference on Informatics in Control, Automation and Robotics, Colmar, France, 21–23 July 2015.
7. Shao, Y.; Tang, X.; Chu, H.; Mei, Y.; Chang, Z.; Zhang, X. Research on Target Tracking System of Quadrotor UAV Based on Monocular Vision. In Proceedings of the 2019 Chinese Automation Congress, Hangzhou, China, 22–24 November 2019; pp. 4772–4775.
8. Sun, X.; Zhang, W. Implementation of Target Tracking System Based on Small Drone. In Proceedings of the 2019 IEEE 4th Advanced Information Technology, Electronic and Automation Control Conference, Chengdu, China, 20–22 December 2019; pp. 1863–1866.
9. RYZE. *Tello SDK 2.0 User Guide*, 1st ed.; RYZE: Shenzhen, China, 2018; p. 5.
10. Song, J.-B.; Byun, K.-S. Design and Control of an Omnidirectional Mobile Robot with SteerableOmnidirectional Wheels. In *Mobile Robots*; Moving Intelligence: Zaltbommel, The Netherlands, 2006; p. 576.
11. Carlisle, B. An Omnidirectional Mobile Robot. In *Development in Robotics*; Kempston: Bedford, UK, 1983; pp. 79–87.
12. Pin, F.; Killough, S. A New Family of Omnidirectional and Holonomic Wheeled Platforms for Mobile Robot. *IEEE Trans. Robot. Autom.* **1999**, *15*, 978–989. [CrossRef]
13. Liu, Y.; Jiang, N.; Wang, J.; Zhao, Y. Vision-based Moving Target Detection and Tracking Using a Quadrotor UAV. In Proceedings of the 11th World Congress on Intelligent Control and Automation, Shenyang, China, 29 June–4 July 2014; pp. 2358–2368.
14. Salih, A.L.; Moghavvemi, M.; Mohamed, H.A.F.; Geaid, K.S. Flight PID Controller Design for a UAV Quadrotor. *Sci. Res. Essays* **2010**, *5*, 3660–3667.
15. Ziegler, J.G.; Nichols, N.B. Optimum settings for automatic controllers. *Trans. ASME* **1942**, *64*, 759–768. [CrossRef]

*Article*

# A Multi-Colony Social Learning Approach for the Self-Organization of a Swarm of UAVs

Muhammad Shafiq [1], Zain Anwar Ali [1,*], Amber Israr [1], Eman H. Alkhammash [2] and Myriam Hadjouni [3]

[1] Electronic Engineering Department, Sir Syed University of Engineering & Technology, Karachi 75300, Pakistan; muhshafiq@ssuet.edu.pk (M.S.); aisrar@ssuet.edu.pk (A.I.)
[2] Department of Computer Science, College of Computers and Information Technology, Taif University, P.O. Box 11099, Taif 21944, Saudi Arabia; eman.kms@tu.edu.sa
[3] Department of Computer Sciences, College of Computer and Information Science, Princess Nourah Bint Abdulrahman University, P.O. Box 84428, Riyadh 11671, Saudi Arabia; mfhaojouni@pnu.edu.sa
* Correspondence: zaali@ssuet.edu.pk

**Abstract:** This research offers an improved method for the self-organization of a swarm of UAVs based on a social learning approach. To start, we use three different colonies and three best members i.e., unmanned aerial vehicles (UAVs) randomly placed in the colonies. This study uses max-min ant colony optimization (MMACO) in conjunction with social learning mechanism to plan the optimized path for an individual colony. Hereinafter, the multi-agent system (MAS) chooses the most optimal UAV as the leader of each colony and the remaining UAVs as agents, which helps to organize the randomly positioned UAVs into three different formations. Afterward, the algorithm synchronizes and connects the three colonies into a swarm and controls it using dynamic leader selection. The major contribution of this study is to hybridize two different approaches to produce a more optimized, efficient, and effective strategy. The results verify that the proposed algorithm completes the given objectives. This study also compares the designed method with the Non-Dominated Sorting Genetic Algorithm II (NSGA-II) to prove that our method offers better convergence and reaches the target using a shorter route than NSGA-II.

**Keywords:** social learning; ant colony optimization; multi-agent system

**Citation:** Shafiq, M.; Ali, Z.A.; Israr, A.; Alkhammash, E.H.; Hadjouni, M. A Multi-Colony Social Learning Approach for the Self-Organization of a Swarm of UAVs. *Drones* **2022**, *6*, 104. https://doi.org/10.3390/drones6050104

Academic Editor: Xiwang Dong

Received: 11 April 2022
Accepted: 20 April 2022
Published: 23 April 2022

**Publisher's Note:** MDPI stays neutral with regard to jurisdictional claims in published maps and institutional affiliations.

## 1. Introduction

In the last decade, research has been exponentially increasing in the domains of flight control, path planning, and obstacle avoidance of unmanned aerial vehicles (UAVs) [1–3]. The analyses get increasingly complex when dealing with multiple UAVs in different formations. The natural behaviors of birds, ants, and fishes have been proven to be significant in formulating successful bio-inspired algorithms for the formation control, route planning, and trajectory tracking of a swarm of multiple UAVs [4–6]. Some of the important algorithms inspired by nature include ant colony optimization [7], pigeon-inspired optimization [8], and particle swarm optimization [9].

The primary inspiration for this study is to utilize the knowledge obtained from studying the natural flocking and swarming activities of ants and use them for controlling UAVs. Researchers have used these nature-inspired algorithms for numerous purposes including cooperative path planning of multiple UAVs [10], distributed UAV flocking among obstacles [11], and forest fire fighting missions [12]. We also find multiple studies that hybridize a bio-inspired algorithm with another method to increase its efficiency [13–15].

There are many existing solutions regarding the problems of path planning and multi-UAV cooperation. One such research study [16] deals with the inspection of an oilfield using multiple UAVs while avoiding obstacles. The researchers achieve this using an improved version of Non-Dominated Sorting Genetic Algorithm (NSGA). Another existing solution to tackle a multi-objective optimization is addressed in reference [17]. In [17], the researchers use a hybrid of NSGA and local fruit fly optimization to solve interval multi-objective optimization problems. In reference [18], academics use a modified particle swarm optimization algorithm for the dynamic target tracking of multiple UAVs.

Our proposed method consists of many concepts, which are explained as follows:

Ant colony optimization (ACO) is an optimization technique used by ant colonies to find the shortest route to take to get to their food [19]. Using pheromones left behind from earlier ants, the ACO mimics ants looking for food. The path used by the most ants contains the most pheromones, which aids the next ant in choosing the shortest way [20]. The ACO technique is used in a variety of applications, such as the routing of autonomous vehicles and robots, which need to find the shortest path to a destination, and the design of computer algorithms, which need to find the optimal solution for a given problem.

Sometimes, however, the ACO is slow to converge and falls into the local optimum. To help solve these issues, researchers introduced a modified version of ACO called max-min ant colony optimization (MMACO). It operates by controlling the maximum and minimum amounts of pheromone that can be left on each possible trial [21]. The range of possible pheromone amounts on each possible route is limited to avoid stagnation in the search process.

In social animals, social learning plays a significant part in behavior learning. Social learning, as opposed to asocial (individual) learning, allows individuals to learn from the actions of others without experiencing the costs of individual trials and errors [22]. That is why this study incorporates social learning mechanisms into MMACO. Unlike traditional MMACO variations, which update ants based on past data, each ant in the proposed SL-MMACO learns from any better ants (called demonstrators) in the present swarm.

Some of the state-of-the-art work in the field of optimization algorithms include research [23] that proposes the use of social learning-based particle swarm optimization (SL-PSO) for integrated circuits manufacturing. The SL-PSO is used to increase the imaging performance in extreme ultraviolet lithography. Results showed that the errors were reduced significantly compared to conventional methods. Similarly, another recent study [24] uses improved ant colony optimization (IACO) for human gait recognition. The IACO is used to enhance the extracted features, which are then passed on to the classifier. Compared with current methods, the IACO technique in [24] is more accurate and takes less time to compute.

A multi-agent system (MAS) is a collection of agents that interact with each other and the environment to achieve a common goal. The main function of the MAS is to tackle issues that a single agent would find difficult to solve. To achieve its objective, another important function of the MAS is to be able to interact with each agent and respond accordingly. In MAS, each agent can determine its state and behavior based on the state and behavior of its neighbors [25]. MAS has multiple uses in fields such as robotics, computer vision, and transportation [26–28]. In most MAS scenarios, an external entity is required to direct the agents toward the destination [29].

A leader is an agent that can alter the states of the follower agents. A leader can be outside or inside the MAS or can even be virtual. A large swarm of UAVs can be controlled more efficiently by selecting fewer leaders than follower agents. For example, when we want to control a fleet of UAVs, we can select a few leader agents, let the remaining UAVs follow the leader, and use the leading UAVs to control the state of the system.

The major contributions of this study are as follows:

1. Designing a novel hybrid algorithm that combines MMACO with the social learning mechanism to enhance its performance.
2. Using the designed algorithm in tandem with the MAS and dynamic leader selection to connect the three colonies.
3. Achieving the self-organization of the three colonies and synchronizing them into one swarm.
4. Demonstrating the validity of the designed method by performing computer simulations that mimic real-life scenarios.

The paper is organized into seven sections. Section 1 presents the introduction and the literature review. In Section 2, we break down the problem into three scenarios and describe each one in detail. Section 3 provides the framework of the proposed solution. Section 4 defines the proposed method with each of its constituent parts discussed at length and offers the algorithm and the flowchart. In Section 5, we discuss the simulations and their outcomes. Section 6 presents the conclusion of the study.

## 2. Research Design

We broke our problem into three different scenarios to help make it easier. In the first scenario, the UAVs are in random positions and then using our proposed algorithm, they organize themselves into a formation. In the second scenario, we navigate the newly organized formations through some obstacles. In the last scenario, we combine the three formations into one swarm and then navigate it through the same environment. Below, we describe each scenario in detail:

Scenario 1:

Figure 1 presents the first scenario. In this scenario, there are three UAVs in three different colonies and the UAVs are placed at random positions within each colony. The environment contains different obstacles like mountains and rough terrain. The goal is to reach the target using the shortest route possible without colliding with other UAVs. The main objective here is to maintain the formation throughout the journey.

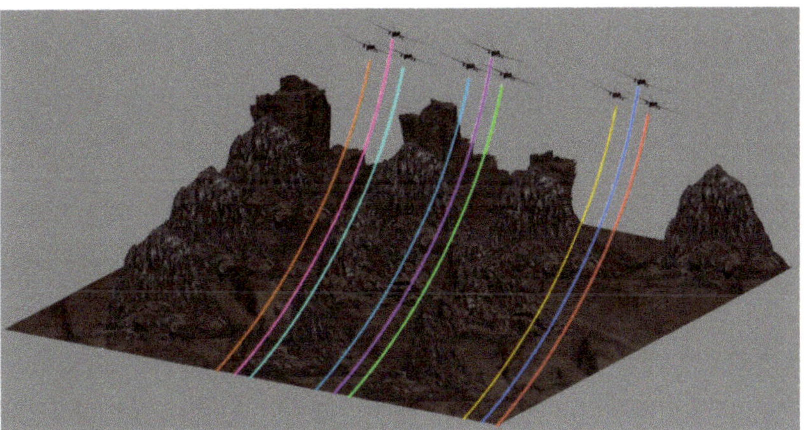

**Figure 1.** Illustration of the first scenario.

Scenario 2:

Figure 2 illustrates the second scenario. In this scenario, there are again three UAVs in three different colonies and the UAVs are placed at random positions within each colony. The environment is also the same. The goal is to reach the target using the shortest route possible without colliding with the obstacles or other UAVs. The main distinction between the first and the second scenario is that, in the first task, we only had to demonstrate the ability of the algorithm to maintain a formation. However, the second task also requires navigating through the obstacles without any collision.

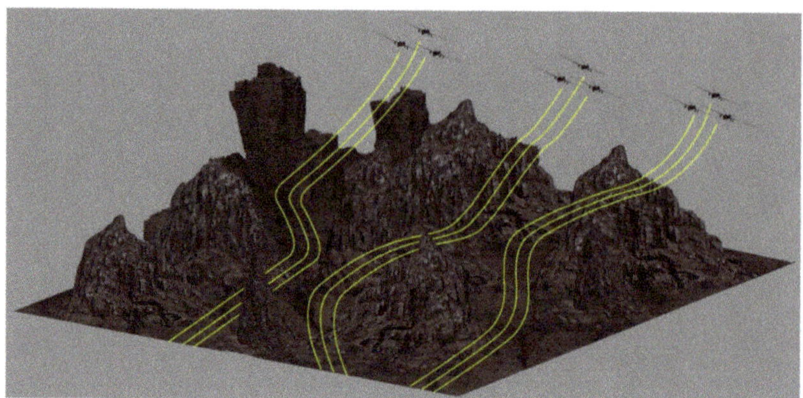

**Figure 2.** Illustration of the second scenario.

Scenario 3:

Figure 3 illustrates the third scenario. In this scenario, we pick up where the second scenario left off, i.e., the three colonies are now in the desired formations. The environment is the same as in the second scenario. The goal is to first synchronize the three colonies into one big swarm, and then, while maintaining the swarm, reach the target using the shortest route possible without colliding with the obstacles or other UAVs.

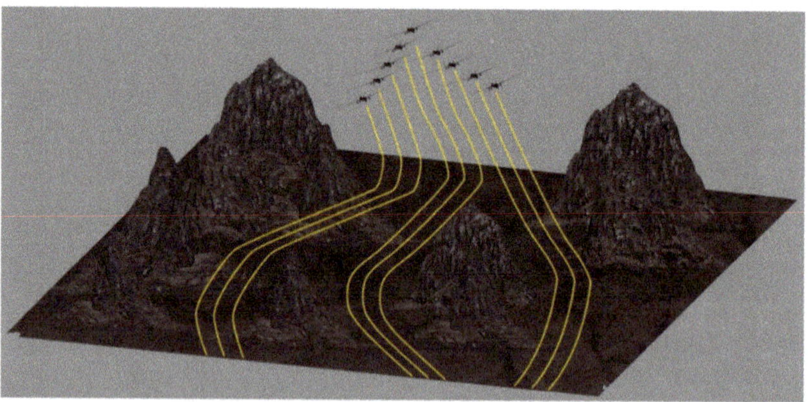

**Figure 3.** Illustration of the third scenario.

## 3. Solution Architecture

Figure 4 presents the framework of our proposed solution for the aforementioned problems. It is clear from the figure that, initially, the UAVs in the three different colonies are at random positions. Here, we apply the social learning-based max-min ant colony optimization (SL-MMACO) to each colony. SL-MMACO works by first finding the most optimal routes for each colony, and then the social learning mechanism sorts the ants from best to worst. Afterward, the multi-agent system (MAS) appoints the best ant as the leader and the remaining ants as agents. In the next step, we see that all the colonies are now synchronized to avoid any collision between the UAVs. Finally, we connect all three colonies into one big swarm and then select its leader dynamically according to the mission requirements.

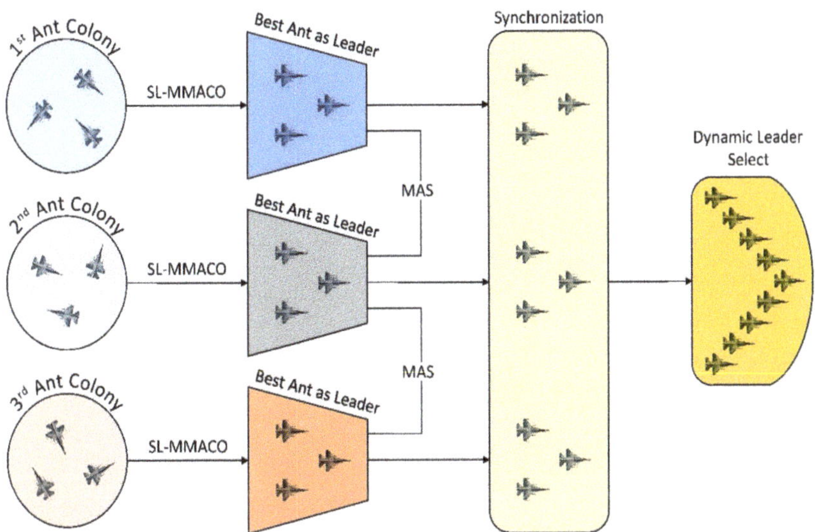

**Figure 4.** Solution architecture.

## 4. Proposed Algorithm

This section introduces the different concepts and theories that we used for the development of our proposed algorithm. In this research paper, we are using a graph-based approach.

### 4.1. Ant Colony Optimization

Here, we are using a graph-based approach. The concept of nodes, edges, and legs is important to understand path planning using ant colony optimization (ACO). Figure 5 presents the relationship between the nodes, edges, and legs. The ACO generates intermediary points between the initial and final positions. These intermediate points are called the nodes. An edge is a link between two nodes. For instance, edge (b,c) is the length from edge b to c, whereas a leg is generated whenever a UAV turns.

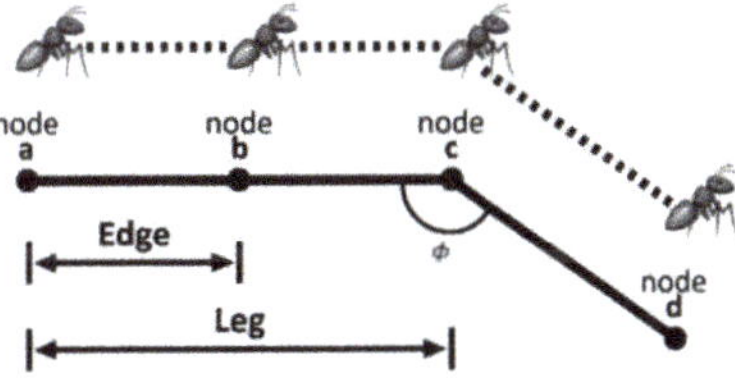

**Figure 5.** Relationship between nodes, edges, and legs.

Suppose that the $m$th ant is at node $i$ on time $t$, the probability of transition can be written as:

$$p_{ij}^{m}(t) = \frac{\tau_{ij}^{\alpha}\eta_{ij}^{\beta}}{\sum_{c \in allowed_i} \tau_{ic}^{\alpha}\eta_{ic}^{\beta}} \tag{1}$$

where the probability of transition from node $i$ to node $j$ of the $m$th ant is $p_{ij}^{m}(t)$, the pheromone on the edge $(i, j)$ is $\tau_{ij}(t)$, the transit feasibility from node $i$ to node $j$ is $\eta_{ij}(t)$, the set of nodes that are neighboring $i$ is $allowed_i$, the constant influencing the $\tau_{ij}(t)$ is $\alpha$, and the constant influencing the $\eta_{ij}(t)$ is $\beta$.

After the method begins, the starting pheromone rate varies according to the edges. The pheromone rate is then reset by each ant that generated the result, which starts the next cycle of the process. $\tau_{ij}(t)$ on the edge $(i, j)$ is:

$$\tau_{ij}(t+1) = (1 - \rho) \times \tau_{ij}(t) + \sum_{m=1}^{k} \Delta\tau_{ij}^{m}(t) \tag{2}$$

where the rate of pheromone evaporation is $\rho$ $(0 \leq \rho \leq 1)$, total ants are represented by $k$, and the pheromone rate of the edge $(i, j)$ is $\Delta\tau_{ij}^{m}(t)$. $\Delta\tau_{ij}^{m}(t)$ can be further defined as

$$\Delta\tau_{ij}^{m}(t) = \begin{cases} Q/L_m; & ant\ 'm'\ uses\ edges\ of\ (i,j) \\ 0; & otherwise \end{cases} \tag{3}$$

where the length of the route built by the $m$th ant is $L_m$ and $Q$ is the constant.

### 4.2. Max-Min Ant Colony Optimization

We need to improve the traditional method to ensure that the ACO converges quickly. MMACO delivers some remarkable results in this area by restricting the pheromones on each route. To understand MMACO, we must first examine the path's cost. The average cost of path $J_{a,k}(t)$ can be given as:

$$J_{a,k}(t) = \frac{1}{k} \sum_{m=1}^{k} J_{a,m}(t) \tag{4}$$

Note that the $m$th ant only updates the pheromone when the cost of path of the $m$th ant in the $t$th iteration fulfill $J_{a,k}(t) \geq J_{a,m}(t)$.

The MMACO updates the route using Equation (3) after every iteration. After each iteration, the algorithm determines the most optimum and least optimal paths. To improve the probability of discovering the global best route, it discards the least optimum route. As a result, Equation (3) may be revised as follows:

$$\Delta\tau_{ij}^{m}(t) = \begin{cases} Q/L_o; & route\ (i,j)\ refer\ to\ the\ optimal\ route \\ -Q/L_w; & route\ (i,j)\ refer\ to\ the\ worst\ route \\ 0; & otherwise \end{cases} \tag{5}$$

In the above equation, $L_O$ is the most optimal route and $L_w$ is the current iteration's worst route. The quantity of pheromone produced by MMACO is limited to specified values. This limitation aids in accelerating convergence and avoiding stagnation.

The algorithm restricts the pheromone on each route to a specified minimum and maximum value, denoted by $\tau_{min}$ and $\tau_{max}$, respectively. This can be represented mathematically as:

$$\tau_{ij}(t) = \begin{cases} \tau_{max}; & \tau_{i,j}(t) \geq \tau_{max} \\ \tau_{ij}(t); & \tau_{min} < \tau_{i,j}(t) < \tau_{max} \\ \tau_{min}; & \tau_{i,j}(t) \leq \tau_{min}(t) \end{cases} \tag{6}$$

### 4.3. Social Learning Mechanism

The conduct of a person to learn from their surroundings is referred to as social learning. You should learn not just from the top students in class, but also from students who are better than you. Most biological groups follow the same idea. Initially, for the map and compass process, locations and velocities of the ants are produced at random and are indicated as $X_i$ and $V_i$ $(i = 1, 2, \ldots, m)$. At the next iteration, the new location Xi and velocity Vi are calculated by the formula [30]:

$$X_a^{N_c} = X_a^{N_c-1} + V_a^{N_c} \tag{7}$$

$$V_a^{N_c} = V_a^{N_c-1} \times e^{-R.N_c} + rand \times c_1 \times (X_{mod} - X_a^{N_c-1}) \tag{8}$$

$$c_1 = 1 - \log\left(\frac{N_c}{m}\right) \tag{9}$$

where $R$ represents the map and compass factor, which is between 0 and 1, $N_c$ is the current number of iterations, $c_1$ the learning factor, $X_{mod}$ the demonstrator ant superior to the current ant, and $m$ is the total number of ants. Each follows the ant better than itself, and this is known as the learning behavior. Figure 6 illustrates the process for the selection of demonstrator $X_{mod}$.

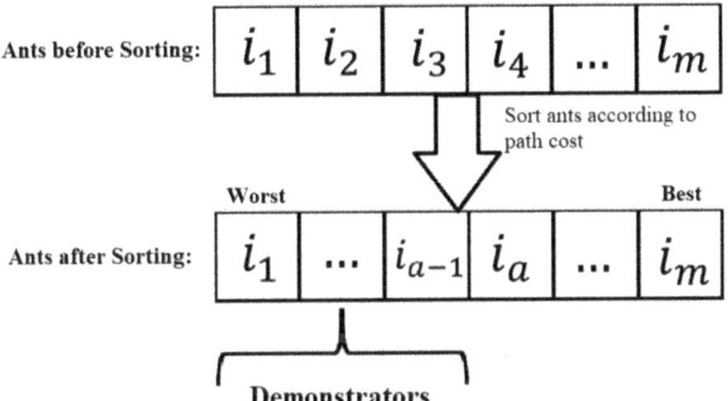


**Figure 6.** Social learning mechanism.

For the landmark operation, the social behavior occurs when ants from the center are removed, and other ants migrate toward the center. The procedure can be given as:

$$X_{cent}^{N_c-1} = \sum_{a=1}^{N_c-1} \frac{X_a^{N_c-1}}{N^{N_c-1}} \tag{10}$$

$$X_a^{N_c} = X_a^{N_c-1} + rand \times c_2 \times \left(X_{cent}^{N_c-1} - X_a^{N_c-1}\right) \tag{11}$$

$$c_2 = \alpha_s\left(\frac{N_c}{m}\right) \tag{12}$$

where the social influence factor is $c_2$, and $\alpha_s$ is called the social coefficient.

### 4.4. Multiple Agent Systems

The multi-agent system comprises $n$ independent agents that move with the same absolute velocity. Every agent's direction is updated to its neighbor's status. At time $t$, the neighbors of an agent $A$ $(1 \leq A \leq n)$ are those who are located within a circle of radius $r$ $(r > 0)$ centered on the position of agent $a$. At time $t$, the neighbor of the agent $a$ is $N_a(t)$,

$$N_a(t) = \{b | d_{ab}(t) < r\} \tag{13}$$

where the Pythagorean Theorem can be used to compute $d_{ab}(t)$.

The coordinates of the agent at time $t$ are $(x_a(t), y_a(t))$, where agent $a$ is a neighbor to agent $b$. The absolute velocity $v$ $(v > 0)$ of each agent in the system is the same.

$$x_a(t+1) = x_a(t) + v\cos\theta_a(t) \tag{14}$$

$$y_a(t+1) = y_a(t) + v\sin\theta_a(t) \tag{15}$$

At time $t$, the heading angle of agent $a$ is $\theta_a(t)$. The following equation is used by the algorithm to update the heading angle:

$$\theta_a(t+1) = \tan^{-1}\frac{\sum_{b\in N_a(t)}\sin\theta_b(t)}{\sum_{b\in N_a(t)}\cos\theta_b(t)} \tag{16}$$

The equation above is used to discover obstructions by examining their surroundings. If a missing node in the neighbor (i.e., a hurdle) exists, the heading angle will be changed to prevent colliding with the obstruction.

We can analyze this algorithm using basic graph theory. Please note that each agent's neighbors are not always the same. The undirected graph set $\mathbb{G}_t = \{\mathcal{V}, \varepsilon_t\}$ is used for agent coordination. Where the set containing every agent is $\mathcal{V} = \{1, 2, \cdots, N\}$, and the time-varying edge set is $\varepsilon_t$. A graph is connected if any two of its vertices are connected.

Equation (16) can be modified as:

$$\tan \theta_a(t+1) = \sum_{b \in N_a(t)} \frac{\cos \theta_b(t)}{\sum_{m \in N_a(t)} \cos \theta_m(t)} \tan \theta_a(t) \tag{17}$$

To further simplify Equation (17), we use a matrix,

$$\tan \theta(t+1) = I(t) \tan \theta(t) \tag{18}$$

where $\tan \theta(t) \triangleq (\tan \theta_1(t), \ldots, \tan \theta_N(t))^\tau$. For the graph $\mathbb{G}_t$, the weighted average matrix is $I(t) \triangleq (i_{ab}(t))$.

$$i_{ab}(t) = \begin{cases} \frac{\cos \theta_b(t)}{\sum_{m \in N_a(t)} \cos \theta_m(t)} & \textit{if } (a,b) \in \varepsilon_t \\ 0, & \textit{otherwise} \end{cases} \tag{19}$$

For the synchronization, we study the linear model of Equation (16) as follows:

$$\theta_a(t+1) = \frac{1}{n_a(t)} \sum_{b \in N_a(t)} \theta_b(t) \tag{20}$$

where the number of elements in $N_a(t)$ is $n_a(t)$. Equation (18) can be rewritten as,

$$\tan \theta(t+1) = \tilde{I}(t) \theta(t) \tag{21}$$

where $\theta(t) \triangleq (\theta_1(t), \ldots, \theta_N(t))^\tau$, and the entries of the matrix $\tilde{I}(t)$ are,

$$\tilde{i}_{ab}(t) = \begin{cases} \frac{1}{n_a(t)}, & \textit{if } (a,b) \in \varepsilon_t \\ 0, & \textit{otherwise} \end{cases} \tag{22}$$

### 4.5. Synchronization and Connectivity

To continue the study of the synchronization of the designed algorithm and the connectivity of the associated neighboring graphs, we should formally describe synchronization. If the headings of each agent match the following criteria, the system will achieve synchronization.

$$\lim_{t \to \infty} \theta_a(t) = \theta, \, a = 1, \ldots, N \tag{23}$$

where $\theta$ varies to the starting values $\{\theta_a(0), x_a(0), y_a(0), a = 1, \ldots, N\}$ and the system parameters $v$, and $r$.

Considering the model in Equations (14)–(16), let $\{\theta_a(0) \in \left(-\frac{\pi}{2}, \frac{\pi}{2}\right), a = 1, \ldots, N\}$, and assume that the neighbor at the start, $\mathbb{G}_0 = \{\mathcal{V}, \varepsilon_0\}$ is connected. Therefore, to achieve synchronization, the system will have to satisfy,

$$v \le \frac{d}{\Delta_0} \left(\frac{\cos \bar{\theta}}{N}\right)^N \tag{24}$$

whereas the number of agents is represented by $N$. Meanwhile,

$$\bar{\theta} = \max_a |\theta_a(0)| \tag{25}$$

$$d = r - \max_{a,b\in\varepsilon_0} d_{ab}(0) \tag{26}$$

$$\Delta_0 = \max_{a,b}\{\tan\theta_a(0) - \tan\theta_b(0)\} \tag{27}$$

Considering the models in Equations (14), (15) and (20), let $\theta_a(0) \in [0, 2\pi)$, and let us assume that the starting neighbor graph is connected. Therefore, to achieve synchronization, the system will have to satisfy,

$$v \leq \frac{d\left(\frac{1}{N}\right)^N}{2\pi} \tag{28}$$

whereas $d$ is the same as described in Equation (26).

### 4.6. Dynamic Leader Selection

Due to communication problems between agents, the formation structure of multi-agent systems might occasionally change. Considering a random failure of communication, each connection $(a,b)\in E$ fails independently with probability $p$. Let $\hat{G}$ be the graph topology and $E_{\hat{G}}(t_{conv})$ be the expected convergence time for this model of communication failure, while $E_{\hat{G}}()$ denotes the argument's expectation across the group of network structures represented by $\hat{G}$. The convergence rate will be maximized by reducing $E_{\hat{G}}(t_{conv})$. As a result, the formula for choosing a leader $k$ to maximize the convergence rate is as follows:

$$\max_Y E_{\hat{G}}\left(\min_{x(0)}x(0)^T(Y\hat{L} + \hat{L}Y)x(0)\right) \tag{29}$$

So that it meets the below conditions,

$$\begin{cases} \boldsymbol{tr}(Y) \geq n - k \\ Y_{aa} \in \{0,1\}\forall_a \neq V \\ Y_{ab} = 0\forall_a \neq b \end{cases} \tag{30}$$

As such, the objective function is in line with the projected convergence rate for potential network topologies. Where $\min_{x(0)}x(0)^T(Y\hat{L} + \hat{L}Y)x(0)$ is a convex function of $Y$.

### 4.7. B-Spline Path Smoothing

The hybrid algorithm's route consists mostly of a combination of line segments. The B-spline curve method is utilized to ensure the smoothness of the route created. The B-spline method is an improvement over the Bezier approach that preserves the convexity and geometrical invariability.

The B-spline path smoothing can be written as:

$$P(u) = \sum_{i=0}^{n} d_i N_{ij}(u) \tag{31}$$

Considering Equation (31), $d_i(i = 0, 1, \ldots, n)$ are control points, and $N_{ij}(u)$ are the normalized b-order functions of the B-spline. These can be described as:

$$\begin{cases} N_{ij}(u) = \begin{cases} 1, & \text{if } u_i \leq u \leq u_{i+1} \\ 0, & \text{otherwise} \end{cases} \\ N_{ij}(u) = \frac{u-u_i}{u_{i+j}-u_i}N_{i,j-1}(u) + \frac{u_{i+j+1}-u}{u_{i+j+1}-u_{i+1}}N_{i+1,j+1}(u) \\ define\frac{0}{0} = 0 \end{cases} \tag{32}$$

The essential functions of the B-spline curve are determined by the parametric knots $\{u_0 \leq u_1 \leq \ldots \leq u_{n+j}\}$. In contrast to the Bezier curve, the B-spline curve is unaffected by altering a single control point. Another benefit of the control points over the Bezier curve is that the degree of polynomials does not increase when the control points are increased.

*4.8. Flowchart and Algorithm*

Figure 7 illustrates the flowchart of the whole system. As we can see, the system first initializes the parameters in all colonies and places ants at the starting random positions. Then, the ants keep moving to the next node until they reach the destination. Hereinafter, they compute the path cost and update the pheromone. However, the system only updates the path if the pheromone is within the desired range. This process is repeated for the predefined number of iterations. Next, the social learning mechanism sorts the ants from best to worst according to the path cost, and the algorithm assigns the best ant to be the leader of each colony and the remaining ants to be agents. Finally, all the colonies are synchronized into one big swarm, and the system dynamically selects its leader.

---

**Algorithm 1** Designed Strategy is Given As.

---

Parameter Initialization of the algorithm
**procedure** algorithm1
  **while** ($N_c < N_{c_{max}}$) **do**
    Put virtual ants on initial positions
    **If** (Destination_Arrived) **then**
      Compute the average cost of the path using Equation (4) and update the phenomenon using Equations (2) and (3) and (5) and (6)
      **else**
      Move to the next node using Equation (1)
    **end if**
    Sort the ants according to the Social Learning mechanism using Equations (7)–(12)
    **Output** the best ant
  **end while**
  **while** (destination) **do**
    **if** (Found_Obstacle) **then**
      Update heading angle using Equation (16)
      **else**
      Keep moving towards the destination
    **end if**
    **while** (Total_UAVs)
      Dynamically select the leader using Equations (29) and (30)
    **end while**
  **end while**
**end procedure**

---

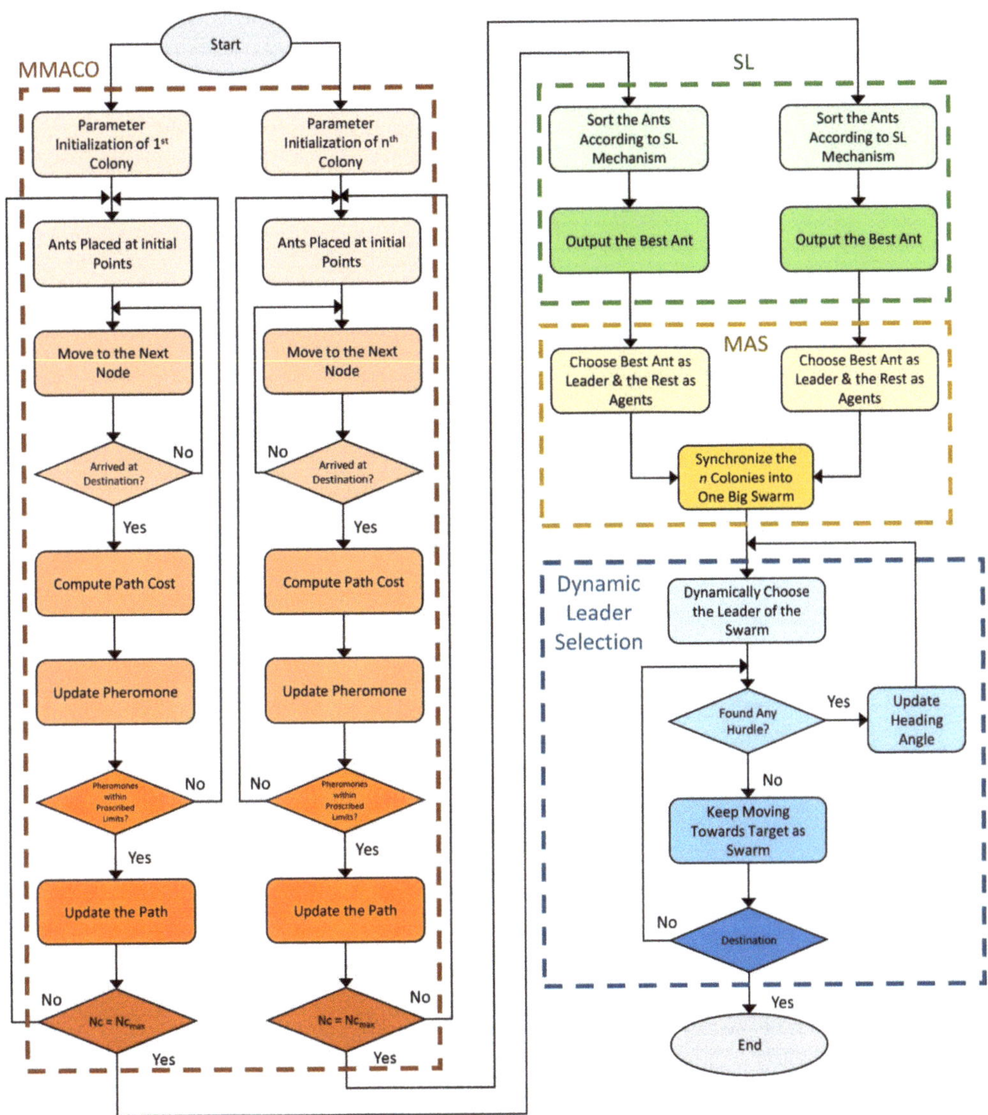


**Figure 7.** Illustration of the flowchart.

## 5. Simulation Results

In this section, we verify the effectiveness of the proposed strategy by applying it to a MATLAB simulation. The conclusions that we want to verify are the following: firstly, we apply the algorithm to three colonies with each having three UAVs, and see if they can organize themselves into desired formations. For the second scenario, we navigate these newly organized formations through some obstacles to see if they can maintain their formations. Lastly, we validate if the designed strategy can successfully synchronize and connect the three colonies into one swarm.

First Scenario:

For the first scenario, the UAVs are at random positions within each colony. The objective is to arrange the UAVs into the desired formations and then reach the target using the shortest possible path. Figure 8 presents the simulation result of the first scenario. As we can see, the algorithm successfully arranges the UAVs into formations, and they maintain these formations throughout the journey. The environment contains different obstacles like mountains and rough terrain. Please note that in this scenario, we are only testing the capability of the algorithm to maintain formations, and hence, we fly the UAV formations in an upward trajectory and not through the obstacles. The second and the third scenarios deal with passing the formations through the obstacles.

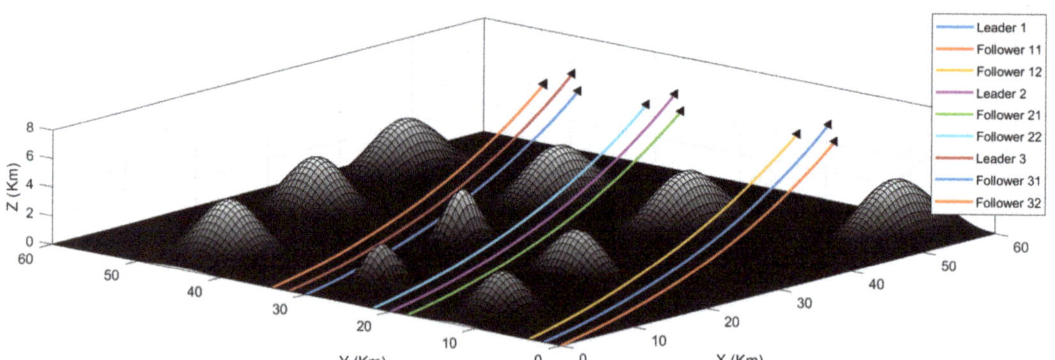

**Figure 8.** Simulation result of the first scenario.

Second Scenario:

In this scenario, after the algorithm successfully maintains the formation, and we check whether it is also capable of obstacle avoidance. The environment contains different obstacles like mountains and rough terrain. The goal is to reach the target using the shortest route possible without colliding with the obstacles or other UAVs. Figure 9 illustrates the simulation result of the second scenario. It is evident from the result that the algorithm achieved the desired goal, and we can see that the three colonies navigated the obstacles while maintaining formations.

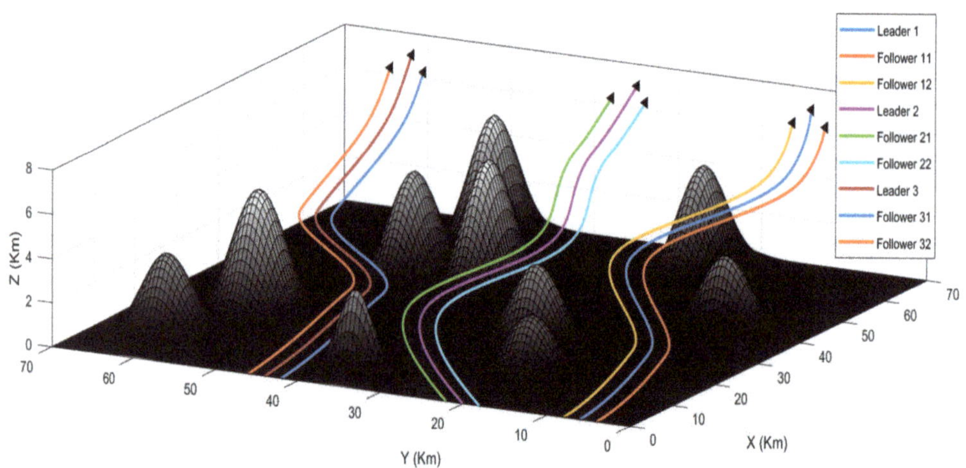

**Figure 9.** Simulation result of the second scenario.

Third Scenario:

In this scenario, we pick up where the second scenario left off, i.e., the three colonies are now in the desired formations. The environment contains different obstacles like mountains and rough terrain. The goal is to first synchronize the three colonies into one big swarm, and then while maintaining the swarm, reach the target using the shortest route possible without colliding with the obstacles or other UAVs. Figure 10 illustrates the simulation result of the third scenario. Again, we see that the algorithm successfully synchronized the three colonies into one swarm, and it reaches its target without any collision.

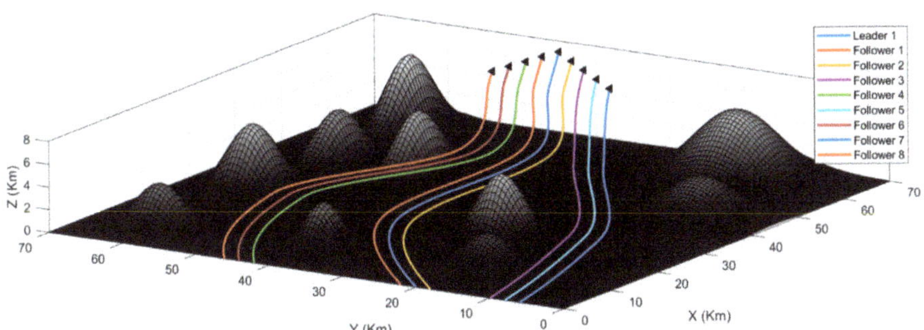


**Figure 10.** Simulation result of the third scenario.

Comparison with NSGA-II:

Lastly, we also compare our proposed algorithm with the Non-Dominated Sorting Genetic Algorithm II (NSGA II). We compare our proposed method with the NSGA II because it is a fast multi-objective genetic algorithm and is a highly regarded evolutionary algorithm. Figure 11 compares our proposed method with NSGA-II. Here, we can see that our designed strategy stays close to the reference while NSGA-II sometimes strays too far. Additionally, our strategy follows a shorter and quicker path to the target.

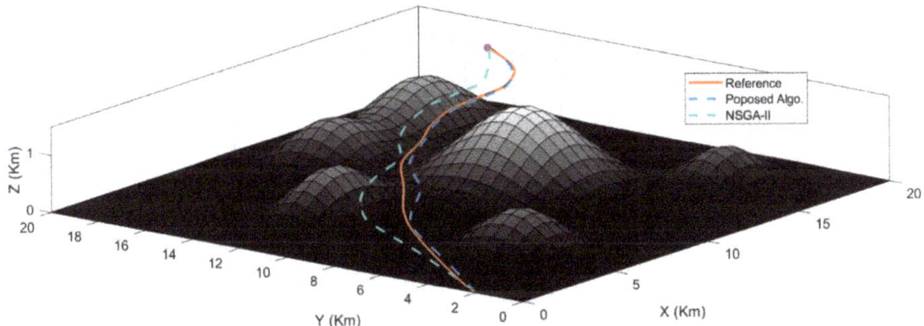


**Figure 11.** Comparison of our proposed method with NSGA-II.

## 6. Conclusions

This research presents a strategy for the self-organization of a swarm of UAVs consisting of three colonies with three UAVs each. To plan the path of each colony to the target, we used max-min ant colony optimization (MMACO), and we used the social learning mechanism to sort the ants from worst to best. To organize the randomly positioned UAVs into different formations, this study used the multi-agent system (MAS). The designed algorithm also synchronized and connected the three colonies into a swarm with the help of dynamic leader selection. The proposed algorithm completed the given objectives in the simulation results.

The salient results and the findings in this research include successfully maintaining the formations in the first scenario. In the second scenario, the algorithm not only maintained the formations, but also navigated them through the obstacles. In the third scenario, the algorithm merged the three formations into one big swarm and then successfully navigated the swarm through the obstacles. By comparing the proposed method with the Non-Dominated Sorting Genetic Algorithm II (NSGA-II), it was clear that our strategy offered better convergence, optimized routes, and reached the destination using a shorter route than NSGA-II.

**Author Contributions:** Investigation, M.S.; supervision, Z.A.A.; validation, A.I.; methodlogy, E.H.A.; writing-original draft preparation, M.S.; software, Z.A.A.; funding acquision, E.H.A.; writing-review and editing, A.I. and M.H. All authors have read and agreed to the published version of the manuscript.

**Funding:** This work is supported by Taif University Researchers Supporting Project number (TURSP-2020/292) Taif University, Taif, Saudi Arabia and this work is also supported by Princess Nourah bint Abdulrahman University Researchers Supporting Project number (PNURSP2022R193), Princess Nourah bint Abdulrahman University, Riyadh, Saudi Arabia.

**Data Availability Statement:** All data used to support the findings of this study are included within the article.

**Acknowledgments:** The authors would like to acknowledge Taif University Researchers Supporting Project number (TURSP-2020/292) Taif University, Taif, Saudi Arabia. The authors would like also to acknowledge, Princess Nourah bint Abdulrahman University Researchers Supporting Project number (PNURSP2022R193), Princess Nourah bint Abdulrahman University, Riyadh, Saudi Arabia.

**Conflicts of Interest:** The authors declare that they have no conflict of interest.

## References


1. Radmanesh, M.; Kumar, M.; Guentert, P.H.; Sarim, M. Overview of path-planning and obstacle avoidance algorithms for UAVs: A comparative study. *Unmanned Syst.* **2018**, *6*, 95–118. [CrossRef]
2. Yao, P.; Wang, H.; Su, Z. Real-time path planning of unmanned aerial vehicle for target tracking and obstacle avoidance in complex dynamic environment. *Aerosp. Sci. Technol.* **2015**, *47*, 269–279. [CrossRef]
3. Seo, J.; Kim, Y.; Kim, S.; Tsourdos, A. Collision avoidance strategies for unmanned aerial vehicles in formation flight. *IEEE Trans. Aerosp. Electron. Syst.* **2017**, *53*, 2718–2734. [CrossRef]
4. Oh, H.; Shirazi, A.R.; Sun, C.; Jin, Y. Bio-inspired self-organising multi-robot pattern formation: A review. *Robot. Auton. Syst.* **2017**, *91*, 83–100. [CrossRef]
5. Ganesan, R.; Raajini, X.M.; Nayyar, A.; Sanjeevikumar, P.; Hossain, E.; Ertas, A.H. Bold: Bio-inspired optimized leader election for multiple drones. *Sensors* **2020**, *20*, 3134. [CrossRef] [PubMed]
6. Xie, Y.; Han, L.; Dong, X.; Li, Q.; Ren, Z. Bio-inspired adaptive formation tracking control for swarm systems with application to UAV swarm systems. *Neurocomputing* **2021**, *453*, 272–285. [CrossRef]
7. Dorigo, M.; Birattari, M.; Stutzle, T. Ant colony optimization. *IEEE Comput. Intell. Mag.* **2006**, *1*, 28–39. [CrossRef]
8. Duan, H.; Qiao, P. Pigeon-inspired optimization: A new swarm intelligence optimizer for air robot path planning. *Int. J. Intell. Comput. Cybern.* **2014**, *7*, 24–37. [CrossRef]
9. Poli, R.; Kennedy, J.; Blackwell, T. Particle swarm optimization. *Swarm Intell.* **2007**, *1*, 33–57. [CrossRef]
10. Ali, Z.A.; Han, Z.; Wang, B.H. Cooperative path planning of multiple UAVs by using max–min ant colony optimization along with Cauchy mutant operator. *Fluct. Noise Lett.* **2021**, *20*, 2150002. [CrossRef]
11. Qiu, H.; Duan, H. A multi-objective pigeon-inspired optimization approach to UAV distributed flocking among obstacles. *Inf. Sci.* **2020**, *509*, 515–529. [CrossRef]
12. Ghamry, K.A.; Kamel, M.A.; Zhang, Y. Multiple UAVs in forest fire fighting mission using particle swarm optimization. In Proceedings of the 2017 IEEE International Conference on Unmanned Aircraft Systems (ICUAS), Miami, FL, USA, 13–16 June 2017; pp. 1404–1409.
13. Vijh, S.; Gaurav, P.; Pandey, H.M. Hybrid bio-inspired algorithm and convolutional neural network for automatic lung tumor detection. *Neural Comput. Appl.* **2020**, 1–14. [CrossRef]
14. Domanal, S.G.; Guddeti, R.M.R.; Buyya, R. A hybrid bio-inspired algorithm for scheduling and resource management in cloud environment. *IEEE Trans. Serv. Comput.* **2017**, *13*, 3–15. [CrossRef]
15. Dhiman, G. ESA: A hybrid bio-inspired metaheuristic optimization approach for engineering problems. *Eng. Comput.* **2021**, *37*, 323–353. [CrossRef]
16. Li, K.; Yan, X.; Han, Y.; Ge, F.; Jiang, Y. Many-objective optimization based path planning of multiple UAVs in oilfield inspection. *Appl. Intell.* **2022**, 1–16. [CrossRef]
17. Ge, F.; Li, K.; Han, Y. Solving interval many-objective optimization problems by combination of NSGA-III and a local fruit fly optimization algorithm. *Appl. Soft Comput.* **2022**, *114*, 108096. [CrossRef]

18. Wang, Y.; Li, K.; Han, Y.; Yan, X. Distributed multi-UAV cooperation for dynamic target tracking optimized by an SAQPSO algorithm. *ISA Trans.* 2021, in press. [CrossRef]

19. Shetty, A.; Shetty, A.; Puthusseri, K.S.; Shankaramani, R. An improved ant colony optimization algorithm: Minion Ant (MAnt) and its application on TSP. In Proceedings of the 2018 IEEE Symposium Series on Computational Intelligence (SSCI), Bangalore, India, 18–21 November 2018; pp. 1219–1225.

20. Shi, B.; Zhang, Y. A novel algorithm to optimize the energy consumption using IoT and based on Ant Colony Algorithm. *Energies* **2021**, *14*, 1709. [CrossRef]

21. Shafiq, M.; Ali, Z.A.; Alkhammash, E.H. A cluster-based hierarchical-approach for the path planning of swarm. *Appl. Sci.* **2021**, *11*, 6864. [CrossRef]

22. Cheng, R.; Jin, Y. A social learning particle swarm optimization algorithm for scalable optimization. *Inf. Sci.* **2015**, *291*, 43–60. [CrossRef]

23. Zhang, Z.; Li, S.; Wang, X.; Cheng, W.; Qi, Y. Source mask optimization for extreme-ultraviolet lithography based on thick mask model and social learning particle swarm optimization algorithm. *Opt. Express* **2021**, *29*, 5448–5465. [CrossRef] [PubMed]

24. Khan, A.; Javed, M.Y.; Alhaisoni, M.; Tariq, U.; Kadry, S.; Choi, J.; Nam, Y. Human Gait Recognition Using Deep Learning and Improved Ant Colony Optimization. *Comput. Mater. Cont.* **2022**, *70*, 2113–2130. [CrossRef]

25. Kamdar, R.; Paliwal, P.; Kumar, Y. A state of art review on various aspects of multi-agent system. *J. Circuits Syst. Comput.* **2018**, *27*, 1830006. [CrossRef]

26. Gulzar, M.M.; Rizvi, S.T.H.; Javed, M.Y.; Munir, U.; Asif, H. Multi-agent cooperative control consensus: A comparative review. *Electronics* **2018**, *7*, 22. [CrossRef]

27. González-Briones, A.; Villarrubia, G.; De Paz, J.F.; Corchado, J.M. A multi-agent system for the classification of gender and age from images. *Comput. Vis. Image Underst.* **2018**, *172*, 98–106. [CrossRef]

28. Hamidi, H.; Kamankesh, A. An approach to intelligent traffic management system using a multi-agent system. *Int. J. Intell. Transp. Syst. Res.* **2018**, *16*, 112–124. [CrossRef]

29. Herrera, M.; Pérez-Hernández, M.; Parlikad, A.K.; Izquierdo, J. Multi-agent systems and complex networks: Review and applications in systems engineering. *Processes* **2020**, *8*, 312. [CrossRef]

30. Ruan, W.; Duan, H. Multi-UAV obstacle avoidance control via multi-objective social learning pigeon-inspired optimization. *Front. Inf. Technol. Electron. Eng.* **2020**, *21*, 740–748. [CrossRef]

*drones*

*Article*

# Mathematical Modeling and Stability Analysis of Tiltrotor Aircraft

Hanlin Sheng *[iD], Chen Zhang and Yulong Xiang

College of Energy and Power Engineering, Nanjing University of Aeronautics and Astronautics (NUAA), Nanjing 210016, China; billowzc@nuaa.edu.cn (C.Z.); xyl1913249177@nuaa.edu.cn (Y.X.)
* Correspondence: dreamshl@nuaa.edu.cn; Tel.: +86-18963646736

**Abstract:** The key problem in the development process of a tiltrotor is its mathematical modeling. Regarding that, this paper proposes a dividing modeling method which divides a tiltrotor into five parts (rotor, wing, fuselage, horizontal tail, and vertical fin) and to develop aerodynamic models for each of them. In that way, force and moment generated by each part are obtained. Then by blade element theory, we develop the rotor's dynamic model and rotor flapping angle expression; by mature lifting line theory, the build dynamic models of the wings, fuselage, horizontal tail and vertical fin and the rotors' dynamic interference on wings, as well as nacelle tilt's variation against center of gravity and moment of inertia, are taken into account. In MATLAB/Simulink simulation environment, a non-linear tiltrotor simulation model is built, Trim command is applied to trim the tiltrotor, and the XV-15 tiltrotor is taken as an example to validate rationality of the model developed. In the end, the non-linear simulation model is linearized to obtain a state-space matrix, and thus the stability analysis of the tiltrotor is performed.

**Keywords:** tiltrotor; blade element theory; flight mechanical model; flight simulation; stability analysis

**Citation:** Sheng, H.; Zhang, C.; Xiang, Y. Mathematical Modeling and Stability Analysis of Tiltrotor Aircraft. *Drones* **2022**, *6*, 92. https://doi.org/10.3390/drones6040092

Academic Editor: Mostafa Hassanalian

Received: 15 March 2022
Accepted: 6 April 2022
Published: 8 April 2022

**Publisher's Note:** MDPI stays neutral with regard to jurisdictional claims in published maps and institutional affiliations.

## 1. Introduction

A tiltrotor is an important derivative of short take-off and vertical landing (STOVL) aircraft. As a hybrid vehicle, it combines the merits of both a helicopter and a fixed-wing aircraft. It can hover and vertically take off and land like a helicopter and fly forward fast like a fixed-wing aircraft. A tiltrotor has a nacelle installed at every wingtip. The nacelle can tilt 0–90° and switch from helicopter mode and transition mode to flight mode. A tiltrotor has many flight modes as Figure 1 and thus is granted with wide flight envelope, which brings great development potential but also many technical issues [1–6]. For example, there is airflow disturbance between the rotor and the wing; the aerodynamic parameters of the aircraft change sharply in the transition section of the tiltrotor aircraft; and there are problems such as redundancy of multiple control surfaces, large aerodynamic coupling between different channels of the tilt rotor aircraft, and poor aerodynamic stability during high-speed flight. Due to its complexity in terms of dynamics, building a complete mathematical model becomes significant for designing the flight control system, which justifies its difficulty as well. This is why mathematical modeling is a critical technical issue for a tiltrotor.

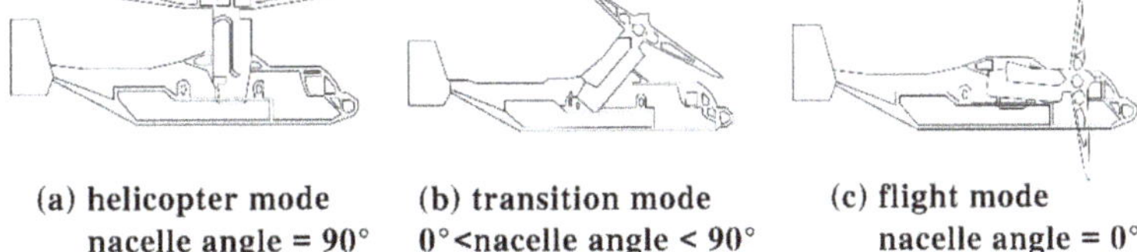



**Figure 1.** Three flight modes of tilt rotor aircraft.

Reference [7] investigates optimal tiltrotor flight trajectories considering the possibility of engine failure. A two-dimensional longitudinal rigid body model of a tiltrotor aircraft is used. It has certain limitations in implementation. References [8,9] built a tiltrotor's 3DOF kinematic model in which the tiltrotor's dynamics was over-simplified. Reference [10] built a more complete 6DOF kinematic model, but the blade's moment of inertia was over-treated, the aerodynamic parts were over-simplified, and aerodynamic interferences between the nacelle dynamics and aerodynamic parts were not considered. The tiltrotor model of V-22 built in reference [11] was good for flight property calculation, flight simulation and the aircraft's stability analysis. Reference [12] build a general parametric flight dynamics model which can be used for online identification purpose in the three flight modes of tiltrotor aircraft (i.e., helicopter mode, conversion mode, and fixed-wing mode) is developed, and an unideal noise model is also introduced in order to minimize the parameter identification error caused by measurement noise. It can be seen that with the development of system identification, the research on identifying the stability derivative, maneuverability derivative and linear model of tilt rotor aircraft through flight data is also developing rapidly in references [13–15]; the modified blade element analysis theory is used to model the rotor, and the modified momentum theory is used to calculate the induced velocity in reference [16]. The model built in reference [17] considers the variation of each blade's flapping due to the elasticity of the blades. In model of reference [18], Primary dynamic equations of the model are developed considering nacelles tilting dynamics. they laid more focus on rotor dynamics or nacelles tilting dynamics. However, they did not conduct further studies concerning the other components and interference characteristics.

Reference [19] built an even better basic dynamic model by studying dynamics issues like aerodynamic interference, change of center of gravity, and gyroscopic moment's interference on the airframe in transition mode. Apart from the aerodynamic effects, the flight dynamic model also includes a model of the air data system and the feedback control laws. However, this study lacked an analysis of the relative force, forming principles of moment, and control plane.

This paper adopts the dividing modeling method, which breaks down a tiltrotor into five parts, rotor, wing, fuselage, horizontal tail and vertical fin, develops aerodynamic models for each part, and thus obtains force and moment generated by each part. In the MATLAB/Simulink simulation environment, a non-linear tiltrotor simulation model is built. The trim command is applied to trim the tiltrotor and an XV-15 tiltrotor is taken as an example to validate the accuracy and rationality of the model developed. In the end, the non-linear simulation model is linearized to obtain a state-space matrix. The stability derivative and eigenvalue of the tiltrotor are analyzed and furthermore the tiltrotor's stability in each flight mode is studied.

## 2. Tiltrotor Aircraft Aerodynamic Model

The main aerodynamic components of a tiltrotor aircraft are its rotor, wing, fuselage, horizontal tail (incl. horizontal rudder) and vertical fin (incl. vertical rudder). This paper is going to develop aerodynamic models for each component. The aerodynamic force of each component will find its solution in wind axis system and finally be converted to body axis system through the conversion matrix. Figure 2 shows the components of the tiltrotor aircraft and the schematic diagram of some axis coordinate systems.

In order to establish an accurate flight dynamics model of tiltrotor aircraft and avoid being complex, the following basic assumptions are made in this paper

(1)  The earth axis system is assumed to be an inertial reference system;
(2)  Assuming that the earth is flat, the curvature of the earth is not considered;
(3)  Tiltrotor aircraft are considered as rigid bodies;
(4)  The rotor blades are bending rigid and linear torsion;
(5)  The blade waving motion is calculated by taking the first-order harmonic;
(6)  The angle of attack and sideslip angle are small angles;
(7)  Ignore the influence of rotor downwash flow on the fuselage;

(8)    Aerodynamic interference between left and right rotors is not considered;

(9)    Tiltrotor aircraft is left-right symmetrical, and its longitudinal axis is its plane of symmetry.

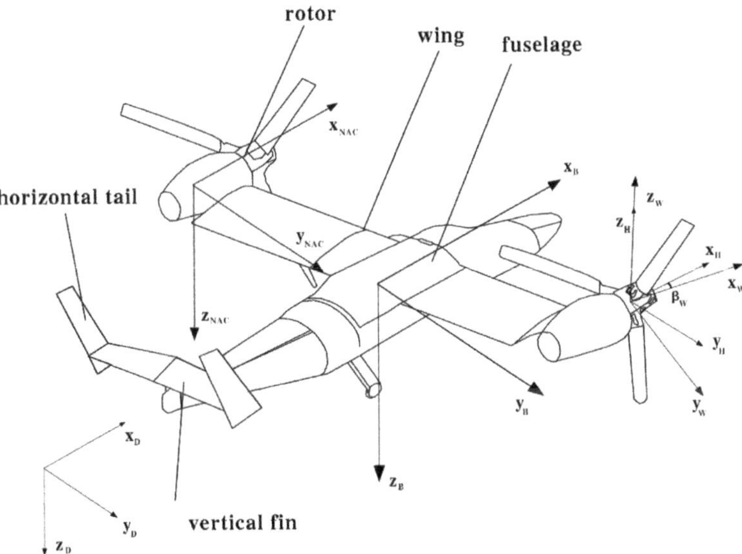


**Figure 2.** Tiltrotor components and coordinate system.

### 2.1. Center of Gravity and Moment of Inertia

The center of gravity of a tiltrotor changes in its longitudinal plane as the nacelle tilts, which causes the change of the moment of inertia as well. Center-of-gravity position and moment of inertia are functions of the nacelle angle.

The nacelle angle $\beta_M = 0$, center-of-gravity position is the initial position. The variation of center-of-mass position as the nacelle angle changes is expressed as:

$$\Delta x = \frac{m_{NAC} R_H \sin \beta_M}{m} \tag{1}$$

$$\Delta z = \frac{m_{NAC} R_H (1 - \cos \beta_M)}{m} \tag{2}$$

where, $R_H$ is the rotor's height against the wing, $m_{NAC}$ is the mass of the nacelle system, $m$ is the gross weight of the airframe.

Moment of inertia changes along with the nacelle angle $\beta_M$ change, which is formulated as [10]:

$$I = I_0 - KI\beta_M \tag{3}$$

where $I_0$ is the moments of inertia of any axis in helicopter mode. $KI$ is the coefficient of the moment of corresponding inertia.

### 2.2. Rotor Aerodynamic Model

The rotor is the most important component of a tiltrotor. In helicopter mode, the rotor is the main lifting and control surface. In fixed-wing mode, the rotor is the propeller. While in transition mode, the rotor plays the aforesaid three roles together. That explains why it is one of the critical technologies to build a precise rotor model for a tiltrotor's modeling.

Compared with a helicopter, a tiltrotor's mathematical model is far more complicated thanks to the aerodynamic interferences between rotors and wings. Rotors' downwash flow confluences at wings, and expands along wings to the fuselage, forming a "fountain flow effect", which boosts rotors' induced velocity. However, on the other hand, the blocking effect from wings to rotors is similar to ground effect and reduces that induced velocity. As a tiltrotor is flying at a low speed, fountain flow effect and blocking effect are generating equivalent induced velocities, which imply the aerodynamic interference from wings to rotors is negligible. Therefore, a tiltrotor can be considered equal to the aerodynamic model of an isolated rotor.

A tiltrotor is composed of left and right two rotors. They tilt in the opposite direction. The right one tilts counter-clockwise while the left one tilts clockwise. Since two rotors use the same modeling method but only differ in terms of some symbols, this paper is going to take the right rotor as an example and to build its aerodynamic model.

According to the helicopter flight dynamics theory [20], external forces on blades in flapping plane are aerodynamic force, centrifugal force, gravity force, inertia force, etc. The resultant moment of above forces against flapping hinge is 0, that is, $\sum M = 0$.

Thus, the rotor flapping motion equation is:

$$I_b \ddot{\beta} + I_b \Omega^2 \beta = M_T - M_s g \tag{4}$$

where, $\Delta x$, $\Delta z$ is the variation of center of gravity as the nacelle tilts.

The velocity being converted to the rotor hub wind axis system at the rotor hub center in the aircraft-body axis system is:

$$\begin{bmatrix} u_h \\ v_h \\ w_h \end{bmatrix} = C_H^{HW} C_{NAC}^{H} C_B^{NAC} \begin{bmatrix} u_h \\ v_h \\ w_h \end{bmatrix}_B \tag{5}$$

Tangential velocity and vertical velocity of the rotor profile, respectively, are:

$$U_T = \Omega R \left( \frac{r}{R} + \mu \sin \psi \right) \tag{6}$$

$$U_P = \Omega R (\lambda_0 - \mu \beta \cos \psi) - v_1 - r\dot{\beta} \tag{7}$$

By blade element theory, take blade element of radial position on propeller and width $dr$, and chord $b$, therefore lift force and resistance force of blade element are:

$$dY = \frac{1}{2} C_y \rho W^2 b dr = \frac{1}{2} C_y \rho (U_T^2 + U_P^2) b dr \tag{8}$$

$$dX = \frac{1}{2} C_x \rho W^2 b dr = \frac{1}{2} C_x \rho (U_T^2 + U_P^2) b dr \tag{9}$$

Component force converted to flapping plane is:

$$dT = dY \cos \beta_* - dX \sin \beta_* \tag{10}$$

$$dQ = dX \cos \beta_* + dY \sin \beta_* \tag{11}$$

The projection of blade element's aerodynamics onto the rotor structure axis system constitutes the rotor's elemental force and moment $dT_s$, $dH_s$, $dS_s$, $dM_k$.

Integrate the above blade element elemental aerodynamic force and moment along the propeller, take its average value against position angle, and then multiply with the number of blades to get the force and moment generated by rotors: $T_s$, $H_s$, $S_s$, $M_k$.

In calculating the rotor's elemental force, its induced velocity cannot be calculated directly by explicit formula. This paper applies the iteration method. Given the initial value of induced velocity in the rotor's vertical velocity, the rotor's thrust is obtained by above equation and the new induced velocity by momentum theory is calculated as below:

$$v_1' = v_{1d} \left( 1 + \frac{r}{R} \cos \psi \right) \tag{12}$$

Equivalent induced velocity is:

$$v_{1d} = \frac{\Omega R C_T}{4\sqrt{\lambda_1{}^2 + \mu^2}} \tag{13}$$

where $\lambda_1$ is inflow ratio: $\lambda_1 = \lambda_0 - v_{1d}/\Omega R$, $C_T$ is the rotor thrust coefficient.

If $v_1$ and $v_1'$ are close enough, iteration will exit, otherwise renew $v_1$ with $(v_1 + v_1')/2$. It will work out the rotor's induced velocity as well as its force and moment.

Convert the force and moment of the rotor into aircraft-body axis system to get its force and moment in that system:

$$\begin{bmatrix} F_{xR} \\ F_{yR} \\ F_{zR} \end{bmatrix} = C_{NAC}^B C_H^{NAC} C_{HW}^H \begin{bmatrix} H_s \\ S_s \\ T_s \end{bmatrix} \tag{14}$$

$$\begin{bmatrix} M_{xR} \\ M_{yR} \\ M_{zR} \end{bmatrix} = C_{NAC}^B C_H^{NAC} C_{HW}^H \begin{bmatrix} 0 \\ 0 \\ M_k \end{bmatrix} + \begin{bmatrix} 0 & -z_h & y_h \\ z_h & 0 & -x_h \\ -y_h & x_h & 0 \end{bmatrix} \begin{bmatrix} F_{xR} \\ F_{yR} \\ F_{zR} \end{bmatrix} \tag{15}$$

Follow the same principle to work out the force and moment of the left rotor in aircraft-body axis system.

### 2.3. Wing Aerodynamic Model

Wing aerodynamic model of a tiltrotor is the most complex among all other components. In helicopter and transition modes, rotors' downwash flow causes complicated aerodynamic interference on wings. This paper assumes wings are rigid and have no elastic distortion, and its aerodynamic center is the working point of force and moment.

A tiltrotor has a left wing and a right wing. Take the left wing as an example to work out its force and moment.

As a tiltrotor is flying at a low speed, rotors' downwash flow confluences at wings and forms "fountain effect". To precisely develop the wing model, the wing is divided into two parts: the first part is slipstream zone affected by rotor wake disturbance while the other part is free flow zone free from rotor disturbance. The two zones are shown as Figure 3:

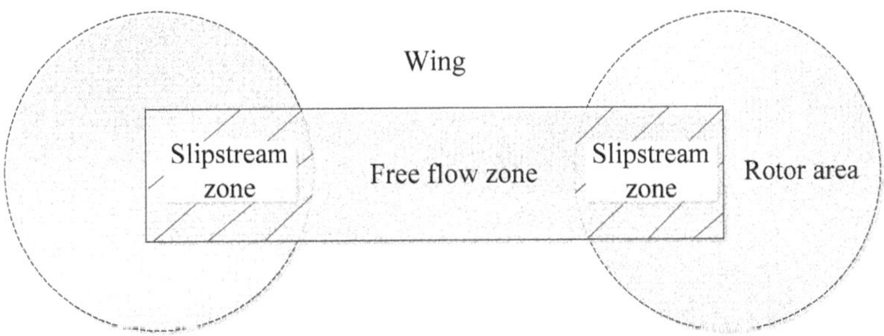


**Figure 3.** Rotor slipstream zone and free flow zone diagram.

It's hard to precisely calculate the area of slipstream zone, but the below formula can approximately estimate that area [7]:

$$S_{wss} = S_{wss_{max}} \left[ \sin(1.386(\frac{\pi}{2} - \beta_M)) + \cos(3.114(\frac{\pi}{2} - \beta_M)) \right] \frac{\mu_{max} - \mu}{\mu_{max}} \tag{16}$$

where, $S_{wss_{max}}$ is the maximum area of slipstream zone, $\mu_{max}$ is advance ratio in helicopter mode when the wing is free from impact of rotor wake.

The area of free flow zone $S_{wfs}$ is the result of wing area deducting slipstream zone area.

(1)  Force and moment in slipstream zone

In slipstream zone, the wing's air velocity is the sum of the rotor's induced velocity at wings and inflow ahead.

$$
\begin{bmatrix} u_{wsl} \\ v_{wsl} \\ w_{wsl} \end{bmatrix} = \begin{bmatrix} u \\ v \\ w \end{bmatrix} + \begin{bmatrix} 0 & z_{wsl} & -y_{wsl} \\ -z_{wsl} & 0 & x_{wsl} \\ y_{wsl} & -x_{wsl} & 0 \end{bmatrix} \begin{bmatrix} p \\ q \\ r \end{bmatrix} + \begin{bmatrix} v_{1d} \sin \beta_M \\ 0 \\ -v_{1d} \cos \beta_M \end{bmatrix} \tag{17}
$$

In helicopter mode, the left wing aerodynamic center has the position against the airframe center of gravity as $P_{wsl0} = \begin{bmatrix} x_{wsl0} & y_{wsl0} & z_{wsl0} \end{bmatrix}^T$. As the nacelle tilts, the wing's aerodynamic center position is:

$$
P_{wsl} = \begin{bmatrix} x_{wsl} \\ y_{wsl} \\ z_{wsl} \end{bmatrix} = \begin{bmatrix} x_{wsl0} - \Delta x \\ y_{wsl0} \\ z_{wsl0} - \Delta z \end{bmatrix} \tag{18}
$$

where, $\Delta x, \Delta z$ is the variation of center of gravity as the nacelle tilts.

Force and moment in the aircraft body axis system are:

$$
\begin{bmatrix} F_{xwsl} \\ F_{ywsl} \\ F_{zwsl} \end{bmatrix} = \begin{bmatrix} \cos \alpha_{wsl} & 0 & -\sin \alpha_{wsl} \\ 0 & 1 & 0 \\ \sin \alpha_{wsl} & 0 & \cos \alpha_{wsl} \end{bmatrix} \begin{bmatrix} -D_{wsl} \\ 0 \\ -L_{wsl} \end{bmatrix} \tag{19}
$$

$$
\begin{bmatrix} M_{xwsl} \\ M_{ywsl} \\ M_{zwsl} \end{bmatrix} = \begin{bmatrix} 0 & -z_{wsl} & y_{wsl} \\ z_{wsl} & 0 & -x_{wsl} \\ -y_{wsl} & x_{wsl} & 0 \end{bmatrix} \begin{bmatrix} F_{xwsl} \\ F_{ywsl} \\ F_{zwsl} \end{bmatrix} + \begin{bmatrix} 0 \\ M_{wsl} \\ 0 \end{bmatrix} \tag{20}
$$

(2)  Force and moment in free flow zone

Free flow zone is dynamic. In helicopter mode, its area is the smallest. Then as the nacelle tilts, part of slipstream zone turns into free flow zone. Therefore, a free flow zone can be considered as two parts: the first part is the zone always being free flow zone (incl. wing flap), and the other part is the zone turned from slipstream zone due to nacelle tilting (incl. aileron).

In free flow zone, wing air velocity is only related to front incident flow. In calculating these two parts, aerodynamic center position deserves more attention.

First, calculate the air velocity $\begin{bmatrix} u_{wfl1} & v_{wfl1} & w_{wfl1} \end{bmatrix}^T$, $\begin{bmatrix} u_{wfl1} & v_{wfl1} & w_{wfl1} \end{bmatrix}^T$ in the two parts free flow zone of the wing, respectively, and then calculate the aerodynamic center dynamic pressure $q_{wfl1}, q_{wfl2}$ and angle of attack $\alpha_{wfl1}, \alpha_{wfl2}$, respectively, in the free flow zone of the left wing according to the air velocity in the free flow zone.

Then, calculate the Lift force $L_{wfl1}, L_{wfl2}$, resistance force $D_{wfl1}, D_{wfl2}$ and moment $M_{wfl1}, M_{wfl2}$ of left wing in free flow zone are:

Aerodynamics in free flow zone in the aircraft-body axis system are:

$$
\begin{bmatrix} F_{xwfl} \\ F_{ywfl} \\ F_{zwfl} \end{bmatrix} = \begin{bmatrix} \cos \alpha_{wfl} & 0 & -\sin \alpha_{wfl} \\ 0 & 1 & 0 \\ \sin \alpha_{wfl} & 0 & \cos \alpha_{wfl} \end{bmatrix} \begin{bmatrix} -D_{wfl} \\ 0 \\ -L_{wfl} \end{bmatrix} \tag{21}
$$

Moment in free flow zone in the aircraft-body axis system are:

$$
\begin{bmatrix} M_{xwfl} \\ M_{ywfl} \\ M_{zwfl} \end{bmatrix} = \begin{bmatrix} 0 & -z_{wfl} & y_{wfl} \\ z_{wfl} & 0 & -x_{wfl} \\ -y_{wfl} & x_{wfl} & 0 \end{bmatrix} \begin{bmatrix} F_{xwfl} \\ F_{ywfl} \\ F_{zwfl} \end{bmatrix} + \begin{bmatrix} 0 \\ M_{wfl} \\ 0 \end{bmatrix} \tag{22}
$$

Take the sum of force and moment of the left wing in slipstream zone and free flow zone, which is the force and moment received by the left wing. Follow the same principle to get the force and moment of the right wing. The total force and moment on left and right wings are the gross sum of forces and moments of all wings.

### 2.4. Fuselage Aerodynamic Model

The fuselage is complex in structure and subject to aerodynamic disturbance from rotors and wings. This paper is going to neglect aerodynamic disturbance from rotors and wings and apply a simplified approach to fuselage modelling. Force and moment of the fuselage are calculated in local wind axis system, working point is the airframe aerodynamic center, which means the calculated force and moment need converted to the body axis system.

First, calculate air velocity $\begin{bmatrix} u_f & v_f & w_f \end{bmatrix}^T$ at fuselage aerodynamic center $(x_f, y_f, z_f)$, and then calculate dynamic pressure $q_f$, angle of attack $\alpha_f$ and sideslip angle of the fuselage $\beta_f$ according to the air velocity at fuselage aerodynamic center.

Then, calculate the aerodynamic force $D_f, S_f, L_f$ and moment $M_{fx}, M_{fy}, M_{fz}$ of the fuselage in wind axis system.

At the end, work out aerodynamic force and moment of the fuselage in body axis system through the conversion matrix from wind axis frame to body axis frame.

$$\begin{bmatrix} F_{xF} \\ F_{yF} \\ F_{zF} \end{bmatrix} = \begin{bmatrix} \cos\beta_f & -\sin\beta_f & 0 \\ \sin\beta_f & \cos\beta_f & 0 \\ 0 & 0 & 1 \end{bmatrix} \begin{bmatrix} \cos\alpha_f & 0 & -\sin\alpha_f \\ 0 & 1 & 0 \\ \sin\alpha_f & 0 & \cos\alpha_f \end{bmatrix} \begin{bmatrix} -D_f \\ S_f \\ -L_f \end{bmatrix} \tag{23}$$

$$\begin{bmatrix} M_{xF} \\ M_{yF} \\ M_{zF} \end{bmatrix} = \begin{bmatrix} \cos\beta_f & -\sin\beta_f & 0 \\ \sin\beta_f & \cos\beta_f & 0 \\ 0 & 0 & 1 \end{bmatrix} \begin{bmatrix} \cos\alpha_f & 0 & -\sin\alpha_f \\ 0 & 1 & 0 \\ \sin\alpha_f & 0 & \cos\alpha_f \end{bmatrix} \begin{bmatrix} M_{fx} \\ M_{fy} \\ M_{fz} \end{bmatrix} + \begin{bmatrix} 0 & -z_f & y_f \\ z_f & 0 & -x_f \\ -y_f & x_f & 0 \end{bmatrix} \begin{bmatrix} F_{xF} \\ F_{yF} \\ F_{zF} \end{bmatrix} \tag{24}$$

### 2.5. Horizontal Tail Aerodynamic Model

Rotor wake has minor impact on horizontal tail, so rotor wake will be overlooked in this paper. Calculate aerodynamic force of horizontal tail and elevator by following the way of treating a fixed-wing. Airflow of horizontal tail pressure center $(x_{HT}, y_{HT}, z_{HT})$ is the sum of airframe linear velocity and angular velocity.

First, calculate air velocity $\begin{bmatrix} u_{HT} & v_{HT} & w_{HT} \end{bmatrix}^T$ at horizontal tail pressure center, and then calculate dynamic pressure $q_{HT}$, angle of attack $\alpha_{HT}$ and sideslip angle of the fuselage $\beta_{HT}$ according to the air velocity at horizontal tail pressure center.

Then, calculate the lift force $L_{HT}$ and resistance force $D_{HT}$ of a horizontal tail in local wind axis system.

Force and torque in wind axis system are converted into body axis system:

$$\begin{bmatrix} F_{HT} \\ F_{HT} \\ F_{HT} \end{bmatrix} = \begin{bmatrix} \cos\beta_{HT} & -\sin\beta_{HT} & 0 \\ \sin\beta_{HT} & \cos\beta_{HT} & 0 \\ 0 & 0 & 1 \end{bmatrix} \begin{bmatrix} \cos\alpha_{HT} & 0 & -\sin\alpha_{HT} \\ 0 & 1 & 0 \\ \sin\alpha_{HT} & 0 & \cos\alpha_{HT} \end{bmatrix} \begin{bmatrix} -D_{HT} \\ 0 \\ -L_{HT} \end{bmatrix} \tag{25}$$

$$\begin{bmatrix} M_{HT} \\ M_{HT} \\ M_{HT} \end{bmatrix} = \begin{bmatrix} 0 & -z_{HT} & 0 \\ z_{HT} & 0 & -x_{HT} \\ 0 & x_{HT} & 0 \end{bmatrix} \begin{bmatrix} F_{HT} \\ F_{HT} \\ F_{HT} \end{bmatrix} \tag{26}$$

### 2.6. Vertical Fin Aerodynamic Model

The vertical fin and rudder have a similar modelling method as horizontal tail. Calculate aerodynamic force and moment of vertical fin in wind axis system and then convert them to body axis system. A tiltrotor has two vertical fins, so it's necessary to develop models for both left and right fins. This paper will take the left fin as an example.

Calculate air velocity $\begin{bmatrix} u_{VT} & v_{VT} & w_{VT} \end{bmatrix}^T$ at vertical fin aerodynamic center $(x_{VT}, y_{VT}, z_{VT})$, and then calculate dynamic pressure $q_{VT}$, angle of attack $\alpha_{VT}$ and sideslip angle of the fuselage $\beta_{VT}$ according to the air velocity at vertical fin aerodynamic center. Thus, aerodynamic force of the vertical fin in wind axis system is:

$$L_{VT} = q_{VT}A_{VT}(C_{L,VT} + \Delta C_{L,VT}) = q_{VT}A_{VT}\left[ a_{VT}(\alpha_{VT} - \alpha_0) + \frac{\partial C_{l,V}}{\partial \delta_{rud}}\delta_{rud}\right] \tag{27}$$

$$D_{VT} = q_{VT}A_{VT}C_{D,VT} \tag{28}$$

Aerodynamic force and moment converted to body axis system are:

$$\begin{bmatrix} F_{xVT} \\ F_{yVT} \\ F_{zVT} \end{bmatrix} = \begin{bmatrix} \cos\beta_{VT} & -\sin\beta_{VT} & 0 \\ \sin\beta_{VT} & \cos\beta_{VT} & 0 \\ 0 & 0 & 1 \end{bmatrix} \begin{bmatrix} \cos\alpha_{VT} & -\sin\alpha_{VT} & 0 \\ \sin\alpha_{VT} & \cos\alpha_{VT} & 0 \\ 0 & 0 & 1 \end{bmatrix} \begin{bmatrix} -D_{VT} \\ -L_{VT} \\ 0 \end{bmatrix} \tag{29}$$

$$\begin{bmatrix} M_{VT} \\ M_{VT} \\ M_{VT} \end{bmatrix} = \begin{bmatrix} 0 & -z_{VT} & y_{VT} \\ z_{VT} & 0 & -x_{VT} \\ -y_{VT} & x_{VT} & 0 \end{bmatrix} \begin{bmatrix} F_{xVT} \\ F_{yVT} \\ F_{zVT} \end{bmatrix} \tag{30}$$

In the same way, work out force and moment of the right vertical fin. The sum of force and moment of the right and left fins is the gross force and moment of vertical fins in aircraft-body axis system.

### 3. Nonlinear Simulation Modelling

The resultant external force and moment exerted on a tiltrotor is:

$$\begin{bmatrix} F_x \\ F_y \\ F_z \end{bmatrix} = \begin{bmatrix} F_{xR} \\ F_{yR} \\ F_{zR} \end{bmatrix}_L + \begin{bmatrix} F_{xR} \\ F_{yR} \\ F_{zR} \end{bmatrix}_R + \begin{bmatrix} F_{xW} \\ F_{yW} \\ F_{zW} \end{bmatrix}_L + \begin{bmatrix} F_{xW} \\ F_{yW} \\ F_{zW} \end{bmatrix}_R + \begin{bmatrix} F_{xF} \\ F_{yF} \\ F_{zF} \end{bmatrix} + \begin{bmatrix} F_{HT} \\ F_{HT} \\ F_{HT} \end{bmatrix} + \begin{bmatrix} F_{VT} \\ F_{VT} \\ F_{VT} \end{bmatrix}_L + \begin{bmatrix} F_{VT} \\ F_{VT} \\ F_{VT} \end{bmatrix}_R \tag{31}$$

$$\begin{bmatrix} M_x \\ M_y \\ M_z \end{bmatrix} = \begin{bmatrix} M_{xR} \\ M_{yR} \\ M_{zR} \end{bmatrix}_L + \begin{bmatrix} M_{xR} \\ M_{yR} \\ M_{zR} \end{bmatrix}_R + \begin{bmatrix} M_{xW} \\ M_{yW} \\ M_{zW} \end{bmatrix}_L + \begin{bmatrix} M_{xW} \\ M_{yW} \\ M_{zW} \end{bmatrix}_R + \begin{bmatrix} M_{xF} \\ M_{yF} \\ M_{zF} \end{bmatrix} + \begin{bmatrix} M_{HT} \\ M_{HT} \\ M_{HT} \end{bmatrix} + \begin{bmatrix} M_{VT} \\ M_{VT} \\ M_{VT} \end{bmatrix}_L + \begin{bmatrix} M_{VT} \\ M_{VT} \\ M_{VT} \end{bmatrix}_R \tag{32}$$

According to momentum theorem, we can get:

$$\begin{bmatrix} \dot{u} \\ \dot{v} \\ \dot{w} \end{bmatrix} = \frac{1}{m}\begin{bmatrix} F_x \\ F_y \\ F_z \end{bmatrix} + C_D^B\begin{bmatrix} 0 \\ 0 \\ g \end{bmatrix} - \begin{bmatrix} 0 & -r & q \\ r & 0 & -p \\ -q & p & 0 \end{bmatrix}\begin{bmatrix} u \\ v \\ w \end{bmatrix} \tag{33}$$

According to theorem of moment of momentum, we can get:

$$\begin{bmatrix} \dot{p} \\ \dot{q} \\ \dot{r} \end{bmatrix} = I^{-1}\begin{bmatrix} M_x \\ M_y \\ M_z \end{bmatrix} - I^{-1}\begin{bmatrix} 0 & -r & q \\ r & 0 & -p \\ -q & p & 0 \end{bmatrix}I \tag{34}$$

where

$$I = \begin{bmatrix} I_{xx} & -I_{xy} & -I_{xz} \\ -I_{yx} & I_{yy} & -I_{yz} \\ -I_{zx} & -I_{zy} & I_{zz} \end{bmatrix} \tag{35}$$

Supplementary kinematic equation group is:

$$\begin{bmatrix} \dot{\phi} \\ \dot{\theta} \\ \dot{\psi} \end{bmatrix} = \begin{bmatrix} 1 & \sin\phi\tan\theta & \cos\phi\tan\theta \\ 0 & \cos\phi & -\sin\phi \\ 0 & \sin\phi\sec\theta & \cos\phi\sec\theta \end{bmatrix}\begin{bmatrix} p \\ q \\ r \end{bmatrix} \tag{36}$$

The above is a tiltrotor's 6DOF flight dynamics equation. The above equation group, if solved, will determine the aircraft's flight condition. This paper uses Function S of MATLAB to develop a tiltrotor's non-linear simulation model. From Section 1, it's known that respective aerodynamic models for the rotor, wing, fuselage, horizontal tail and vertical fin will be developed. Substitute the calculated force and moment from each component to 6DOF flight dynamics equation, which will get us the nonlinear mathematical modelling of a tiltrotor aircraft. The model has the structure diagram as described in Figure 4:

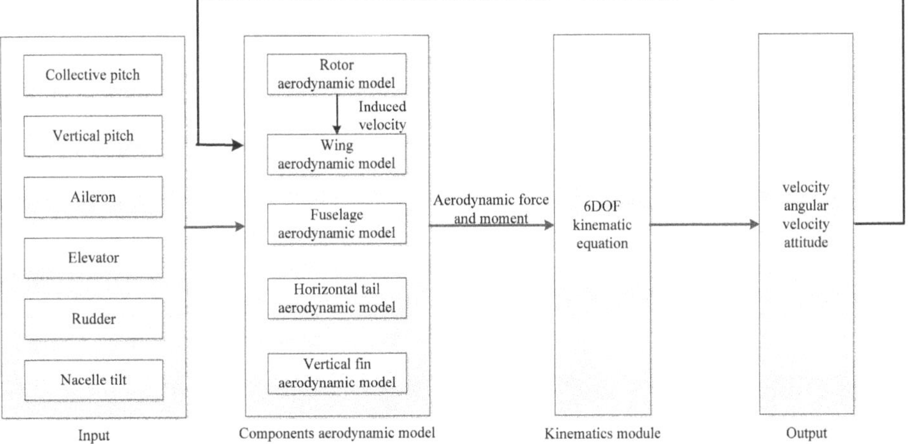


**Figure 4.** Non-linear dynamic model structure diagram.

## 4. Trimming and Result Analysis

After the non-linear simulation model is built, it is necessary to validate its effectiveness and accuracy. Thus, this paper is going to compare trimming results of the built model and actual trimming results. Considering the inevitability of errors, if the change trend is aligned and values are close, it can be deduced that the built model reflects flight characteristics of a tiltrotor, which means the model built in this paper is accurate and effective.

Steady flight is the case where the aircraft's linear and angular velocity are constant. Then all external forces and moments exerted on the aircraft are zero, the aircraft stays in state of equilibrium. The maneuver applied to reach equilibrium is called trim control. The trimming calculation is based the condition that force and moment acting on the aircraft in steady flight condition is balanced, and then use appropriate mathematical algorithms to define control inputs in trimmed condition. Mathematically speaking, trimming is the point to make sure the system status derivative as zero. The most popular trimming methods are the Newton iteration method, simplex method, steepest descent method, genetic algorithm, particle swarm optimization algorithm, etc. Meanwhile, MATLAB/Simulink toolkit also provides trim function, which also uses optimization algorithms. Thus, this paper is going to perform trimming with trim function.

This paper takes an XV-15 tiltrotor as an example, builds its mathematical model in Simulink, and then trims it with the trim function. Trimming results in helicopter mode, flight mode and transition mode where nacelle tilt angles are 15° and 75° are given hereafter.

Trimming results in helicopter mode and flight mode is as Figure 5, Trimming results at transition mode when nacelle tilt angle is 75° and 15° is as Figure 6:

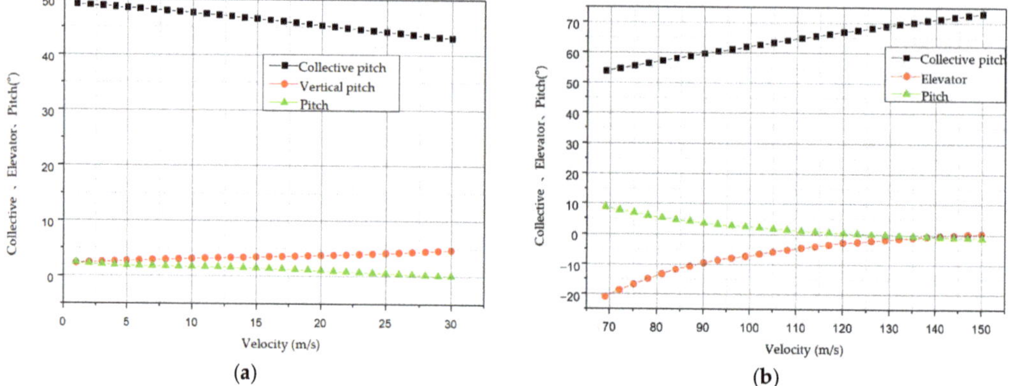

Figure 5. (a) Trimming results in helicopter mode; (b) Trimming results in flight mode.

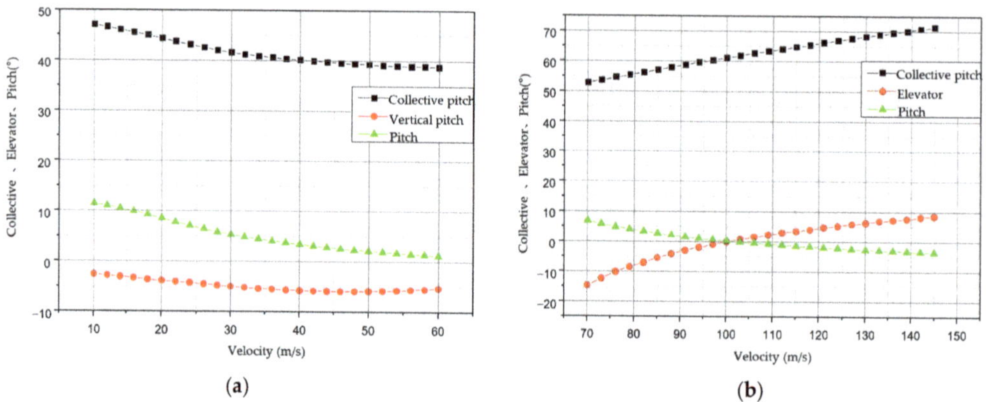

**Figure 6.** (a) Trimming results in transition mode with nacelle tilt angle 15°; (b) trimming results in transition mode with nacelle tilt angle 75°.

Trimming results are analyzed in below:

In helicopter mode: an aircraft starts forward flight from hovering. With the velocity increasing, disturbance from wings to rotors reduces, lift force provided by wings starts to increase, collective pitch input starts to reduce. As the aircraft flight velocity increases, a forward force is necessary to supply, thus need increase vertical pitch. The bigger the flight velocity is, the bigger vertical pitch is necessary to be supplied. At the meantime of vertical pitch increase, nose-down pitch is generated, which means the aircraft's angle of pitch will continuously reduce.

In-flight mode: rotors are equivalent to propellers. As the flight velocity increases, resistance on the airframe and propeller plane increases. To overcome the resistance, it is necessary to increase collective pitch. While with the velocity increases, rudder effectiveness reinforces, so the rudder maneuver decreases continuously.

Transition mode: when the nacelle tilt angle is 15°, use helicopter manipulation method, compared to helicopter mode, forward flight velocity is the same and collective pitch maneuver is quite small, which is mainly caused by nacelle tilt leading to the dramatic reduction of blocking effect of wings on rotors. In the meantime, flight velocity increases and lift force provided by wings increases, so the collective pitch control keeps on reducing. Forward force generated by nacelle tilt, backward force generated by propeller plane moving backward, velocity increasing, pitch angle decreasing, more nose-down and vertical cyclic feathering reduction all together lead to a negative vertical pitch.

When the nacelle tilt angle is 75°, use fixed-wing aircraft control. Compared to aircraft mode, forward flight velocity is the same and collective pitch maneuver is quite small, which is due to the fact that the rotors have been equaled to propellers. As the nacelle angle increases, propeller plane resistance increases along, which needs a bigger collective pitch to generate thrust to obtain balance.

To test accuracy and effectiveness of the model, this paper will take flight model as an example. Given the comparison between the balancing results of the mathematical model, the GTRS results and the actual balancing results of XV-15 is given. The comparison result is as Figure 7.

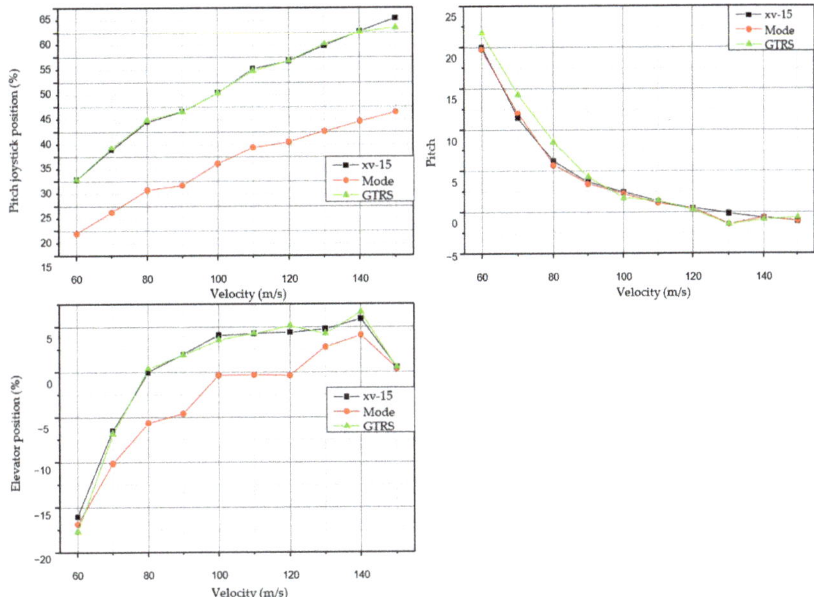

**Figure 7.** Comparison of trimming results.

The balancing results are completely consistent in trend, which shows that the model is feasible and reasonable. However, due to the lack of the data of XV-15, there will inevitably be some differences in the results. the model does not need a lot of experimental data to look up the table, which makes the model have good universality.

## 5. Tiltrotor Maneuvering Stability Characteristics Analysis

Flight dynamic linear differential equations, to simplify analysis and equation solution. Non-linear model can be linearized by following steps as Figure 8:

The above figure describes how to turn a non-linear model to linear model. Linear model is obtained by linearizing a steady flight condition and by trimming kinematic equations of that steady flight condition to obtain corresponding steady-state values.

Balance point is shown as Equation (37):

$$f(x_e, u_e) = 0 \tag{37}$$

where, $x_e$ is the trimmed value of state vector. $u_e$ is the control vector in steady state.

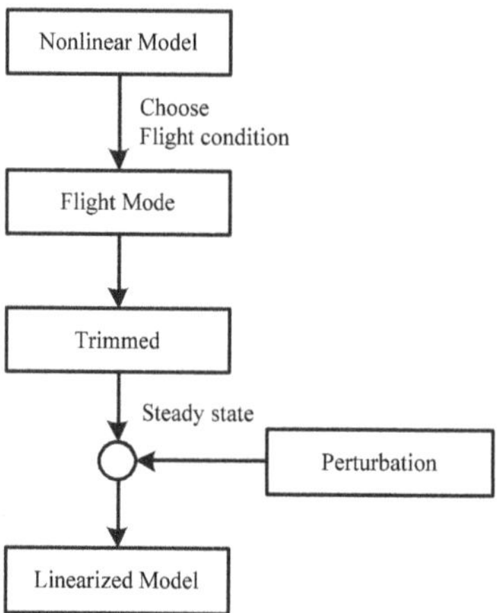


**Figure 8.** Linearizing steps for non-linear model.

At balance point, a small disturbance on flight kinematic equation being assumed and trimmed will get the linearized model, which is written to state space as:

$$\dot{x}(t) = Ax(t) + Bu(t) \tag{38}$$

where, $x(t)$ is state variable, $u(t)$ is control variable. $A$ and $B$ are matrixes of coefficients.

Where, state variable is:

$$x = \begin{bmatrix} u & v & w & p & q & r & \phi & \theta & \psi \end{bmatrix}^T \tag{39}$$

Control variable is:

$$u = \begin{bmatrix} \delta_c & \delta_{cc} & \delta_e & \delta_{ec} & \delta_{ail} & \delta_{ele} & \delta_{rud} \end{bmatrix}^T \tag{40}$$

In Equation (38):

$$A = \begin{bmatrix}
F_x^u & F_x^v & F_x^w & F_x^p & F_x^q & F_x^r & 0 & -mg\cos\theta\cos\phi & mg\cos\theta\sin\phi \\
F_y^u & F_y^v & F_y^w & F_y^p & F_y^q & F_y^r & mg\cos\theta\cos\phi & -mg\sin\theta & 0 \\
F_z^u & F_z^v & F_z^w & F_z^p & F_z^q & F_z^r & -mg\cos\theta\sin\phi & -mg\sin\theta & 0 \\
M_x^u & M_x^v & M_x^w & M_x^p & M_x^q & M_x^r & 0 & 0 & 0 \\
M_y^u & M_y^v & M_y^w & M_y^p & M_y^q & M_y^r & 0 & 0 & 0 \\
M_z^u & M_z^v & M_z^w & M_z^p & M_z^q & M_z^r & 0 & 0 & 0 \\
0 & 0 & 0 & 1 & 0 & \tan\theta & 0 & 0 & 0 \\
0 & 0 & 0 & 0 & 1 & 0 & 0 & 0 & 0 \\
0 & 0 & 0 & 0 & 0 & \sec\theta & 0 & 0 & 0
\end{bmatrix} \tag{41}$$

$$B = \begin{bmatrix}
F_x^{\delta_c} & F_x^{\delta_{cc}} & F_x^{\delta_e} & F_x^{\delta_{ec}} & F_x^{\delta_{ail}} & F_x^{\delta_{ele}} & F_x^{\delta_{rud}} \\
F_y^{\delta_c} & F_y^{\delta_{cc}} & F_y^{\delta_e} & F_y^{\delta_{ec}} & F_y^{\delta_{ail}} & F_y^{\delta_{ele}} & F_y^{\delta_{rud}} \\
F_z^{\delta_c} & F_z^{\delta_{cc}} & F_z^{\delta_e} & F_z^{\delta_{ec}} & F_z^{\delta_{ail}} & F_z^{\delta_{ele}} & F_{xz}^{\delta_{rud}} \\
M_x^{\delta_c} & M_x^{\delta_{cc}} & M_x^{\delta_e} & M_x^{\delta_{ec}} & M_x^{\delta_{ail}} & M_x^{\delta_{ele}} & M_x^{\delta_{rud}} \\
M_y^{\delta_c} & M_y^{\delta_{cc}} & M_y^{\delta_e} & M_y^{\delta_{ec}} & M_y^{\delta_{ail}} & M_y^{\delta_{ele}} & M_y^{\delta_{rud}} \\
M_z^{\delta_c} & M_z^{\delta_{cc}} & M_z^{\delta_e} & M_z^{\delta_{ec}} & M_z^{\delta_{ail}} & M_z^{\delta_{ele}} & M_z^{\delta_{rud}} \\
0 & 0 & 0 & 0 & 0 & 0 & 0 \\
0 & 0 & 0 & 0 & 0 & 0 & 0 \\
0 & 0 & 0 & 0 & 0 & 0 & 0
\end{bmatrix} \tag{42}$$

In the previous section, the aircraft's balance point is trimmed. By the function Linmod in the MATLAB/Simulink environment, one can obtain a non-linear model near the balance point linearized and obtain an *A* matrix and *B* matrix.

*5.1. Helicopter Mode*

5.1.1. Derivative Analysis

The parameters of matrix A and matrix B in helicopter mode are shown in Tables 1 and 2. Stability derivative analysis on Matrix A is performed as below:

**Table 1.** Linearized state matrix A in helicopter mode.

| Matrix A | $\partial u$ | $\partial v$ | $\partial w$ | $\partial p$ | $\partial q$ | $\partial r$ | $\partial \phi$ | $\partial \theta$ | $\partial \psi$ |
|---|---|---|---|---|---|---|---|---|---|
| $\partial F_x$ | −0.0580 | 0 | 0.0187 | 0 | 0.4600 | 0 | 0 | −9.8046 | 0 |
| $\partial F_y$ | 0 | 0.5453 | 0 | 0.7229 | 0 | −0.2858 | 9.8046 | 0 | 0 |
| $\partial F_z$ | −0.1099 | −0.0003 | −0.0944 | 0 | 0.3296 | 0 | 0 | −0.4429 | 0 |
| $\partial M_x$ | 0 | 0.1097 | 0 | 0.0719 | 0 | 0.0423 | 0 | 0 | 0 |
| $\partial M_y$ | 0.0078 | 0 | −0.0136 | 0 | −0.1815 | 0 | 0 | 0 | 0 |
| $\partial M_z$ | 0 | 0.4464 | 0 | 0.9910 | 0 | −0.0660 | 0 | 0 | 0 |
| $\partial \phi$ | 0 | 0 | 0 | 1 | 0 | 0 | 0 | 0 | 0 |
| $\partial \theta$ | 0 | 0 | 0 | 0 | 1 | 0.0452 | 0 | 0 | 0 |
| $\partial \psi$ | 0 | 0 | 0 | 0 | 0 | 1.0010 | 0 | 0 | 0 |

**Table 2.** Linearized input matrix B in helicopter mode.

| Matrix B | $\partial \delta_c$ | $\partial \delta_{cc}$ | $\partial \delta_e$ | $\partial \delta_{ec}$ | $\partial \delta_{ail}$ | $\partial \delta_{ele}$ | $\partial \delta_{rud}$ |
|---|---|---|---|---|---|---|---|
| $\partial F_x$ | 2.1930 | 0 | 11.3653 | 0 | 0 | 0 | 0 |
| $\partial F_y$ | 0 | 3.3550 | 0 | 0.1902 | 0 | 0 | 0 |
| $\partial F_z$ | −41.4224 | 0 | −0.0880 | 0 | 0 | −0.0001 | 0 |
| $\partial M_x$ | 0 | −17.0922 | 0 | −0.0929 | 0 | 0 | 0 |
| $\partial M_y$ | −0.1171 | 0 | −4.4119 | 0 | 0 | −0.0001 | 0 |
| $\partial M_z$ | 0 | 0.9384 | 0 | −3.6585 | −3.6585 | 0 | 0 |
| $\partial \phi$ | 0 | 0 | 0 | 0 | 0 | 0 | 0 |
| $\partial \theta$ | 0 | 0 | 0 | 0 | 0 | 0 | 0 |
| $\partial \psi$ | 0 | 0 | 0 | 0 | 0 | 0 | 0 |

$F_x^u$, $F_z^u$: forward flight speed $u$ increases, rotors flap backward, rotor thrust vector is backward, which results in $F_x^u$ as negative. At the meantime, upward thrust of wings and horizontal tail increases, $F_z^u$ as negative.

$F_x^w$, $F_z^w$: $F_x^w$: being small explains that disturbance of vertical speed has little impact on force in X direction. Increase of will enlarge the rotor's angle of attack, thrust of rotor increases, thus $F_x^w$ is positive and $F_z^w$ is negative.

$F_x^q$, $F_z^q$: as an airframe has pitching movement, passive flapping will happen, $a_{1w} = \left(-\frac{16}{\gamma_b}\frac{q}{\Omega} + \frac{p}{\Omega}\right)/\left(1 - \frac{1}{2}\mu^2\right)$ decreases, thus $F_x^q$ and $F_z^q$ are positive.

$F_x^\theta$, $F_z^\theta$: $\theta$ increase leads to the rotor's angle of attack and thrust of rotor increasing, thus $F_z^\theta$ is negative. And, analogous to instability of helicopter angle of attack, rotors reverse backward, $F_x^\theta$ is negative.

$M_y^u$, $M_y^w$ and $M_y^q$: forward flight speed increases, rotor tip plane is reversing, backward force increases, pitch-up moment is generated, $M_y^u$ is positive. $M_y^w$ being zero is since the vertical distance from the rotor hub to the aircraft center is short, moment change is small. As analyzed afore, as there is pitching movement, passive flapping happens, the increased force in X and Y directions leads to nose-down moment, and thus $M_y^q$ is negative.

5.1.2. Eigenvalue Analysis

Eigenvalue is a very important characteristic of analyzing stability of the model. It demonstrates a vehicle's motion modes under different flight conditions. Since longitudinal and lateral coupling of a helicopter is severe, for easier analysis, this paper is going to discuss longitudinal module and lateral module in helicopter mode separately.

By observing eigenvalues as Table 3, it can be found that a tiltrotor's right half plane in helicopter mode has roots, which means its stability in such mode is poor and control system must be applied to improve stability.

**Table 3.** Eigenvalue of longitudinal module and lateral module in helicopter mode.

| Helicopter Hovering Status | Eigenvalue |
|---|---|
| Longitudinal | $0.1850 + 0.3665i$ <br> $0.1850 - 0.3665i$ <br> $-0.3520 + 0.0643i$ <br> $-0.3520 - 0.0643i$ |
| Lateral | $-0.2862 - 0.9332i$ <br> $-0.2862 + 0.9332i$ <br> $-0.2257$ <br> $1.3495$ <br> $0$ |

Longitudinal eigenvalue: motion modes of velocity and angle of attack corresponding to a pair of positive complex conjugate roots are similar, with long period and divergent. Motion modes of angle of attack and angle of pitch corresponding to a pair of negative complex conjugate roots converge fast.

Lateral eigenvalue: the mode of complex conjugate roots is similar to the oscillation mode of longitudinal hovering. Large negative real root corresponds to rolling convergence mode. Since rotors rotate behind the airframe, rotors have larger rolling aerodynamic damping and converge faster. Small negative root represents spiral mode. Zero root means level flight in any heading course has no difference.

*5.2. Flight Mode*

5.2.1. Derivative Analysis

The parameters of matrix A and matrix B in flight mode are shown in Tables 4 and 5. Stability derivative analysis on Matrix A is performed as below:

**Table 4.** Linearized state matrix A in flight mode.

| Matrix A | $\partial u$ | $\partial v$ | $\partial w$ | $\partial p$ | $\partial q$ | $\partial r$ | $\partial \phi$ | $\partial \theta$ | $\partial \psi$ |
|---|---|---|---|---|---|---|---|---|---|
| $\partial F_x$ | $-0.3112$ | 0 | 0.1477 | 0 | $-0.9607$ | $-0.0717$ | 0 | $-9.8143$ | 0 |
| $\partial F_y$ | 0 | $-0.5236$ | $-0.0001$ | 0.7131 | 0 | $-99.0077$ | 9.8143 | 0 | 0 |
| $\partial F_z$ | $-0.1191$ | $-0.8137$ | $-1.1330$ | 0 | 97.3686 | $-0.2118$ | 0 | $-0.0723$ | 0 |
| $\partial M_x$ | 0 | 0.1425 | 0 | $-0.6207$ | 0 | $-0.1000$ | 0 | 0 | 0 |
| $\partial M_y$ | 0.0241 | 0 | $-0.3377$ | 0 | $-1.4017$ | 0.0693 | 0 | 0 | 0 |
| $\partial M_z$ | 0 | 0.0685 | 0 | $-0.1094$ | 0 | $-0.9424$ | 0 | 0 | 0 |
| $\partial \phi$ | 0 | 0 | 0 | 1 | 0 | 0.0074 | 0 | 0 | 0 |
| $\partial \theta$ | 0 | 0 | 0 | 0 | 1 | 0 | 0 | 0 | 0 |
| $\partial \psi$ | 0 | 0 | 0 | 0 | 0 | 1 | 0 | 0 | 0 |

**Table 5.** Linearized input matrix B in flight mode.

| Matrix B | $\partial \delta_c$ | $\partial \delta_{cc}$ | $\partial \delta_e$ | $\partial \delta_{ec}$ | $\partial \delta_{ail}$ | $\partial \delta_{ele}$ | $\partial \delta_{rud}$ |
|---|---|---|---|---|---|---|---|
| $\partial F_x$ | 71.0156 | 0 | $-6.3152$ | 0 | 0 | 0.0794 | 0 |
| $\partial F_y$ | 0 | $-2.9203$ | 0 | 2.5071 | 0 | 0 | $-5.4356$ |
| $\partial F_z$ | $-18.1784$ | 0 | 14.7210 | 0 | 0 | $-10.7707$ | 0 |
| $\partial M_x$ | 0 | $-18.8236$ | 0 | 8.3808 | $-1.0303$ | 0 | $-0.4183$ |
| $\partial M_y$ | $-4.2089$ | 0 | $-3.3294$ | 0 | 0 | $-15.7326$ | 0 |
| $\partial M_z$ | 0 | $-23.4353$ | 0 | 2.3839 | $-0.0223$ | 0 | 2.4838 |
| $\partial \phi$ | 0 | 0 | 0 | 0 | 0 | 0 | 0 |
| $\partial \theta$ | 0 | 0 | 0 | 0 | 0 | 0 | 0 |
| $\partial \psi$ | 0 | 0 | 0 | 0 | 0 | 0 | 0 |

$F_x^u$, $F_z^u$: forward flight speed $u$ increases, airframe resistance and propeller resistance increase, and thrust generated by wings increase, which leads to $F_x^u$ as negative. Thrusts of wings and horizontal tail increase, $F_z^u$ as negative.

$F_x^w$, $F_z^w$: increases, vertical upward air flow increases, angle of attack of wings increases, lift component in X direction increases, so $F_x^w$ is positive. increase leads to the increase of thrust of wings, $F_z^w$ is negative.

$F_x^q$, $F_z^q$: $q$ increase, downward airflow at wings is generated, thrust of wings reduces, $F_z^q$ is positive. Backward thrust component of wings increases, $F_x^q$ is negative.

$F_x^\theta$, $F_z^\theta$: $\theta$ increase, thrust vector of wings is backward, $F_x^\theta$ is negative. Since the aircraft has pitch-up already, with the increase of $\theta$, it pitches up more. Thrust is reducing in Z direction, $F_z^\theta$ is negative.

$M_y^u$, $M_y^w$ and $M_y^q$: $u$ increase, airframe resistance and propeller resistance increase, which leads to the tiltrotor pitching up, $M_y^u$ is positive. $w$ increase, thrust of wings is forwarding, pitching-up moment reduces, $M_y^w$ is negative. $q$ increases, significant upward airflow at horizontal tail is generated, thrust of horizontal tail in Z direction increases, which leads to bigger nose-down moment, and $M_y^q$ is negative.

5.2.2. Eigenvalue Analysis

Eigenvalues in flight mode is shown as Table 6. Different from the helicopter mode, the flight mode has less severe longitudinal and lateral coupling than that in helicopter mode and thus has much better stability. Eigenvalue analysis at forward speed of 100 m/s in flight mode is performed.

**Table 6.** Eigenvalue of longitudinal module and lateral module in flight mode.

| Flight Mode 100 m/s | Eigenvalue |
| --- | --- |
| Longitudinal | $-0.5295$<br>$-0.2232$<br>$-1.2688 + 5.7348i$<br>$-1.2688 - 5.7348i$ |
| Lateral | $-0.9300 + 2.6358i$<br>$-0.9300 - 2.6358i$<br>$0.3028$<br>$-0.5295$<br>$0$ |

Longitudinal eigenvalue: the longitudinal mode has long period and short period (two modes). The external force generated after receiving disturbance makes it hard to change flight speed but easy to change the angle of attack (incl. angle of pitch). The long period mode corresponds to speed mode while the short period mode corresponds to the angle of attack variation.

Lateral motion: large complex roots represent rolling convergence mode, since wings converge quickly for large aerodynamic damping. Small complex roots correspond to the spiral mode. A pair of complex conjugate roots represent oscillating motion mode, also known as Dutch roll mode, a motion mode in which heading course and rolling are recurrent.

## 6. Conclusions

This paper adopts the dividing modeling method, which breaks down a tiltrotor into five parts, rotor, wing, fuselage, horizontal tail and vertical fin, develops aerodynamic models for each part and thus obtains force and moment generated by each part. The force and moment then is converted to the airframe coordinate frame. By blade element theory, the rotor's dynamic model and rotor flapping angle expression are built. Then, according to mature lifting line theory, dynamic models of the wing, horizontal tail, and vertical fin are built. At the meantime, the rotor's dynamic interference on wings and nacelle tilt's variance against center of gravity and moment of inertia are considered. A more perfect mathematical model of the tilt rotor aircraft is established.

In the MATLAB/Simulink simulation environment, a non-linear tiltrotor simulation model is built, Trim command is applied to trim the tiltrotor and an XV-15 tiltrotor is taken as an example to validate accuracy and rationality of the model developed. Due to the lack of complete XV-15 data, there will inevitably be some differences in the results. The results show that the trend of the model is completely consistent with the actual data and GTRS model, which validate the accuracy and rationality of the model developed.

In the end, the non-linear simulation model is linearized to get State-space matrix, the stability derivative and eigenvalue of the tiltrotor are analyzed, and furthermore the tiltrotor's stability in each flight mode is studied. The stability of the aircraft in the helicopter mode is poor, which must be improved through the control system. The vertical and horizontal coupling of the flight mode is not as serious as that of the helicopter mode, and the stability is much better than that of the helicopter mode. It therefore provides a theoretical basis for the subsequent controller setting.

**Author Contributions:** Conceptualization, H.S.; methodology, H.S.; validation, H.S.; formal analysis, C.Z.; investigation, C.Z.; data curation, Y.X.; writing—original draft preparation, H.S.; writing—review and editing, C.Z.; supervision, H.S. All authors have read and agreed to the published version of the manuscript.

**Funding:** This work was supported by Funding of National Key Laboratory of Rotorcraft Aeromechanics (No. 61422202108), National Natural Science Foundation of China (No. 51906103, No. 52176009).

**Institutional Review Board Statement:** Not applicable.

**Informed Consent Statement:** Not applicable.

**Data Availability Statement:** The data presented in this study are available on request from the corresponding author.

**Conflicts of Interest:** The authors declare no conflict of interest. The funders had no role in the design of the study; in the collection, analyses, or interpretation of data; in the writing of the manuscript; or in the decision to publish the results.

## Nomenclature

| | |
|---|---|
| $m_{NAC}$ | mass of the nacelle system |
| $\beta_M$ | nacelle angle |
| $R_H$ | rotor's height against the wing |
| $m$ | gross weight of the airframe |
| $I$ | moment of inertia |
| $I_0$ | moment of inertia in helicopter mode |
| $KI$ | moment of inertia coefficient |
| $I_b$ | blade mass moment of inertia |
| $\Omega$ | angular velocity of blade |
| $\beta$ | sideslip angle |
| $M_T$ | blade flapping aerodynamic moment |
| $M_s$ | mass moment of the blade to the swinging hinge |
| $g$ | gravitational acceleration |
| $r$ | radius of blade |
| $R$ | hub center height |
| $\phi$ | roll angle |
| $\theta$ | pitch angle |
| $\psi$ | yaw angle |
| $u$ | linear velocity of the aircraft's $x$-axis |
| $v$ | linear velocity of the aircraft's $y$-axis |
| $w$ | linear velocity of the aircraft's $z$-axis |
| $p$ | angular velocity of the aircraft's $x$-axis |

| | |
|---|---|
| $q$ | angular velocity of the aircraft's $y$-axis |
| $r$ | angular velocity of the aircraft's $z$-axis |
| $\beta_*$ | inflow angle |
| $C_y$ | lift coefficient of blade airfoil |
| $C_x$ | drag coefficient of blade airfoil |
| $W$ | absolute velocity of air |
| $\rho$ | air density |
| $T_s$ | elemental thrust |
| $H_s$ | elemental backward force |
| $S_s$ | elemental side force |
| $M_k$ | elemental moment |
| $C_L$ | lift coefficient |
| $C_D$ | resistance coefficient |
| $C_M$ | moment coefficient |
| $A$ | stress area |
| $L$ | lift force |
| $D$ | resistance force |
| $T_s$ | elemental thrust |
| $H_s$ | elemental backward force |
| $S_s$ | elemental side force |
| $M_k$ | elemental moment |
| $S_{wss_{max}}$ | maximum area of slipstream zone |
| $\mu_{max}$ | advance ratio in helicopter mode |
| $S_{wfs}$ | result of wing area deducting slipstream zone area |
| $P_{wsl0}$ | wing's aerodynamic center position |
| $\omega$ | rotational speed in the international system of units |
| $\lambda$ | propeller's aerodynamic efficiency |
| $\delta$ | manipulation quantity |
| $\delta_c$ | rotor collective pitch |
| $\delta_e$ | longitudinal cyclic pitch control amount |
| $\delta_{ail}$ | aileron control amount |
| $\delta_{rud}$ | rudder control amount |
| $\delta_{cc}$ | collective differential control amount |
| $\delta_{ec}$ | longitudinal period variable pitch differential |
| $\delta_{ele}$ | elevator control amount |
| **Subscript** | |
| $x$ | vector component corresponding to the $x$-axis of the coordinate system |
| $y$ | vector component corresponding to the $y$-axis of the coordinate system |
| $z$ | vector component corresponding to the $z$-axis of the coordinate system |
| $h$ | parameter at the rotor hub center |
| $R$ | parameter of the right rotor |
| $W$ | wing |
| $B$ | body shafting |
| $NAC$ | nacelle shafting |
| $H$ | hub shafting |
| $HW$ | hub wind shaft system |
| $Wfl$ | parameter in left wing free flow |
| $ail$ | parameter in aileron |
| $Wsl$ | parameter in left wing slipstream zone |
| $wfs$ | parameter in free flow zone |
| $wss$ | parameter in slipstream zone |
| $f$ | parameter at fuselage aerodynamic center |
| $Ht$ | parameter of a horizontal tail |
| $VT$ | parameter at vertical fin aerodynamic center |

## References


1. Harendra, P.B.; Joglekar, M.J.; Gaffey, T.M.; Marr, R.L. V/STOL Tilt Rotor Study. Volume 5: A Mathematical Model for Real Time Flight Simulation of the Bell Model 301 Tilt Rotor Research Aircraft. NASA CR-114614. 1973. Available online: https://ntrs.nasa.gov/citations/19730022217 (accessed on 10 March 2022).
2. Foster, M.; Textron, B.H. Evolution of Tiltrotor Aircraft. In Proceedings of the AIAA/ICAS International Air and Space Symposium and Exposition: The Next 100 Years, Dayton, OH, USA, 14–17 July 2003.
3. Alli, P. Erica: The European Tiltrotor. Design and Critical Technology Projects. In Proceedings of the AiAA International Air & Space Symposium & Exposition: The Next 100 Years, Dayton, OH, USA, 14–17 July 2003.
4. Choi, S.W.; Kang, Y.; Chang, S.; Koo, S.; Kim, J.M. Development and conversion flight test of a small tiltrotor unmanned aerial vehicle. *J. Aircr.* **2010**, *47*, 730–732. [CrossRef]
5. Belardo, M.; Marano, A.D.; Beretta, J.; Diodati, G.; Graziano, M.; Capasso, M.; Ariola, P.; Orlando, S.; Di Caprio, F.; Paletta, N.; et al. Wing Structure of the Next-Generation Civil Tiltrotor: From Concept to Preliminary Design. *Aerospace* **2021**, *8*, 102. [CrossRef]
6. Kim, T.; Shin, S.J. Advanced Analysis on Tiltrotor Aircraft Flutter Stability, Including Unsteady Aerodynamics. *AIAA J.* **2008**, *46*, 1002–1012. [CrossRef]
7. Carlson, E.B.; Zhao, Y.J. Optimal city-center takeoff operation of tiltrotor aircraft in one engine failure. *J. Aerosp. Eng.* **2004**, *17*, 26–39. [CrossRef]
8. Carlson, E.B.; Zhao, Y.J. Prediction of tiltrotor height-velocity diagrams using optimal control theory. *J. Aircr.* **2003**, *40*, 896–905. [CrossRef]
9. Wang, X.; Cai, L. Mathematical modeling and control of a tilt-rotor aircraft. *Aerosp. Sci. Technol.* **2015**, *47*, 473–492. [CrossRef]
10. Kleinhesselink, K.M. Stability and Control Modeling of Tiltrotor Aircraft. Ph.D. Thesis, University of Maryland, College Park, MD, USA, 2007.
11. Miller, M.; Narkiewicz, J. Tiltrotor modelling for simulation in various flight conditions. *J. Theor. Appl. Mech.* **2006**, *44*, 881–906.
12. Wu, W.; Chen, R. An improved online system identification method for tiltrotor aircraft-ScienceDirect. *Aerosp. Sci. Technol.* **2021**, *110*, 106491. [CrossRef]
13. Jategaonkar, R.V.; Fischenberg, D.; Grünhagen, W.V. Aerodynamic Modeling and System Identification from Flight Data-Recent Applications at DLR. *J. Aircr.* **2004**, *41*, 651–691. [CrossRef]
14. Bicker, A. Quadrotor Comprehensive Identification from Frequency Responses. *Sci. Eng. Res.* **2014**, *5*, 1438–1447.
15. Lichota, P. Multi-Axis Inputs for Identification of a Reconfigurable Fixed-Wing UAV. *Aerospace* **2020**, *76*, 113. [CrossRef]
16. Yang, X.; Zhu, J.; Huang, X.; Hu, C.; Sun, Z. Modeling and simulation for tiltrotor airplane. *Acta Aeronaut. Astronaut. Sin.-Ser. A B-* **2006**, *27*, 584.
17. Abà, A.; Barra, F.; Capone, P.; Guglieri, G. Mathematical Modelling of Gimballed Tilt-Rotors for Real-Time Flight Simulation. *Aerospace* **2020**, *7*, 124. [CrossRef]
18. Wang, Q.; Wu, W. Modelling and Analysis of Tiltrotor Aircraft for Flight Control Design. *Inf. Technol. J.* **2014**, *13*, 885–894.
19. Haixu, L.; Xiangju, Q.; Weijun, W. Multi-body Motion Modeling and Simulation for Tilt Rotor Aircraft. *Chin. J. Aeronaut.* **2010**, *23*, 415–422. [CrossRef]
20. Gao, Z.; Chen, R.L. *Helicopter Flight Dynamics*; Science Press: Beijing, China, 2003.

*drones*

MDPI

*Article*

# Design and Implementation of Sensor Platform for UAV-Based Target Tracking and Obstacle Avoidance

Abera Tullu [1], Mostafa Hassanalian [2] and Ho-Yon Hwang [3,*]

1    Department of Aerospace Engineering, Sejong University, 209, Neungdong-ro, Gwangjin-gu, Seoul 05006, Korea; tuab@sejong.ac.kr

2    Department of Mechanical Engineering, New Mexico Institute of Mining and Technology, Socorro, NM 87801, USA; mostafa.hassanalian@nmt.edu

3    Department of Aerospace Engineering, and Convergence Engineering for Intelligence Drone, Sejong University, Seoul 05006, Korea

*    Correspondence: hyhwang@sejong.edu; Tel.: +82-10-6575-2282

**Abstract:** Small-scale unmanned aerial vehicles are being deployed in urban areas for missions such as ground target tracking, crime scene monitoring, and traffic management. Aerial vehicles deployed in such cluttered environments are required to have robust autonomous navigation with both target tracking and obstacle avoidance capabilities. To this end, this work presents a simple-to-design but effective steerable sensor platform and its implementation techniques for both obstacle avoidance and target tracking. The proposed platform is a 2-axis gimbal system capable of roll and pitch/yaw. The mathematical model that governs the dynamics of this platform is developed. The performance of the platform is validated through a software-in-the-loop simulation. The simulation results show that the platform can be effectively steered to all regions of interest except backward. With its design layout and mount location, the platform can engage sensors for obstacle avoidance and target tracking as per requirements. Moreover, steering the platform in any direction does not induce aerodynamic instability on the unmanned aerial vehicle in mission.

**Keywords:** autonomous navigation; gimbal design; obstacle avoidance; target tracking; unmanned aerial vehicles; law enforcement

**Citation:** Tullu, A.; Hassanalian, M.; Hwang, H.-Y. Design and Implementation of Sensor Platform for UAV-Based Target Tracking and Obstacle Avoidance. *Drones* **2022**, *6*, 89. https://doi.org/10.3390/drones6040089

Academic Editors: Daobo Wang and Zain Anwar Ali

Received: 10 March 2022
Accepted: 27 March 2022
Published: 29 March 2022



## 1. Introduction

The emergence of small-scale unmanned aerial vehicles (UAVs) revolutionizes the way missions are conducted in various sectors. Being small in size and agile in their operation, these UAVs can operate in cluttered and confined environments. With the advent of miniature sensors, as well as computer vision and machine learning technologies [1–3], these small-scale UAVs can acquire artificial intelligence and be deployed to monitor production processing in complex industrial facilities [4].

The feasibility of deploying small-scale UAVs for missions in urban areas is already proven in various scenarios [5]. Douglas and James [6] discussed the implementation and operational feasibility of such UAVs for law enforcement. As reported in their conclusion, the use of small UAVs in both urban environments and open areas is feasible. Romeo [7] reported that, in 2020 alone, more than 1500 public safety departments of the United States implemented small UAVs as situation awareness tools.

However, small-scale UAVs have various technological constraints [8–11], including insufficient power sources for long flight endurance and limited payload weight. Perry et al. [12] reported that, of all challenges and limitations of small-scale UAVs, the weight constraint is the most serious setback to their technological advancement. Contrary to this, such UAVs are required to have multiple sensors on board to have robust autonomous navigation in urban areas where access to GPS-based navigation is unreliable. The requirement of multiple sensors onboard a small UAVs also poses challenges, such as sensors data fusion and the high purchase cost of the sensors. Jixian [13] reported that acquiring effective methodology for multi-sensor data fusion and interpretation is not yet realized. Bahador et al. [14] also reported various reasons for the challenges in multi-sensor data fusion, including data imperfection, inconsistent data, and variation in operational timing of sensors.

The hurdle to overcome is, therefore, not only the weight but also the complexity of multi-sensor data fusion. In resolving the aforementioned setbacks—flight endurance, weight constraint, and sensors data fusion—reducing the number of sensors is a promising approach. In closing the gap between the need to scan the surrounding environment and reducing number of sensors, it is necessary to implement a movable sensor platform that enables few sensors to scan the environment around the UAV.

There are various movable sensor platforms (gimbals) which are operating successfully in the objectives they are designed and deployed for [15–17]. Steerable degrees of freedom (DoF) and mount locations dictate the design objectives of the sensor platforms. Most, if not all, of the existing platforms are UAV-belly-mounted platforms; hence, their feasibility for obstacle avoidance, which is critical for operating in urban areas [18–20], is controvertible. When obstacles such as high, multi-storey buildings are encountered, there is the possibility that the UAV has to pass over the buildings. In such cases, scanning the environment above the UAV is required. Belly-mounted gimbals can not engage sensors for obstacle detection in such scenarios. Consequently, this work presents a simple-to-design but effective solution for improving the performance of the sensor platforms with a different mount location and orientation for small-scale, fixed-wing UAVs.

For the designed gimbals to operate as desired, mathematical models that control their dynamics are essential. Mohd et al. [21] developed a mathematical model for a two-axis gimbal system with pan-and-tilt DoF. The developed mathematical model is specific to the design, mount location, and orientation of the gimbal system. For the same DoF gimbal system, but with different orientation, Alexander et al. [22] developed a specific mathematical model for the gimbal. For the control of the dynamics of a three-axis gimbal system, Aytaç and Rıfat [23] formulated a unique mathematical model for the system. The proposed sensor platform has unique design layout, mount location, and orientation. Therefore, a mathematical model specific to this platform is formulated and a control algorithm is developed based on the mathematical model.

Constrained to these requirements, this paper proposes steerable sensor platform design and its implementation techniques, as described in subsequent sections. A problem statement and the methodology are presented in Section 2. The custom sensor platform design and its operational modes and techniques are described in Section 3. In Section 4, the mathematical model that governs dynamics of the platform is derived. In Section 5, the designed platform performance testing methods are discussed. Results and discussions are given in Section 6. Conclusions and future works are given in Section 7.

## 2. Problem Statement and Methodology

The use of small-scale UAVs in urban areas incurs required but incompatible features. A small-scale UAV is required to operate in urban areas because its potential danger in the event of a crash is low. To autonomously navigate in urban areas, this UAV requires multiple sensors onboard. However, small-scale UAV is highly weight-constrained. Therefore, the plausible approach to alleviate the incompatibility is to reduce the number of sensors onboard the UAV. The approach to reduce the number of sensors onboard the UAV remarkably resolves multiple issues: payload weight, technical challenge of sensors integration, computational burdens of data fusion, and sensors purchase costs.

Rather than rigidly mounting multiple sensors on different sides of the UAV, it is feasible to mount few sensors on a movable platform, so that the sensors can be steered to scan regions of interest. There are various movable sensor platforms available for small-scale UAVs. However, they are designed to be mounted under the belly of the UAVs. This can be conceivable for target tracking mission. However, since obstacle avoidance is one of the critical requirements for missions in urban areas, such belly mounting is not feasible. Moreover, belly-mounted platforms induce aerodynamic instability to certain extent. The flight control surfaces have to counteract this instability, which drains the power source.

The aforementioned studies imply that, for civilian UAVs operating in urban areas, there has to be a new design of sensor platform that can engage sensors for obstacle avoidance, in addition to other requirements. The design and mount location of the sensor platform should enable sensors for obstacle avoidance and for monitoring a region of interest or track moving target.

Taking these into consideration, this research work presents a design and technical implementation of low-cost and light-weight steerable sensor platform that can be mounted on the nose of fixed-wing VTOL UAVs. The design and mount location of the proposed sensor platform are such that the platform can engage the sensors for both obstacle avoidance and target tracking. In addition to the tracking and avoidance capabilities, the platform is designed with the intention that it does not induce any aerodynamic instability while engaging sensors in different direction or being at idle state.

### 3. Custom Platform Design

Prior to the design of a sensor platform, the type of the intended UAV on which the platform is to be mounted should be known. The mount location of the platform is also determined by the type of mission the UAV is design for.

#### 3.1. Airframe Selection

The selection of UAV type depends on the mission requirement. As indicated in the title of this work, the UAV is required to have target tracking capability. For target tracking, a UAV has to cruise with high speed that exceeds the speed of the potential target. A fixed-wing UAV is appropriate for this mission. However, the challenge that comes with fixed-wing UAVs is the requirement of runway for take-off and landing, as well as the very low stall speed in the case when the target being tracked is moving slowly. As law enforcement operations are often conducted in urban areas, constructing runways in every law enforcement compound is infeasible. Moreover, law enforcement missions, such as crime scene monitoring, need a UAV with hovering capability. Such UAVs often have a multicopter-type airframe, which lacks high cruise speed.

Referring to the aforementioned requirements, a hybrid of the multicopter and fixed-wing UAVs is compulsory. A VTOL UAV can change mode from multicopter to fixed-wing or vice versa depending on the status of the target being tracked, mission type, and the environment it is flying in. Furthermore, VTOL UAV does not require runway or launch pad for take-off or trap net for landing as the law enforcement offices are often in highly populated cities. As a result, a fixed-wing UAV with vertical take-off and landing (VTOL) capabilities is selected as preferable airframe.

#### 3.2. Platform Design

To the knowledge of the authors, there is no extant steerable sensor platform designed to be nose-mounted on a fixed-wing VTOL UAV for missions requiring obstacle avoidance and target tracking. To fill this gap, a custom sensor platform is designed and its implementation techniques are explained. The sensor platform, presented in this work, is designed in such a way that it can be mounted on the nose of a UAV and, hence, has better aerodynamic efficiency and capabilities to scan environments around the UAV except the backside, which is not important for either target tracking or collision avoidance. Moreover, a simple platform layout that avoids complex design philosophy and control system is preferred to reduce the platform design and manufacturing burdens as well as the required material purchase costs. This is appropriate for making the purchase cost of small-scale UAVs affordable. For the design of the sensors platform presented in this study, blender version 2.91—an open source 3D modeling software—was used. The designed components of the platform and their assembly are shown in Figure 1. The two quarter spherical shells (canopies) and the central bay make a full spherical shell of diameter 138.6 mm. The two canopies, the central bay, boom, and ring components, shown in Figure 1a, form the complete sensor platform. This platform is designed in such a way that it carries a Sony FDR-X3000 camera for imagery data input and LIDAR lite v3 sensor for obstacle ranging. Both the camera and the LiDAR are mounted on the central bay of the platform. They are mounted on sensor case, as shown in Figure 1b.

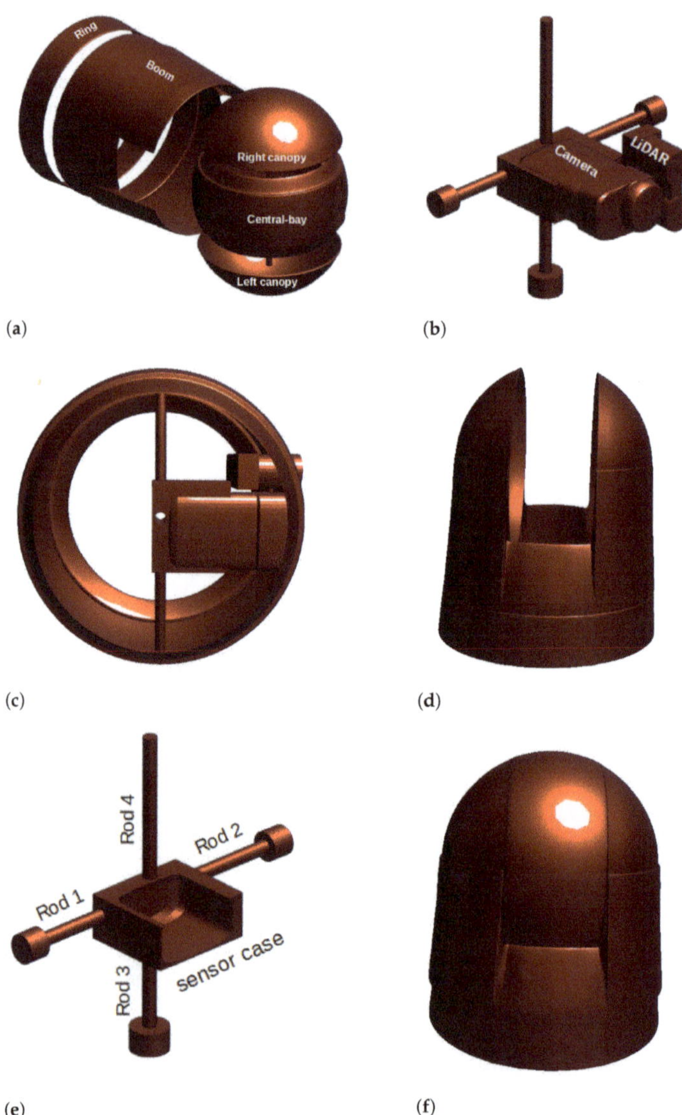


**Figure 1.** Platform components and their configuration. (**a**) Components of the platform; (**b**) sensors configuration in the platform; (**c**) inner gimbal; (**d**) outer gimbal; (**e**) sensor case and connection rods; (**f**) assembled platform.

The inner gimbal, comprising the central bay, the camera, the camera case, LIDAR, and the rods, is shown in Figure 1c. These components are fixed to one another and the pitch/yaw as a single rigid body. The outer gimbal is composed of the boom and the two canopies, as shown in Figure 1d. The four rods, shown in Figure 1e are used to connect the sensor case to the inner gimbal as well as the inner gimbal to the outer gimbal. Rod 1 and rod 2 connect the sensor case to the inner gimbal, and rod 3 connects the inner gimbal to the outer gimbal. The servo motor (not shown here) is connected to rod 4 to control the pitch/yaw motion of the inner gimbal. The inner has grooves that maintain the pitch/yaw smoothly on the rims of the two canopies. The roll motion of the outer gimbal is controlled by a servo motor fixed to the nose of a UAV (not shown). The outer gimbal rolls with its groove, sliding inside the ring. The ring is the part of the platform that is fixed to the nose of UAV. The assembled platform is shown in Figure 1f. The detailed specification of the designed sensor platform is given in Table 1.

**Table 1.** Specifications of the designed sensor platform.

| Parameter | Unit | Value |
| --- | --- | --- |
| platform thickness | mm | 3 |
| canopy diameter | mm | 128.3 |
| boom—front diameter | mm | 140.1 |
| boom—rear diameter | mm | 153.9 |
| boom—length | mm | 99.4 |
| ring—front diameter | mm | 156.0 |
| ring—rear diameter | mm | 156.9 |

*3.3. Platform Implementation Techniques*

The platform is designed with the intention that it enables onboard sensors scan the environment around a UAV for both obstacle avoidance and target tracking purposes. To this end, the appropriate mount location of the platform is on the nose of the fixed-wing VTOL UAV. The platform operation technique relies on two servo motors capable of steering the platform to all possible directions. Servo motors are implemented for gimbal control, though it induces vibration, unlike the brushless motors often used for gimbal control. The implementation of servos reduces both payload weight and purchase cost as compared with brushless motors. Moreover, since the objective of this platform design is for target tracking and collision avoidance, the image quality is not as big a concern as it is for other mission objectives such as aerial photography. Dampers are used to reduce the vibration issue. One servo is fixed to the nose of the UAV and the other servo is fixed to the right canopy of the platform. Both the right and the left canopies are fixed to the outer gimbal of the platform. The servo fixed to the nose of the UAV controls the rolling movement of the platform about the x-axis and the servo—fixed to the right canopy of the platform—controls the pitch or yaw movement of the inner gimbal of the platform about the y-axis. These roll and pitch/yaw directions of the platform are shown in Figure 2.

The two servos control the movement of the platform independently. The roll in the platform tilts inner gimbal. If the platform roll angle is $0°$, then the inner gimbal can pitch either up or down. If the platform rolls by $\pm90°$ (clockwise or anti-clockwise), then the inner gimbal can yaw either left or right. With such configuration, the platform is capable of rolling in the range of $[-90°, +90°]$ and pitching or yawing in the range of $[-90°, +90°]$.

To monitor the environment above or below the UAV, only the inner gimbal pitches up or down, while the other components of the platform remain tied to the UAV. To monitor the environment either to the left or to the right of the UAV, the whole platform has to first roll, followed by the yawing of the inner gimbal. The combination of rolling of the whole platform and pitching/yawing of the inner gimbal enables the UAV to monitor any direction. In such way, the regions of interest, except the rear side of the UAV, can be scanned making use of only two servos. In all these engagements, there is no aerodynamic instability due to the movement of the platform.

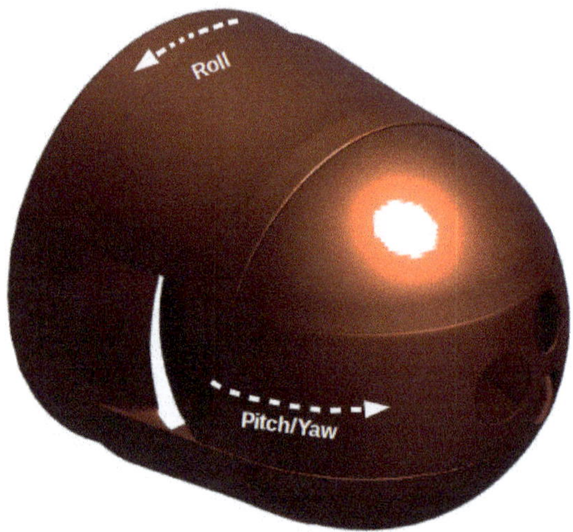



**Figure 2.** Operational techniques of the platform.

The designed platform has generic applications that include the mission objectives of all belly-mounted platforms. Furthermore, although the design requirements of the sensor platform are constrained to target tracking and obstacle avoidance capabilities, with its wide range of angle of view, the platform has potential applications for various missions, such as door-to-door package delivery, infrastructure monitoring, sewer inspection, and as a visual guide for aerial-robot-based repair and painting of high, multi-storey buildings.

## 4. Platform Motion Control

### 4.1. Kinematics of the Platform

The kinematics of the platform deals with determination of linear and angular positions, velocities, and acceleration of the two gimbals—the inner and outer gimbals—in such a way that the attached sensors focus on a region of interest. The orientations of the two gimbals in their idle states are shown in Figure 3. The origin of the camera frame ($C$) is on the rim of gimbal 2 and the axes of the gimbals and the camera are as shown in the figure. The hidden axes of these frames are generated by right-hand rule.

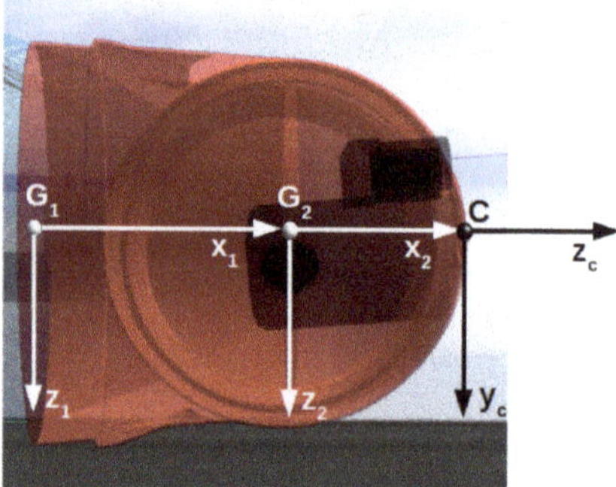



**Figure 3.** Frames of reference for the gimbals and camera.

The two gimbals are considered as rigid bodies whose motions are constrained through joints, and the motion of one affects the other. As shown in Figure 4a, the origin of the body frame $b$ is located at center of gravity of the UAV, the relative position of frame $G_1$ with respect to body frame is $(d_{1x}, 0, d_{1z})$, and the relative position of frame $G_2$ with respect to frame $G_1$ is $(d_{2x}, 0, 0)$. Gimbal 2 is attached to gimbal 1 with a revolute joint at frame $G_2$ and constrained to rotate about an axis perpendicular to $x_2 z_2$-plane. Gimbal 1 is attached to the UAV (henceforth: body) frame with revolute joint at frame $G_1$ and rotates in the $y_1 z_1$-plane, as shown in Figure 4b. $\alpha$ and $\beta$ represent arbitrary roll and pitch angles made by gimbal 1 about the body frame and gimbal 2 about the gimbal 1 frame, respectively.

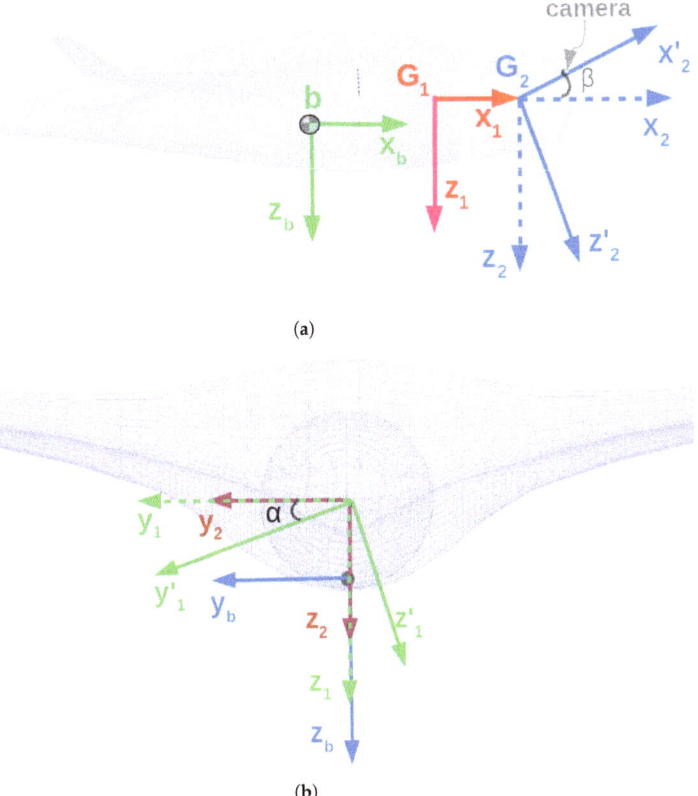


**Figure 4.** Coordinate frame transformation. (**a**) Side view; (**b**) front view.

### 4.1.1. Coordinated Frame Transformation

Coordinated frame transformation is necessary, where any value obtained in one frame can be derived in another frame. Let the coordinate frame axes of the camera are represented by $(x_c, y_c, z_c)$ where $z_c$-axis is parallel to $x_2$-axis and $x_c$-axis is parallel to $y_2$-axis. Therefore, any known information in the camera frame can be transformed to the gimbal 2 frame, as follows:

$$^2T_c = \begin{bmatrix} ^2R_c & ^2d_c \\ 0_{1\times3} & 1 \end{bmatrix} = \begin{bmatrix} 0 & 0 & 1 & r \\ 1 & 0 & 0 & 0 \\ 0 & 1 & 0 & 0 \\ 0 & 0 & 0 & 1 \end{bmatrix} \tag{1}$$

where $^2T_c$ is a homogeneous coordinate transformation from camera frame to frame $G_2$. $^2R_c$ and $^2d_c$ are the orientation and translation of the camera frame, with respect to gimbal 2 frame, respectively, and $r$ is the radius of gimbal 2. Similarly, the homogeneous coordinate transformation from $G_2$ frame to $G_1$ frame and from $G_1$ to body frame, respectively, are as follows:

$$^1T_2 = \begin{bmatrix} \cos\beta & 0 & -\sin\beta & d_{2x} \\ 0 & 1 & 0 & 0 \\ \sin\beta & 0 & \cos\beta & 0 \\ 0 & 0 & 0 & 1 \end{bmatrix} \quad \text{and} \quad {}^bT_1 = \begin{bmatrix} 1 & 0 & 0 & d_{1x} \\ 0 & \cos\alpha & \sin\alpha & 0 \\ 0 & -\sin\alpha & \cos\alpha & d_{1z} \\ 0 & 0 & 0 & 1 \end{bmatrix} \tag{2}$$

Following the two transformations given in Equation (2), the transformation from gimbal 2 to body frame is as follows:

$$^bT_2 = {}^bT_1 {}^1T_2 = \begin{bmatrix} \cos\beta & 0 & -\sin\beta & d_{1x}+d_{2x} \\ \sin\alpha\sin\beta & \cos\alpha & \sin\alpha\cos\beta & 0 \\ \cos\alpha\sin\beta & -\sin\alpha & \cos\alpha\cos\beta & d_{1z} \\ 0 & 0 & 0 & 1 \end{bmatrix} \tag{3}$$

$^bT_2$ and $^bT_1$ shall be utilized in Section 4.2 to transform the center of mass of each gimbal into body frame. Using to Equations (1) and (3), the transformation from the camera frame to body frame is as follows:

$$^bT_c = {}^bT_2 {}^2T_c = \begin{bmatrix} 0 & \sin\beta & \cos\beta & r\cos\beta+d_{1x}+d_{2x} \\ \cos\alpha & \sin\alpha\cos\beta & -\sin\alpha\sin\beta & r\sin\alpha\sin\beta \\ \sin\alpha & \cos\alpha\cos\beta & -\cos\alpha\sin\beta & r\cos\alpha\sin\beta+d_{1z} \\ 0 & 0 & 0 & 1 \end{bmatrix} \tag{4}$$

Information such as target location in camera image frame are transformed to body frame using Equation (4).

### 4.1.2. Jacobian Transformation

The ultimate objective of camera gimbals is to keep the camera focused on the region of interest while a carrier UAV undergoes different flight maneuvers. To keep the camera in the desired position and orientation, the angular positions and velocities of joints of the gimbals have to vary accordingly. That means that the position and orientation of the origin of the camera frame can be defined in terms of the joint angular positions $\alpha$ and $\beta$. The Jacobian transformation handles such interdependence. Let the position vector of the origin of camera frame—with respect to the $i = (G_2, G_1, b)$ frame–be represented as $^i\vec{r}_c$. As the gimbals undergo roll, pitch, and yaw motions, the origin of the camera frame traces the 3D space. Hence, the position vector $^i\vec{r}_c$ can be decomposed as follows:

$$^i\vec{r}_c = \begin{bmatrix} {}^ix_c \\ {}^iy_c \\ {}^iz_c \end{bmatrix} \tag{5}$$

The same position of the camera frame can be expressed in $j$ frame as follows:

$$^j\vec{r}_c = {}^jT_i \, {}^i\vec{r}_c \tag{6}$$

where $^jT_i$ is the homogeneous coordinate transformation from the $i$ frame to the $j$ frame.

Referring to Figure 4a, the position of the camera origin with respect to the $G_2$ frame is given as follows:

$$^2X_c = \begin{bmatrix} r \\ 0 \\ 0 \\ 1 \end{bmatrix} \tag{7}$$

This position can be transformed to the $G_1$ and $b$ frames as follows:

$$^1X_c = {}^1T_2\,{}^2X_c = \begin{bmatrix} r\cos\beta + d_{1x} \\ 0 \\ -r\sin\beta \\ 1 \end{bmatrix} \quad \text{and} \quad {}^bX_c = {}^bT_2\,{}^2X_c = \begin{bmatrix} r\cos\beta + d_{1x} + d_{2x} \\ r\sin\alpha\sin\beta \\ r\cos\alpha\sin\beta + d_{2z} \\ 1 \end{bmatrix} \tag{8}$$

where the transformation matrices $^1T_2$ and $^bT_2$, given in Equations (2) and (3), are used. The position vectors $^1X_c$ and $^bX_c$ are functions of the angular positions $\alpha$ and $\beta$, as shown below.

$$^1X_c = \begin{bmatrix} {}^1x_c(\beta) \\ {}^1y_c(\beta) \\ {}^1z_c(\beta) \\ 1 \end{bmatrix} = \begin{bmatrix} r\cos\beta + d_{2x} \\ 0 \\ -r\sin\beta \\ 1 \end{bmatrix} \quad \text{and} \quad {}^bX_c = \begin{bmatrix} {}^bx_c(\beta) \\ {}^by_c(\beta) \\ {}^bz_c(\beta) \\ 1 \end{bmatrix} = \begin{bmatrix} r\cos\beta + d_{1x} + d_{2x} \\ r\sin\alpha\sin\beta \\ r\cos\alpha\sin\beta + d_{2z} \\ 1 \end{bmatrix} \tag{9}$$

Referring to Equation (9), translational Jacobian matrices of the following forms are derived as follows:

$$J_{{}^iX_c} := \begin{bmatrix} \frac{\partial^i x_c}{\partial\alpha} & \frac{\partial^i x_c}{\partial\beta} \\ \frac{\partial^i y_c}{\partial\alpha} & \frac{\partial^i y_c}{\partial\beta} \\ \frac{\partial^i z_c}{\partial\alpha} & \frac{\partial^i z_c}{\partial\beta} \\ 0 & 0 \end{bmatrix} \quad i = 1, b \tag{10}$$

This implies that

$$J_{{}^1X_c} = \begin{bmatrix} 0 & -r\sin\beta \\ 0 & 0 \\ 0 & -r\cos\beta \\ 0 & 0 \end{bmatrix} \quad \text{and} \quad J_{{}^bX_c} = \begin{bmatrix} 0 & -r\sin\beta \\ r\cos\alpha\sin\beta & r\sin\alpha\cos\beta \\ -r\sin\alpha\sin\beta & r\cos\alpha\cos\beta \\ 0 & 0 \end{bmatrix} \tag{11}$$

where $J_{{}^1X_c}$ is translational Jacobian matrix that relates the angular position ($\beta$) of gimbal 2 to the camera position, and $J_{{}^bX_c}$ is translation Jacobian matrix that relates the combined variation of angular positions ($\alpha$) and ($\beta$) of gimbal 1 and gimbal 2, respectively, to the camera position.

In addition to the position of the origin of camera frame, its orientation is important. The angular rotations of the origin of the camera frame with respect to $G_1$ and $b$ frames, respectively, are as follows:

$$^1\Theta_c = \begin{bmatrix} {}^1\theta_{xc} \\ {}^1\theta_{yc} \\ {}^1\theta_{zc} \\ 1 \end{bmatrix} = \begin{bmatrix} 0 \\ \beta \\ 0 \\ 1 \end{bmatrix} \quad \text{and} \quad {}^b\Theta_c = \begin{bmatrix} {}^b\theta_{xc} \\ {}^b\theta_{yc} \\ {}^b\theta_{zc} \\ 1 \end{bmatrix} = \begin{bmatrix} \alpha \\ \beta \\ 0 \\ 1 \end{bmatrix} \tag{12}$$

Applying the differential relations shown in Equation (10) into Equation (12), the rotational Jacobian matrices of these angular rotations are as follows:

$$J_{{}^1\Theta_c} = \begin{bmatrix} 0 & 0 \\ 0 & 1 \\ 0 & 0 \\ 0 & 0 \end{bmatrix} \quad \text{and} \quad J_{{}^b\Theta_c} = \begin{bmatrix} 1 & 0 \\ 0 & 1 \\ 0 & 0 \\ 0 & 0 \end{bmatrix} \tag{13}$$

The linear and angular velocities of the origin of camera frame can be obtained from the time derivatives of linear position and angular rotation, given by Equations (9) and (12), respectively. These velocities of the origin of the camera frame can be expressed as functions of the angular velocities $\dot{\alpha}$ and $\dot{\beta}$ of gimbal 1 and 2, respectively.

$$^i\dot{\Lambda}_c = \frac{\partial^i\Lambda_c}{\partial\alpha}\dot{\alpha} + \frac{\partial^i\Lambda_c}{\partial\beta}\dot{\beta} \quad \Lambda = X, \Theta \quad \text{and} \quad i = 1, b \tag{14}$$

Using Equation (14), the linear velocities of the origin of the camera frame with respect to $G_1$ and $b$ frames are the following:

$$\begin{bmatrix} {}^1\dot{x}_c(\beta) \\ {}^1\dot{y}_c(\beta) \\ {}^1\dot{z}_c(\beta) \\ 0 \end{bmatrix} = J_{1X_c}\begin{bmatrix} \dot{\alpha} \\ \dot{\beta} \end{bmatrix} \quad \text{and} \quad \begin{bmatrix} {}^b\dot{x}_c(\beta) \\ {}^b\dot{y}_c(\beta) \\ {}^b\dot{z}_c(\beta) \\ 0 \end{bmatrix} = J_{bX_c}\begin{bmatrix} \dot{\alpha} \\ \dot{\beta} \end{bmatrix} \tag{15}$$

where $J_{1X_c}$ and $J_{bX_c}$ are given in Equation (11).

Similarly, applying Equation (14), the angular velocities with respect to $G_1$ and $b$ are as follows:

$$\begin{bmatrix} {}^1\dot{\theta}_{xc} \\ {}^1\dot{\theta}_{yc} \\ {}^1\dot{\theta}_{zc} \\ 1 \end{bmatrix} = J_{1\Theta_c}\begin{bmatrix} \dot{\alpha} \\ \dot{\beta} \end{bmatrix} \quad \text{and} \quad \begin{bmatrix} {}^b\dot{\theta}_{xc} \\ {}^b\dot{\theta}_{yc} \\ {}^b\dot{\theta}_{zc} \\ 1 \end{bmatrix} = J_{b\Theta_c}\begin{bmatrix} \dot{\alpha} \\ \dot{\beta} \end{bmatrix} \tag{16}$$

where $J_{1\Theta_c}$ and $J_{b\Theta_c}$ are given in Equation (13), respectively.

### 4.2. Dynamics of the Platform

The platform is considered as constituent of two rigid bodies: gimbal 1 and gimbal 2. To control the dynamics of these gimbals, a governing mathematical model is developed. Both gimbals are set into rotation by the desired torques applied through their respective servo motors so as to focus the sensors on a given region of interest. Each gimbal has its own inertia with both static and dynamic mass unbalance. The dynamics of one of the gimbal affects the other. Moreover, the change in attitude of a UAV on which the platform is attached also affects the dynamics of the gimbals. The platform controller to be developed from the mathematical model has to take all these effects into consideration and keep the sensors focused on region of interest.

The gimbal 2 rotates about $y_2$-axis that passes through the gimbal's geometric center and gimbal 1 rotates about $x_1$-axis that, also, passes through the gimbal's geometric center. Due to the sensors mounted on gimbal 2, the center of mass is offset by certain amount from the geometric center. Similarly, due to gimbal 2 and servo motors attached to gimbal 1, the center of mass of gimbal 1 is shifted off the center. Therefore, the off-diagonal elements of inertia matrices of the two gimbals are non-zero. Let, the mass moment of inertia about the respective mass centers of the gimbals are given by the following:

$$ {}^iI_j = \begin{bmatrix} {}^iI_{jx} & {}^iI_{jxy} & {}^iI_{jxz} \\ {}^iI_{jxy} & {}^iI_{jy} & {}^iI_{jyz} \\ {}^iI_{jxz} & {}^iI_{jyz} & {}^iI_{jz} \end{bmatrix} \tag{17}$$

where $j = 1,2$ represents the mass moment of inertia for gimbal 1 and gimbal 2, respectively, and $i$ represents the coordinate frame with respect to which the moment of inertia is measured. The Lagrangian equation of motion of the sensor platform is given as follows:

$$\frac{d}{dt}\left(\frac{\partial \mathcal{L}}{\partial \dot{\vartheta}_j}\right) - \frac{\partial \mathcal{L}}{\partial \vartheta_j} = Q_j \quad \text{where} \quad \mathcal{L} = K - V \tag{18}$$

$K$ and $V$ are the kinetic and potential energies, $Q$ is non-conservative force that includes the desired force applied by the servo motors on the gimbals and the undesired external forces. The joint angular position $\vartheta = (\alpha, \beta)$ and $\dot{\vartheta}$ represents their time derivative.

Kinetic energy of gimbal $j$ is given as follows:

$$K_j = \frac{1}{2}{}^b\dot{X}_j^T m_j\,{}^b\dot{X}_j + \frac{1}{2}{}^b\dot{\Theta}_j^{Tb}I_j\,{}^b\dot{\Theta}_j^T \tag{19}$$

where the first and the second terms on the right hand side of Equation (19) represent translational and rotational kinetic energies of center of mass of the gimbal with respect to body frame. ${}^b\dot{X}_j$ and ${}^b\dot{\Theta}_j$ are the linear and angular velocities of the center of mass with respect to body frame. These velocities can be expressed in terms of joint angular velocities $\dot{\vartheta}$ by applying translational and rotational Jacobian matrices of the form given in Equations (15) and (16). Applying the corresponding Jacobian matrices to Equation (19), the total kinetic energy of the gimbals, in terms of joint angular velocities, is expressed as follows:

$$K = \frac{1}{2}\dot{\vartheta}^T D\dot{\vartheta} \tag{20}$$

where $D$ is the $n \times n$ inertial-type matrix composed of translational and rotational inertia of the following form:

$$D(\vartheta) = \sum_{j}^{2} \left( J_{Xj}^{T} \, m_j \, J_{Xj} + \frac{1}{2} J_{\Theta j}^{T} \, {}^{b}I_j \, J_{\Theta j} \right) \tag{21}$$

where $J_{Xj}$ and $J_{\Theta j}$ are Jacobian matrices that transform joint angular velocities $\dot{\vartheta}_j$ of gimbal $j$ into translational and rotational velocities of the gimbal's center of mass. ${}^{b}I_j$ is the mass moment of inertia of the gimbal about its center of mass as expressed in the body frame and $m_j$ is the mass of the gimbal.

To determine the potential energy ($V$) of a gimbal, one must calculate the acceleration due to gravity of the center of mass of the gimbal. Acceleration due to gravity is a vector pointing into the Earth's center. In gimbal 2 frame, the acceleration due to gravity is as follows:

$$^{2}\mathcal{G} = \begin{bmatrix} -g\sin\beta \\ 0 \\ g\cos\beta \end{bmatrix} \tag{22}$$

where $\beta$ is the current angular position of gimbal 2 as shown in Figure 4a. The gravity vector acts at the center of mass of the gimbal. The gravity vector in body frame can be obtained using the coordinate transformation matrix given in Equation (3). However, the first three yaws and the three columns of the matrix are used to transform the gravity vector, since the translation vector has no effect on the gravity vector. Therefore,

$$^{b}\mathcal{G} = \begin{bmatrix} -g\sin 2\beta \\ g\sin\alpha\left(\cos^2\beta - \sin^2\beta\right) \\ g\cos\alpha\left(\cos^2\beta - \sin^2\beta\right) \end{bmatrix} \tag{23}$$

The potential energy of gimbal $j$, with respect to the body frame, is given as follows:

$$V_j = -m_j \, {}^{b}\mathcal{G}^{T} \, {}^{b}r_j \tag{24}$$

where ${}^{b}r_j$ is the position vector of the center of mass of gimbal $j$ with respect to the body frame. The total potential energy of the sensor platform is, thus, given as follows:

$$V_j = -\sum_{i=1}^{2} m_i \, {}^{b}\mathcal{G}^{T} \, J_{Xi}^{j} \, \vartheta \tag{25}$$

where $i$ represents the coordinate frame in which inertia measurement was made before it is transformed into body frame. For $j > i$, $j$'s column of matrix $J$ is zero. The position of the center of mass for each gimbal is obtained using the CATIA inertia measuring tool.

Substituting the expressions of Equations (20) and (25) into Equation (18), and after rearranging the terms, the dynamic equation of motion of the platform is as follows:

$$\sum_{i=1}^{2} D_{ji}\ddot{\vartheta}_i + \sum_{k=1}^{2}\sum_{l=1}^{2} \left( \frac{\partial D_{jk}}{\partial \vartheta_l} - \frac{1}{2}\frac{\partial D_{kl}}{\partial \vartheta_j} \right) \dot{\vartheta}_k\dot{\vartheta}_l + \sum_{i=1}^{2} m_i \, {}^{b}\mathcal{G}^{T} \, J_{Xi}^{j} = \mathcal{Q}_j \tag{26}$$

where the first term represents the reaction of gimbals to external force $\mathcal{Q}_j$, the second term is a velocity coupling force, and the third term represents gravitational force. Let

$$H_{jkl} := \sum_{k=1}^{2}\sum_{l=1}^{2} \left( \frac{\partial D_{jk}}{\partial \vartheta_l} - \frac{1}{2}\frac{\partial D_{kl}}{\partial \vartheta_j} \right) \dot{\vartheta}_k\dot{\vartheta}_l \tag{27}$$

and

$$\Gamma_j := \sum_{i=1}^{2} m_i \, {}^{b}\mathcal{G}^{T} \, J_{Xi}^{j} \tag{28}$$

With the definitions given in Equations (27) and (28), Equation (26) can be written in compact form as

$$\ddot{\vartheta} = D^{-1}[\mathcal{Q} - (H + \Gamma)] \tag{29}$$

Let the state–space model be given as follows:

$$\chi = \begin{bmatrix} \vartheta \\ \dot{\vartheta} \end{bmatrix} = \begin{bmatrix} \alpha \\ \beta \\ \dot{\alpha} \\ \dot{\beta} \end{bmatrix} \implies \dot{\chi} = \begin{bmatrix} \dot{\vartheta} \\ \ddot{\vartheta} \end{bmatrix} = \begin{bmatrix} \dot{\alpha} \\ \dot{\beta} \\ \ddot{\alpha} \\ \ddot{\beta} \end{bmatrix} \tag{30}$$

The state–space model can be expressed in terms of Equation (29), as follows:

$$\begin{bmatrix} \dot{\vartheta} \\ \ddot{\vartheta} \end{bmatrix} = \begin{bmatrix} \vartheta_{2\times1} \\ D(\vartheta_{2\times1})^{-1}[Q - (H + \Gamma(\vartheta_{2\times1}))] \end{bmatrix} \tag{31}$$

The platform control algorithm makes use of the above equations to determine the required angular positions and velocities of the gimbals. The inertia matrices of the two gimbals are obtained from their CATIA models. The inertial matrices of the gimbals are obtained with respect to their respective coordinate frame and transformed into body frame.

## 5. Platform Performance Validation Test

To test the performance of the proposed sensor platform, the platform is nose-mounted on a fixed-wing UAV, which is commanded to randomly change its flight status. Under such random changes in flight status, the responses of the platform in its collision avoidance and target tracking operational modes are tested. In this phase of performance validation, the tests are conducted in a virtual environment using SITL simulation.

*SITL-Based Performance Tests*

Prior to the actual deployment of a new model, conducting a simulation-based performance test is common practice. There are various SITL simulation frameworks available for use. For the proposed sensor platform performance tests, PX4 flight control firmware-based SITL simulation framework was selected. This PX4 firmware has ready-made models of various vehicles, LiDar Lite, and camera sensors required by the proposed sensor platform.

To reproduce the actual platform control techniques described in Section 3.3, the designed platform model is nose-mounted on a standard VTOL UAV model of PX4 firmware, as shown in Figure 5. The outer gimbal of the platform is attached to the nose of the UAV with revolute joint in roll and the inner gimbal of the platform is attached to the right canopy of the platform with revolute joint in pitch/yaw. The two revolute joints represent the two servo motors that control the sensor platform in real flight scenario. Both joints can rotate in the range of $[-90°, +90°]$. The PX4 firmware models of camera and LIDAR sensors are mounted on the inner gimbal of the platform. Inertia measurement unit (IMU) sensor is mounted on the inner gimbal and the attitude angles of the sensor platform obtained through IMU are used to determine the gravity vector on the gimbals of the platform.

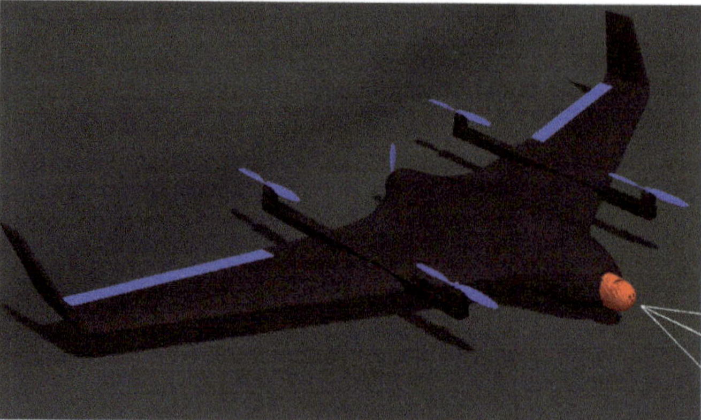

**Figure 5.** Platform nose-mount on fixed-wing VTOL UAV.

Gazebo Simulation Environment

Gazebo is an open source three dimensional environment simulator that is rich in realistic features of both indoor and outdoor environments. It implements open dynamics engine (ODE) that handles rigid body dynamics simulation and collision detection. The dynamics of UAV and its sensor platform system are governed by this ODE. Sensor models are attached on the UAV and sensor platform to acquire their dynamics information. These sensor information are sent to flight control firmware. Based on the received information, the flight control firmware decides actuator commands and controls the dynamics of the UAV and the sensor platform. PX4 firmware is already integrated to Gazebo simulation environment and have been widely used by many researchers over the years [24–26].

Access to the ODE and other functionalities of gazebo is through gazebo plugins. The plugin is a code from which a shared library is generated. Communication between PX4 flight control firmware and gazebo simulation environment are enabled through the generated plugin libraries. A custom gazebo plugin that enables the PX4 firmware to acquire information about the sensor platform and send actuator commands to the platform servos is written by the authors and implemented in the SITL simulation process. The custom plugin takes the current angular positions and velocities of the gimbals as inputs and applies a force required to steer the platform to a desired region of interest. A simple PID controller is used in the plugin to control the motion of the gimbals. The position, mass, and inertial properties of all components of sensor platform, including that of FDR-X3000 camera and LiDAR sensors, are incorporated to simulation description format (SDF) file of UAV model and their dynamics are simulated by gazebo ODE.

A custom gazebo plugin module that controls the revolute joints was written and included to SITL_gazebo plugins of PX4 firmware. To send actuator outputs to the joints (servos), a custom mixer was defined and included to SITL_mixers of the firmware. For obstacle avoidance, the custom mixer takes normalized velocity vector of the UAV and provides actuator outputs. For target tracking, a robot operating system (ROS)-based node is written in ROS workspace. This node subscribes to current location of UAV in the gazebo simulation environment and generates simulated target locations so that the platform steers the sensors to lock-on/track a virtual target on those locations. The relative position vector of the virtual target with respect to the UAV is determined from the current UAV location and the generated virtual target location. Based on the relative position vector, the required roll, and the pitch/yaw angles are calculated and published to actuator control topic of PX4 firmware. The custom mixer takes the actuator control values and produces corresponding actuator output that controls the joints.

The platform's operational mode switching from obstacle avoidance to target tracking or vice versa is enabled through parameter tuning and flight status of the UAV. A parameter is defined in the mc_att_control module of the PX4 firmware and mapped to radio control (RC) transmitter for tuning. The mc_att_control module is, also, customized with conditional statements. For instance, the following:

- ⊙ The platform engages the sensors for obstacle avoidance if the parameter is tuned to be in a certain range of values and UAV flight status is in either of the following flight modes:
    - ▷ automatic take-off;
    - ▷ automatic landing;
    - ▷ fly to a known location of interest (e.g., crime scene);
    - ▷ return-to-launch.

- ⊙ Under the condition that the UAV is in hover or altitude control mode, manual override of the mode is disabled and the platform is set to ready to be manually steered by RC transmitter to search for an intended target. Although manual override is disabled during a UAV's hover and altitude control mode, flight mode switching is active and can be carried out through either aground control unit or an RC transmitter.

- ⊙ If a target is identified and the parameter is tuned to certain range of values, then the platform locks on the target and pursues it. The flight mode is then switched to mission so that the UAV tracks the target.

## 6. Results and Discussion

### 6.1. Obstacle Avoidance

When the UAV is on the ground, the platform disengages the sensors and acquires a horizontal orientation, as shown in Figure 6a. When a take-off command is given to the UAV, it takes-off with the sensor platform pitched by 90° upward, as shown in Figure 6b, till it attains its designated altitude.

Soon, the UAV attains the intended altitude and enters hover mode, and the platform automatically disengages pitching, as shown in Figure 6c, and is ready for either manual steering to search for target or cruises to the region of interest in the obstacle avoidance mode. During landing, as shown in Figure 6d, the platform pitches the sensors down by 90° to scan the environment below the UAV.

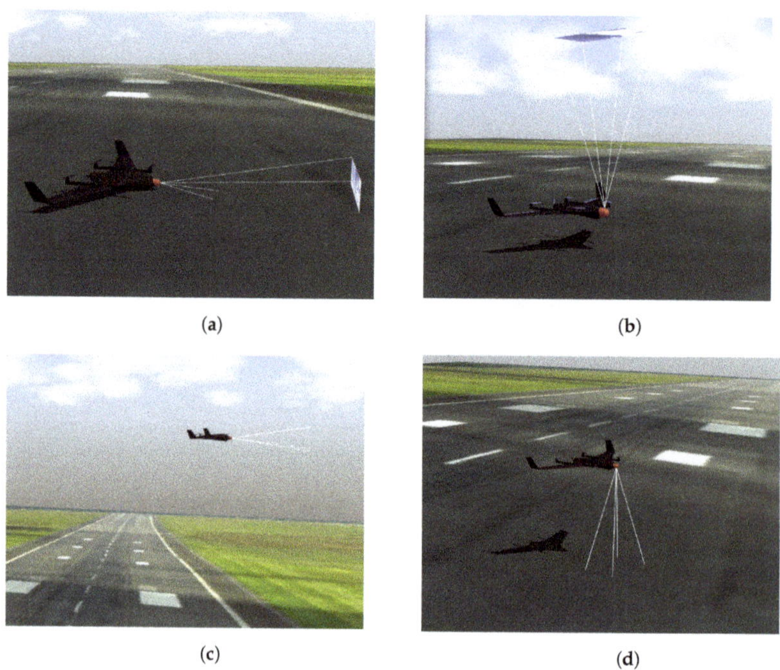

Figure 6. SITL-based platform operation onboard UAV. (a) Platform during UAV arming; (b) platform during UAV take-off; (c) platform during UAV hovering; (d) platform during UAV landing.

In Figure 7, an inertial measurement unit (IMU) is mounted on inner gimbal of the platform and the pitch responses of the platform, to take-off and land commands, are compared to the IMU reading. There is a complete overlap between the IMU reading and the pitch angle variation of the sensor platform.

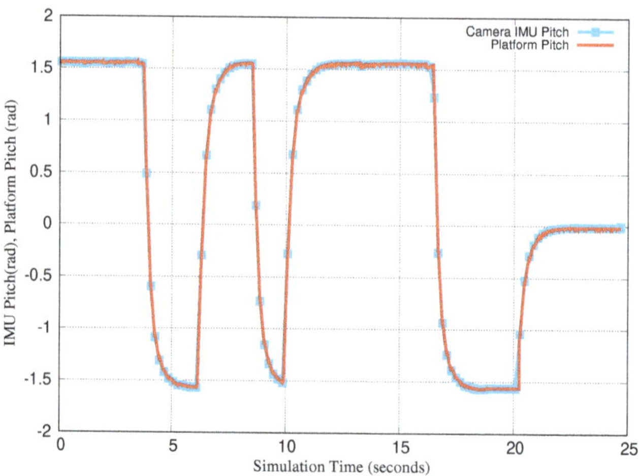


Figure 7. Camera IMU vs. platform pitch.

In order to further test the pitch response of the platform to random changes in the flight status of the UAV, take-off and land commands are given to the UAV at random altitudes. Therefore, the UAV aborts its flight status randomly and the platform has to respond to that random changes. The pitch response of the platform is checked with respect to the velocity vector of the UAV as shown in Figure 8.

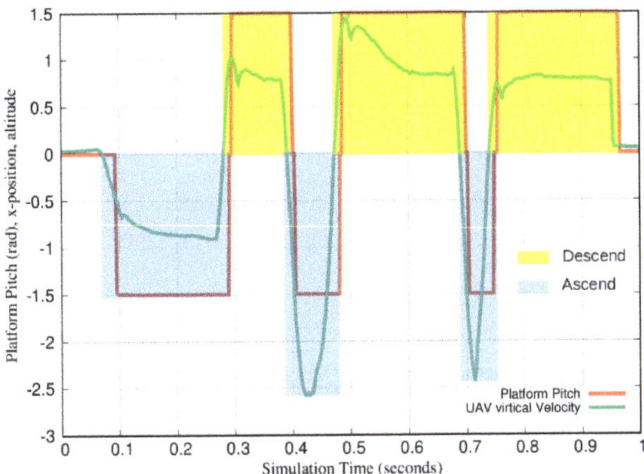


**Figure 8.** Obstacle avoidance test during random change in flight status.

The negative velocity of the UAV corresponds to the ascending mode, whereas the positive velocity is for the descending mode. Likewise, the negative and positive pitch angles correspond to the pitch up and down, respectively, of the inner gimbal of the platform. As can be seen in the figure, the platform responds to the change in the direction, but not the magnitude, of the velocity of the UAV. This shows that the platform is not bothered by how fast or slow the UAV is flying but by the change in the flight course.

The performance of the platform is also tested while a UAV is navigating through waypoints. The waypoints shown in Figure 9 are sent to the UAV. In its mission to fly to the destination, the UAV is commanded to take-off to altitude of 10 m, descend to 5 m, roll to the left for 30 m (along x-axis), pitch forward about 15 m (along y-axis), and then roll and pitch (simultaneously) towards the destination, which is 15 m along the y-axis and 15 m along the x-axis in the negative direction. Therefore, in this flight path, the performance of the sensor platform in roll, pitch, and combination of roll–yaw can be tested.

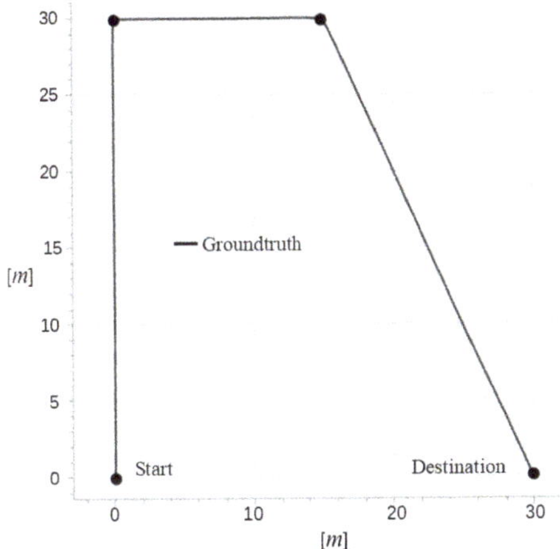


**Figure 9.** Path followed by the UAV.

For obstacle avoidance, the sensor platform has to be steered towards the velocity vector of the UAV so that the sensors focus on the flight course. As shown in Figure 10, the UAV is commanded to fly to the destination along the given waypoints, while the roll, pitch, or yaw movement of the platform remains aligned with the velocity vector of the UAV. Referring to the figure, during take-off and descend, the platform roll angle remains zero and the pitch angle is $\mp 1.5$ rads ($\mp 90°$). At 19.05 s of simulation time, the UAV starts to roll side-way along x-axis with velocity $V_x$ while other components of the velocity ($V_y$ and $V_z$) remain zero. The platform rolls right and yaws left by $+90°$s to orient the sensors along flight course of the UAV. After 31.71 s of simulation time, the UAV completes rolling and changes its flight course towards the y-axis (forward). Following this change in flight course, the platform orients the sensors forward ($0°$) until the UAV completes 15 m of forward flight. The forward flight is completed at 43.76 s of simulation time where the $V_y$ velocity component drops to zero. The remaining mission is to fly to the destination that requires the UAV to simultaneously fly forward and roll sideways. In this flight course, the platform has to roll left and yaw right as shown in the simulation time range from around 0.47–0.6 s

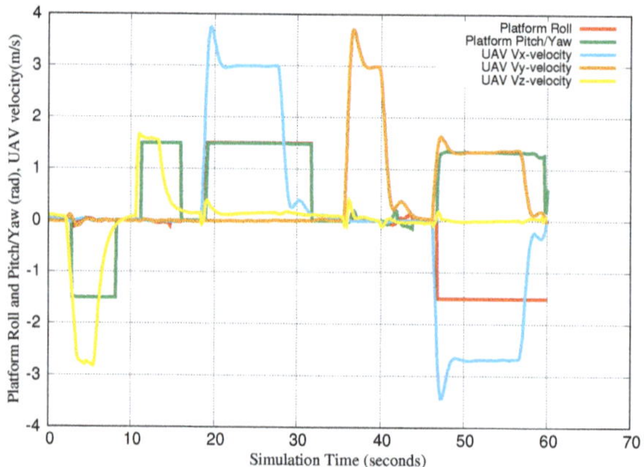

**Figure 10.** Obstacle avoidance mode.

### 6.2. Target Tracking

Target tracking or lock on target responses of the sensor platform are shown in Figure 11, where the UAV is commanded to randomly ascend and descend. The platform remains focused on a target located 30 m in front of the UAV. The response of the sensor platform to UAV's altitude change is immediate. The altitudes at which the UAV is commanded to change its flight mode are normalized, where 1 corresponds to altitude of 10 m, to magnify the variation in the platform pitch angle.

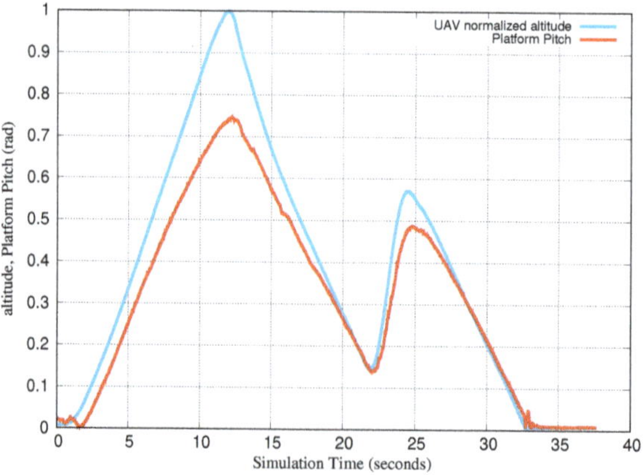

**Figure 11.** Lock-on-target capability.

The lock-on-target capability of the platform is visualized using the YOLO object detection algorithm that runs in the ROS workspace. As shown in Figure 12, the UAV locks on a target located 30 m in front of it.

**Figure 12.** Lock-on-target mode.

The aforementioned waypoints are used to test the target tracking or lock on target performance of the platform. In the lock-on-target operation mode, wherever the UAV is heading, the platform has to keep focusing on the target (destination). This is depicted by the variations in the roll and pitch/yaw angles of the sensor platform as compared to the location of the UAV along the flight path, as shown in Figure 13. The UAV takes-off to altitude of 10 m and descends to 5 m before flying to waypoints. The locations of the UAV in the given waypoints are scaled down by 10% to magnify the variation of roll and pitch/yaw angles of the platform. During take-off and descend modes, the platform pitches so as to lock the sensors on a virtual target located at destination point (30 m forward). After 15 s of simulation time, the UAV rolls left along x-axis keeping its altitude at 5 m. To lock the sensors on the destination point, the platform has to roll left and yaw right, while the UAV is rolling along the x-axis, it recedes from the destination, and hence the roll and yaw angles of the platform increase up to 28.24 s of simulation time. The UAV then heads forward for 15 m up to 35.64 s of simulation time at which the UAV completes rolling left. During this forward flight course, the platform roll angle remains fixed while the yaw angle keeps on increasing so that the sensors remains locked on the target. Then, the UAV rolls right and fly forward, simultaneously, towards the destination up to 45.13 s of simulation time in which the platform yaw angle increases to +1.5 rads and the roll angle reduces to 0 as the UAV approaches and hovers over the target. At the destination, the UAV hovers over the target with the platform pitching down +1.5 to remain locked on the target.

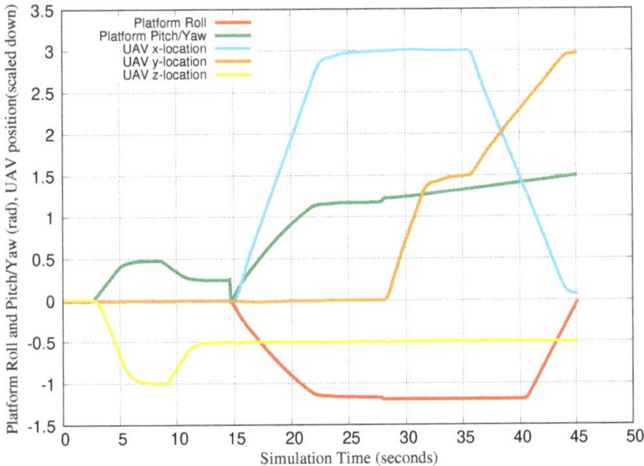


**Figure 13.** Target tracking mode.

In the hover or position hold flight mode of the UAV, the incorporated custom parameters and conditions given in mc_att_control enable a user to manually roll or pitch/yaw the platform, using RC transmitter to search for a target without affecting the attitude of the UAV. Figure 14 shows random manual steering of the sensor platform in search for a target, while the UAV remains level in hover mode. The moment the target is obtained, the user toggles a switch on the RC transmitter that is mapped to a parameter to lock on the target, and the UAV starts to pursue the target.

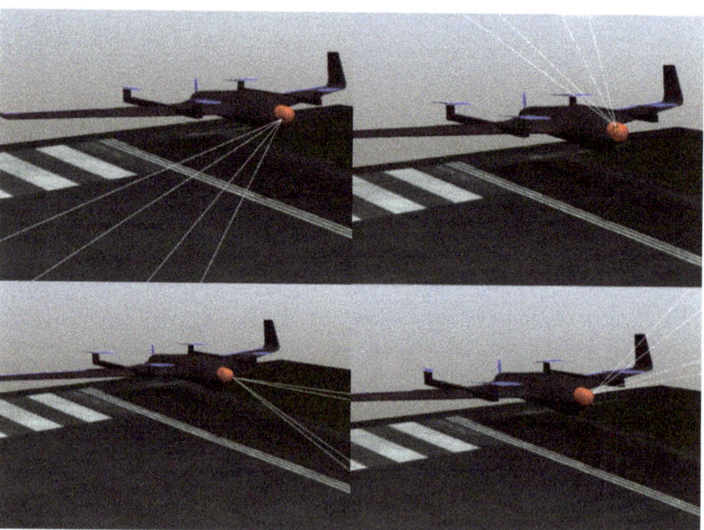

**Figure 14.** Manual search for target.

## 7. Conclusions and Future Work

The use of small-scale UAVs for law enforcement missions is significantly increasing in cities and towns. However, these UAVs are required to have robust autonomy in executing missions in such areas. This means that the UAVs have to have enough information about the environment they are operating in so as to execute their missions while avoiding collision with potential dangers. Information about the surrounding environment is obtained through sensors onboard a UAV. Often, multiple sensors are rigidly mounted on different sides of a UAV to scan the surrounding environment. However, this practice is not feasible for various reasons, including the following: small-scale UAVs are highly weight-constrained, synchronizing and fusing data of multiple sensors are challenging tasks, and the purchase cost of sensors is high. Therefore, to avoid these setbacks, utilizing movable sensor platforms that can carry few sensors is an indispensable solution. To this end, a movable sensor platform is designed, its technical implementation is described, and the mathematical model that governs its dynamics is derived. The proposed sensor platform design is unique in its capability to engage sensors for collision avoidance and target tracking tasks. Moreover, the design layout and mount location of the platform do not induce aerodynamic instability during all modes of its operation.

The performance of the platform was tested using software-in-the-loop simulations. The simulation tests were based on the two operational modes of the sensor platform: collision avoidance and target tracking. To test the responses of the sensor platform, the UAV was commanded to randomly change its flight modes and cruise in different directions. The results show that the platform effectively steers the sensors in roll, pitch, or yaw directions in response to random change in UAV's flight mode and flight course.

In our future work, the custom gazebo plugin that controls the dynamics of the platform will be modified to be incorporated to PX4 flight control software. To validate the successful performance tests obtained through software-in-the-loop simulations, the platform, with its sensors, shall be nose-mounted on a fixed-wing VTOL UAV and real flight experiments will be conducted. If required, further improvements shall be carried out on the platform control module to make sure that the operation of the sensor platform is accurate and stable in both of its operational modes.

**Author Contributions:** In this manuscript, the sensor platform design philosophy and its implementation techniques were performed by A.T.; customization of the PX4 flight control firmware and integration of the sensor platform to software-in-the-loop simulation were performed by M.H.; the preliminary platform design and its performance test methodologies were performed by H.-Y.H. All authors have read and agreed to the published version of the manuscript.

**Funding:** This work was supported by the Korea Agency for Infrastructure Technology Advancement (KAIA) grant funded by the Ministry of Land, Infrastructure, and Transport (Grant 21CTAP-C157731-02).

**Institutional Review Board Statement:** Not applicable.

**Informed Consent Statement:** Not applicable.

**Data Availability Statement:** Not applicable.

**Conflicts of Interest:** All authors mentioned in this manuscript were involved in the study from beginning to end. The manuscript was thoroughly reviewed by the authors before being submitted to *Drones*. This manuscript has not been submitted to another journal for publication.

## References

1. Burgués, J.; Marco, S. Environmental chemical sensing using small drones: A review. *Sci. Total Environ.* **2020**, *748*, 141172. [CrossRef] [PubMed]
2. Smith, M.L.; Smith, L.N.; Hansen, M.F. The quiet revolution in machine vision—A state-of-the-art survey paper, including historical review, perspectives, and future directions. *Comput. Ind.* **2021**, *130*, 103472. [CrossRef]
3. Agarwal, A.; Kumar, S.; Singh, D. Development of Neural Network Based Adaptive Change Detection Technique for Land Terrain Monitoring with Satellite and Drone Images. *Def. Sci.* **2019**, *69*, 474. [CrossRef]
4. Salhaoui, M.; Guerrero-González, A.; Arioua, M.; Ortiz, F.J.; El Oualkadi, A.; Torregrosa, C.L. Smart Industrial IoT Monitoring and Control System Based on UAV and Cloud Computing Applied to a Concrete Plant. *Sensors* **2019**, *19*, 3316. [CrossRef] [PubMed]
5. Glaser, A. Police Departments Are Using Drones to Find and Chase down Suspects. Vox. 2017. Available online: https://www.vox.com/2017/4/6/15209290/police-fire-department-acquired-drone-us-flying-robot-law-enforcement (accessed on 28 December 2021).
6. Murphy, D.W.; Cycon, J. Application for mini VTOL UAV for law enforcement. In Proceedings of the Volume 3577, Sensors, C3I, Information, and Training Technologies for Law Enforcement, Boston, MA, USA, 7 January 1999. [CrossRef]
7. Durscher, R. How Law Enforcement Has Been Using Drones. Government Fleet. 2020. Available online: https://www.government-fleet.com/359403/how-law-enforcement-has-been-utilizing-drones (accessed on 28 December 2021).
8. Hardin, P.J.; Jensen, R.R. Small-Scale Unmanned Aerial Vehicles in Environmental Remote Sensing: Challenges and Opportunities. *J. GISci. Remote Sens.* **2011**, *48*, 99–111. [CrossRef]
9. Geuther, S.; Capristan, F.; Kirk, J.; Erhard, R. A VTOL small unmanned aircraft system to expand payload capabilities. In Proceedings of the 31st Congress of the International Council of the Aeronautical Sciences, Belo Horizonte, Brazil, 9–14 September 2018.
10. Chand, B.N.; Mahalakshmi, P.; Naidu, V.P.S. Sense and Avoid Technology in Unmanned Aerial Vehicles: A Review. In Proceedings of the International Conference on Electrical, Electronics, Communication, Computer and Optimization Techniques, Mysuru, India, 15–16 December 2017.
11. Mukhamediev, R.I.; Symagulov, A.; Kuchin, Y.; Zaitseva, E.; Bekbotayeva, A.; Yakunin, K.; Assanov, I.; Levashenko, V.; Popova, Y.; Akzhalova, A.; et al. Review of Some Applications of Unmanned Aerial Vehicles Technology in the Resource-Rich Country. *Appl. Sci.* **2021**, *11*, 10171. [CrossRef]
12. Hardin, P.J.; Lulla, V.; Jensen, R.R.; Jensen, J.R. Small Unmanned Aerial Systems (sUAS) for environmental remote sensing: Challenges and opportunities revisited. *GISci. Remote Sens.* **2019**, *56*, 309–322. [CrossRef]
13. Zhang, J. Multi-source remote sensing data fusion: Status and trends. *Int. J. Image Data Fusion* **2010**, *1*, 5–24. [CrossRef]
14. Khaleghi, B.; Khamis, A.; Karray, F.O.; Razavi, S.N. Multisensor data fusion: A review of the state-of-the-art. *Inf. Fusion* **2013**, *14*, 28–44. [CrossRef]
15. Quigley, M.; Goodrich, M.A.; Griffiths, S.; Eldredge, A.; Beard, R.W. Target Acquisition, Localization, and Surveillance Using a Fixed-Wing Mini-UAV and Gimbaled Camera. In Proceedings of the 2005 IEEE International Conference on Robotics and Automation, Barcelona, Spain, 18–22 April 2005.
16. Kuzey, B.; Yemenicioğlu, E.; Kuzucu, A. 2 Axis Gimbal Camera Design. *ResearchGate* **2007**, *1*, 32–39. [CrossRef]
17. Gremsy. Stabilizing Gimbals & Stabilized Camera Mounts for Drones & UAVs. Unmanned System Technology. Available online: https://www.unmannedsystemstechnology.com/company/gremsy/ (accessed on 12 October 2021).
18. Sánchez, P.; Casado, R.; Bermúdez, A. Real-Time Collision-Free Navigation of Multiple UAVs Based on Bounding Boxes. *Electronics* **2020**, *9*, 1632. [CrossRef]
19. Shakhatreh, H.; Sawalmeh, A.; Al-Fuqaha, A.I.; Dou, Z.; Almaita, E.; Khalil, I.M.; Othman, N.S.; Khreishah, A.; Guizani, M. Unmanned Aerial Vehicles: A Survey on Civil Applications and Key Research Challenges. *arXiv* **2018**, arXiv:1805.00881.
20. Lancovs, D. Broadcast transponders for low flying unmanned aerial vehicles. *Transp. Res. Procedia* **2017**, *24*, 370–376. [CrossRef]

21. Ahmad, M.H.; Osman, K.; Zakeri, M.F.M.; Samsudin, S.I. Mathematical Modelling and PID Controller Design for Two DOF Gimbal System. In Proceedings of the 2021 IEEE 17th International Colloquium on Signal Processing & Its Applications (CSPA), Langkawi, Malaysia, 5–6 March 2021. [CrossRef]
22. Isaev, A.M.; Adamchuk, A.S.; Amirokov, S.R.; Isaev, M.A.; Grazhdankin, M.A. Mathematical Modelling of the Stabilization System for a Mobile Base Video Camera Using Quaternions. Available online: http://ceur-ws.org/Vol-2254/10000051.pdf (accessed on 17 February 2022).
23. Aytaç, A.; Rıfat, H. Model predictive control of three-axis gimbal system mounted on UAV for real-time target tracking under external disturbances. *Mech. Syst. Signal Process.* **2020**, *138*, 106548. [CrossRef]
24. Nguyen, K.D.; Nguyen, T.T. Vision-based software-in-the-loop-simulation for Unmanned Aerial Vehicle Using Gazebo and PX4 Open Source. In Proceedings of the 2019 International Conference on System Science and Engineering (ICSSE), Dong Hoi, Vietnam, 20–21 July 2019.
25. Nguyen, K.D.; Ha, C. Development of Hardware-in-the-Loop Simulation Based on Gazebo and Pixhawk for Unmanned Aerial Vehicles. *Int. J. Aeronaut. Space Sci.* **2018**, *19*, 238–249. [CrossRef]
26. Omar, H.M. Hardware-In-the-Loop Simulation of Time-Delayed Anti-SwingController for Quadrotor with Suspended Load. *Appl. Sci.* **2022**, *12*, 1706. [CrossRef]

 *drones*

*Article*

# Research on Modeling and Fault-Tolerant Control of Distributed Electric Propulsion Aircraft

Jiacheng Li [1,*], Jie Yang [2] and Haibo Zhang [1]

[1] College of Energy and Power Engineering, Nanjing University of Aeronautics and Astronautics, Nanjing 210016, China; zhhbjason@nuaa.edu.cn
[2] College of Aerospace Engineering, Nanjing University of Aeronautics and Astronautics, Nanjing 210016, China; nuaa_yj@nuaa.edu.cn
* Correspondence: ileejc@nuaa.edu.cn; Tel.: +86-18662706520

**Abstract:** Distributed electric propulsion (DEP) aircrafts have high propulsion efficiency and low fuel consumption, which is very promising for propulsion. The redundant thrusters of DEP aircrafts increase the risk of fault in the propulsion system, so it is necessary to study fault-tolerant control to ensure flight safety. There has been little research on coordinated thrust control, and research on fault-tolerant control of the propulsion system for DEP aircrafts is also in the preliminary stage. In this study, a mathematical model of DEP aircrafts was built. Aiming at the lateral and longitudinal control of DEP aircrafts, a coordinated thrust control method based on total energy control and total heading control was designed. Furthermore, a fault-tolerant control strategy and control method was developed for faults in the propulsion system. Simulation results showed that the controller could control the thrust to the prefault level. The correctness and effectiveness of the designed coordinated thrust control method and the fault-tolerant control method for DEP aircrafts were theoretically verified. This study provides a theoretical basis for future engineering application and development of the control system for DEP aircrafts.

**Keywords:** distributed electric propulsion; coordinated thrust control; fault-tolerant control; flight simulation

**Citation:** Li, J.; Yang, J.; Zhang, H. Research on Modeling and Fault-Tolerant Control of Distributed Electric Propulsion Aircraft. *Drones* **2022**, *6*, 78. https://doi.org/10.3390/drones6030078

Academic Editor: Abdessattar Abdelkefi

Received: 26 February 2022
Accepted: 11 March 2022
Published: 17 March 2022

**Publisher's Note:** MDPI stays neutral with regard to jurisdictional claims in published maps and institutional affiliations.



## 1. Introduction

A distributed electric propulsion aircraft is a new type of aircraft that converts mechanical energy into electrical energy through an engine-driven generator. It is used in conjunction with energy storage devices, such as lithium batteries, to power multiple electric propulsion devices distributed on the wings or fuselage. The DEP aircraft studied in this work is presented in Figure 1. With distributed propulsion, an aircraft's propeller slipstream can significantly increase the airflow velocity behind its propeller disks, which will improve the aircraft performance in flight [1], enhance the stability of the wing structure [2], and realize short take off. Electric propulsion can increase efficiency of the propulsion system [3] and reduce noise [4]. The fuel consumption and pollution emission of an aircraft diminish as the DEP system improves the working condition of the gas turbines and aerodynamic efficiency of the vehicle, which satisfies the green requirements for the future [5,6]. In addition, the DEP system has multiple redundancy of a power system, which is safer and labeled as a very promising propulsion type.

In the study of methods for modeling of DEP aircraft, Joseph W. Connolly et al. of the NASA Glenn Research Center developed a nonlinear dynamic model with full flight envelope controller for the propulsion system of a partially turboelectric single-aisle aircraft. Optimization strategies for efficiency of the aircraft were investigated by adjusting the power between the energy for turbofan thrust and the extracted energy used to power the tail fan [7]. Nhan T. Nguyen et al. from the NASA Ames Research Center proposed an adaptive aeroelastic shape control framework for distributed propulsion aircrafts, which allows the wing-mounted distributed propulsion system to twist the wing shape in flight to improve aerodynamic efficiency through the flexibility of an elastic wing. In addition, an aero-propulsive-elastic model of a highly flexible wing distributed propulsion transport aircraft was established, and analysis of the initial simulation results showed that the scheme could solve the potential flutter problem and effectively improve the aerodynamic efficiency quantity of the lift-to-drag ratio [8]. Zhang Jing et al. from Beihang University systematically investigated the integrated flight/propulsion modeling and optimal control of distributed propulsion configuration with boundary layer ingestion and supercirculation features and proposed an integrated flight/propulsion optimal control scheme to deal with the strong coupling effects and to implement comprehensive control of redundant control surfaces as well as the distributed engines [9]. Lei Tao et al. from Northwestern Polytechnic University built a complete simulation model of the DEP aircraft power system and comparatively analyzed the pros and cons of three evaluation indexes, namely the propulsion power, the propulsion efficiency, and the range in pure electric propulsion and turboelectric propulsion architectures, based on a flight profile [10]. Da Xingya et al. from the High-Speed Aerodynamics Research Institute under China Aerodynamics Research and Development Center introduced the power-to-thrust ratio as a parameter. They analyzed the effects of the state of a boundary layer and propulsion system parameters on system performance through a numerical analysis method based on the integral equation of boundary layer and verified the reliability of the calculation method by comparing the baseline state with N3-X [11]. For future electric airliners, Shanghai Jiao Tong University and the NASA Glenn Research Center developed a design method and a propulsion electric grid simulator for a turboelectric distributed propulsion (TeDP) system, explored the influence of the motor size and the spread length and air inlet conditions on the number of thrusters, and established a simulation system of a generator driven by a gas turbine engine and a system constituting two permanent magnet motors to simulate the drive of motor propelling fans. These techniques can convert a common motor system into a unique TeDP electric grid simulation program [12,13]. P.M. Rothhaar, a research engineer from the NASA Langley Research Center, developed the full process of testing, modeling, simulation, control, and flight test of a distributed propulsion vertical takeoff and landing (VTOL) tilt-wing aircraft and established methods for self-adaptive control architectures, control distribution research and design, trajectory optimization and analysis, flight system identification, and incremental flight testing [14]. J.L. Freeman performed a dynamic flight simulation of directional control authority-oriented spreading DEP and developed a linear time-invariant state space model to simulate the six-degree-of-freedom flight dynamics of a DEP aircraft controlled by a throttle lever. The study showed that further development of this technology could reduce or eliminate the vertical tail of an aircraft [15].

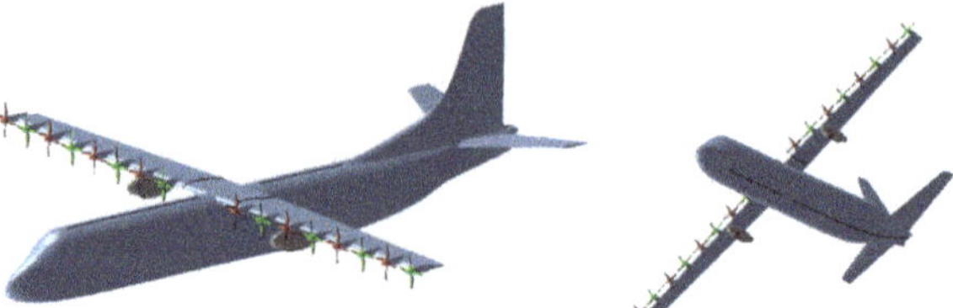

**Figure 1.** The DEP aircraft studied in this paper.

In the research of the coordinated control technology of thrust and fault-tolerant control technology for DEP aircrafts, Jonathan L. Kratz et al. from the NASA Glenn Research Center designed a flight control plan for a single-aisle turboelectric aircraft with aft boundary layer thruster, and the designed controller was validated by simulation within the flight envelope. The results also showed that the engine efficiency was greatly improved [16]. Eric Nguyen Van et al. proposed a method to calculate a motor's bandwidth and control law for an active DEP aircraft with designed longitudinal/lateral control law and distribution modules, and the results demonstrated that the method can reduce the surface area of a vertical tail by 60% [17,18]. The NASA Glenn Research Center developed an 11 kw lightweight and efficient motor controller for X-57 DEP aircrafts. The controller includes a control processor and a three-phase power inverter weighing 1 kg and not requiring a heat sink, and its efficiency is over 97% [19]. Garrett T. Klunk et al. considered the stability and control effectiveness in the event of engine fault. An active thruster-based control system can redistribute thrust to offer dynamic directional stability when a thruster is unable to recover symmetric thrust. This capability satisfies the function of a vertical tail in an aircraft and, if permitted during certification, can completely replace the vertical tail [20]. The University of Michigan investigated the fault detection and control of DEP aircraft engines. For thruster faults in DEP aircrafts, Kalman filtering was adopted to detect motor faults, and a model predictive controller was leveraged to recover the altitude of cruising flight and redistribute thrust to a properly operating motor [21]. In recent years, the development of artificial intelligence provides a new technical way for fault-tolerant control. R. Shah from Cornell University proposed adaptive and learning methods and compared them to control DC motors actuating control surfaces of unmanned underwater vehicles. The result showed that deterministic artificial intelligence (DAI) outperformed the model-following approach in minimal peak transient value by approximately 2–70% [22]. S.M. Koo from Cornell University determined the threshold for the computational rate of actuator motor controllers for unmanned underwater vehicles necessary to accurately follow discontinuous square wave commands. The results showed that continuous DAI surpassed all modeling approaches, making it the safest and most viable solution to future commercial applications in unmanned underwater vehicles [23]. It can be seen that DAI has broad application prospects in the field of fault-tolerant control of DEP aircraft actuator in the future and should be deeply studied.

The redundant thrusters of DEP aircrafts also increase the risk of fault in the propulsion system, so it is necessary to study fault-tolerant control to ensure flight safety. At present, there is little research on coordinated thrust control, and research on fault-tolerant control of propulsion system for DEP aircrafts is also in the preliminary stage. In this context, a power system model for DEP aircrafts, including the engine module, the generator and energy storage system module, and the thruster module, is established in Section 2. A mathematical model of a six-degree-of-freedom DEP aircraft was built based on the principles of aerodynamics and flight dynamics. In Section 3, research on control methods to coordinate thrust from multiple thrusters is discussed based on the mathematical model of DEP aircrafts. The lateral and longitudinal control loops of DEP aircrafts were set up based on the principles of total energy control and total heading control, and a fault-tolerant control method was developed for the case where a thruster of a DEP aircraft has failed. In Section 4, experiments simulating flight tests and fault-tolerant control within the mission segment are outlined, and the experimental results are used to verify the effectiveness of the designed coordinated thrust control system and the fault-tolerant control method. Finally, all the major results are summarized and discussed in Section 5. In this study, the correctness and effectiveness of the designed coordinated thrust control method and the fault-tolerant control method for DEP aircrafts were theoretically verified, providing a theoretical basis for future engineering application and development of the control system for DEP aircrafts.

## 2. Modelling of the DEP Aircraft

Unlike traditional aircrafts, a DEP aircraft is powered by electrical energy converted from the mechanical energy of its engine, so the energy flow of its propulsion system differs from that of traditional aircraft. In this study, a mathematical model of the DEP aircraft's propulsion system was established, including its engine, generator, energy storage, thruster, and other modules. Then, a mathematical model of the DEP aircraft was built according to aerodynamics and flight dynamics to deepen understanding of the drive mode and flight mechanism of DEP aircrafts and lay the foundation for flight control and simulation research.

*2.1. Mathematical Model of the DEP Aircraft's Propulsion System*

2.1.1. Engine Module

In this study, two turboshaft engines were adopted to convert mechanical energy into electrical energy stored in the energy storage system. The turboshaft engines follow the ideal Brayton cycle.

Flow in the inlet was considered as an isentropic process with no total pressure loss and temperature loss, so the isentropic flow equation is as follows:

$$\frac{P_t}{P_s} = \left(1 + \frac{k-1}{2} M_a{}^2\right)^{\frac{k}{k-1}} \tag{1}$$

$$\frac{T_t}{T_s} = 1 + \frac{k-1}{2} M_a{}^2 \tag{2}$$

where $P_t$ is the total pressure, $T_t$ is the total temperature, $P_s$ is the static pressure, $T_s$ is the static temperature, $\mu$ is the specific heat ratio of the ideal gas, and $M_a$ is the Mach number.

The pressure ratio of a compressor is as follows, where $P_{t2}$ is the total inlet pressure of the compressor, and $P_{t3}$ is the total outlet pressure of the compressor.

$$P_{\text{ratio}} = \frac{P_{t3}}{P_{t2}} \tag{3}$$

It was assumed that the compressor is ideal and therefore provides isentropic compression. The temperature ratio can be calculated from the isentropic relations, where $T_{t2}$ is the total inlet temperature of the compressor, and $T_{t3}$ is the total outlet temperature of the compressor.

$$\frac{T_{t3}}{T_{t2}} = \left(\frac{P_{t3}}{P_{t2}}\right)^{\frac{k-1}{k}} \tag{4}$$

The increase in heat in the airflow within the combustor is proportional to the fuel consumption rate and the fuel heat value, as described below:

$$dm_0 Q = dm_f H_V \tag{5}$$

where $dm_0$ is the mass flow of air, $dm_f$ is the mass flow of fuel, $Q$ is the heat exchanged with the system, and $H_V$ is the heat value of fuel.

With the ideal burner efficiency and constant specific heat, the equation is as follows:

$$\left(dm_0 + dm_f\right) C_p T_{t4} - (dm_0) C_p T_{t3} = dm_f H_V \tag{6}$$

The maximum mass flow of fuel $dm_{f\text{max}}$ can be calculated using the highest temperature of the turbine inlet temperature $TIT$ at a constant-pressure specific heat $C_p$.

$$dm_{f\text{max}} = \frac{-(TIT_{\text{max}}) - T_{t3} C_p dm_0}{C_p TIT_{\text{max}} - H_V} \tag{7}$$

The turbine provides enough power to drive the compressor. Therefore, there is a condition to be satisfied, namely the turbine power should be equal to the compressor power. Under ideal conditions, the equation for this condition is as follows, where $T_{t4}$ is the total inlet temperature of the gas turbine, and $T_{t41}$ is the total inlet temperature of the power turbine.

$$dm_0 C_p (T_{t3} - T_{t2}) = \left(dm_0 + dm_f\right) C_p (T_{t4} - T_{t41}) \tag{8}$$

It was assumed that the turbine is ideal and is therefore isentropically depressurized. The temperature ratio can be calculated based on the isentropic relations. The isentropic relations were then adopted to change the pressure ratio of the turbine according to the following equation, where $P_{t4}$ is the total inlet pressure of the gas turbine, and $P_{t41}$ is the total inlet pressure of the power turbine.

$$\frac{P_{t41}}{P_{t4}} = \left[ 1 - \frac{T_{t2}}{T_{t4}} \frac{1}{\left(1 + \frac{dm_f}{dm_0}\right)} \left\{ \left(\frac{P_{t3}}{P_{t2}}\right)^{\frac{k-1}{k}} - 1 \right\} \right]^{\frac{k}{k-1}} \tag{9}$$

The power turbine extends the flow to ambient pressure to obtain the maximum power. It was assumed that the turbine is ideal and therefore it is isentropically depressurized. The isentropic relations were adopted to change the temperature ratio as follows:

$$\frac{T_{t3}}{T_{t2}} = \left(\frac{p_{t3}}{p_{t2}}\right)^{\frac{k-1}{k}} \tag{10}$$

The nozzle works isentropically, and there is no loss of total pressure and temperature. The total inlet pressure of nozzle $P_{t5}$ is equal to the total outlet pressure of nozzle $P_{t7}$.

$$P_{t5} = P_{t7} \tag{11}$$

Power recovery of the turboshaft engine is a function of the total enthalpy change of the turbine:

$$P_{Recovery} = \left(dm_0 + dm_f\right) C_p (T_{t4} - T_{t41}) \tag{12}$$

The specific fuel consumption $SFC$ is shown below:

$$SFC = \frac{dm_f}{P_{Recovery}} \tag{13}$$

### 2.1.2. Electric Power Generation and Energy Storage Module

Mechanical energy generated by the turboshaft engine is mechanically connected to a generator through the reduction gear box, and the generator then stores the generated electrical energy in the energy storage battery. Ports of the generator and the energy storage system were defined as presented in Figure 2.

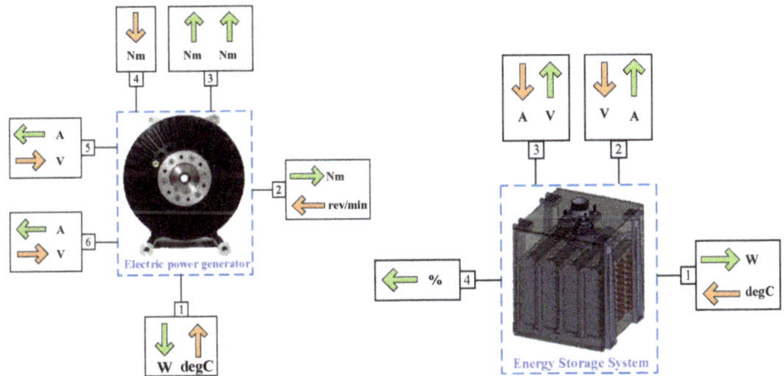


**Figure 2.** Definition of ports of the electric power generation and energy storage system.

The mechanical power of the generator $P_{mec}$ is calculated based on the following equation:

$$P_{mec} = T_m w_s \tag{14}$$

In the equation, $T_m$ and $w_s$ are the torque and the rotational speed of the shaft at the generator's Port 2, respectively.

The lost power $P_{lost}$ is calculated according to the following equation:

$$P_{lost} = (1 - \eta) \cdot |P_{mec}| \tag{15}$$

where $\eta$ is the efficiency defined by the motor's characteristics.

The electrical energy generated is as follows:

$$P_{elec} = P_{mec} - P_{lost} \tag{16}$$

The load of the charge $q_l$ extracted from the energy storage system for use is calculated as follows:

$$\frac{dq_l}{dt} = -I_3 \tag{17}$$

where $I_3$ represents the current of a battery at Port 3 of the energy storage system. When $I_3$ is negative, the battery is in a discharged state. The state of charge $SOC$ is a state variable, and its derivative is calculated as follows:

$$\frac{dSOC}{dt} = -\frac{dq}{dt} \cdot \frac{100}{C_{norn}} \tag{18}$$

where $C_{norn}$ is the rated capacity of a battery. The output power $P_{bat}$ of the energy storage system at Port 1 is calculated as follows:

$$P_{bat} = R_{cell} I_{cell}^2 S_{cell} P_{cell} \tag{19}$$

where $R_{cell}$ is the internal resistance of a battery cell, $I_{cell}$ is the battery current, $S_{cell}$ is the number of cells in series in a battery, and $P_{cell}$ is the number of cells in parallel in a battery.

### 2.1.3. Thruster Module

The thruster module of a DEP aircraft consists of 16 sets of motors connected to propellers through a reduction gear box. The thrust $F_P$ and the torque $T_P$ of a single propeller are calculated as follows:

$$F_P = C_{Thrust} \rho n_T^2 D_p^4 \tag{20}$$

$$T_P = \frac{C_{power} \rho n_T^3 D_p^5}{\omega} \tag{21}$$

where $\rho$ is the air density, $n_T$ is the rotational speed, $\omega$ is the rotational speed in the international system of units, $D_p$ is the propeller's diameter calculated from the propeller's radius, $C_{Thrust}$ is the thrust coefficient, and $C_{power}$ is the power coefficient.

The thrust coefficient $C_{Thrust}$ and power coefficient $C_{power}$ of the propeller are related to the geometric characteristics of the propeller, such as diameter, number of blades, blade area, rotating area, blade angle, theoretical pitch angle, etc. The $C_{Thrust}$ and $C_{power}$ map of the propeller can be generated by the propeller performance map generator tool. According to the propeller shaft speed, aircraft speed, and actual pitch angle, the value of $C_{Thrust}$ and $C_{power}$ at this time can be interpolated.

The propulsion ratio $J$ of the propeller is calculated as follows:

$$J = \frac{V_a}{n_T D_p} \tag{22}$$

In the equation, $V_a$ is the norm of the airspeed vector $\vec{V_a}$. The thrust and the torque coefficients are equal to zero when the rotational speed is opposite to the rotation (counterclockwise or clockwise) direction.

The propeller's aerodynamic efficiency $\lambda$ is defined as follows:

$$\lambda = J\frac{C_{Thrust}}{C_{power}} \tag{23}$$

Based on the position of the propeller relative to the body, the thrust and the moment of the propulsion system acting on the aircraft can be calculated.

### 2.2. Mathematical Model of the DEP Aircraft

In this study, a mathematical model of DEP aircrafts was established based on the principles of aerodynamics and aircraft dynamics. $(u, v, w)$ is the linear velocity of the aircraft, $(p, q, r)$ is its angular velocity, and $(\phi, \theta, \psi)$ represents its roll angle, pitch angle, and yaw angle.

#### 2.2.1. Earth-Surface Reference Frame $O_E x_E y_E z_E$

The Earth-surface reference frame was defined to obtain a transformational relationship between the aircraft body and Earth and used to determine the attitude and heading of the aircraft. The selected takeoff point is the origin $O_E$, the axis $z_E$ is vertical to the horizontal plane and points to the Earth's core, and the axis $x_E$ is located in the horizontal plane and points to the direction of the nose when the aircraft takes off. The axis $y_E$ is also located in the horizontal plane and is perpendicular to the axis $x_E$, whose direction is determined by the right-hand rule.

#### 2.2.2. Aircraft-Body Coordinate Frame $O_B x_B y_B z_B$

The selected mass center of the aircraft is the origin $O_B$ of the coordinates. The coordinate system is fixed to the aircraft body, and the axis $x_B$ is along the axis of the aircraft's symmetry plane, which points to the nose of the aircraft. $y_B$ points to the starboard side of the aircraft, while $z_B$ is perpendicular to $x_B$ in the aircraft's symmetry plane, which points to the bottom of the body.

The Earth-surface reference frame was converted to the aircraft-body coordinate frame as follows:

$$R_E^B = \begin{pmatrix} \cos\theta\cos\psi & \cos\theta\sin\psi & -\sin\theta \\ \sin\phi\sin\theta\cos\psi - \cos\phi\sin\psi & \sin\phi\sin\theta\sin\psi + \cos\phi\cos\psi & \sin\phi\cos\theta \\ \cos\phi\sin\theta\cos\psi + \sin\phi\sin\psi & \cos\phi\sin\theta\sin\psi - \sin\phi\cos\psi & \cos\phi\cos\theta \end{pmatrix} \tag{24}$$

#### 2.2.3. Velocity of Aircraft Relative to the Air

The velocity of the aircraft relative to the air is defined as follows:

$$\vec{V_a} = \begin{pmatrix} V_{ax} \\ V_{ay} \\ V_{az} \end{pmatrix} = \begin{pmatrix} V_{Gx}^E - V_{windx}^E \\ V_{Gy}^E - V_{windy}^E \\ V_{Gz}^E - V_{windz}^E \end{pmatrix} \tag{25}$$

where $\vec{V_a}$ is the airspeed of the aircraft, $V_{Gx}^E, V_{Gy}^E, V_{Gz}^E$ is the relative velocities to the Earth on the axis $x, y, z$, and $V_{windx}^E, V_{windy}^E, V_{windz}^E$ is the relative wind speed to the Earth on the axis $x, y, z$.

#### 2.2.4. Angle of Attach and Sideslip Angle

The aircraft should make its wings fly at a positive angle with respect to the airspeed vector in order to rise. The positive angle is the angle of attack, noted as $\alpha$; the angle between the velocity vector and the plane $x_B z_B$ is deemed as the sideslip angle, noted as $\beta$.

The velocity under the aircraft-body coordinate frame was adopted to calculate the angle of attach and the sideslip angle:

$$\begin{cases} \alpha = \arctan\left(\dfrac{V^B_{az}}{V^B_{ax}}\right) \\[2ex] \beta = \arcsin\left(\dfrac{V^B_{ay}}{\left|\overrightarrow{V_a}\right|}\right) \end{cases} \tag{26}$$

#### 2.2.5. Force and Moment

Based on aerodynamic principles, forces and moments acting on a DEP aircraft can be summarized as follows:

$$\begin{pmatrix} F_x \\ F_y \\ F_z \end{pmatrix} = \begin{pmatrix} -mg\sin\theta \\ mg\cos\theta\sin\phi \\ mg\cos\theta\cos\phi \end{pmatrix} + \begin{pmatrix} -\frac{1}{2}\rho SV_a^2 C_x \\ -\frac{1}{2}\rho SV_a^2 C_y \\ -\frac{1}{2}\rho SV_a^2 C_z \end{pmatrix} + \begin{pmatrix} F_{P_x} \\ F_{P_y} \\ F_{P_z} \end{pmatrix} \tag{27}$$

$$\begin{pmatrix} M_x \\ M_y \\ M_z \end{pmatrix} = \begin{pmatrix} \frac{1}{2}\rho SbV_a^2 C_l \\ \frac{1}{2}\rho S\bar{c}V_a^2 C_m \\ \frac{1}{2}\rho SbV_a^2 C_n \end{pmatrix} + \begin{pmatrix} T_{P_x} \\ T_{P_y} \\ T_{P_z} \end{pmatrix} \tag{28}$$

where $m$ is the mass of aircraft, $\rho$ is the air density, $S$ is the wing area for reference, $V_a$ is the norm of the airspeed vector $\overrightarrow{V_a}$, $b$ is the wingspan for reference, and $\bar{c}$ is the mean aerodynamic wing chord. $F_P$ and $T_P$ are the thrust and torque generated by the thrusters above. $C_x$ represents the drag coefficient, $C_y$ is the lateral force coefficient, and $C_z$ is the lift coefficient. $C_l$ is the roll moment coefficient, $C_m$ is the pitch moment coefficient, and $C_n$ is the yaw moment coefficient, as calculated and shown below.

The aerodynamic coefficient is as follows:

$$\begin{cases} C_x = C_{x0} + C_{x\alpha}\alpha + C_{xq}\frac{c}{2V_a}q + C_{x\delta_E}\left|\delta_E\right| \\ C_y = C_{y\beta}\beta + C_{yp}\frac{c}{2V_a}p + C_{yr}\frac{c}{2V_a}r + C_{y\delta_A}\delta_A + C_{y\delta_R}\delta_R \\ C_z = C_{z0} + C_{z\alpha}\alpha + C_{zq}\frac{c}{2V_a}q + C_{z\delta_E}\delta_E \end{cases} \tag{29}$$

The pneumatic moment coefficient is as follows:

$$\begin{cases} C_l = C_{l\beta}\beta + C_{lp}\frac{c}{2V_a}p + C_{lr}\frac{c}{2V_a}r + C_{l\delta_A}\delta_A + C_{l\delta_R}\delta_R \\ C_m = C_{m0} + C_{m\alpha}\alpha + C_{mq}\frac{c}{2V_a}q + C_{m\delta_E}\delta_E \\ C_n = C_{n\beta}\beta + C_{np}\frac{c}{2V_a}p + C_{nr}\frac{c}{2V_a}r + C_{n\delta_A}\delta_A + C_{n\delta_R}\delta_R \end{cases} \tag{30}$$

In the abovementioned equation, coefficients such as $C_{x0}, C_{x\alpha}, C_{xq}, C_{x\delta_E}$ are derived from the partial derivatives in a Taylor series approximation process and are dimensionless values, which are determined by the aircraft's parameters.

#### 2.2.6. Flight Dynamics Equations

From the momentum theorem, the following can be obtained:

$$\begin{pmatrix} \dot{u} \\ \dot{v} \\ \dot{w} \end{pmatrix} = \frac{1}{m}\begin{pmatrix} F_x \\ F_y \\ F_z \end{pmatrix} + R^B_E\begin{pmatrix} 0 \\ 0 \\ g \end{pmatrix} - \begin{pmatrix} 0 & -r & q \\ r & 0 & -p \\ -q & p & 0 \end{pmatrix}\begin{pmatrix} u \\ v \\ w \end{pmatrix} \tag{31}$$

From the moment of momentum theorem, the following can be obtained:

$$\begin{pmatrix} \dot{p} \\ \dot{q} \\ \dot{r} \end{pmatrix} = I^{-1}\begin{pmatrix} M_x \\ M_y \\ M_z \end{pmatrix} - I^{-1}\begin{pmatrix} 0 & -r & q \\ r & 0 & -p \\ -q & p & 0 \end{pmatrix}I \tag{32}$$

In the equations, $I$ is the moment of inertia of the aircraft:

$$I = \begin{pmatrix} I_{xx} & -I_{xy} & -I_{xz} \\ -I_{yx} & I_{yy} & -I_{yz} \\ -I_{zx} & -I_{zy} & I_{zz} \end{pmatrix} \tag{33}$$

The set of supplementary kinematic equations is presented as follows:

$$\begin{pmatrix} \dot{\phi} \\ \dot{\theta} \\ \dot{\psi} \end{pmatrix} = \begin{pmatrix} 1 & \sin\phi\tan\theta & \cos\phi\tan\theta \\ 0 & \cos\phi & -\sin\phi \\ 0 & \sin\phi\sec\theta & \cos\phi\sec\theta \end{pmatrix} \begin{pmatrix} p \\ q \\ r \end{pmatrix} \tag{34}$$

Equations (31), (32), and (34) are the six-degree-of-freedom flight dynamics equations of DEP aircrafts, and the flight state of an aircraft can be obtained by solving the above equations.

Simulation was carried out in order to verify the correctness of the mathematical model of DEP aircrafts. The inputs of the model were the target roll angle and pitch angle, and the aircraft was controlled to fly in a steady state with zero sideslip angle. The attitude response inputs of the aircraft are shown in Figures 3 and 4.

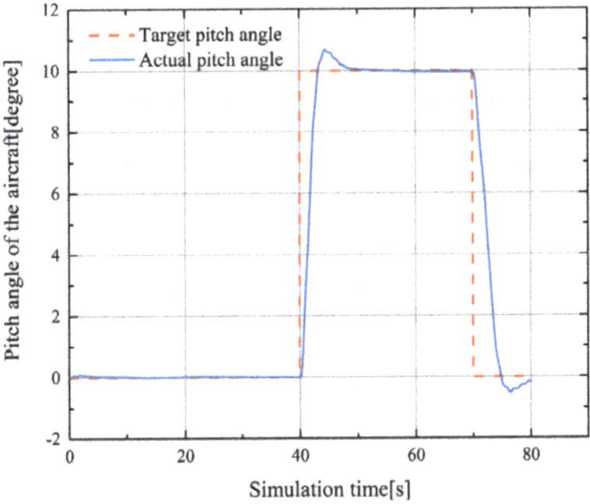


**Figure 3.** The pitch angle response of the DEP aircraft.

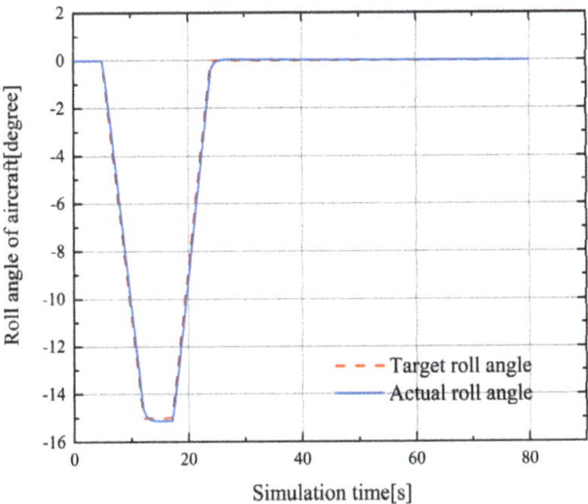


**Figure 4.** The roll angle response of the DEP aircraft.

At the fifth second, the input command of the roll angle changed, and the sideslip angle of the aircraft changed accordingly. As the roll channel is coupled with the yaw channel, in order to ensure zero sideslip angle flight, the yaw channel must respond to meet the control requirements. Figure 5 shows the response curve of the aircraft's sideslip angle. It can be seen that the sideslip angle caused by the roll channel only changed slightly. The aircraft returned to the steady flight with zero sideslip angle quickly, meaning it had a good control effect.

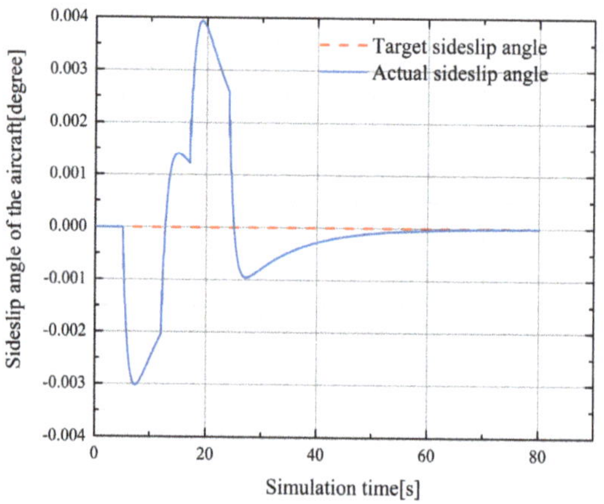


**Figure 5.** The sideslip angle response of the DEP aircraft.

## 3. Coordinated Thrust Control and Fault-Tolerant Control of the DEP Aircraft

### 3.1. Coordinated Thrust Control of the DEP Aircraft

### 3.1.1. Longitudinal Control Loop

In this study, a total energy control system (TECS) was designed for the longitudinal control of DEP aircrafts, which controls the entire flight of climb, cruise and descent with the best goal of minimizing the aircraft's energy consumption [24]. The system solves the coupling problems concerning the power lever angle and the elevator. Based on the total aircraft energy, the throttle directly corresponds to the increase or decrease in the overall aircraft energy, the rise and fall directly correspond to the distribution of the aircraft's kinetic and potential energy, and the altitude and airspeed are the results produced by the joint action of the power lever angle and the elevator. TECS was derived as follows:

$$E_{tot} = E_{kin} + E_{pot} = \frac{1}{2}mV_a^2 + mgH \tag{35}$$

$$\frac{\dot{E}_{tot}}{mg} = \frac{mV_a\dot{V}_a}{mg} + \frac{\dot{H}mg}{mg} = \frac{V_a\dot{V}_a}{g} + \dot{H} \tag{36}$$

$$\dot{E}_{spec} = \frac{\dot{E}_{tot}}{mgV_a} = \frac{\dot{V}_a}{g} + \frac{\dot{H}}{V_a} = \frac{\dot{V}_a}{g} + \sin\gamma \approx \frac{\dot{V}_a}{g} + \gamma \tag{37}$$

$$\dot{E}_{dist} = \gamma - \frac{\dot{V}_a}{g} \tag{38}$$

In the equations, $E_{tot}$ is the total energy of the aircraft, $E_{kin}$ is the kinetic energy of the aircraft, $E_{pot}$ is the potential energy of the aircraft, $m$ is the aircraft mass, $H$ is the altitude, $g$ is the gravitational acceleration, $V_a$ is the airspeed, $\gamma$ is the track angle, and $\dot{E}_{dist}$ is the specific energy distribution rate. In addition, the incremental thrust $\Delta T_c$ is associated with the specific energy gradient $\dot{E}_{spec}$.

The TECS approach connects the change in commanded thrust to the change in specific energy rate as follows:

$$\Delta T_c = \left( K_{TP} + \frac{K_{TI}}{s} \right) \dot{E}_{spec} \tag{39}$$

In the above equation in $s$ domain, $K_{TP}$ is the proportional gain of the thrust control loop, while $K_{TI}$ is the integral gain of the thrust control loop that drives the steady-state error to zero. It was assumed that the elevator control is under energy conservation and the elevator can convert kinetic energy to potential energy, so the specific energy distribution rate is presented as follows:

$$\dot{E}_{dist} = \gamma - \frac{\dot{V}_a}{g} \tag{40}$$

Based on that, changes in pitch angle command $\Delta\theta_c$ are related to changes in $\dot{E}_{dist}$:

$$\Delta\theta_c = \left( K_{EP} + \frac{K_{EI}}{s} \right) \dot{E}_{dist} \tag{41}$$

where $K_{EP}$ is the proportional gain of the pitch angle control loop, and $K_{EI}$ is the integral gain of the pitch angle control loop.

The aircraft's thrust is associated with the thrust command, and the change of elevator deflection angle $\Delta\delta_e$ is related to the pitch command:

$$\Delta T = G_{eng}(s)\Delta T_c, \ \Delta\delta_e = G_{elev}(s)\Delta\theta_c \tag{42}$$

where $G_{thr}(s)$ denotes the combined thrust control function, and $G_{elev}(s)$ is the combined pitch control and elevator actuator dynamics function. Based on the above derivation, the functional block diagram of TECS can be represented as in Figure 6.

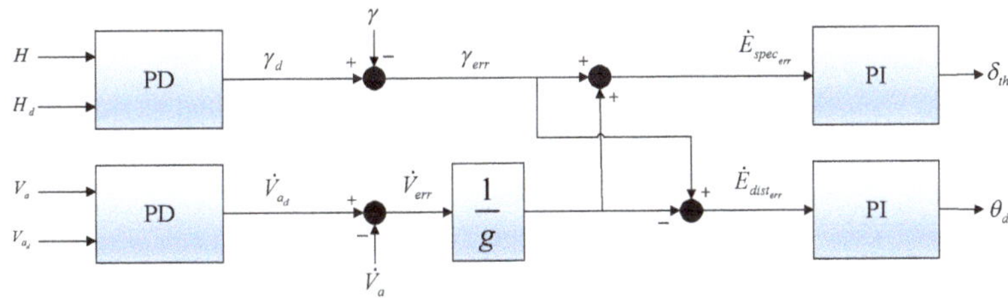

**Figure 6.** Functional block diagram of TECS.

### 3.1.2. Lateral Control Loop

The total heading control system (THCS) is leveraged for the lateral control of DEP aircrafts [24]. The error signals between the commanded and actual rate of change of heading ($\dot{\psi}_c$ and $\dot{\psi}$) and between the commanded and actual rate of change of sideslip ($\dot{\beta}_c$ and $\dot{\beta}$) are computed as follows:

$$\Delta\dot{\psi} = \dot{\psi}_c - \dot{\psi} \tag{43}$$

$$\Delta\dot{\beta} = \dot{\beta}_c - \dot{\beta} \tag{44}$$

The commanded roll angle changes $\Delta\phi_c$ and the yaw rate changes $\Delta r_c$ based on these errors are calculated as follows:

$$\Delta\phi_c = \frac{V_a}{g} \left( K_{RP} + \frac{K_{RI}}{s} \right) \left( \Delta\dot{\psi} + \Delta\dot{\beta} \right) \tag{45}$$

$$\Delta r_c = \frac{V_a}{g} \left( K_{YP} + \frac{K_{YI}}{s} \right) \left( \Delta\dot{\psi} - \Delta\dot{\beta} \right) \tag{46}$$

where $K_{RP}, K_{RI}$ is the proportional gain and integral gain of the roll angle control loop, and $K_{YP}, K_{YI}$ is the proportional gain and integral gain of the yaw rate control loop.

In terms of aircraft-related components, deflection changes in ailerons $\Delta\delta_a$ and the deflection changes in rudder $\Delta\delta_r$ are calculated in response to roll angle commands and roll angle speed variations, respectively.

$$\Delta\delta_a = G_{ail}(s)\Delta\phi_c, \ \Delta\delta_r = G_{rud}(s)\Delta r_c \tag{47}$$

where $G_{ail}(s)$ and $G_{rud}(s)$ are ailerons and the rudder controller and the actuator dynamics function, respectively.

*3.2. Fault Response Strategy and Fault-Tolerant Control of DEP Aircrafts*

This study focused on stuck and failed thrusters. Causes of thruster fault include decreased gain of a brushless motor due to aging of the motor stator coils, excessive friction of the motor rotor's shaft, and degradation of the motor's magnet performance, leading to the output deviating from the normal one. Macroscopically, when the output of a brushless motor is weak during the actual flight, changes in attitude angle of the motor is reduced with the same control amount, and the entire aircraft becomes "sluggish". In this study, 16 electric thrusters were adopted for the model object, with a symmetric distribution of eight thrusters on the left and eight on the right. The thruster near the center was numbered 1, and the outermost thruster was numbered 8. The state matrices of the thrusters on the left and the right were expressed by $X_L$ and $X_R$. In the preliminary design, the total thrust of the system is given by Equation (50), assuming that the thrust of all thrusters on the same side is equal [25].

$$T_{R\_i} = T_{R\_j}, \forall i \in [1,8], j \in [1,8] \tag{48}$$

$$T_{L\_i} = T_{L\_j}, \forall i \in [1,8], j \in [1,8] \tag{49}$$

$$T_{total} = T_L \sum_{i=1}^{8} X_L(i) + T_R \sum_{i=1}^{8} X_R(i) \tag{50}$$

Through this thruster counting method, the total yaw moment provided by this propulsion system is given by Equation (51), where the diameter of each thruster is $D$:

$$M_{tot} = T_L \frac{D}{2} \sum_{i=1}^{8} (2i-1)X_L(i) - T_R \frac{D}{2} \sum_{i=1}^{8} (2i-1)X_R(i) \tag{51}$$

Solving $T_R$ in (50) and (51), $T_{Left}$ and $T_{Right}$ can be obtained as shown in Equations (53) and (54) below:

$$M_{tot} = T_L \frac{D}{2} \sum_{i=1}^{8} (2i-1)X_L(i) - \left( \frac{T_{tot}}{\sum_{i=1}^{8} X_R(i)} - \frac{\sum_{i=1}^{8} X_L(i)}{\sum_{i=1}^{8} X_R(i)} \right) \frac{D}{2} \sum_{i=1}^{8} (2i-1)X_R(i) \tag{52}$$

$$T_{Left} = \frac{\frac{2}{D}M_{tot} + \frac{T_{tot}}{\sum_{i=1}^{8} X_R(i)} \sum_{i=1}^{8} (2i-1)X_R(i)}{\sum_{i=1}^{8} (2i-1)[X_L(i)] + \frac{\sum_{i=1}^{8} X_L(i)}{\sum_{i=1}^{8} X_R(i)} \sum_{i=1}^{8} (2i-1)[X_R(i)]} \tag{53}$$

$$T_{Right} = \frac{T_{tot}}{\sum_{i=1}^{8} X_R(i)} - \frac{\frac{2}{D}M_{tot} + \frac{T_{tot}}{\sum_{i=1}^{8} X_R(i)} \sum_{i=1}^{8} (2i-1)X_R(i)}{\frac{\sum_{i=1}^{8} X_R(i)}{\sum_{i=1}^{8} X_L(i)} \sum_{i=1}^{8} (2i-1)[X_L(i)] + \sum_{i=1}^{8} (2i-1)[X_R(i)]} \tag{54}$$

$T_{Left}$ and $T_{Right}$ of the above equations were input as thrust commands to the electric thrusters on both sides, where the state matrices of the thrusters were considered for monitoring the minimum thrust demand, turbine engine state, generator state, power bus state, and electric thruster state. If any of the components fail, the corresponding variable in the state matrices degrades to 0. In addition to enabling coordinated control of the electric thrusters on both sides, the thrust can be redistributed to maintain stability and maneuverability of the aircraft in case of a component fault. The fault-tolerant controller of the DEP system designed in this study features a fault injection module. The function developed so far allows the remaining thrusters to make corresponding changes to recover the aircraft's thrust to the prefault level when a single thruster on the left/right fails and the torque and the rotational speed fail to reach the normal operational level.

When the *i*th thruster fails, the mathematical form of the rotational speed of the thruster can be expressed as follows:

$$\omega_i^f = \sigma_i \omega_i \tag{55}$$

where $0 \leq \sigma_i < 1$ denotes the fault rate of the *i*th thruster under a fault. When the motor is completely jammed, then $\sigma_i = 0$.

When a simulation test of thruster fault-tolerant control is conducted in this simulation platform, a random fault thruster ID number, that is $n_{Fault}$, will be randomly generated in the $\tau$th second in order to simulate a thruster fault more realistically.

When a thruster numbered $n_{Fault}$ fails in the $\tau$th second, the torque of the thruster corresponding to the failed thruster should be first controlled to the torque value of the failed one, i.e., $T_i * \sigma_i$, at which point the difference between the thrust in a steady-state flight and that of a failed aircraft is deemed as the control error $\psi_{err}$. In this case, the thruster control torque needed to recover the prefault thrust can be calculated by PID control, which will be fed back to the aircraft control input, in order to achieve the fault-tolerant control of thrusters in the DEP system. The functional block diagram of the designed fault-tolerant control methods for DEP aircrafts in this study is shown in Figure 7.

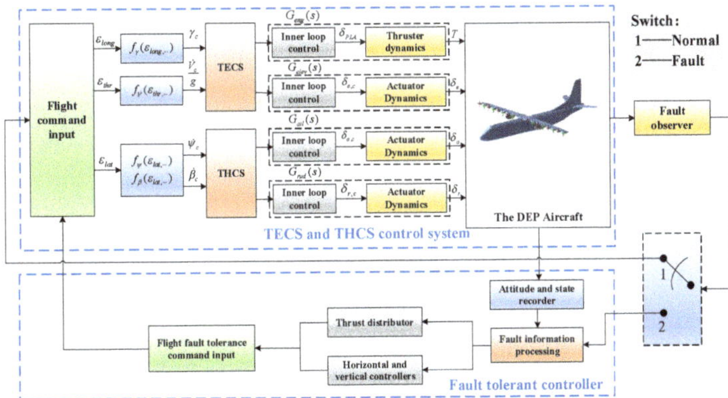


**Figure 7.** Functional block diagram of the fault-tolerant control of the DEP aircraft.

## 4. Simulation Results and Discussion

*4.1. Simulation Tests Carried out within Mission Segments*

Simulation tests of the DEP aircraft on the coordinated and comprehensive control of thrust were conducted during the entire process in the mission profile, namely takeoff, cruise, and descent. The set flight conditions are shown in Table 1.

**Table 1.** Parameter setting in mission segments of takeoff/cruise/descent.

| Flight Phase | Starting Height (m) | Final Height (m) | Mach Number |
| --- | --- | --- | --- |
| Climb | 0 | 10,000 | 0.49 |
| Cruise | 10,000 | 10,000 | 0.79 |
| Descent | 10,000 | 0 | 0.18 |

The flight simulation test results of the control system within the full mission segments are presented in Figure 8, with the response curve of the flight altitude showing good tracking effects.

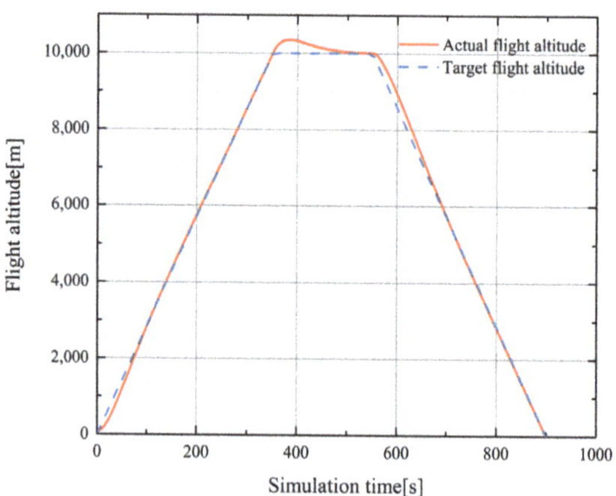


**Figure 8.** Response curve of the flight altitude.

For the altitude control within the flight mission segment, the quantitative description of the control effect is shown in Table 2, including rise time, peak time, settling time, and overshoot.

**Table 2.** The performance index of DEP aircraft's altitude control.

| Performance Index | Value | Unit |
| --- | --- | --- |
| Rise time | 349.75 | seconds |
| Peak time | 375.97 | seconds |
| Settling time | 493.61 | seconds |
| Overshoot | 3.46 | percent |

Figure 9 is the acceleration response curve of the z-axis. As can be seen, there is a change of acceleration when the aircraft's flight state changes. The curve then converges to zero. Figure 10 is the velocity response curve of the z-axis. When the aircraft enters cruise from climb, changes in acceleration results in the aircraft's velocity in the z-axis reaching almost zero in order to maintain a flight state with constant height and uniform speed.

Variation trend of the pitch angle of the aircraft in the corresponding mission segments is shown in Figure 11. During the cruise phase, the pitch angle of the aircraft returns to zero degrees.

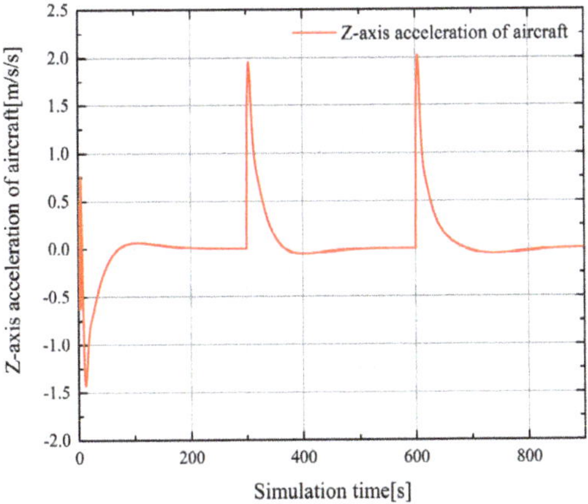

**Figure 9.** Acceleration response curve of the *z*-axis.

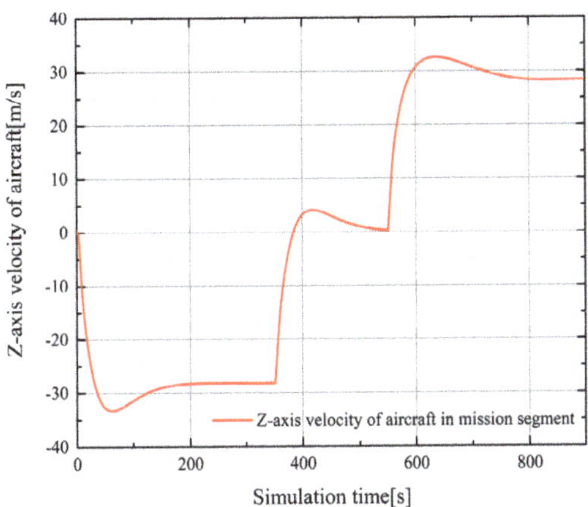

**Figure 10.** Velocity response curve of the *z*-axis.

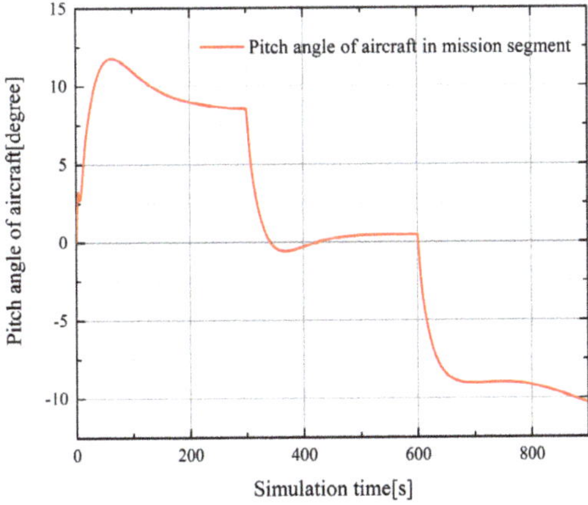

**Figure 11.** Variation trend of the pitch angle in the mission segment.

Figures 12 and 13 show the variation trend of the roll angle and yaw angle of an aircraft in the corresponding mission segments. The roll angle and the yaw angle will witness some small changes at the moment the flight state switches due to changes in the thrust and the attitude of thrusters in the DEP system. They will then return to a flight state without roll and deviation. The test results verifies the stability of the control system designed in this study.

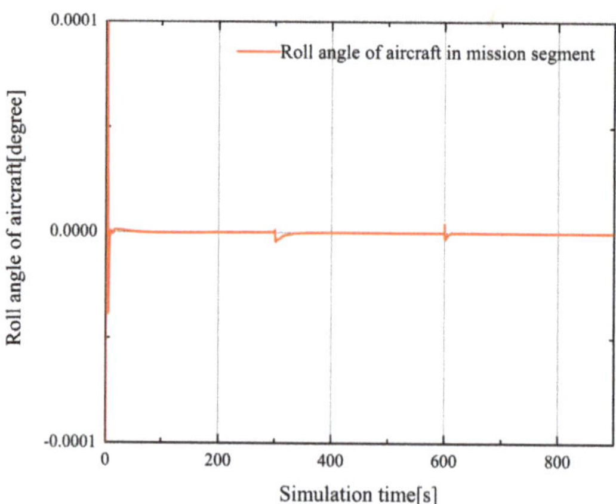


**Figure 12.** Variation trend of the roll angle in the mission segment.

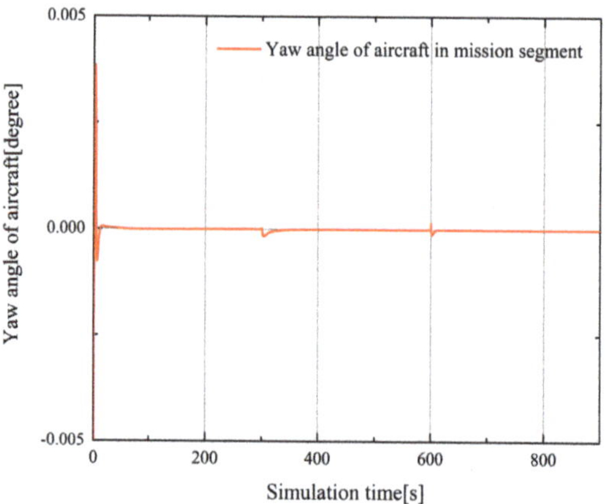


**Figure 13.** Variation trend of the yaw angle in the mission segment.

The thrust controller solves the control input torque of a corresponding single thruster to produce thrust in different stages. The curve of the total thrust variations generated by all thrusters in different mission segments is displayed in Figure 14. When the aircraft enters cruise in 300 s, the thrust required by the aircraft decreases, and the thrust is further reduced after it descends. The total power generated by the thruster module in the entire mission segments is demonstrated in Figure 15.

*4.2. DEP System Thruster Fault-Tolerant Control Simulation Test*

Propellers of thrusters on the left and right wings of an aircraft are designed to be right-handed and left-handed, and the torque direction is also symmetrical. When a thruster fails when $n_{Fault} = 2$ and $\sigma_i = 0.2$ is randomly generated at the 200th second, the torque of the failed thruster instantly drops to the moment value of $T_i * \sigma_i$, as shown by the red curve in Figure 16. Therefore, the torque of the symmetrical thruster No. 15 should change symmetrically in order to first ensure the balance of moment, as shown by the blue curve in Figure 16.

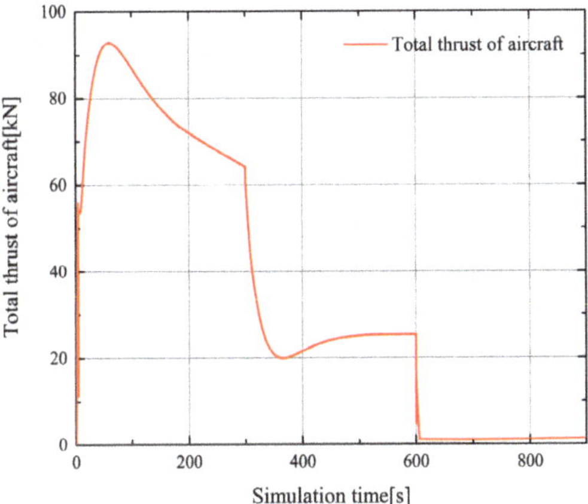

**Figure 14.** Curve of the total thrust variations generated by the thruster module.

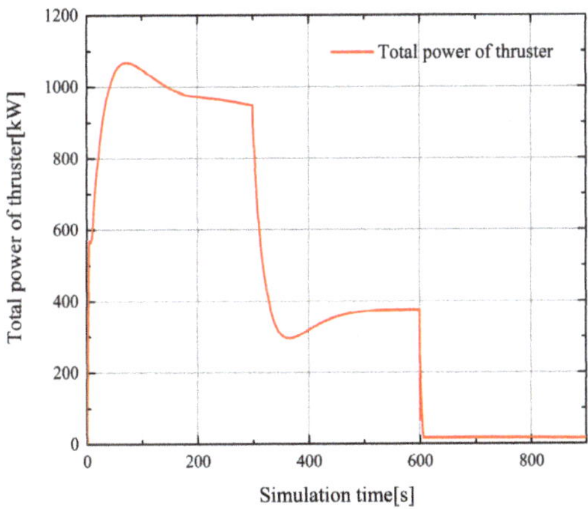

**Figure 15.** Curve of the total power variations generated by the thruster module.

In order to recover a stable flight state, all thrusters, except the thrusters symmetrical to the failed ones, should increase their thrust, so the torque input of the rest thrusters should be up. As the remaining thrusters change in the same way, the response value of the thruster torque can be observed with thruster No. 1 as an example, and the input control torque of thruster No. 1 gradually increases after the fault occurs in the 200th second. The generated thrust also grows at the 200th second, as shown in Figure 17.

The variation curve of the total thrust of the DEP aircraft after the fault is shown in Figure 18. As can be seen, the total thrust decreases after the thruster fault occurs in the 200th second, and the thrust of each thruster on the left and right is then altered by coordinate control to recover the thrust to a level that can maintain a stable flight of the aircraft.

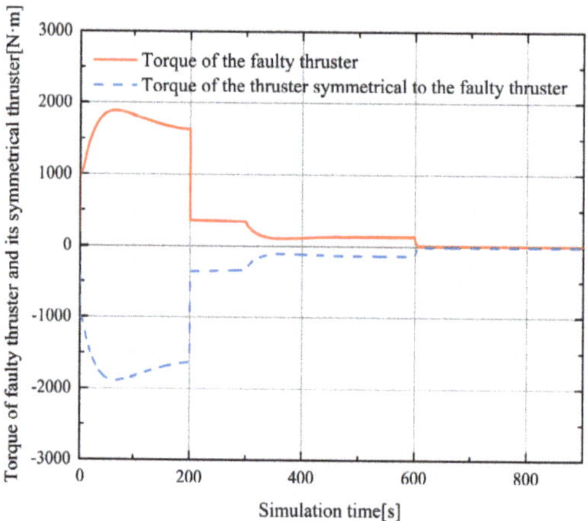

**Figure 16.** Variation curve of torque of the faulty thruster and symmetrical thruster.

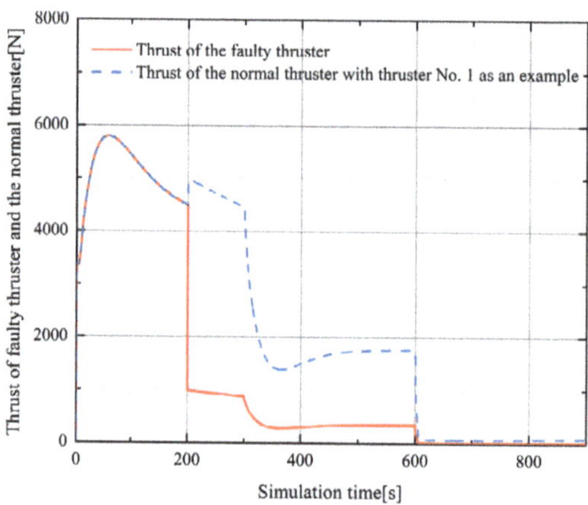

**Figure 17.** Variation curve of the remaining normal thrusters with thruster No. 1 as an example.

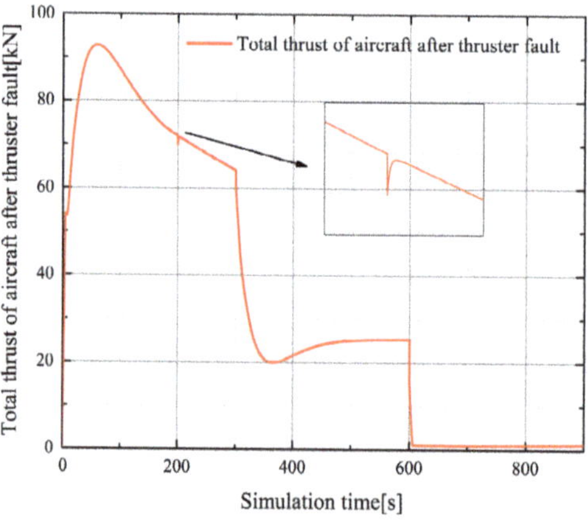

**Figure 18.** Total thrust curve of the DEP aircraft after thruster fault.

The controller can better control the thrust to the prefault level without overshoot after the total thrust changes in the 200th second, taking 0.3 s for adjustment. In other words, thrust generated by the thrusters on the left and right wing of the aircraft is evenly distributed by coordinated control after the fault.

## 5. Conclusions

First, a mathematical model of the DEP aircraft's propulsion system, including the engine module, the generator and energy storage system module, and the thruster module, was established. Then, a mathematical model of the six-degree-of-freedom DEP aircraft was built based on the principles of aerodynamics and flight dynamics, which laid a theoretical foundation for subsequent simulation experiment.

Research on control methods to coordinate thrust from multiple thrusters were carried out based on the mathematical model of DEP aircrafts. The lateral and longitudinal control loops of DEP aircrafts were set up based on the principles of total energy and total heading control, and a simulation experiment was carried out in the mission segment of the DEP aircraft. The effects of aircraft attitude control and altitude control verified the stability and accuracy of the mathematical model of the aircraft.

Furthermore, a fault-tolerant control method was developed for the case where a thruster of a DEP aircraft has failed. Experiments simulating flight tests and fault-tolerant control within the mission segment were conducted, and the experimental results verified the effectiveness of the designed coordinated thrust control system and the fault-tolerant control method. The controller could control the thrust to the prefault level.

The correctness and effectiveness of the designed coordinated thrust control method and fault-tolerant control method for DEP aircrafts were theoretically verified, providing a theoretical basis for future engineering application and development of the control system for DEP aircrafts.

**Author Contributions:** Conceptualization, J.L.; methodology, J.L.; validation, J.L.; formal analysis, J.L.; investigation, J.L.; data curation, J.L.; writing—original draft preparation, J.L.; writing—review and editing, J.Y.; supervision, H.Z. All authors have read and agreed to the published version of the manuscript.

**Funding:** This research received no external funding.

**Institutional Review Board Statement:** Not applicable.

**Informed Consent Statement:** Not applicable.

**Data Availability Statement:** The data presented in this study are available on request from the corresponding author.

**Conflicts of Interest:** The authors declare no conflict of interest. The funders had no role in the design of the study; in the collection, analyses, or interpretation of data; in the writing of the manuscript; or in the decision to publish the results.

## Nomenclature

**Abbreviations**

| | |
|---|---|
| DEP | distributed electric propulsion |
| NASA | National Aeronautics and Space Administration |
| TeDP | turboelectric distributed propulsion |
| VTOL | vertical takeoff and landing |
| TIT | turbine inlet temperature |
| SOC | state of charge |

**Roman letters**

| | |
|---|---|
| $k$ | specific heat ratio of the ideal gas |
| $P_t$ | total pressure |
| $T_t$ | total temperature |

| | |
|---|---|
| $P_s$ | static pressure |
| $T_s$ | static temperature |
| $M_a$ | Mach number |
| $P_{t2}$ | total inlet pressure of compressor |
| $P_{t3}$ | total outlet pressure of compressor |
| $T_{t2}$ | total inlet temperature of compressor |
| $T_{t3}$ | total outlet temperature of compressor |
| $C_p$ | constant pressure specific heat |
| $T_{t4}$ | total inlet temperature of gas turbine |
| $T_{t41}$ | total inlet temperature of power turbine |
| $P_{t4}$ | total inlet pressure of gas turbine |
| $P_{t41}$ | total inlet pressure of power turbine |
| $P_{t5}$ | total inlet pressure of nozzle |
| $P_{t7}$ | total outlet pressure of nozzle |
| $dm_0$ | mass flow of air |
| $Q$ | heat exchanged with the system |
| $dm_f$ | mass flow of fuel |
| $dm_{f\max}$ | maximum mass flow of fuel |
| $H_V$ | heat value of fuel |
| $TIT_{\max}$ | highest temperature of the turbine inlet temperature |
| $P_{Recovery}$ | power recovery of the turboshaft engine |
| $SFC$ | specific fuel consumption |
| $P_{mec}$ | mechanical power of the generator |
| $T_m$ | torque of the shaft at the generator's Port 2 |
| $w_s$ | rotational speed of the shaft at the generator's Port 2 |
| $P_{lost}$ | lost power |
| $P_{elec}$ | electrical energy generated by generator |
| $q_l$ | load of the charge extracted from the energy storage system for use |
| $I_3$ | current of a battery at Port 3 of the energy storage system |
| $C_{norn}$ | rated capacity of a battery |
| $P_{bat}$ | output power of the energy storage system at Port 1 |
| $R_{cell}$ | internal resistance of a battery cell |
| $I_{cell}$ | battery current |
| $S_{cell}$ | number of cells in series in a battery |
| $P_{cell}$ | number of cells in parallel in a battery |
| $F_P$ | thrust of a single propeller |
| $T_P$ | torque of a single propeller |
| $C_{Thrust}$ | thrust coefficient of a single propeller |
| $C_{power}$ | power coefficient of a single propeller |
| $n_T$ | rotational speed |
| $D_p$ | diameter of propeller |
| $J$ | propulsion ratio of the propeller |
| $\vec{V_a}$ | airspeed vector |
| $V_a$ | norm of the airspeed vector |
| $u$ | linear velocity of the aircraft's x-axis |
| $v$ | linear velocity of the aircraft's y-axis |
| $w$ | linear velocity of the aircraft's z-axis |
| $p$ | angular velocity of the aircraft's x-axis |
| $v$ | angular velocity of the aircraft's y-axis |
| $r$ | angular velocity of the aircraft's z-axis |
| $O_E x_E y_E z_E$ | Earth-surface reference frame |
| $O_B x_B y_B z_B$ | aircraft-body coordinate frame |
| $V_G^E$ | relative velocities to the Earth |
| $V_{wind}^E$ | relative wind speed to the Earth |
| $m$ | mass of aircraft |
| $S$ | wing area |
| $b$ | wingspan |

| | |
|---|---|
| $\bar{c}$ | mean aerodynamic wing chord |
| $C_x$ | drag coefficient |
| $C_y$ | lateral force coefficient |
| $C_z$ | lift coefficient |
| $C_l$ | roll moment coefficient |
| $C_m$ | pitch moment coefficient |
| $C_n$ | yaw moment coefficient |
| $I$ | moment of inertia of the aircraft |
| $E_{tot}$ | total energy of the aircraft |
| $E_{kin}$ | kinetic energy of the aircraft |
| $E_{pot}$ | potential energy of aircraft |
| $H$ | altitude of the aircraft |
| $g$ | gravitational acceleration |
| $K_{TP}$ | proportional gain of thrust control loop |
| $K_{TI}$ | integral gain of thrust control loop |
| $\dot{E}_{dist}$ | specific energy distribution rate |
| $\Delta T_c$ | incremental thrust |
| $\dot{E}_{spec}$ | specific energy gradient |
| $\Delta\theta_c$ | commanded pitch angle changes |
| $K_{EP}$ | proportional gain of pitch angle control loop |
| $K_{EI}$ | integral gain of pitch control loop |
| $\Delta\delta_e$ | change of elevator deflection angle |
| $G_{thr}(s)$ | combined thrust control and function |
| $G_{elev}(s)$ | combined pitch control and elevator actuator dynamics function |
| $\Delta\phi_c$ | commanded roll angle changes |
| $\Delta r_c$ | commanded yaw rate changes |
| $K_{RP}$ | proportional gain of roll angle control loop |
| $K_{RI}$ | integral gain of roll angle control loop |
| $K_{YP}$ | proportional gain of yaw rate control loop |
| $K_{YI}$ | integral gain of yaw rate control loop |
| $G_{ail}(s)$ | ailerons controller and the actuator dynamics function |
| $G_{rud}(s)$ | rudder controller and the actuator dynamics function |
| $X$ | state matrices of the thrusters |
| $M_{tot}$ | total yaw moment provided by DEP system |
| $n_{Fault}$ | random fault thruster ID number |

**Greek letters**

| | |
|---|---|
| $\mu$ | specific heat ratio of the ideal gas |
| $\eta$ | efficiency defined by the motor's characteristics |
| $\rho$ | air density |
| $\omega$ | rotational speed in the international system of units |
| $\lambda$ | propeller's aerodynamic efficiency |
| $\phi$ | roll angle of the aircraft |
| $\theta$ | pitch angle of the aircraft |
| $\psi$ | yaw angle of the aircraft |
| $\alpha$ | angle of attack |
| $\beta$ | sideslip angle |
| $\gamma$ | track angle |
| $\sigma$ | fault rate |
| $\tau$ | time of thruster failure in simulation |
| $\psi_{err}$ | control error of yaw angle |

**Subscript**

| | |
|---|---|
| $x$ | vector component corresponding to the $x$-axis of the coordinate system |
| $y$ | vector component corresponding to the $y$-axis of the coordinate system |
| $z$ | vector component corresponding to the $z$-axis of the coordinate system |
| $c$ | the variable control command input into the system |
| $L$ | thruster's variable on the left side of DEP aircraft |
| $R$ | thruster's variable on the right side of DEP aircraft |

|  |  |
|---|---|
| $i$ | count value of left thruster |
| $j$ | count value of right thruster |
| **Superscript** | |
| $B$ | vector or scalar under the aircraft-body coordinate frame |
| $E$ | vector or scalar under the Earth-surface reference frame |
| **Prefix** | |
| $\Delta$ | change value of variable |

## References


1. Serrano, J.R.; García-Cuevas, L.M.; Bares, P.; Varela, P. Propeller Position Effects over the Pressure and Friction Coefficients over the Wing of an UAV with Distributed Electric Propulsion: A Proper Orthogonal Decomposition Analysis. *Drones* **2022**, *6*, 38. [CrossRef]
2. Amoozgar, M.; Friswell, M.; Fazelzadeh, S.; Khodaparast, H.H.; Mazidi, A.; Cooper, J. Aeroelastic Stability Analysis of Electric Aircraft Wings with Distributed Electric Propulsors. *Aerospace* **2021**, *8*, 100. [CrossRef]
3. Serrano, J.; Tiseira, A.; García-Cuevas, L.; Varela, P. Computational Study of the Propeller Position Effects in Wing-Mounted, Distributed Electric Propulsion with Boundary Layer Ingestion in a 25 kg Remotely Piloted Aircraft. *Drones* **2021**, *5*, 56. [CrossRef]
4. Kirner, R.; Raffaelli, L.; Rolt, A.; Laskaridis, P.; Doulgeris, G.; Singh, R. An assessment of distributed propulsion: Part B – Advanced propulsion system architectures for blended wing body aircraft configurations. *Aerosp. Sci. Technol.* **2016**, *50*, 212–219. [CrossRef]
5. Gohardani, A.S.; Doulgeris, G.; Singh, R. Challenges of future aircraft propulsion: A review of distributed propulsion technology and its potential application for the all electric commercial aircraft. *Prog. Aerosp. Sci.* **2011**, *47*, 369–391. [CrossRef]
6. Nickol, C.L.; Haller, W.J. Assessment of the Performance Potential of Advanced Subsonic Transport Concepts for NASA's Environmentally Responsible Aviation Project. In Proceedings of the 54th AIAA Aerospace Sciences Meeting, San Diego, CA, USA, 4–8 January 2016; p. 1030.
7. Connolly, J.W.; Chapman, J.W.; Stalcup, E.J.; Chicatelli, A.; Hunker, K.R. Modeling and Control Design for a Turboelectric Single Aisle Aircraft Propulsion System. In Proceedings of the 2018 AIAA/IEEE Electric Aircraft Technologies Symposium (EATS), Cincinnati, OH, USA, 12–14 July 2018; pp. 1–19.
8. Nguyen, N.T.; Reynolds, K.; Ting, E.; Nguyen, N. Distributed Propulsion Aircraft with Aeroelastic Wing Shaping Control for Improved Aerodynamic Efficiency. *J. Aircr.* **2018**, *55*, 1122–1140. [CrossRef]
9. Zhang, J.; Kang, W.; Li, A.; Yang, L. Integrated flight/propulsion optimal control for DPC aircraft based on the GA-RPS algorithm. *Proc. Inst. Mech. Eng. Part G J. Aerosp. Eng.* **2015**, *230*, 157–171. [CrossRef]
10. Lei, T.; Kong, D.; Wang, R.; Li, W.; Zhang, X. Evaluation and optimization method for power systems of distributed electric propulsion aircraft. *Acta Aeronaut. Et Astronaut. Sin.* **2021**, *42*, 624047. (In Chinese) [CrossRef]
11. Da, X.; Fan, Z.; Xiong, N.; Wu, J.; Zhao, Z. Modeling and analysis of distributed boundary layer ingesting propulsion system. *Acta Aeronaut. Et Astronaut. Sin.* **2018**, *39*, 122048. (In Chinese) [CrossRef]
12. Liu, C.; Si, X.; Teng, J.; Ihiabe, D. Method to Explore the Design Space of a Turbo-Electric Distributed Propulsion System. *J. Aerosp. Eng.* **2016**, *29*, 04016027. [CrossRef]
13. Choi, B.; Brown, G.V.; Morrison, C.; Dever, T. Propulsion Electric Grid Simulator (PEGS) for Future Turboelectric Distributed Propulsion Aircraft. In Proceedings of the 12th International Energy Conversion Engineering Conference, Cleveland, OH, USA, 28–30 July 2014; p. 3644.
14. Rothhaar, P.M.; Murphy, P.C.; Bacon, B.J.; Gregory, I.M.; Grauer, J.A.; Busan, R.C.; Croom, M.A. NASA Langley Distributed Propulsion VTOL TiltWing Aircraft Testing, Modeling, Simulation, Control, and Flight Test Development. In Proceedings of the 14th AIAA Aviation Technology, Integration, and Operations Conference, Atlanta, GA, USA, 16–20 June 2014; p. 2999. [CrossRef]
15. Freeman, J.L.; Klunk, G.T. Dynamic Flight Simulation of Spanwise Distributed Electric Propulsion for Directional Control Authority. In Proceedings of the 2018 AIAA/IEEE Electric Aircraft Technologies Symposium, Cincinnati, OH, USA, 9–11 July 2018; Volume 2018, pp. 1–15. [CrossRef]
16. Kratz, J.L.; Thomas, G.L. Dynamic Analysis of the STARC-ABL Propulsion System. In Proceedings of the AIAA Propulsion and Energy 2019 Forum, Indianapolis, IN, USA, 19–22 August 2019.
17. Van, E.N.; Alazard, D.; Döll, C.; Pastor, P. Co-design of aircraft vertical tail and control laws using distributed electric propulsion. *IFAC-Pap.* **2019**, *52*, 514–519. [CrossRef]
18. Van, E.N.; Alazard, D.; Döll, C.; Pastor, P. Co-design of aircraft vertical tail and control laws with distributed electric propulsion and flight envelop constraints. *CEAS Aeronaut. J.* **2021**, *12*, 101–113. [CrossRef]
19. Garrett, M.; Avanesian, D.; Granger, M.; Kowalewski, S.; Maroli, J.; Miller, W.A.; Jansen, R.; Kascak, P.E. Development of an 11 kW lightweight, high efficiency motor controller for NASA X-57 Distributed Electric Propulsion using SiC MOSFET Switches. In Proceedings of the AIAA (American Institute of Aeronautics and Astronautics) Propulsion and Energy 2019 Forum, Indianapolis, IN, USA, 19–22 August 2019; pp. 1–8.
20. Klunk, G.T.; Freeman, J.L. Vertical Tail Area Reduction for Aircraft with Spanwise Distributed Electric Propulsion. In Proceedings of the 2018 AIAA/IEEE Electric Aircraft Technologies Symposium, Cincinnati, OH, USA, 9–11 July 2018; p. 5022. [CrossRef]

21. Suzuki, Y.; Dunham, W.; Kolmanovsky, I.; Girard, A. Failure Detection and Control of Distributed Electric Propulsion Aircraft Engines. In Proceedings of the AIAA Scitech 2019 Forum, San Diego, CA, USA, 7–11 January 2019; p. 0109.
22. Shah, R.; Sands, T. Comparing Methods of DC Motor Control for UUVs. *Appl. Sci.* **2021**, *11*, 4972. [CrossRef]
23. Koo, S.M.; Travis, H.D.; Sands, T. Evaluation of Adaptive and Learning in Unmanned Systems. *Preprints* **2022**. [CrossRef]
24. Chakraborty, I.; Ahuja, V.; Comer, A.; Mulekar, O. Development of a Modeling, Flight Simulation, and Control Analysis Capability for Novel Vehicle Configurations. In Proceedings of the AIAA Aviation 2019 Forum, Dallas, TX, USA, 17–21 June 2019; p. 3112.
25. Armstrong, M.; Ross, C.; Phillips, D.; Blackwelder, M. *Stability, Transient Response, Control, and Safety of a High-Power Electric Grid for Tur-Boelectric Propulsion of Aircraft*; NASA/CR 2013-217865; National Aeronautics and Space Administration: Cleveland, OH, USA, 2013.

MDPI
St. Alban-Anlage 66
4052 Basel
Switzerland
Tel. +41 61 683 77 34
Fax +41 61 302 89 18
www.mdpi.com

*Drones* Editorial Office
E-mail: drones@mdpi.com
www.mdpi.com/journal/drones